T0419817

Springer Series in

MATERIALS SCIENCE 137

Springer Series in
MATERIALS SCIENCE

The Springer Series in Materials Science covers the complete spectrum of materials physics, including fundamental principles, physical properties, materials theory and design. Recognizing the increasing importance of materials science in future device technologies, the book titles in this series reflect the state-of-the-art in understanding and controlling the structure and properties of all important classes of materials.

Please view available titles in *Springer Series in Materials Science* on series homepage http://www.springer.com/series/856

Kamakhya Prasad Ghatak
Sitangshu Bhattacharya

Thermoelectric Power in Nanostructured Materials

Strong Magnetic Fields

With 174 Figures

Professor Dr. Kamakhya Prasad Ghatak
University of Calcutta
Deptartment of Electronic Science
Acharya Prafulla Chandra Rd. 92
Kolkata, 700 009, India
E-mail: kamakhyaghatak@yahoo.co.in

Dr. Sitangshu Bhattacharya
Indian Institute of Science
Center of Electronics Design and Technology
Nano Scale Device Research Laboratory
Bangalore, 560 012, India
E-mail: isbsin@yahoo.co.in

Series Editors:

Springer Series in Materials Science ISSN 0933-033X
ISBN 978-3-642-10570-8 e-ISBN 978-3-642-10571-5
DOI 10.1007/978-3-642-10571-5
Springer Heidelberg Dordrecht London New York

Library of Congress Control Number: 2010931384

Cover design: eStudio Calamar Steinen

Printed on acid-free paper

Springer is part of Springer Science+Business Media (www.springer.com)

Dedicated to the Sweet Memories of Late Professor Sushil Chandra Dasgupta, D. Sc., Formerly Head of the Department of Mathematics of the then Bengal Engineering College (Presently Bengal Engineering and Science University), Shibpur, West Bengal, India, Late Professor Biswaranjan Nag, D. Sc., Formerly Head of the Departments of Radiophysics and Electronics and Electronic Science, University of Calcutta, Kolkata, West Bengal, India, and Late Professor Sankar Sebak Baral, D. Sc., Formerly Founding Head of the Department of Electronics and Telecommunication Engineering of the then Bengal Engineering College (Presently Bengal Engineering and Science University), Shibpur, West Bengal, India, for their pioneering contributions in research and teaching of Applied Mathematics, Semiconductor Science, And Applied Electronics, respectively, to which the first author remains ever grateful as a student and research worker

Preface

The merging of the concept of introduction of asymmetry of the wave vector space of the charge carriers in semiconductors with the modern techniques of fabricating nanostructured materials such as MBE, MOCVD, and FLL in one, two, and three dimensions (such as ultrathin films, nipi structures, inversion and accumulation layers, quantum well superlattices, carbon nanotubes, quantum wires, quantum wire superlattices, quantum dots, magneto inversion and accumulation layers, quantum dot superlattices, etc.) spawns not only useful quantum effect devices but also unearth new concepts in the realm of nanostructured materials science and related disciplines. It is worth remaking that these semiconductor nanostructures occupy a paramount position in the entire arena of low-dimensional science and technology by their own right and find extensive applications in quantum registers, resonant tunneling diodes and transistors, quantum switches, quantum sensors, quantum logic gates, heterojunction field-effect, quantum well and quantum wire transistors, high-speed digital networks, high-frequency microwave circuits, quantum cascade lasers, high-resolution terahertz spectroscopy, superlattice photo-oscillator, advanced integrated circuits, superlattice photocathodes, thermoelectric devices, superlattice coolers, thin film transistors, intermediate-band solar cells, microoptical systems, high-performance infrared imaging systems, bandpass filters, thermal sensors, optical modulators, optical switching systems, single electron/molecule electronics, nanotube based diodes, and other nanoelectronic devices. Mathematician Simmons rightfully tells us [1] that the mathematical knowledge is said to be doubling in every 10 years, and in this context, we can also envision the extrapolation of the Moore's law by projecting it in the perspective of the advancement of new research and analyses, in turn, generating novel concepts particularly in the area of nanoscience and technology [2].

With the advent of Seebeck effect in 1821 [3–6], it is evident that the investigations regarding the thermoelectric materials, the subset of the generalized set materials science have unfathomable proportions with respect to accumulated knowledge and new research in multidimensional aspects of thermoelectrics in general [7–17]. The timeline of thermoelectric and related research during the 200 years spanning from 1800 to 2000 is given in [18], and with great dismay, we admit that the citation of even pertinent references in this context is placed permanently in the gallery of impossibility theorems. It is rather amazing to observe from the detailed survey of almost the whole spectrum of the literature in this particular aspect that the available monographs, hand books, and review articles on thermoelectrics and related topics have not included any detailed investigations on the thermoelectric power in nanostructured materials under strong magnetic field (TPSM).

It is well known that the TPSM is a very important quantity [19], since the change in entropy (a vital concept in thermodynamics) can be known from this relation by determining the experimental values of the change of electron concentration. The analysis of TPSM generates information regarding the effective mass of the carriers in materials, which occupies a central position in the whole field of materials science in general [20]. The classical TPSM (G_0) equation is valid only under the nondegenerate carrier concentration, and the magnitude of the TPSM is given by ($G_0 = (\pi^2 k_B/3e)$ (k_B and e are Boltzmann's constant and the magnitude of the carrier charge, respectively; [21]). From this equation, it is readily inferred that this conventional form is a function of three fundamental constants only, being independent of the signature of the charge carriers in materials. The significant work of Zawadzki [22–24] reflects the fact that the TPSM for materials having degenerate electron concentration is independent of scattering mechanisms and is exclusively determined by the dispersion laws of the respective carriers. It will, therefore, assume different values for different systems and varies with the doping, the magnitude of the reciprocal quantizing magnetic field under magnetic quantization, the nano thickness in ultrathin films, quantum wires and dots, the quantizing electric field as in inversion layers, the carrier statistics in various types of quantum-confined superlattices having different carrier energy spectra, and other types of low-dimensional field assisted systems.

This monograph, which is based on our 20 years of continuous and ongoing research, is divided into four parts. The first part deals with the thermoelectric power under large magnetic field in quantum-confined materials and it contains four chapters. In Chap. 1, we have investigated the TPSM for quantum dots of nonlinear optical, III–V, II–VI, n-GaP, n-Ge, Te, Graphite, $PtSb_2$, zerogap, II–V, Gallium Antimonide, stressed materials, Bismuth, IV–VI, lead germanium telluride, Zinc and Cadmium diphosphides, Bi_2Te_3, and Antimony on the basis of respective carrier energy spectrum. In Chap. 2, the TPSM in ultrathin films and quantum wires of nonlinear optical, Kane type III–V, II–VI, Bismuth, IV–VI, stressed materials, and carbon nanotubes (a very important quantum material) have been investigated. In Chap. 3, the TPSM in quantum dot III–V, II–VI, IV–VI, HgTe/CdTe superlattices with graded interfaces and quantum dot effective mass superlattices of the aforementioned materials have been investigated. In Chap. 4, the TPSM in quantum wire superlattices of the said materials have been studied.

The second part of this monograph deals with the thermoelectric power under magnetic quantization in macro and micro electronic materials. In Chap. 5, the thermoelectric power in nonlinear optical, Kane type III–V, II–VI, Bismuth, IV-VI, and stressed materials has been investigated in the presence of quantizing magnetic field. In Chap. 6, the thermoelectric power under magnetic quantization in III–V, II–VI, IV–VI, HgTe/CdTe superlattices with graded interfaces and effective mass superlattices of the aforementioned materials together with the quantum wells of said superlattices have been investigated. In Chap. 7, the thermoelectric power under magnetic quantization in ultrathin films of nonlinear optical, Kane type III–V, II–VI, Bismuth, IV–VI, and stressed materials has been investigated.

The third part deals with the thermoelectric power under large magnetic field in quantum-confined optoelectronic materials in the presence of light waves. In Chap. 8, the influence of light on the thermoelectric power under large magnetic field in ultrathin films and quantum wires of optoelectronic materials has been investigated. In Chap. 9, the thermoelectric power under large magnetic field in quantum dots of optoelectronic materials has been studied in the presence of external light waves. In Chap. 10, the same has been studied for III–V quantum wire and quantum dot superlattices with graded interfaces and III–V quantum wire and quantum dot effective mass superlattices, respectively.

The last part of this monograph deals with thermoelectric power under magnetic quantization in macro and micro optoelectronic materials in the presence of light waves. In Chap. 11, the optothermoelectric power in macro optoelectronic materials under magnetic quantization has been investigated. In Chap. 12, the optothermoelectric power in ultrathin films of optoelectronic materials under magnetic quantization has been studied. In Chap. 13, the magneto thermo power in III–V quantum well superlattices with graded interfaces and III–V quantum well effective mass superlattices have been studied. Chapter 14 discusses eight applications of our results in the realm of quantum effect devices and also discusses very briefly the experimental results, and additionally, we have proposed a single multidimensional open research problem for experimentalists regarding the thermoelectric power in nanostructured materials having various carrier energy spectra under different physical conditions. Chapter 15 contains the conclusion and scope of future research. Appendix A contains the TPSM for bulk specimens of few technologically important materials. Each chapter except the last two contains a table highlighting the basic results pertaining to it in a summarized form.

It is well known that the errorless first edition of any book is virtually impossible from the perspective of academic reality and the same stands very true for this monograph in spite of the Herculean joint effort of not only the authors but also the seasoned Editorial team of Springer. Naturally, we are open to accept constructive criticisms for the purpose of their inclusion in the future edition. From Chap. 1 till end, this monograph presents to its esteemed readers 150 open research problems, which will be useful in the real sense of the term for the researchers in the fields of solid state sciences, materials science, computational and theoretical nanoscience and technology, nanostructured thermodynamics and condensed matter physics in general in addition to the graduate courses on modern thermoelectric materials in various academic departments of many institutes and universities. We strongly hope that the alert readers of this monograph will not only solve the said problems by removing all the mathematical approximations and establishing the appropriate uniqueness conditions, but will also generate new research problems, both theoretical and experimental and, thereby, transforming this monograph into a monumental superstructure.

It is needless to say that this monograph exposes only the tip of the iceberg, the rest of which will be worked upon by the researchers of the appropriate fields whom we would like to believe are creatively superior to us. It is an amazing fact to observe that the experimental investigations of the thermoelectric power under strong magnetic field in nanostructured materials have been relatively less investigated in the literature, although such studies will throw light on the understanding of the band structures of nanostructured materials, which, in turn, control the transport phenomena in such low-dimensional quantized systems. Various mathematical analyses and few chapters of this monograph are appearing for the first time in printed form. We hope that our esteemed readers will enjoy the investigations of TPSM in a wide range of nanostructured materials having different energy-wave vector dispersion relation of the carriers under various physical conditions as presented in this book. Since a monograph on the thermoelectric power in nanostructured materials under strong magnetic field is really nonexistent to the best of our knowledge even in the field of nanostructured thermoelectric materials, we earnestly hope our continuous effort of 20 years will be transformed into a standard reference source for creatively enthusiastic readers and researchers engaged either in theoretical or applied research in connection with low-dimensional thermal electronics in general to probe into the in-depth investigation of this extremely potential and promising research area of materials science.

Acknowledgements

Acknowledgement by Kamakhya Prasad Ghatak

I am grateful to T. Roy of the Department of Physics of Jadavpur University for creating the interest in the applications of quantum mechanics in diverse fields when I was in my late teens pursuing the engineering degree course. I am indebted to M. Mitra, S. Sarkar, and P. Chowdhury for creating the passion in me for number theory. I express my gratitude to D. Bimberg, W. L. Freeman, H. L. Hartnagel, and W. Schommers for various academic interactions spanning the last two decades. The renowned scientist P. N. Butcher has been a driving hidden force since 1985 before his untimely demise with respect to our scripting the series in band structure-dependent properties of nanostructured materials. He insisted me repeatedly regarding it and to tune with his high rigorous academic frequency, myself with my truly able students and later on colleagues wrote the Einstein Relation in Compound Semiconductors and their Nanostructures, Springer Series in Materials Science, Vol. 116 as the first one, Photoemission from Optoelectronic Materials and their Nanostructures, Springer Series in Nanostructure Science and Technology as the second one, and the present monograph as the third one.

I am grateful to N. Guho Chowdhury of Jadavpur University, mentor of many academicians including me and a very pivotal person in my academic career, for instigating me to carry out extensive research of the first order bypassing all the difficulties. My family members and especially my beloved parents deserve a very special mention for really forming the backbone of my long unperturbed research career. I must express my gratitude to P. K. Sarkar of Semiconductor Device Laboratory, S. Bania of Digital Electronics Laboratory, and B. Nag of Applied Physics Department for motivating me at rather turbulent times. I must humbly concede the Late V. S. Letokhov with true reverence for inspiring me with the ebullient idea that the publication of research papers containing innovative concepts in eminent peer-reviewed international journals is the central cohesive element to excel in creative research activity, and at the same time being a senior research scientist he substantiated me regarding the unforgettable learning for all the scientists in general from the eye opening and alarming articles by taking a cue from [25–27] immediately after their publications. Besides, this monograph has been completed under the grant (8023/BOR/RID/RPS-95/2007–08) as jointly sanctioned with D. De of the Department of Computer Science and Engineering, West Bengal University of Technology by the All India Council for Technical Education in their research promotion scheme 2008.

Acknowledgement by Sitangshu Bhattacharya

It is virtually impossible to express my gratitude to all the admirable persons who have influenced my academic and social life from every point of view; nevertheless, a short memento to my teachers S. Mahapatra, at the Centre for Electronics Design and Technology and R. C. Mallick at Department of Physics at Indian Institute of Science, Bangalore, for their fruitful academic advices and guidance remains everlasting. I am indebted to my friend Ms. A. Sood for standing by my side at difficult times of my research life, which still stimulates and amplifies my efficiency for performing in-depth research. Besides, my sister Ms. S. Bhattacharya, my father I. P. Bhattacharya and my mother B. Bhattacharya rinsed themselves for my academic development starting from my childhood till date, and for their joint speechless and priceless contribution, no tears of gratitude seems to be enough at least when myself is considered. I am also grateful to the Department of Science and Technology, India, for sanctioning the project and the fellowship under "SERC Fast Track Proposal of Young Scientist" scheme-2008–2009 (SR/FTP/ETA-37/08) under which this monograph has been completed. As usual I am grateful to my friend, philosopher and guide, the first author.

Joint Acknowledgements

We are grateful to Dr. C. Ascheron, Executive Editor Physics, Springer Verlag in the real sense of the term for his inspiration and priceless technical assistance. We owe a lot to Ms. A. Duhm, Associate Editor Physics, Springer and Mrs. E. Suer, assistant to Dr. Ascheron. We are grateful to Ms. S.M. Adhikari and N. Paitya for their kind help. We are indebted to our beloved parents from every fathomable point of view and also to our families for instilling in us the thought that the academic output = ((desire × determination × dedication) − (false enhanced self ego pretending like a true friend although a real unrecognizable foe)). We firmly believe that our Mother Nature has propelled this joint collaboration in her own unseen way in spite of several insurmountable obstacles.

Kolkata
Bangalore
April 2010

K.P. Ghatak
S. Bhattacharya

References

1. G.E. Simmons, *Differential Equations with Application and Historical Notes*, International Series in Pure and Applied Mathematics (McGraw-Hill, USA, 1991)
2. H. Huff (ed.), *Into the Nano Era – Moore's Law beyond Planar Silicon CMOS*, Springer Series in Materials Science, vol 106 (Springer-Verlag, Germany, 2009)
3. T.J. Seebeck, Abh. K. Akad. Wiss. Berlin, 289 (1821)
4. T.J. Seebeck, Abh. K. Akad. Wiss. Berlin, 265 (1823)
5. T.J. Seebeck, Ann. Phys. (Leipzig) **6**, 1 (1826)
6. T.J. Seebeck, Schweigger's J. Phys. **46**, 101 (1826)
7. A.F. Ioffe, *Semiconductor Thermoelements and Thermoelectric Cooling* (Inforsearch, London, 1957)
8. R.W. Ure, R.R. Heikes (eds.), *Thermoelectricity: Science and Engineering* (Interscience, London, 1961)
9. T.C. Harman, J.M. Honig, *Thermoelectric and Thermomagnetic Effects and Applications* (McGraw-Hill, New York, 1967)
10. R. Kim, S. Datta, M.S. Lundstrom, J. Appl. Phys. **105**, 034506 (2009)
11. C. Herring, Phys. Rev. **96**, 1163 (1954)
12. T.M. Tritt (ed.), *Semiconductors and Semimetals, Vols. 69, 70 and 71: Recent Trends in Thermoelectric Materials Research I, II and III* (Academic Press, USA, 2000)
13. D.M. Rowe (ed.), *CRC Handbook of Thermoelectrics* (CRC Press, USA, 1995)
14. D.M. Rowe, C.M. Bhandari, *Modern Thermoelectrics* (Reston Publishing Company, Virginia, 1983)
15. D.M. Rowe (ed.), *Thermoelectrics Handbook: Macro to Nano* (CRC Press, USA, 2006)
16. I.M. Tsidil'kovski, *Thermomagnetic Effects in Semiconductors* (Academic Press, New York, 1962), p. 290
17. J. Tauc, *Photo and Thermoelectric Effects in Semiconductors* (Pergamon Press, New York, 1962)
18. G.S. Nolas, J. Sharp, J. Goldsmid, *Thermoelectrics: Basic Principles and New Materials Developments*, Springer Series in Materials Science, vol 45 (Springer-Verlag, Germany, 2001)
19. J. Hajdu, G. Landwehr, in *Strong and Ultrastrong Magnetic Fields and Their Applications*, Topics in Applied Physics, vol 57, ed. by F. Herlach (Springer-Verlag, Germany, 1985), p. 97
20. I.M. Tsidilkovskii, *Band Structures of Semiconductors* (Pergamon Press, London, 1982)
21. K.P. Ghatak, S. Bhattacharya, S. Bhowmik, R. Benedictus, S. Choudhury, J. Appl. Phys. **103**, 034303 (2008)
22. W. Zawadzki, in *Two-Dimensional Systems, Heterostructures and Superlattices*, Springer Series in Solid-State Science, Vol. 53, ed. by G. Bauer, F. Kuchar, H. Heinrich (Springer-Verlag, Germany, 1984)

23. W. Zawadzki, *11th International Conference on the Physics of Semiconductors*, vol 1 (Elsevier Publishing Company, Netherlands, 1972)
24. S.P. Zelenin, A.S. Kondrat'ev, A.E. Kuchma, Sov. Phys. Semiconduct. **16**, 355 (1982)
25. M. Gad-el-Hak, Phys. Today **57**(3), 61 (2004)
26. P. Gwynne, Phys. World **15**(5), 5, (2002)
27. T. Choy, M. Stoneham, Mater. Today **7**(4) 64 (2004)

Contents

List of Symbols

α	Band nonparabolicity parameter
α_{11}, α_{12}	Energy-band constants
$\overline{\alpha}_{11}, \alpha_{22}, \alpha_{33}, \alpha_{23}$	Spectrum constants
$\overline{\alpha}_{11}, \overline{\alpha}_{22}, \overline{\alpha}_{33}, \overline{\alpha}_{23}$	
$\beta_1, \beta_2, \beta_4, \beta_5$	System constants
γ_{11}, γ_{12}	Energy-band constants
$\beta_{11}, \beta_{12}, \gamma_1, \gamma_5$	
δ	Crystal field splitting constant
δ'	Dirac's delta function
δ_0	Band constant
$\Delta_{\parallel}$	Spin–orbit splitting constant parallel to the C-axis
$\Delta_{\perp}$	Spin–orbit splitting constant perpendicular to the C-axis
Δ	Isotropic spin–orbit splitting constant
Δ_0	Interface width in superlattices
Δ_1	Energy-band constant
Δ_c', Δ_c''	Spectrum constants
λ	Wavelength
λ_0	Band constant
$\hat{\varepsilon}$	Strain tensor
ε	Trace of the strain tensor
$\overline{\varepsilon}$	Energy as measured from the center of the band gap E_{g_0}
ε_{sc}	Semiconductor permittivity
ε_0	Permittivity of vacuum
$\zeta(2r)$	Zeta function of order $2r$
$\overline{\zeta}_0$	Constant of the spectrum
$\Gamma(j+1)$	Complete Gamma function
ω_0	Cyclotron resonance frequency
υ	Frequency
$\theta_1(k)$	Warping of the Fermi surface
$\theta_2(k)$	Inversion asymmetry splitting of the conduction band
Γ_i	Broadening parameter
Ω	Thermodynamic potential
$\overline{a}$	Constant of the spectrum

$\overline{\overline{a}}$	Lattice constant
a_c	Nearest neighbor C–C bonding distance
a_{13}	Nonparabolicity constant
a_{15}	Warping parameter
a_0	The width of the barrier for superlattice structures
$\overline{A}$	Spectrum constant
A_i	Energy band constants
b_0	The width of the well for superlattice structures
B	Quantizing magnetic field
B_2	Momentum matrix element
c	Velocity of light
$\bar{c}$	Constant of the spectrum
C_0	Splitting of the two-spin states by the spin orbit coupling and the crystalline field
C_1	Conduction band deformation potential
C_2	Strain interaction between the conduction and valance bands
d_x, d_y, d_z	Nano thickness along the x, y, and z-directions
e	Magnitude of electron charge
E	Total energy of the carrier
E'	Sub-band energy
$\overline{\overline{E}}$	Energy of the hole as measured from the top of the valance band in the vertically downward direction
E_F	Fermi energy
E_{F_1}	Fermi energy as measured from the mid of the band gap in the vertically upward direction in connection with nanotubes
E_{FB}	Fermi energy in the presence of magnetic field
$\bar{E}_i$	Energy-band constant
E_{n_z}	Energy of the nth sub-band
E_{FQD}	Fermi energy in the presence of 3D quantization as measured from the edge of the conduction band in the vertically upward direction in the absence of any quantization
E_{F1D}	Fermi energy in the presence of two-dimensional quantization as measured from the edge of the conduction band in the vertically upward direction in the absence of any quantization
E_{F2D}	Fermi energy in the presence of size quantization as measured from the edge of the conduction band in the vertically upward direction in the absence of any quantization
$E_{FQDSLGI}$	Fermi energy in quantum dot superlattices with graded interfaces as measured from the edge of the conduction band in the vertically upward direction in the absence of any quantization
$E_{FQDSLEM}$	Fermi energy in quantum dot effective mass superlattices as measured from the edge of the conduction band in the vertically upward direction in the absence of any quantization

E_{FQWSLGI}	Fermi energy in quantum wire superlattices with graded interfaces as measured from the edge of the conduction band in the vertically upward direction in the absence of any quantization
E_{FQWSLEM}	Fermi energy in quantum wire effective mass superlattices as measured from the edge of the conduction band in the vertically upward direction in the absence of any quantization
E_{FQWSLEML}	Fermi energy in quantum wire effective mass superlattices in the presence of light waves as measured from the edge of the conduction band in the vertically upward direction in the absence of any quantization
E_{FQDSLEML}	Fermi energy in quantum dot effective mass superlattices in the presence of light waves as measured from the edge of the conduction band in the vertically upward direction in the absence of any quantization
E_{go}	Band gap in the absence of any field
E_{B}	Bohr electron energy
E_{QD}	Totally quantized energy
E_{FQD}	Fermi energy in quantum dots as measured from the edge of the conduction band in the vertically upward direction in the absence of any quantization
$E_{\text{F2}DL}$	Fermi energy in ultrathin films in the presence of light waves as measured from the edge of the conduction band in the vertically upward direction in the absence of any quantization
$E_{\text{F0}DL}$	Fermi energy in quantum dots in the presence of light waves as measured from the edge of the conduction band in the vertically upward direction in the absence of any quantization
E_{FBL}	Fermi energy under quantizing magnetic field in the presence of light waves as measured from the edge of the conduction band in the vertically upward direction in the absence of any quantization
E_{F2DBL}	Fermi energy in ultrathin films under quantizing magnetic field in the presence of light waves as measured from the edge of the conduction band in the vertically upward direction in the absence of any quantization
$E_{\text{FBQWSLEML}}$	Fermi energy in quantum well effective mass superlattices under magnetic quantization in the presence of light waves as measured from the edge of the conduction band in the vertically upward direction in the absence of any quantization
$E_{\text{FBQWSLGIL}}$	Fermi energy in quantum well superlattices with graded interfaces under magnetic quantization in the presence of light waves as measured from the edge of the conduction band in the vertically upward direction in the absence of any quantization
$E_{n_x}, E_{n_y}, E_{n_z}$	The quantized energy levels due to infinity deep potential well along the x, y and z-directions
$f(E)$	Fermi–Dirac occupation probability factor

$F_j(\eta)$	One parameter Fermi–Dirac integral of order j
$F_0(\eta_{n_z})$	Special case of the one parameter Fermi–Dirac integral of order j
$F_1(\eta_{n_z})$	Special case of the one parameter Fermi–Dirac integral of order j
g_v	Valley degeneracy
G_0	Thermoelectric power under strong magnetic field
h	Planck's constant
$\hbar$	Dirac's constant ($\equiv h/(2\pi)$)
H	Heaviside step function
i	Integer
I_0	Light intensity
k_0	Constant of the energy spectrum
$\bar{k}_0$	Inverse Bohr radius
k_B	Boltzmann's constant
k	Electron wave vector
l_x	Sample length along x direction
$\bar{l}, \bar{m}, \bar{n}$	Matrix elements of the strain perturbation operator
l	Band constant
L_0	Period of the superlattices
m_0	Free electron mass
m^*	Isotropic effective electron mass at the edge of the conduction band
$m^*_{\parallel}$	Longitudinal effective electron mass at the edge of the conduction band
$m^*_{\perp}$	Transverse effective electron mass at the edge of the conduction band
m_1	Effective carrier mass at the band-edge along x direction
m_2	Effective carrier mass at the band-edge along y direction
m_3	The effective carrier mass at the band-edge along z direction
m'_2	Effective-mass tensor component at the top of the valence band (for electrons) or at the bottom of the conduction band (for holes)
m^*_t	The transverse effective mass at $k = 0$
m^*_l	The longitudinal effective mass at $k = 0$
$m^*_{\perp,1}, m^*_{\parallel,1}$	Transverse and longitudinal effective electron mass at the edge of the conduction band for the first material in superlattice
m_r	Reduced mass
m_v	Effective mass of the heavy hole at the top of the valance band in the absence of any field
m^*_v	Effective mass of the holes at the top of the valence band
$m^{\pm}_t$	Contributions to the transverse effective mass of the external L_6^+ and L_6^- bands arising from the $\vec{k} \cdot \vec{p}$ perturbations with the other bands taken to the second order

$m_l^{\pm}$	Contributions to the longitudinal effective mass of the external L_6^+ and L_6^- bands arising from the $\vec{k} \cdot \vec{p}$ perturbations with the other bands taken to the second order
m_{tc}	Transverse effective electron mass of the conduction electrons at the edge of the conduction band
m_{lc}	Longitudinal effective electron mass of the conduction electrons at the edge of the conduction band
m_{tv}	Transverse effective hole mass of the holes at the edge of the valence band
m_{lv}	Longitudinal effective hole mass of the holes at the edge of the valence band
n_x, n_y, n_z	Size quantum numbers along the x, y, and z-directions
n_0	Carrier degeneracy
n	Landau quantum number/chiral indices
$\bar{\bar{n}}$	Band constant
$N(E)$	Density of states in bulk specimens
N_c	Effective number of states in the conduction band
$N_{2D}(E)$	2D density-of-states function per sub-band
$N_{2DT}(E)$	Total 2D density-of-states function
P_0	Momentum matrix element
$(\bar{P})$	Energy-band constant
$P_{\parallel}, P_{\perp}$	Momentum matrix elements parallel and perpendicular to the direction of C-axis
$\bar{Q}, \bar{R}$	Spectrum constants
r_0	Radius of the nanotube
$\bar{s}$	Spectrum constant
s_0	Upper limit of the summation
$\overline{S}_0$	Entropy per unit volume
t_c	Tight binding parameter
t_i	Energy band constants
T	Temperature
v	Band constant
$\bar{v}_0, \bar{w}_0$	Constants of the spectrum
V_0	Potential barrier
$\|V_G\|$	Constants of the energy spectrum
x, y	Alloy compositions

Part I
Thermoelectric Power Under Large Magnetic Field in Quantum Confined Materials

Chapter 1
Thermoelectric Power in Quantum Dots Under Large Magnetic Field

1.1 Introduction

In recent years, with the advent of Quantum Hall Effect (QHE) [1,2], there has been considerable interest in studying the thermoelectric power under strong magnetic field (TPSM) in various types of nanostructured materials having quantum confinement of their charge carriers in one, two, and three dimensions of the respective wave-vector space leading to different carrier energy spectra [3–38]. The classical TPSM equation as mentioned in the preface is valid only under the condition of carrier nondegeneracy, is being independent of carrier concentration, and reflects the fact that the signature of the band structure of any material is totally absent in the same.

Zawadzki [8] demonstrated that the TPSM for electronic materials having degenerate electron concentration is essentially determined by their respective energy band structures. It has, therefore, different values in different materials and changes with the doping; with the magnitude of the reciprocal quantizing magnetic field under magnetic quantization, quantizing electric field as in inversion layers, and nanothickness as in quantum wells, wires, and dots; and with the superlattice period as in quantum-confined semiconductor superlattices with graded interfaces having various carrier energy spectra and also in other types of field-assisted nanostructured materials. Some of the significant features that have emerged from these studies are:

(a) The TPSM decreases with the increase in carrier concentration.
(b) The TPSM decreases with increasing doping in heavily doped semiconductors forming band tails.
(c) The nature of variations is significantly influenced by the spectrum constants of various materials having different band structures.
(d) The TPSM exhibits oscillatory dependence with inverse quantizing magnetic field because of the Shubnikov–de Haas effect.
(e) The TPSM decreases with the magnitude of the quantizing electric field in inversion layers.
(f) The TPSM exhibits composite oscillations with significantly different values in superlattices and various other quantized field aided structures.

In this chapter, an attempt is made to investigate the TPSM in quantum dots of nonlinear optical, III–V, II–VI, GaP, Ge, Te, Graphite, $PtSb_2$, zerogap, II–V, GaSb, stressed materials, Bismuth, IV–VI, Lead Germanium Telluride, Zinc and Cadmium diphosphides, Bi_2Te_3, and Antimony from Sects. 1.2.1 to 1.2.18, respectively. In this context, it may be noted that with the advent of fine line lithography [39], molecular beam epitaxy [40, 41], organometallic vapor-phase epitaxy [42], and other experimental techniques, low-dimensional structures [43–55] having quantum confinement of the charge carriers in one, two, and three dimensions [such as ultrathin films (UFs), nipi structures, inversion and accumulation layers, quantum well superlattices, carbon nanotubes, quantum wires (QWs), quantum wire superlattices, quantum dots (QDs), magnetoinversion and accumulation layers, quantum dot superlattices, etc.] have, in the last few years, created tremendous passion among the interdisciplinary researchers not only for the potential of these quantized structures in uncovering new phenomena in nanostructured science but also for their new diverse technological applications. As the dimension of the UFs increases from one dimension to three dimension, the degree of freedom of the free carriers decreases drastically and the density-of-states function changes from the Heaviside step function in UFs to the Dirac's delta function in QDs [56, 57].

The QDs can be used for visualizing and tracking molecular processes in cells using standard fluorescence microscopy [58–61]. They display minimal photobleaching [62], thus allowing molecular tracking over prolonged periods, and consequently, single molecule can be tracked by using optical fluorescence microscopy [63, 64]. The salient features of quantum dot lasers [65–67] include lower threshold currents, higher power, and greater stability compared with that of the conventional one, and the QDs find extensive applications in nanorobotics [68–71], neural networks [72–74], and high density memory or storage media [75]. The QDs are also used in nanophotonics [76] because of their theoretically high quantum yield and have been suggested as implementations of qubits for quantum information processing [77]. The QDs also find applications in diode lasers [78], amplifiers [79, 80], and optical sensors [81, 82]. High-quality QDs are well suited for optical encoding [83, 84] because of their broad excitation profiles and narrow emission spectra. The new generations of QDs have far-reaching potential for the accurate investigations of intracellular processes at the single-molecule level, high-resolution cellular imaging, long-term in vivo observation of cell trafficking, tumor targeting, and diagnostics [85, 86]. The QD nanotechnology is one of the most promising candidates for use in solid-state quantum computation [87, 88]. It may also be noted that the QDs are being used in single electron transistors [89, 90], photovoltaic devices [91, 92], photoelectrics [93], ultrafast all-optical switches and logic gates [94–97], organic dyes [98–100], and in other types of nanodevices.

Section 1.2.1 investigates the TPSM in QDs of nonlinear optical materials (taking n-$CdGeAs_2$ as an example), which find applications in nonlinear optics and light-emitting diodes [101]. The quasicubic model can be used to investigate the symmetric properties of both the bands at the zone center of wave-vector space of the same compound [102]. Including the anisotropic crystal potential in the Hamiltonian and the special features of the nonlinear optical compounds, Kildal [103] formulated the electron dispersion law under the assumptions of isotropic momentum matrix and the isotropic spin–orbit splitting constant, respectively, although the anisotropies in the two aforementioned band constants are the significant physical features of the said materials [104–106].

In this context, it may be noted that the III–V compounds find potential applications in infrared detectors [107], quantum dot light-emitting diodes [108], quantum cascade lasers [109], quantum well wires [110], optoelectronic sensors [111], high electron mobility transistors [112], etc. The III–V, ternary and quaternary materials are called the Kane-type compounds, since their electron energy spectra are being defined by the three-band model of Kane [113]. In Sect. 1.2.2, the TPSM from QDs of III–V materials has been studied, and the simplified results for two-band model of Kane and that of wide gap materials have further been demonstrated as special cases. Besides Kane, the conduction electrons of III–V materials also obey another six different types of electron dispersion laws as given in the literature. The TPSM has also been investigated for all the cases for the purpose of complete presentation and relative assessment among the energy band models of III–V compounds.

The II–VI compounds are being extensively used in nanoribbons, blue green diode lasers, photosensitive thin films, infrared detectors, ultrahigh-speed bipolar transistors, fiber-optic communications, microwave devices, photovoltaic and solar cells, semiconductor gamma-ray detector arrays, and semiconductor detector gamma camera and allow for a greater density of data storage on optically addressed compact discs [114–121]. The carrier energy spectra in II–VI materials are defined by the Hopfield model [122], where the splitting of the two-spin states by the spin–orbit coupling and the crystalline field has been taken into account. Section 1.2.3 contains the investigation of the TPSM in QDs of II–VI compounds, taking p-CdS as an example.

The n-Gallium Phosphide (n-GaP) is being used in quantum dot light-emitting diode [123], high-efficiency yellow solid state lamps, light sources, and high peak current pulse for high gain tubes. The green and yellow light-emitting diodes made of nitrogen-doped n-GaP possess a longer device life at high drive currents [124–126]. In Sect. 1.2.4, the TPSM in QDs of n-GaP is studied. The importance of Germanium is already well known since the inception of transistor technology, and in recent years, memory circuits, single photon detectors, single photon avalanche diode, ultrafast all-optical switch, THz lasers, and THz spectrometers [127–130] are made of Ge. In Sect. 1.2.5, the TPSM has been studied in QDs of Ge.

Tellurium (Te) is also an elemental semiconductor which has been used as the semiconductor layer in thin-film transistors (TFT) [131]. Te also finds extensive applications in CO_2 laser detectors [132], electronic imaging, strain sensitive devices [133, 134], and multichannel Bragg cell [135]. Section 1.2.6 contains the investigation of TPSM in QDs of Tellurium. The importance of graphite is already well known in the whole spectrum of materials science, and the low-dimensional graphite is used instead of carbon wire in many practical applications. Graphite intercalation compounds are often used as suitable model for investigation of low-dimensional systems and, in particular, for investigation of phase transition in such systems [136–139]. In Sect. 1.2.7, the TPSM in QDs of graphite has been explored. Platinum Antimonide ($PtSb_2$) finds applications in device miniaturization, colloidal nanoparticle synthesis, sensors and detector materials, and thermo-photovoltaic devices [140–142]. The TPSM in QDs of p-$PtSb_2$ has been investigated in Sect. 1.2.8.

Zerogap compounds are used in optical waveguide switch or modulators that can be fabricated by using the electro-optic and thermo-optic effects for facilitating optical communications and signal processing. The gapless materials also find extensive applications in infrared detectors and night vision cameras [143–147]. Section 1.2.9 contains the study of TPSM in QDs of the same taking p-HgTe as an example.

The II–V materials are used in photovoltaic cells constructed of single crystal materials in contact with electrolyte solutions. Cadmium selenide shows an open-circuit voltage of 0.8 V and power conservation coefficients of nearly 6% for 720-nm light [148]. They are also used in ultrasonic amplification [149]. The thin film transistor using cadmium selenide as the semiconductor has been developed [150, 151]. In Sect. 1.2.10, the TPSM in QDs of II–V materials has been studied taking CdSb as an example. Gallium antimonide (GaSb) finds applications in the fiber-optic transmission window, heterojunctions, and quantum wells. A complementary heterojunction field effect transistor (CHFET) in which the channels for the p-FET device and the n-FET device forming the complementary FET are formed from GaSb. The band gap energy of GaSb makes it suitable for low power operation [152–157]. In Sect. 1.2.11, the TPSM in QDs of GaSb has been studied.

It may be noted that the stressed materials are being widely investigated for strained silicon transistors, quantum cascade lasers, semiconductor strain gages, thermal detectors, and strained-layer structures [158–161]. The TPSM in QDs of stressed materials (taking stressed n-InSb as an example) has been investigated in Sect. 1.2.12. In recent years, Bismuth (Bi) nanolines are fabricated, and Bi also finds use in array of antennas which leads to the interaction of electromagnetic waves with such Bi nanowires [162, 163]. Several dispersion relations of the carriers have been proposed for Bi. Shoenberg [164, 165] experimentally verified that the de Haas–Van Alphen and cyclotron resonance experiments supported the ellipsoidal parabolic model of Bi, although, the magnetic field dependence of many physical properties of Bi supports the two-band model [166]. The experimental investigations on the magneto-optical [167] and the ultrasonic quantum oscillations [168] support the Lax ellipsoidal nonparabolic model [166]. Kao [169], Dinger and Lawson [170], and Koch and Jensen [171] demonstrated that the Cohen model [172] is in conformity with the experimental results in a better way. Besides, the Hybrid model of bismuth, as developed by Takoka et al., also finds use in the literature [173]. McClure and Choi [174] devised a new model of Bi and they showed that it can explain the data for a large number of magneto-oscillatory and resonance experiments. In Sect. 1.2.13, we have formulated the TPSM in QDs of Bi in accordance with the aforementioned energy band models for the purpose of relative assessment.

Lead chalcogenides (PbTe, PbSe, and PbS) are IV–VI compounds whose studies over several decades have been motivated by their importance in infrared IR detectors, lasers, light-emitting devices, photovoltaics, and high-temperature thermoelectric [175–179]. PbTe, in particular, is the end compound of several ternary and quaternary high-performance high-temperature thermoelectric materials [180–184]. It has been used not only as bulk but also as films [185–188], quantum wells [189], superlattices [190, 191], nanowires [192], and colloidal and embedded nanocrystals [193–196]. PbTe films doped with various impurities have also been investigated [197–200]. These studies revealed some of the interesting features that have been observed in bulk PbTe, such as Fermi level pinning, and in the case of superconductivity [201]. In Sects. 1.2.14 and 1.2.15, the TPSM in QDs of IV–VI materials $Pb_{1-x}Ge_xTe$ has been studied.

The diphosphides find prominent role in biochemistry where the folding and structural stabilization of many important extracellular peptide and protein molecules, including hormones, enzymes, growth factors, toxins, and immunoglobulin, are concerned [202–204]. Besides, artificial introduction of extra diphosphides into peptides or proteins can improve biological activity [205, 206] or confer thermostability [207]. The asymmetric diphosphide bond formation in peptides containing a free thiol group takes place over a wide pH range in aqueous buffers and can be crucially monitored by spectrophotometric titration of the released 3-nitro-2-pyridinethiol [208, 209]. In Sect. 1.2.16, the TPSM in QDs of zinc and cadmium diphosphides has been investigated.

Bismuth telluride (Bi_2Te_3) was first identified as a material for thermoelectric refrigeration in 1954 [210] and its physical properties were later improved by the addition of bismuth selenide and antimony telluride to form solid solutions [211–215]. The alloys of Bi_2Te_3 are very important compounds for the thermoelectric industry and have extensively been investigated in the literature [211–215]. In Sect. 1.2.17, the TPSM in QDs of Bi_2Te_3 has been considered. In recent years, antimony has emerged to be very promising, since glasses made from antimony are being extensively used in near infrared spectral range for third- or second-order nonlinear processes. The chalcogenide glasses are in general associated with high nonlinear properties for their Infrared transmission from 0.5–1 μm to 12–18 μm [216–221]. Alloys of Sb are used as ultrahigh-frequency indicators and in thin-film thermocouple [216–221]. In Sect. 1.2.18, the TPSM in QDs of Sb has been studied. Section 1.3 contains results and discussion for this chapter. Section 1.4 contains the open research problems pertinent to this chapter.

1.2 Theoretical Background

1.2.1 Magnetothermopower in Quantum Dots of Nonlinear Optical Materials

The form of **k.p** matrix for nonlinear optical compounds can be expressed extending Bodnar [222] as

$$H = \begin{bmatrix} H_1 & H_2 \\ H_2^+ & H_1 \end{bmatrix}, \qquad (1.1)$$

where

$$H_1 \equiv \begin{bmatrix} E_{g0} & 0 & P_{\|}k_z & 0 \\ 0 & (-2\Delta_{\|}/3) & \left(\sqrt{2}\Delta_{\perp}/3\right) & 0 \\ P_{\|}k_z & \left(\sqrt{2}\Delta_{\perp}/3\right) & -\left(\delta + \frac{1}{3}\Delta_{\|}\right) & 0 \\ 0 & 0 & 0 & 0 \end{bmatrix}$$

and

$$H_2 \equiv \begin{bmatrix} 0 & -f_{,+} & 0 & f_{,-} \\ f_{,+} & 0 & 0 & 0 \\ 0 & 0 & 0 & 0 \\ f_{,+} & 0 & 0 & 0 \end{bmatrix}$$

in which E_{g0} is the band gap in the absence of any field; $P_{\|}$ and $P_{\perp}$ the momentum matrix elements parallel and perpendicular to the direction of crystal axis, respectively; δ the crystal field splitting constant; and $\Delta_{\|}$ and $\Delta_{\perp}$ are the spin–orbit splitting constants parallel and perpendicular to the C-axis, respectively, $f_{,\pm} \equiv (P_{\perp}/\sqrt{2})\left(k_x \pm ik_y\right)$ and $i = \sqrt{-1}$. Thus, neglecting the contribution of the higher bands and the free electron term, the diagonalization of the above matrix leads to the dispersion relation of the conduction electrons in bulk specimens of nonlinear optical compounds [223] as

$$\gamma(E) = f_1(E)k_s^2 + f_2(E)k_z^2, \tag{1.2}$$

where

$$\gamma(E) \equiv E(E+E_{g0})\Big[\left(E+E_{g0}\right)\left(E+E_{g0}+\Delta_{\|}\right) + \delta\left(E+E_{g0}+\frac{2}{3}\Delta_{\|}\right) + \frac{2}{9}\left(\Delta_{\|}^2 - \Delta_{\perp}^2\right)\Big], \quad k_s^2 = k_x^2 + k_y^2,$$

$$f_1(E) \equiv \frac{\hbar^2 E_{g0}\left(E_{g0}+\Delta_{\perp}\right)}{\left[2m_{\perp}^*\left(E_{g0}+\frac{2}{3}\Delta_{\perp}\right)\right]}\Big[\delta\left(E+E_{g0}+\frac{1}{3}\Delta_{\|}\right) + \left(E+E_{g0}\right)\left(E+E_{g0}+\frac{2}{3}\Delta_{\|}\right) + \frac{1}{9}\left(\Delta_{\|}^2 - \Delta_{\|}^2\right)\Big],$$

$$f_2(E) \equiv \frac{\hbar^2 E_{g0}\left(E_{g0}+\Delta_{\|}\right)}{\left[2m_{\|}^*\left(E_{g0}+\frac{2}{3}\Delta_{\|}\right)\right]}\left[\left(E+E_{g0}\right)\left(E+E_{g0}+\frac{2}{3}\Delta_{\|}\right)\right]$$

and $m_{\|}^*$ and $m_{\perp}^*$ are the longitudinal and transverse effective electron masses at the edge of the conduction band, respectively.

Let E_{n_i} ($i = x, y$ and z) be the quantized energy levels due to infinitely deep potential well along ith axis with $n_i = 1, 2, 3\ldots$ as the size quantum numbers. Therefore, from (1.2), one can write

$$\gamma\left(E_{n_x}\right) = f_1\left(E_{n_x}\right)\left(\frac{\pi n_x}{d_x}\right)^2 \tag{1.3}$$

$$\gamma\left(E_{n_y}\right) = f_1\left(E_{n_y}\right)\left(\frac{\pi n_y}{d_y}\right)^2 \tag{1.4}$$

$$\gamma\left(E_{n_z}\right) = f_2\left(E_{n_z}\right)\left(\frac{\pi n_z}{d_z}\right)^2 \tag{1.5}$$

From (1.2), the totally quantized energy $\left(E_{\mathrm{QD1}}\right)$ in this case can be expressed as

$$\gamma\left(E_{\mathrm{QD1}}\right) = f_1\left(E_{\mathrm{QD1}}\right)\left[\left(\frac{\pi n_x}{d_x}\right)^2 + \left(\frac{\pi n_y}{d_y}\right)^2\right] + f_2\left(E_{\mathrm{QD1}}\right)\left[\left(\frac{\pi n_z}{d_z}\right)^2\right] \tag{1.6}$$

The total electron concentration per unit volume in this case assumes the form

$$n_0 = \frac{2g_{\mathrm{v}}}{d_x d_y d_z}\sum_{n_x=1}^{n_{x_{\max}}}\sum_{n_y=1}^{n_{y_{\max}}}\sum_{n_z=1}^{n_{z_{\max}}}\frac{L_{11}}{M_{11}}, \tag{1.7}$$

where g_{v} is the valley degeneracy,

$$L_{11} = \left[1 + A_1 \cos H_1\right] \tag{1.8}$$

$$M_{11} = 1 + A_1^2 + 2A_1 \cos H_1 \tag{1.9}$$

in which

$$A_1 = \exp\left[\frac{E_{\mathrm{QD1}} - E_{\mathrm{FQD}}}{k_{\mathrm{B}}T}\right],$$

E_{FQD} is the Fermi energy in the presence of three-dimensional quantization as measured from the edge of the conduction band in the vertically upward direction in the absence of any quantization; T the temperature; $H_1 = \Gamma_1/k_{\mathrm{B}}T$; and Γ_1 is the broadening parameter in this case.

The TPSM (G_0) can, in general, be expressed as [3]

$$G_0 = \frac{1}{e}\left(\frac{\partial \overline{S_0}}{\partial n_0}\right)_{E_F,T}, \tag{1.10}$$

where E_{F} is the Fermi energy corresponding to the electron concentration n_0 and $\overline{S_0}$ is the entropy per unit volume which can be written as

$$\overline{S_0} = -\left.\frac{\partial \Omega}{\partial T}\right|_{E=E_{\mathrm{F}}} \tag{1.11}$$

in which Ω is the thermodynamic potential which, in turn, can be expressed in accordance with the Fermi–Dirac statistics as

$$\Omega = -k_{\mathrm{B}}T \sum \ln \left| 1 + \exp \left[\frac{E_{\mathrm{F}} - E_{\delta_0}}{k_{\mathrm{B}}T} \right] \right|, \tag{1.12}$$

where the summation is carried out over all the possible δ_0 states.

Thus, combining (1.10)–(1.12), the magnitude of the TPSM can be written in a simplified form as [10]

$$G_0 = \left(\pi^2 k_{\mathrm{B}}^2 T / 3en_0\right) \left(\frac{\partial n_0}{\partial E_{\mathrm{F}}} \right) \tag{1.13}$$

It should be noted that being a thermodynamic relation and temperature-induced phenomena, the TPSM as expressed by (1.13), in general, is valid for electronic materials having arbitrary dispersion relations and their nanostructures. In addition to bulk materials in the presence of strong magnetic field, (1.13) is valid under one-, two-, and three-dimensional quantum confinement of the charge carriers (such as quantum wells in ultrathin films, nipi structures, inversion and accumulation layers, quantum well superlattices, carbon nanotubes, quantum wires, quantum wire superlattices, quantum dots, magnetoinversion and accumulation layers, quantum dot superlattices, magneto nipis, quantum well superlattices under magnetic quantization, ultrathin films under magnetic quantization, etc.). The formulation of G_0 requires the relation between the electron statistics and the corresponding Fermi energy which is basically the band-structure-dependent quantity and changes under different physical conditions. It is worth remarking to note that the number $\left(\pi^2/3\right)$ has occurred as a consequence of mathematical analysis and is not connected with the well-known Lorenz number. For quantum wells in ultrathin films, nipi structures, inversion and accumulation layers, quantum well superlattices, magnetoinversion and accumulation layers, magneto nipis, quantum well superlattices under magnetic quantization and magnetosize quantization, the carrier concentration is measured per unit area, whereas, for quantum wires, quantum wires under magnetic field, quantum wire superlattices, and such allied systems, the same can be measured per unit length. Besides, for bulk materials under strong magnetic field, quantum dots, quantum dots under magnetic field, quantum dot superlattices, and quantum dot superlattices under magnetic field, the carrier concentration is expressed per unit volume.

The TPSM in this case using (1.7) and (1.13) can be written as

$$G_0 = \frac{\pi^2 k_{\mathrm{B}}}{3e} \left[\sum_{n_x=1}^{n_{x_{\max}}} \sum_{n_y=1}^{n_{y_{\max}}} \sum_{n_z=1}^{n_{z_{\max}}} \frac{L_{11}}{M_{11}} \right]^{-1} \left[\sum_{n_x=1}^{n_{x_{\max}}} \sum_{n_y=1}^{n_{y_{\max}}} \sum_{n_z=1}^{n_{z_{\max}}} \frac{Q_{11}}{(M_{11})^2} \right], \tag{1.14}$$

where

$$Q_{11} = A_1 \left[\left(1 + A_1^2\right) \cos H_1 + 2A_1 \right]. \tag{1.15}$$

1.2.2 *Magnetothermopower in Quantum Dots of III–V Materials*

The dispersion relation of the conduction electrons of III–V compounds are described by the models of Kane (both three and two bands) [224, 225], Stillman et al. [226], Newson and Kurobe [227], Rossler [228], Palik et al. [229], Johnson and Dickey [230], and Agafonov et al. [231], respectively. For the purpose of complete and coherent presentation, the TPSM in QDs of III–V compounds has also been investigated in accordance with the aforementioned different dispersion relations for the purpose of relative comparison as follows.

1.2.2.1 The Three Band Model of Kane

Under the conditions, $\delta = 0$, $\Delta_{\parallel} = \Delta_{\perp} = \Delta$ (isotropic spin–orbit splitting constant), and $m_{\parallel}^{*} = m_{\perp}^{*} = m^{*}$ (isotropic effective electron mass at the edge of the conduction band), (1.2) gets simplified into the form

$$\frac{\hbar^2 k^2}{2m^*} = I(E),\ I(E) \equiv \frac{E\left(E + E_{\mathrm{go}}\right)\left(E + E_{\mathrm{go}} + \Delta\right)\left(E_{\mathrm{go}} + \frac{2}{3}\Delta\right)}{E_{\mathrm{go}}\left(E_{\mathrm{go}} + \Delta\right)\left(E + E_{\mathrm{go}} + \frac{2}{3}\Delta\right)} \tag{1.16}$$

which is known as the three band model of Kane [224, 225] and is often used to study the electronic properties of III–V materials.

The totally quantized energy $\left(E_{\mathrm{QD2}}\right)$in this case assumes the form

$$I\left(E_{\mathrm{QD2}}\right) = \frac{\hbar^2 \pi^2}{2m^*}\left[\left(\frac{n_x}{d_x}\right)^2 + \left(\frac{n_y}{d_y}\right)^2 + \left(\frac{n_z}{d_z}\right)^2\right]. \tag{1.17}$$

The electron concentration is given by

$$n_0 = \frac{2g_{\mathrm{v}}}{d_x d_y d_z} \sum_{n_x=1}^{n_{x_{\max}}} \sum_{n_y=1}^{n_{y_{\max}}} \sum_{n_z=1}^{n_{z_{\max}}} \frac{L_{12}}{M_{12}}, \tag{1.18}$$

where $L_{12} = [1 + A_2 \cos H_2]$, $A_2 = \exp[E_{\mathrm{QD2}} - E_{\mathrm{FQD}}/k_{\mathrm{B}}T]$, $H_2 = \Gamma_2/k_{\mathrm{B}}T$, Γ_2 is the broadening parameter in this case, and $M_{12} = 1 + A_2^2 + 2A_2 \cos H_2$.

The TPSM in this case, using (1.13) and (1.18), can be expressed as

$$G_0 = \frac{\pi^2 k_{\mathrm{B}}}{3e}\left[\sum_{n_x=1}^{n_{x_{\max}}} \sum_{n_y=1}^{n_{y_{\max}}} \sum_{n_z=1}^{n_{z_{\max}}} \frac{L_{12}}{M_{12}}\right]^{-1}\left[\sum_{n_x=1}^{n_{x_{\max}}} \sum_{n_y=1}^{n_{y_{\max}}} \sum_{n_z=1}^{n_{z_{\max}}} \frac{Q_{12}}{(M_{12})^2}\right], \tag{1.19}$$

where $Q_{12} = A_2\left[\left(1 + A_2^2\right) \cos H_2 + 2A_2\right]$.

1.2.2.2 The Two Band Model of Kane

Under the inequalities $\Delta \gg E_{g_0}$ or $\Delta \ll E_{g_0}$, (1.16) assumes the form

$$E(1+\alpha E) = \left(\hbar^2 k^2 / 2m^*\right), \; \alpha \equiv 1/E_{g_0}. \tag{1.20}$$

Equation (1.20) is known as the two-band model of Kane and should be as such for studying the electronic properties of the materials whose band structures obey the above inequalities [224, 225].

The totally quantized energy E_{QD3} in this case is given by

$$E_{QD3}\left(1+\alpha E_{QD3}\right) = \frac{\hbar^2 \pi^2}{2m^*}\left[\left(\frac{n_x}{d_x}\right)^2 + \left(\frac{n_y}{d_y}\right)^2 + \left(\frac{n_z}{d_z}\right)^2\right]. \tag{1.21}$$

The electron concentration can be written as

$$n_0 = \frac{2g_v}{d_x d_y d_z} \sum_{n_x=1}^{n_{x_{max}}} \sum_{n_y=1}^{n_{y_{max}}} \sum_{n_z=1}^{n_{z_{max}}} \frac{L_{13}}{M_{13}}, \tag{1.22}$$

where $L_{13} = [1 + A_3 \cos H_3]$, $A_3 = \exp[E_{QD3} - E_{FQD}/k_B T]$, $H_3 = \Gamma_3/k_B T$, Γ_3 is the broadening parameter in this case, and $M_{13} = 1 + A_3^2 + 2A_3 \cos H_3$.

The TPSM in this case, using (1.13) and (1.22), can be expressed as

$$G_0 = \frac{\pi^2 k_B}{3e}\left[\sum_{n_x=1}^{n_{x_{max}}} \sum_{n_y=1}^{n_{y_{max}}} \sum_{n_z=1}^{n_{z_{max}}} \frac{L_{13}}{M_{13}}\right]^{-1}\left[\sum_{n_x=1}^{n_{x_{max}}} \sum_{n_y=1}^{n_{y_{max}}} \sum_{n_z=1}^{n_{z_{max}}} \frac{Q_{13}}{(M_{13})^2}\right], \tag{1.23}$$

where $Q_{13} = A_3\left[\left(1 + A_3^2\right) \cos H_3 + 2A_3\right]$.

1.2.2.3 The Model of Stillman et al.

In accordance with the model of Stillman et al. [226], the electron dispersion law of III–V materials assumes the form

$$E = t_{11}k^2 - t_{12}k^4 \tag{1.24}$$

where $t_{11} \equiv \hbar^2/2m^*$ and

$$t_{12} \equiv \left(1 - \frac{m^*}{m_0}\right)^2 \left(\frac{\hbar^2}{2m^*}\right)^2 \times\left[\left(3E_{g_0} + 4\Delta + \frac{2\Delta^2}{E_{g_0}}\right) . \left\{\left(E_{g_0} + \Delta\right)\left(2\Delta + 3E_{g_0}\right)\right\}^{-1}\right]$$

Equation (1.24) can be expressed as

$$\frac{\hbar^2 k^2}{2m^*} = I_{11}(E), \tag{1.25}$$

where

$$I_{11}(E) \equiv a_{11}\left[1-(1-a_{12}E)^{1/2}\right],$$
$$a_{11} \equiv \left(\frac{\hbar^2 t_{11}}{4m^* t_{12}}\right), a_{12} \equiv \frac{4t_{12}}{t_{11}^2}.$$

The E_{QD4} in this case can be defined as

$$I_{11}\left(E_{\mathrm{QD4}}\right) = \frac{\hbar^2\pi^2}{2m^*}\left[\left(\frac{n_x}{d_x}\right)^2 + \left(\frac{n_y}{d_y}\right)^2 + \left(\frac{n_z}{d_z}\right)^2\right]. \tag{1.26}$$

The electron concentration is given by

$$n_0 = \frac{2g_{\mathrm{v}}}{d_x d_y d_z}\sum_{n_x=1}^{n_{x_{\max}}}\sum_{n_y=1}^{n_{y_{\max}}}\sum_{n_z=1}^{n_{z_{\max}}}\frac{L_{14}}{M_{14}}, \tag{1.27}$$

where $L_{14} = [1 + A_4 \cos H_4]$, $A_4 = \exp[E_{\mathrm{QD4}} - E_{\mathrm{FQD}}/k_{\mathrm{B}}T]$, $H_4 = \Gamma_4/k_{\mathrm{B}}T$, Γ_4 is the broadening parameter in this case, and $M_{14} = 1 + A_4^2 + 2A_4 \cos H_4$.

The TPSM in this case, using (1.13) and (1.27), can be expressed as

$$G_0 = \frac{\pi^2 k_{\mathrm{B}}}{3e}\left[\sum_{n_x=1}^{n_{x_{\max}}}\sum_{n_y=1}^{n_{y_{\max}}}\sum_{n_z=1}^{n_{z_{\max}}}\frac{L_{14}}{M_{14}}\right]^{-1}\left[\sum_{n_x=1}^{n_{x_{\max}}}\sum_{n_y=1}^{n_{y_{\max}}}\sum_{n_z=1}^{n_{z_{\max}}}\frac{Q_{14}}{(M_{14})^2}\right], \tag{1.28}$$

where $Q_{14} = A_4\left[\left(1 + A_4^2\right)\cos H_4 + 2A_4\right]$.

1.2.2.4 The Model of Newson and Kurobe

In accordance with the model of Newson and Kurobe, the electron dispersion law in this case assumes the form as [227]

$$E = a_{13}k_z^4 + \left[\frac{\hbar^2}{2m^*} + a_{14}k_s^2\right]k_z^2 + \frac{\hbar^2}{2m^*}k_s^2 + a_{14}k_x^2k_y^2 + a_{13}\left(k_x^4 + k_y^4\right), \tag{1.29}$$

where a_{13} is the nonparabolicity constant, a_{14} ($\equiv 2a_{13} + a_{15}$) and a_{15} is known as the warping parameter.

The totally quantized energy E_{QD5} in this case can be written as

$$E_{\mathrm{QD5}} = a_{13}\left(\frac{\pi n_z}{d_z}\right)^4 + \left[\frac{\hbar^2}{2m^*} + a_{14}\left(\left(\frac{\pi n_x}{d_x}\right)^2 + \left(\frac{\pi n_y}{d_y}\right)^2\right)\right]\cdot\left(\frac{\pi n_z}{d_z}\right)^2 + \frac{\hbar^2}{2m^*}\left[\left(\frac{\pi n_x}{d_x}\right)^2 + \left(\frac{\pi n_y}{d_y}\right)^2\right] + a_{14}\pi^4\left(\frac{n_x n_y}{d_x d_y}\right)^2 + a_{13}\pi^4\left[\left(\frac{n_x}{d_x}\right)^4 + \left(\frac{n_y}{d_y}\right)^4\right]. \quad (1.30)$$

The electron concentration is given by

$$n_0 = \frac{2g_{\mathrm{v}}}{d_x d_y d_z}\sum_{n_x=1}^{n_{x_{\max}}}\sum_{n_y=1}^{n_{y_{\max}}}\sum_{n_z=1}^{n_{z_{\max}}}\frac{L_{15}}{M_{15}}, \quad (1.31)$$

where $L_{15} = [1 + A_5 \cos H_5]$, $A_5 = \exp[E_{\mathrm{QD5}} - E_{\mathrm{FQD}}/k_{\mathrm{B}}T]$, $H_5 = \Gamma_5/k_{\mathrm{B}}T$, Γ_5 is the broadening parameter in this case, and $M_{15} = 1 + A_5^2 + 2A_5 \cos H_5$.

The TPSM in this case, using (1.13) and (1.31), can be expressed as

$$G_0 = \frac{\pi^2 k_{\mathrm{B}}}{3e}\left[\sum_{n_x=1}^{n_{x_{\max}}}\sum_{n_y=1}^{n_{y_{\max}}}\sum_{n_z=1}^{n_{z_{\max}}}\frac{L_{15}}{M_{15}}\right]^{-1}\left[\sum_{n_x=1}^{n_{x_{\max}}}\sum_{n_y=1}^{n_{y_{\max}}}\sum_{n_z=1}^{n_{z_{\max}}}\frac{Q_{15}}{(M_{15})^2}\right], \quad (1.32)$$

where $Q_{15} = A_5\left[\left(1 + A_5^2\right)\cos H_5 + 2A_5\right]$.

1.2.2.5 The Model of Rossler

The dispersion relation of the conduction electrons in accordance with the model of Rossler can be written as [228]

$$E = \frac{\hbar^2 k^2}{2m^*} + [\alpha_{11} + \alpha_{12}k]\,k^4 + (\beta_{11} + \beta_{12}k)\left[k_x^2 k_y^2 + k_y^2 k_z^2 + k_z^2 k_x^2\right] \pm [\gamma_{11} + \gamma_{12}k]\left[k^2\left(k_x^2 k_y^2 + k_y^2 k_z^2 + k_z^2 k_x^2\right) - 9k_x^2 k_y^2 k_z^2\right]^{1/2}, \quad (1.33)$$

where $\alpha_{11}, \alpha_{12}, \beta_{11}, \beta_{12}, \gamma_{11}$, and γ_{12} are energy-band constants.

$E_{\mathrm{QD6},\pm}$ in this case assumes the form

$$\begin{aligned}
E_{\mathrm{QD6},\pm} \equiv & \frac{\hbar^2\pi^2}{2m^*}\left[\left(\frac{n_x}{d_x}\right)^2+\left(\frac{n_y}{d_y}\right)^2+\left(\frac{n_z}{d_z}\right)^2\right] \\
& +\left[\alpha_{11}+\alpha_{12}\left[\left(\frac{\pi n_x}{d_x}\right)^2+\left(\frac{\pi n_y}{d_y}\right)^2+\left(\frac{\pi n_z}{d_z}\right)^2\right]^{1/2}\right] \\
& \times\left[\left(\frac{\pi n_x}{d_x}\right)^2+\left(\frac{\pi n_y}{d_y}\right)^2+\left(\frac{\pi n_z}{d_z}\right)^2\right]^2 \\
& +\left[\beta_{11}+\beta_{12}\left[\left(\frac{\pi n_x}{d_x}\right)^2+\left(\frac{\pi n_y}{d_y}\right)^2+\left(\frac{\pi n_z}{d_z}\right)^2\right]^{1/2}\right] \\
& \times\left[\pi^4\left(\frac{n_x n_y}{d_x d_y}\right)^2+\pi^4\left(\frac{n_y n_z}{d_y d_z}\right)^2+\pi^4\left(\frac{n_z n_x}{d_z d_x}\right)^2\right] \\
& \pm\left[\gamma_{11}+\gamma_{12}\left[\left(\frac{\pi n_x}{d_x}\right)^2+\left(\frac{\pi n_y}{d_y}\right)^2+\left(\frac{\pi n_z}{d_z}\right)^2\right]^{1/2}\right] \\
& \times\left[\left[\left(\frac{\pi n_x}{d_x}\right)^2+\left(\frac{\pi n_y}{d_y}\right)^2+\left(\frac{\pi n_z}{d_z}\right)^2\right]\right. \\
& \times\left[\pi^4\left(\frac{n_x n_y}{d_x d_y}\right)^2+\pi^4\left(\frac{n_y n_z}{d_y d_z}\right)^2+\pi^4\left(\frac{n_z n_x}{d_z d_x}\right)^2\right] \\
& \left.-9\pi^6\left(n_x n_y n_z/d_x d_y d_z\right)^6\right]
\end{aligned} \tag{1.34}$$

The electron concentration is given by

$$n_0=\frac{g_v}{d_x d_y d_z}\sum_{n_x=1}^{n_{x_{\max}}}\sum_{n_y=1}^{n_{y_{\max}}}\sum_{n_z=1}^{n_{z_{\max}}}\frac{L_{16,\pm}}{M_{16,\pm}}, \tag{1.35}$$

where $L_{16,\pm}=\left[1+A_{6,\pm}\cos H_6\right]$, $A_{6,\pm}=\exp[E_{\mathrm{QD6},\pm}-E_{\mathrm{FQD}}/k_B T]$, $H_6=\Gamma_6/k_B T$, Γ_6 is the broadening parameter in this case, and $M_{16,\pm}=1+A_{6,\pm}^2+2A_{6,\pm}\cos H_6$.

The TPSM in this case, using (1.13) and (1.35), can be expressed as

$$G_0=\frac{\pi^2 k_B}{3e}\left[\sum_{n_x=1}^{n_{x_{\max}}}\sum_{n_y=1}^{n_{y_{\max}}}\sum_{n_z=1}^{n_{z_{\max}}}\frac{L_{16,\pm}}{M_{16,\pm}}\right]^{-1}\left[\sum_{n_x=1}^{n_{x_{\max}}}\sum_{n_y=1}^{n_{y_{\max}}}\sum_{n_z=1}^{n_{z_{\max}}}\frac{Q_{16,\pm}}{\left(M_{16,\pm}\right)^2}\right], \tag{1.36}$$

where $Q_{16,\pm}=A_{6,\pm}\left[\left(1+A_{6,\pm}^2\right)\cos H_6+2A_{6,\pm}\right]$.

1.2.2.6 The Model of Palik et al.

In accordance with the model of Palik et al. [229], the energy spectrum of the conduction electrons in III–V materials, up to the fourth order in effective mass theory, taking into account the interactions the heavy hole, the light hole, and the split-off bands can be written as [229]

$$E = \frac{\hbar^2 k^2}{2m^*} - b_{11} k^4, \tag{1.37}$$

where

$$b_{11} \equiv \left[\frac{\hbar^4}{4E_{\mathrm{go}}\,(m^*)^2}\right]\left[\frac{1+\frac{x_{11}^2}{2}}{1+\frac{x_{11}}{2}}\right](1-y_1)^2,$$

$$x_{11} \equiv \left[1+\left(\frac{\Delta}{E_{\mathrm{go}}}\right)\right]^{-1},$$

and $y_1 \equiv m^*/m_0$.

From (1.37) we get

$$\frac{\hbar^2 k^2}{2m^*} = I_{12}(E), \tag{1.38}$$

where

$$I_{12}(E) \equiv b_{12}\left[a_{12} - \sqrt{a_{12}^2 - 4Eb_{11}}\right],$$

$b_{12} \equiv a_{12}/2b_{11}$, and $a_{12} \equiv \hbar^2/2m^*$.

The totally quantized energy (E_{QD7}) in this case can be defined as

$$I_{12}\left(E_{\mathrm{QD7}}\right) = \frac{\hbar^2}{2m^*}\left[\left(\frac{\pi n_x}{d_x}\right)^2 + \left(\frac{\pi n_y}{d_y}\right)^2 + \left(\frac{\pi n_z}{d_z}\right)^2\right]. \tag{1.39}$$

The electron concentration is given by

$$n_0 = \frac{2g_{\mathrm{v}}}{d_x d_y d_z} \sum_{n_x=1}^{n_{x_{\max}}} \sum_{n_y=1}^{n_{y_{\max}}} \sum_{n_z=1}^{n_{z_{\max}}} \frac{L_{17}}{M_{17}}, \tag{1.40}$$

where $L_{17} = [1 + A_7 \cos H_7]$, $A_7 = \exp[E_{\mathrm{QD7}} - E_{\mathrm{FQD}}/k_{\mathrm{B}}T]$, $H_7 = \Gamma_7/k_{\mathrm{B}}T$, Γ_7 is the broadening parameter in this case, and $M_{17} = 1 + A_7^2 + 2A_7 \cos H_7$.

The TPSM in this case, using (1.13) and (1.40), can be expressed as

$$G_0 = \frac{\pi^2 k_B}{3e} \left[\sum_{n_x=1}^{n_{x_{\max}}} \sum_{n_y=1}^{n_{y_{\max}}} \sum_{n_z=1}^{n_{z_{\max}}} \frac{L_{17}}{M_{17}} \right]^{-1} \left[\sum_{n_x=1}^{n_{x_{\max}}} \sum_{n_y=1}^{n_{y_{\max}}} \sum_{n_z=1}^{n_{z_{\max}}} \frac{Q_{17}}{(M_{17})^2} \right], \tag{1.41}$$

where $Q_{17} = A_7 \left[\left(1 + A_7^2\right) \cos H_7 + 2A_7 \right]$.

1.2.2.7 The Model of Johnson and Dickey

In accordance with the model of Johnson and Dickey [230], the electron dispersion law in III–V materials assumes the form

$$E = -\frac{E_{g0}}{2} + \frac{\hbar^2 k^2}{2} \left[\frac{1}{m^*} - \frac{1}{m_0} \right] + \frac{E_{g0}}{2} \left[1 + a_{15}(E) k^2 \right]^{1/2}, \tag{1.42}$$

where

$$a_{15}(E) \equiv \frac{4\hbar^2}{2m^* E_{g0}} a_{14}(E),$$

$$a_{14}(E) \equiv \frac{\left(E_{g0} + \Delta\right)\left(E + E_{g0} + \frac{2}{3}\Delta\right)}{\left(E_{g0} + \frac{2}{3}\Delta\right)\left(E + E_{g0} + \Delta\right)},$$

$$\frac{m_0}{m^*} = \left(\bar{P}\right)^2 \left[\frac{\left(E_{g0} + \frac{2}{3}\Delta\right)}{\left(E_{g0}\right)\left(E_{g0} + \Delta\right)} \right]$$

and $\left(\bar{P}\right)$ is the energy band constant in this case.

The E_{QD8} in this case is given by

$$E_{QD8} = -\frac{E_{g0}}{2} + \frac{\hbar^2}{2} \left[\frac{1}{m^*} - \frac{1}{m_0} \right] \left[\left(\frac{\pi n_x}{d_x} \right)^2 + \left(\frac{\pi n_y}{d_y} \right)^2 + \left(\frac{\pi n_z}{d_z} \right)^2 \right]$$
$$+ \frac{E_{g0}}{2} \left[1 + a_{15}\left(E_{QD8}\right) \left[\left(\frac{\pi n_x}{d_x} \right)^2 + \left(\frac{\pi n_y}{d_y} \right)^2 + \left(\frac{\pi n_z}{d_z} \right)^2 \right] \right]^{1/2}. \tag{1.43}$$

The electron concentration can be written as

$$n_0 = \frac{2g_v}{d_x d_y d_z} \sum_{n_x=1}^{n_{x_{\max}}} \sum_{n_y=1}^{n_{y_{\max}}} \sum_{n_z=1}^{n_{z_{\max}}} \frac{L_{18}}{M_{18}}, \tag{1.44}$$

where $L_{18} = [1 + A_8 \cos H_8]$, $A_8 = \exp[E_{QD8} - E_{FQD}/k_B T]$, $H_8 = \Gamma_8/k_B T$, Γ_8 is the broadening parameter in this case, and $M_{18} = 1 + A_8^2 + 2A_8 \cos H_8$.

The TPSM in this case, using (1.13) and (1.44), can be expressed as

$$G_0 = \frac{\pi^2 k_B}{3e}\left[\sum_{n_x=1}^{n_{x_{\max}}}\sum_{n_y=1}^{n_{y_{\max}}}\sum_{n_z=1}^{n_{z_{\max}}}\frac{L_{18}}{M_{18}}\right]^{-1}\left[\sum_{n_x=1}^{n_{x_{\max}}}\sum_{n_y=1}^{n_{y_{\max}}}\sum_{n_z=1}^{n_{z_{\max}}}\frac{Q_{18}}{(M_{18})^2}\right], \tag{1.45}$$

where $Q_{18} = A_8\left[\left(1 + A_8^2\right)\cos H_8 + 2A_8\right]$.

1.2.2.8 The Model of Agafonov et al.

In accordance with the model of Agafonov et al. [231], the electron dispersion law in III–V materials can be written as

$$E = \frac{\bar{y} - E_{g0}}{2}\left[1 - T_5\left(\frac{k_x^4 + k_y^4 + k_z^4}{\bar{y}k^2}\right)\right], \tag{1.46}$$

where

$$(\bar{y})^2 = \left[E_{g0}^2 + \frac{8}{3}P_0^2 k^2\right],$$

$$T_5 = \frac{\bar{D}\sqrt{3} - 3\bar{B}}{2},$$

$$\bar{B} = -21\left(\frac{\hbar^2}{2m_0}\right),$$

$$\bar{D} = -40\left(\frac{\hbar^2}{2m_0}\right)$$

and P_0 is the momentum matrix element.

The totally quantized energy (E_{QD9}) in this case can be written as

$$E_{QD9} = \left(\frac{\psi_{30} - E_{g0}}{2}\right)\left[1 - \left[T_5\left[\left(\frac{\pi n_x}{d_x}\right)^4 + \left(\frac{\pi n_y}{d_y}\right)^4 + \left(\frac{\pi n_z}{d_z}\right)^4\right]\right]\right.$$
$$\left.\times\left[\psi_{30}\left[\left(\frac{\pi n_x}{d_x}\right)^2 + \left(\frac{\pi n_y}{d_y}\right)^2 + \left(\frac{\pi n_z}{d_z}\right)^2\right]\right]^{-1}\right], \tag{1.47}$$

where

$$\psi_{30} = \left[E_{g0}^2 + \frac{8}{3}P_0^2\left[\left(\frac{\pi n_x}{d_x}\right)^2 + \left(\frac{\pi n_y}{d_y}\right)^2 + \left(\frac{\pi n_z}{d_z}\right)^2\right]\right]^{1/2}$$

The electron concentration is given by

$$n_0 = \frac{2g_{\mathrm{v}}}{d_x d_y d_z} \sum_{n_x=1}^{n_{x_{\max}}} \sum_{n_y=1}^{n_{y_{\max}}} \sum_{n_z=1}^{n_{z_{\max}}} \frac{L_{19}}{M_{19}}, \tag{1.48}$$

where $L_{19} = [1 + A_9 \cos H_9]$, $A_9 = \exp E_{\mathrm{QD9}} - E_{\mathrm{FQD}}/k_{\mathrm{B}}T$, $H_9 = \Gamma_9/k_{\mathrm{B}}T$, Γ_9 is the broadening parameter in this case, and $M_{19} = 1 + A_9^2 + 2A_9 \cos H_9$.

The TPSM in this case, using (1.13) and (1.48), can be expressed as

$$G_0 = \frac{\pi^2 k_{\mathrm{B}}}{3e} \left[\sum_{n_x=1}^{n_{x_{\max}}} \sum_{n_y=1}^{n_{y_{\max}}} \sum_{n_z=1}^{n_{z_{\max}}} \frac{L_{19}}{M_{19}} \right]^{-1} \left[\sum_{n_x=1}^{n_{x_{\max}}} \sum_{n_y=1}^{n_{y_{\max}}} \sum_{n_z=1}^{n_{z_{\max}}} \frac{Q_{19}}{(M_{19})^2} \right], \tag{1.49}$$

where $Q_{19} = A_9 \left[\left(1 + A_9^2\right) \cos H_9 + 2A_9 \right]$.

1.2.3 Magnetothermopower in Quantum Dots of II–VI Materials

The carrier energy spectra in bulk specimens of II–VI compounds in accordance with Hopfield model can be written as [122]

$$E = A_0 k_s^2 + B_0 k_z^2 \pm C_0 k_s \tag{1.50}$$

where $A_0 \equiv \hbar^2/2m_\perp^*$, $B_0 \equiv \hbar^2 \big/ 2m_\parallel^*$, and C_0 represents the splitting of the two-spin states by the spin–orbit coupling and the crystalline field.

Using (1.50), the totally quantized energy ($E_{\mathrm{QD10},\pm}$) in this case can be expressed as

$$\begin{aligned} E_{\mathrm{QD10},\pm} &= A_0 \left[\left(\frac{\pi n_x}{d_x} \right)^2 + \left(\frac{\pi n_y}{d_y} \right)^2 \right] + B_0 \left(\frac{\pi n_z}{d_z} \right)^2 \\ &\pm C_0 \left[\left(\frac{\pi n_x}{d_x} \right)^2 + \left(\frac{\pi n_y}{d_y} \right)^2 \right]^{1/2} \end{aligned} \tag{1.51}$$

The carrier concentration is given by

$$n_0 = \frac{g_{\mathrm{v}}}{d_x d_y d_z} \sum_{n_x=1}^{n_{x_{\max}}} \sum_{n_y=1}^{n_{y_{\max}}} \sum_{n_z=1}^{n_{z_{\max}}} \frac{L_{20,\pm}}{M_{20,\pm}}, \tag{1.52}$$

where $L_{20,\pm} = \left[1 + A_{10,\pm} \cos H_{10}\right]$, $A_{10,\pm} = \exp[E_{\text{QD}10,\pm} - E_{\text{FQD}}/k_{\text{B}}T]$, $H_{10} = \Gamma_{10}/k_{\text{B}}T$, Γ_{10} is the broadening parameter in this case, and $M_{20,\pm} = 1 + A_{10,\pm}^2 + 2A_{10,\pm} \cos H_{10}$.

The TPSM in this case, using (1.13) and (1.52), can be expressed as

$$G_0 = \frac{\pi^2 k_{\text{B}}}{3e} \left[\sum_{n_x=1}^{n_{x_{\max}}} \sum_{n_y=1}^{n_{y_{\max}}} \sum_{n_z=1}^{n_{z_{\max}}} \frac{L_{20,\pm}}{M_{20,\pm}} \right]^{-1} \left[\sum_{n_x=1}^{n_{x_{\max}}} \sum_{n_y=1}^{n_{y_{\max}}} \sum_{n_z=1}^{n_{z_{\max}}} \frac{Q_{20,\pm}}{\left(M_{20,\pm}\right)^2} \right], \tag{1.53}$$

where $Q_{20,\pm} = A_{10,\pm} \left[\left(1 + A_{10,\pm}^2\right) \cos H_{10} + 2A_{10,\pm} \right]$.

1.2.4 Magnetothermopower in Quantum Dots of n-Gallium Phosphide

The dispersion relation of the conduction electrons in bulk specimens of n-GaP is given by [232]

$$E = \frac{\hbar^2 k_s^2}{2m_\perp^*} + \frac{\hbar^2}{2m_{\|}^*} \left[k_s^2 + k_z^2\right] - \left[\frac{\hbar^4 k_0^2}{m_{\|}^{*2}} \left(k_s^2 + k_z^2\right) + |V_G|^2 \right]^{1/2} + |V_G|, \tag{1.54}$$

where k_0 and $|V_G|$ are constants of the energy spectrum.

Using (1.54), the totally quantized energy $E_{\text{QD}11}$ in this case is given by

$$\begin{aligned} E_{\text{QD}11} = {} & \frac{\hbar^2}{2m_\perp^*} \left[\left(\frac{\pi n_x}{d_x}\right)^2 + \left(\frac{\pi n_y}{d_y}\right)^2 \right] \\ & + \frac{\hbar^2}{2m_{\|}^*} \left[\left(\frac{\pi n_x}{d_x}\right)^2 \left(\frac{\pi n_y}{d_y}\right)^2 + \left(\frac{\pi n_z}{d_z}\right)^2 \right] \\ & - \left[\frac{\hbar^4 k_0^2}{m_{\|}^{*2}} \left[\left(\frac{\pi n_x}{d_x}\right)^2 \left(\frac{\pi n_y}{d_y}\right)^2 + \left(\frac{\pi n_z}{d_z}\right)^2 \right] + |V_G|^2 \right]^{1/2} + |V_G| \end{aligned} \tag{1.55}$$

The electron concentration can be written as

$$n_0 = \frac{2g_{\text{v}}}{d_x d_y d_z} \sum_{n_x=1}^{n_{x_{\max}}} \sum_{n_y=1}^{n_{y_{\max}}} \sum_{n_z=1}^{n_{z_{\max}}} \frac{L_{21}}{M_{21}}, \tag{1.56}$$

where $L_{21} = [1 + A_{11} \cos H_{11}]$, $A_{11} = \exp[E_{\text{QD}11} - E_{\text{FQD}}/k_{\text{B}}T]$, $H_{11} = \Gamma_{11}/k_{\text{B}}T$, Γ_{11} is the broadening parameter in this case, and $M_{21} = 1 + A_{11}^2 + 2A_{11} \cos H_{11}$.

The TPSM in this case, using (1.13) and (1.56), can be expressed as

$$G_0 = \frac{\pi^2 k_B}{3e} \left[\sum_{n_x=1}^{n_{x_{\max}}} \sum_{n_y=1}^{n_{y_{\max}}} \sum_{n_z=1}^{n_{z_{\max}}} \frac{L_{21}}{M_{21}} \right]^{-1} \left[\sum_{n_x=1}^{n_{x_{\max}}} \sum_{n_y=1}^{n_{y_{\max}}} \sum_{n_z=1}^{n_{z_{\max}}} \frac{Q_{21}}{(M_{21})^2} \right], \tag{1.57}$$

where $Q_{21} = A_{11} \left[\left(1 + A_{11}^2\right) \cos H_{11} + 2A_{11} \right]$.

1.2.5 *Magnetothermopower in Quantum Dots of n-Germanium*

It is well known that the conduction electrons of n-Ge obey two different types of dispersion laws since band nonparabolicity has been included in two different ways as given in the literature [233–235].

(a) The energy spectrum of the conduction electrons in bulk specimens of n-Ge can be expressed in accordance with Cardona et al. [233, 234] as

$$E = -\frac{E_{g0}}{2} + \frac{\hbar^2 k_z^2}{2m_{\parallel}^*} + \left[\frac{E_{g0}^2}{4} + E_{g0} k_s^2 \left(\frac{\hbar^2}{2m_{\perp}^*} \right) \right]^{1/2}, \tag{1.58}$$

where, in this case, $m_{\parallel}^*$ and $m_{\perp}^*$ are the longitudinal and transverse effective masses along $\langle 111 \rangle$ direction at the edge of the conduction band, respectively. The totally quantized energy E_{QD12} in this case is given by

$$E_{\text{QD12}} = -\frac{E_{g0}}{2} + \frac{\hbar^2}{2m_{\parallel}^*} \left(\frac{\pi n_z}{d_z} \right)^2 + \left[\frac{E_{g0}^2}{4} + E_{g0} \left(\frac{\hbar^2}{2m_{\perp}^*} \right) \left\{ \left(\frac{\pi n_x}{d_x} \right)^2 + \left(\frac{\pi n_y}{d_y} \right)^2 \right\} \right]^{1/2} \tag{1.59}$$

The electron concentration can be written as

$$n_0 = \frac{2g_v}{d_x d_y d_z} \sum_{n_x=1}^{n_{x_{\max}}} \sum_{n_y=1}^{n_{y_{\max}}} \sum_{n_z=1}^{n_{z_{\max}}} \frac{L_{22}}{M_{22}}, \tag{1.60}$$

where $L_{22} = [1 + A_{12} \cos H_{12}]$, $A_{12} = \exp[E_{\text{QD12}} - E_{\text{FQD}}/k_B T]$, $H_{12} = \Gamma_{12}/k_B T$, Γ_{12} is the broadening parameter in this case, and $M_{22} = 1 + A_{12}^2 + 2A_{12} \cos H_{12}$.

The TPSM in this case, using (1.13) and (1.60), can be expressed as

$$G_0 = \frac{\pi^2 k_B}{3e}\left[\sum_{n_x=1}^{n_{x_{\max}}}\sum_{n_y=1}^{n_{y_{\max}}}\sum_{n_z=1}^{n_{z_{\max}}}\frac{L_{22}}{M_{22}}\right]^{-1}\left[\sum_{n_x=1}^{n_{x_{\max}}}\sum_{n_y=1}^{n_{y_{\max}}}\sum_{n_z=1}^{n_{z_{\max}}}\frac{Q_{22}}{(M_{22})^2}\right], \tag{1.61}$$

where $Q_{22} = A_{12}\left[\left(1 + A_{12}^2\right)\cos H_{12} + 2A_{12}\right]$.

(b) The dispersion relation of the conduction electron in bulk specimens of n-Ge can be expressed in accordance with the model of Wang and Ressler [235] and can be written as

$$E = \frac{\hbar^2 k_z^2}{2m_{\parallel}^*} + \frac{\hbar^2 k_s^2}{2m_{\perp}^*} - \bar{c}_1\left(\frac{\hbar^2 k_s^2}{2m_{\perp}^*}\right)^2 - \bar{d}_1\left(\frac{\hbar^2 k_s^2}{2m_{\perp}^*}\right)\left(\frac{\hbar^2 k_z^2}{2m_{\parallel}^*}\right) - \bar{e}_1\left(\frac{\hbar^2 k_z^2}{2m_{\parallel}^*}\right)^2, \tag{1.62}$$

where

$$\begin{aligned}
\bar{c}_1 &= \bar{C}\left(2m_{\perp}^*/\hbar^2\right)^2,\ \bar{C} = 1.4\bar{A},\\
\bar{A} &= \tfrac{1}{4}\left(\hbar^4/E_{g0}m_{\perp}^{*2}\right)\left(1 - \tfrac{m_{\perp}^*}{m_0}\right)^2,\ \bar{d}_1 = \bar{d}\left(\tfrac{4m_{\perp}^* m_{\parallel}^*}{\hbar^4}\right),\\
\bar{d} &= 0.8\bar{A},\ \bar{e}_1 = \bar{e}_0\left(2m_{\parallel}^*/\hbar^2\right)^2,\ \text{and } \bar{e}_0 = 0.005\bar{A}
\end{aligned} \tag{1.63}$$

Using (1.62), the totally quantized energy E_{QD13} in this case is given by

$$\begin{aligned}
E_{QD13} = {} & \frac{\hbar^2}{2m_{\parallel}^*}\left(\frac{\pi n_z}{d_z}\right)^2 + \frac{\hbar^2}{2m_{\perp}^*}\left\{\left(\frac{\pi n_x}{d_x}\right)^2 + \left(\frac{\pi n_y}{d_y}\right)^2\right\}\\
& - \bar{c}_1\left(\frac{\hbar^2}{2m_{\perp}^*}\right)^2\left\{\left(\frac{\pi n_x}{d_x}\right)^2 + \left(\frac{\pi n_y}{d_y}\right)^2\right\}^2\\
& - \bar{d}_1\left(\frac{\hbar^2}{2m_{\perp}^*}\right)\left\{\left(\frac{\pi n_x}{d_x}\right)^2 + \left(\frac{\pi n_y}{d_y}\right)^2\right\}\left(\frac{\hbar^2}{2m_{\parallel}^*}\right)\left(\frac{\pi n_z}{d_z}\right)^2\\
& - \bar{e}_1\left(\frac{\hbar^2}{2m_{\parallel}^*}\right)^2\left(\frac{\pi n_z}{d_z}\right)^4
\end{aligned} \tag{1.64}$$

The electron concentration can be written as

$$n_0 = \frac{2g_v}{d_x d_y d_z}\sum_{n_x=1}^{n_{x_{\max}}}\sum_{n_y=1}^{n_{y_{\max}}}\sum_{n_z=1}^{n_{z_{\max}}}\frac{L_{23}}{M_{23}}, \tag{1.65}$$

where $L_{23} = [1 + A_{13} \cos H_{13}]$, $A_{13} = \exp[E_{QD13} - E_{FQD}/k_B T]$, $H_{13} = \Gamma_{13}/k_B T$, Γ_{13} is the broadening parameter in this case, and $M_{23} = 1 + A_{13}^2 + 2A_{13} \cos H_{13}$.

The TPSM in this case, using (1.13) and (1.65), can be expressed as

$$G_0 = \frac{\pi^2 k_B}{3e} \left[\sum_{n_x=1}^{n_{x_{\max}}} \sum_{n_y=1}^{n_{y_{\max}}} \sum_{n_z=1}^{n_{z_{\max}}} \frac{L_{23}}{M_{23}} \right]^{-1} \left[\sum_{n_x=1}^{n_{x_{\max}}} \sum_{n_y=1}^{n_{y_{\max}}} \sum_{n_z=1}^{n_{z_{\max}}} \frac{Q_{23}}{(M_{23})^2} \right], \tag{1.66}$$

where $Q_{23} = A_{13} \left[\left(1 + A_{13}^2\right) \cos H_{13} + 2A_{13} \right]$.

1.2.6 Magnetothermopower in Quantum Dots of Tellurium

The carriers of Tellurium find various descriptions for the energy–wave vector dispersion relations in the literature. Among them we shall use the E–k dispersion relations as given by Bouat et al. [236] and Ortenberg and Button [237], respectively.

(a) The dispersion relation of the conduction electrons of Tellurium can be written in accordance with Bouat et al. as [236]

$$E = A_6 k_z^2 + A_7 k_s^2 \pm \left[A_8 k_z^2 + A_9 k_s^2 \right]^{1/2} \tag{1.67}$$

where A_6, A_7, A_8, and A_9 are the energy band constants.

The totally quantized energy can be written as

$$E_{QD14,\pm} = A_6 \left(\frac{\pi n_z}{d_z} \right)^2 + A_7 \left[\left(\frac{\pi n_x}{d_x} \right)^2 + \left(\frac{\pi n_y}{d_y} \right)^2 \right] \\ \pm \left[A_8 \left(\frac{\pi n_z}{d_z} \right)^2 + A_9 \left[\left(\frac{\pi n_x}{d_x} \right)^2 + \left(\frac{\pi n_y}{d_y} \right)^2 \right] \right]^{1/2} \tag{1.68}$$

The electron concentration is given by

$$n_0 = \frac{g_v}{d_x d_y d_z} \sum_{n_x=1}^{n_{x_{\max}}} \sum_{n_y=1}^{n_{y_{\max}}} \sum_{n_z=1}^{n_{z_{\max}}} \frac{L_{24,\pm}}{M_{24,\pm}}, \tag{1.69}$$

where $L_{24,\pm} = [1 + A_{14,\pm} \cos H_{14}]$, $A_{14,\pm} = \exp[E_{QD14,\pm} - E_{FQD}/k_B T]$, $H_{14} = \Gamma_{14}/k_B T$, Γ_{14} is the broadening parameter in this case, and $M_{24,\pm} = 1 + A_{14,\pm}^2 + 2A_{14,\pm} \cos H_{14}$.

The TPSM in this case, using (1.13) and (1.69), can be expressed as

$$G_0 = \frac{\pi^2 k_B}{3e} \left[\sum_{n_x=1}^{n_{x_{\max}}} \sum_{n_y=1}^{n_{y_{\max}}} \sum_{n_z=1}^{n_{z_{\max}}} \frac{L_{24,\pm}}{M_{24,\pm}} \right]^{-1} \left[\sum_{n_x=1}^{n_{x_{\max}}} \sum_{n_y=1}^{n_{y_{\max}}} \sum_{n_z=1}^{n_{z_{\max}}} \frac{Q_{24,\pm}}{(M_{24,\pm})^2} \right], \tag{1.70}$$

where $Q_{24,\pm} = A_{14,\pm} \left[\left(1 + A_{14,\pm}^2 \right) \cos H_{14} + 2A_{14,\pm} \right]$.

(b) The dispersion relation of the conduction electrons of Tellurium can be written in accordance with the model of Ortenberg and Button as [237]

$$E = t_1 + t_2 k_z^2 + t_3 k_s^2 + t_4 k_s^4 + t_5 k_s^2 k_z^2 \pm \left[\left(t_1 + t_6 k_s^2 \right)^2 + t_7 k_z^2 \right]^{1/2}, \tag{1.71}$$

where t_1, t_2, t_3, t_4, t_5, t_6, and t_7 are the energy band constants.

The totally quantized energy can be written as

$$\begin{aligned} E_{\mathrm{QD15},\pm} &= t_1 + t_2 \left(\frac{\pi n_z}{d_z} \right)^2 + t_3 \left[\left(\frac{\pi n_x}{d_x} \right)^2 + \left(\frac{\pi n_y}{d_y} \right)^2 \right] \\ &+ t_4 \left[\left(\frac{\pi n_x}{d_x} \right)^2 + \left(\frac{\pi n_y}{d_y} \right)^2 \right]^2 + t_5 \left[\left(\frac{\pi n_x}{d_x} \right)^2 + \left(\frac{\pi n_y}{d_y} \right)^2 \right] \left(\frac{\pi n_z}{d_z} \right)^2 \\ &\pm \left[\left[t_1 + t_6 \left\{ \left(\frac{\pi n_x}{d_x} \right)^2 + \left(\frac{\pi n_y}{d_y} \right)^2 \right\} \right]^2 + t_7 \left(\frac{\pi n_z}{d_z} \right)^2 \right]^{1/2} \end{aligned} \tag{1.72}$$

The electron concentration is given by

$$n_0 = \frac{g_v}{d_x d_y d_z} \sum_{n_x=1}^{n_{x_{\max}}} \sum_{n_y=1}^{n_{y_{\max}}} \sum_{n_z=1}^{n_{z_{\max}}} \frac{L_{25,\pm}}{M_{25,\pm}}, \tag{1.73}$$

where $L_{25,\pm} = \left[1 + A_{15,\pm} \cos H_{15} \right]$, $A_{15,\pm} = \exp[E_{\mathrm{QD15},\pm} - E_{\mathrm{FQD}}/k_B T]$, $H_{15} = \Gamma_{15}/k_B T$, Γ_{15} is the broadening parameter in this case, and $M_{25,\pm} = 1 + A_{15,\pm}^2 + 2A_{15,\pm} \cos H_{15}$.

The TPSM in this case, using (1.13) and (1.73), can be expressed as

$$G_0 = \frac{\pi^2 k_B}{3e} \left[\sum_{n_x=1}^{n_{x_{\max}}} \sum_{n_y=1}^{n_{y_{\max}}} \sum_{n_z=1}^{n_{z_{\max}}} \frac{L_{25,\pm}}{M_{25,\pm}} \right]^{-1} \left[\sum_{n_x=1}^{n_{x_{\max}}} \sum_{n_y=1}^{n_{y_{\max}}} \sum_{n_z=1}^{n_{z_{\max}}} \frac{Q_{25,\pm}}{(M_{25,\pm})^2} \right], \tag{1.74}$$

where $Q_{25,\pm} = A_{15,\pm} \left[\left(1 + A_{15,\pm}^2 \right) \cos H_{15} + 2A_{15,\pm} \right]$.

1.2.7 Magnetothermopower in Quantum Dots of Graphite

The carrier dispersion law in graphite can be written as [238]

$$E = \frac{1}{2}(E_2 + E_3) \pm \left[\frac{1}{4}(E_2 - E_3)^2 + \eta_{46}^2 k^2\right]^{1/2} \tag{1.75}$$

where $E_2 = \left(\Delta_1 - 2\gamma_1 \cos\phi_0 + 2\gamma_5 \cos^2\phi_0\right)$; Δ_1, γ_1, and γ_5 are energy band constants; $\phi_0 = (\bar{c}k_z/2)$; $E_3 = \left(2\gamma_2 \cos^2\phi_0\right)$; $\eta_{46} = (\sqrt{3}/2)\,\bar{a}\,(\gamma_0 + 2\gamma_4 \cos\phi_0)$; $\bar{c}$ and $\bar{a}$ are constants of the spectrum.

The totally quantized energy can be expressed as

$$\begin{aligned}
E_{\text{QD16},\pm} = {} & \frac{1}{2}\left[\Delta_1 - 2\gamma_1 \cos\left(\frac{\bar{c}(\pi n_z)}{2d_z}\right) + 2\gamma_5 \cos^2\left(\frac{\bar{c}(\pi n_z)}{2d_z}\right)\right. \\
& \left. + 2\gamma_2 \cos^2\left(\frac{\bar{c}(\pi n_z)}{2d_z}\right)\right] \pm \left[\frac{1}{4}\left(\Delta_1 - 2\gamma_1 \cos^2\left(\frac{\bar{c}(\pi n_z)}{2d_z}\right)\right.\right. \\
& \left. + 2\gamma_5 \cos^2\left(\frac{\bar{c}(\pi n_z)}{2d_z}\right) - 2\gamma_2 \cos^2\left(\frac{\bar{c}(\pi n_z)}{2d_z}\right)\right)^2 \\
& + \frac{3}{4}(\bar{a})^2\left(\gamma_0 + 2\gamma_4 \cos\left(\frac{n_z \pi \bar{c}}{2d_z}\right)\right)^2 \\
& \left. \times \left[\left(\frac{\pi n_x}{d_x}\right)^2 + \left(\frac{\pi n_y}{d_y}\right)^2 + \left(\frac{\pi n_z}{d_z}\right)^2\right]\right]^{1/2}
\end{aligned} \tag{1.76}$$

The electron concentration is given by

$$n_0 = \frac{g_v}{d_x d_y d_z} \sum_{n_x=1}^{n_{x_{\max}}} \sum_{n_y=1}^{n_{y_{\max}}} \sum_{n_z=1}^{n_{z_{\max}}} \frac{L_{26,\pm}}{M_{26,\pm}}, \tag{1.77}$$

where $L_{26,\pm} = [1 + A_{16,\pm} \cos H_{16}]$, $A_{16,\pm} = \exp[E_{\text{QD16},\pm} - E_{\text{FQD}}/k_B T]$, $H_{16} = \Gamma_{16}/k_B T$, Γ_{16} is the broadening parameter in this case, and $M_{26,\pm} = 1 + A_{16,\pm}^2 + 2A_{16,\pm} \cos H_{16}$.

The TPSM in this case, using (1.13) and (1.77), can be expressed as

$$G_0 = \frac{\pi^2 k_B}{3e}\left[\sum_{n_x=1}^{n_{x_{\max}}} \sum_{n_y=1}^{n_{y_{\max}}} \sum_{n_z=1}^{n_{z_{\max}}} \frac{L_{26,\pm}}{M_{26,\pm}}\right]^{-1} \left[\sum_{n_x=1}^{n_{x_{\max}}} \sum_{n_y=1}^{n_{y_{\max}}} \sum_{n_z=1}^{n_{z_{\max}}} \frac{Q_{26,\pm}}{(M_{26,\pm})^2}\right], \tag{1.78}$$

where $Q_{26,\pm} = A_{16,\pm}\left[\left(1 + A_{16,\pm}^2\right)\cos H_{16} + 2A_{16,\pm}\right]$.

1.2.8 Magnetothermopower in Quantum Dots of Platinum Antimonide

The dispersion relation of the carriers in $PtSb_2$ can be written as [239]

$$\left(E+\lambda_0\frac{(\bar{\bar{a}})^2}{4}k^2-lk_s^2\frac{(\bar{\bar{a}})^2}{4}\right)\left(E+\delta_0-\upsilon\frac{(\bar{\bar{a}})^2}{4}k^2-\bar{\bar{n}}\frac{(\bar{\bar{a}})^2}{4}k_s^2\right)=I\frac{(\bar{\bar{a}})^4}{16}k^4, \tag{1.79}$$

where λ_0,l, δ_0,υ, and $\bar{\bar{n}}$ are the band constants, and $\bar{\bar{a}}$ is the lattice constant.

Equation (1.79) can be expressed as

$$\left[E+\omega_1k_s^2+\omega_2k_z^2\right]\left[E+\delta_0-\omega_3k_s^2-\omega_4k_z^2\right]=I_1\left(k_z^2+k_s^2\right)^2, \tag{1.80}$$

where

$$\omega_1\equiv\left(\lambda_0\frac{(\bar{\bar{a}})^2}{4}-l\frac{(\bar{\bar{a}})^2}{4}\right),\quad \omega_2\equiv\lambda_0\frac{(\bar{\bar{a}})^2}{4},$$

$$\omega_3\equiv\left(\bar{\bar{n}}\frac{(\bar{\bar{a}})^2}{4}+\upsilon\frac{(\bar{\bar{a}})^2}{4}\right),\quad \omega_4\equiv\upsilon\frac{(\bar{\bar{a}})^2}{4}$$

and

$$I_1\equiv I\left(\frac{(\bar{\bar{a}})^2}{4}\right)^2.$$

The totally quantized energy E_{QD17} is given by

$$\begin{aligned}&\left[E_{QD17}+\omega_1\left[\left(\frac{\pi n_x}{d_x}\right)^2+\left(\frac{\pi n_y}{d_y}\right)^2\right]+\omega_2\left(\frac{\pi n_z}{d_z}\right)^2\right]\\&\times\left[E_{QD17}+\delta_0+\omega_3\left[\left(\frac{\pi n_x}{d_x}\right)^2+\left(\frac{\pi n_y}{d_y}\right)^2\right]-\omega_4\left(\frac{\pi n_z}{d_z}\right)^2\right]\\&=I_1\left[\left(\frac{\pi n_x}{d_x}\right)^2+\left(\frac{\pi n_y}{d_y}\right)^2+\left(\frac{\pi n_z}{d_z}\right)^2\right]^2\end{aligned} \tag{1.81}$$

The hole concentration is given by

$$p_0=\frac{2g_v}{d_xd_yd_z}\sum_{n_x=1}^{n_{x_{max}}}\sum_{n_y=1}^{n_{y_{max}}}\sum_{n_z=1}^{n_{z_{max}}}\frac{L_{27}}{M_{27}}, \tag{1.82}$$

where $L_{27} = [1 + A_{17} \cos H_{17}]$, $A_{17} = \exp[E_{QD17} - E_{FQD}/k_B T]$, $H_{17} = \Gamma_{17}/k_B T$, Γ_{17} is the broadening parameter in this case, and $M_{27} = 1 + A_{17}^2 + 2A_{17} \cos H_{17}$.

The TPSM in this case, using (1.13) and (1.82), can be expressed as

$$G_0 = \frac{\pi^2 k_B}{3e} \left[\sum_{n_x=1}^{n_{x_{\max}}} \sum_{n_y=1}^{n_{y_{\max}}} \sum_{n_z=1}^{n_{z_{\max}}} \frac{L_{27}}{M_{27}} \right]^{-1} \left[\sum_{n_x=1}^{n_{x_{\max}}} \sum_{n_y=1}^{n_{y_{\max}}} \sum_{n_z=1}^{n_{z_{\max}}} \frac{Q_{27}}{(M_{27})^2} \right], \tag{1.83}$$

where $Q_{27} = A_{17}\left[\left(1 + A_{17}^2\right) \cos H_{17} + 2A_{17}\right]$.

1.2.9 Magnetothermopower in Quantum Dots of Zerogap Materials

The dispersion relation of the holes in gapless materials [240] assumes the form

$$\bar{\bar{E}} = \frac{\hbar^2 k^2}{2m_v^*} + \frac{3e^2 k}{128\varepsilon_{sc}} - \frac{2}{\pi} E_B \ln \left| k . \left(\overline{k_0}\right)^{-1} \right|, \tag{1.84}$$

where $\bar{\bar{E}}$ is the energy of the hole as measured from the top of the valence band in the vertically downward direction, m_v^* is the effective mass of the holes at the top of the valence band, ε_{sc} is the semiconductor permittivity, E_B is the Bohr electron energy, and $\overline{k_0}$ is the inverse Bohr radius.

The totally quantized energy E_{QD18} is given by

$$\begin{aligned} E_{QD18} = {} & \frac{\hbar^2 \pi^2}{2m^*} \left[\left(\frac{n_x}{d_x}\right)^2 + \left(\frac{n_y}{d_y}\right)^2 + \left(\frac{n_z}{d_z}\right)^2 \right] \\ & + \frac{3e^2}{128\varepsilon_{sc}} \left[\left(\frac{\pi n_x}{d_x}\right)^2 + \left(\frac{\pi n_y}{d_y}\right)^2 + \left(\frac{\pi n_z}{d_z}\right)^2 \right]^{1/2} \\ & + \frac{E_B}{\pi} \ln \left| \frac{(\pi n_x/d_x)^2 + (\pi n_y/d_y)^2 + (\pi n_z/d_z)^2}{\left(\overline{k_0}\right)^2} \right| \end{aligned} \tag{1.85}$$

The hole concentration can be written as

$$p_0 = \frac{2g_v}{d_x d_y d_z} \sum_{n_x=1}^{n_{x_{max}}} \sum_{n_y=1}^{n_{y_{max}}} \sum_{n_z=1}^{n_{z_{max}}} \frac{L_{28}}{M_{28}}, \tag{1.86}$$

where $L_{28} = [1 + A_{18} \cos H_{18}]$, $A_{18} = \exp[E_{QD18} - E_{FQD}/k_B T]$, $H_{18} = \Gamma_{18}/k_B T$, Γ_{18} is the broadening parameter in this case, and $M_{28} = 1 + A_{18}^2 + 2A_{18} \cos H_{18}$.

The TPSM in this case, using (1.13) and (1.86), can be expressed as

$$G_0 = \frac{\pi^2 k_B}{3e} \left[\sum_{n_x=1}^{n_{x_{max}}} \sum_{n_y=1}^{n_{y_{max}}} \sum_{n_z=1}^{n_{z_{max}}} \frac{L_{28}}{M_{28}} \right]^{-1} \left[\sum_{n_x=1}^{n_{x_{max}}} \sum_{n_y=1}^{n_{y_{max}}} \sum_{n_z=1}^{n_{z_{max}}} \frac{Q_{28}}{(M_{28})^2} \right], \tag{1.87}$$

where $Q_{28} = A_{18} \left[\left(1 + A_{18}^2\right) \cos H_{18} + 2A_{18} \right]$.

1.2.10 Magnetothermopower in Quantum Dots of II–V Materials

The dispersion relation of the holes in II–V compounds in accordance with Yamada [241, 242] can be expressed as

$$\begin{aligned} E = {} & A_{10} k_x^2 + A_{11} k_y^2 + A_{12} k_z^2 + A_{13} k_x \\ & \pm \left[\left(A_{14} k_x^2 + A_{15} k_y^2 + A_{16} k_z^2 + A_{17} k_x \right)^2 + A_{18} k_y^2 + A_{19}^2 \right]^{1/2}, \end{aligned} \tag{1.88}$$

where A_{10}, A_{11}, A_{12}, A_{13}, A_{14}, A_{15}, A_{16}, A_{17}, A_{18}, and A_{19} are energy band constants.

The totally quantized energy $E_{QD19,\pm}$ is given by

$$\begin{aligned} E_{QD19,\pm} = {} & A_{10} \left(\frac{\pi n_x}{d_x} \right)^2 + A_{11} \left(\frac{\pi n_y}{d_y} \right)^2 + A_{12} \left(\frac{\pi n_z}{d_z} \right)^2 + A_{13} \left(\frac{\pi n_x}{d_x} \right) \\ & \pm \left[\left(A_{14} \left(\frac{\pi n_x}{d_x} \right)^2 + A_{15} \left(\frac{\pi n_y}{d_y} \right)^2 + A_{16} \left(\frac{\pi n_z}{d_z} \right)^2 \right. \right. \\ & \left. \left. + A_{17} \left(\frac{\pi n_x}{d_x} \right) \right)^2 + A_{18} \left(\frac{\pi n_y}{d_y} \right)^2 + A_{19}^2 \right]^{1/2} \end{aligned} \tag{1.89}$$

The hole concentration can be written as

$$p_0 = \frac{g_v}{d_x d_y d_z} \sum_{n_x=1}^{n_{x_{\max}}} \sum_{n_y=1}^{n_{y_{\max}}} \sum_{n_z=1}^{n_{z_{\max}}} \frac{L_{29,\pm}}{M_{29,\pm}}, \tag{1.90}$$

where $L_{29,\pm} = \left[1 + A_{19,\pm} \cos H_{19}\right]$, $A_{19,\pm} = \exp[E_{QD19,\pm} - E_{FQD}/k_B T]$, $H_{19} = \Gamma_{19}/k_B T$, Γ_{19} is the broadening parameter in this case, and $M_{29,\pm} = 1 + A_{19,\pm}^2 + 2A_{19,\pm} \cos H_{19}$.

The TPSM in this case, using (1.13) and (1.90), can be expressed as

$$G_0 = \frac{\pi^2 k_B}{3e} \left[\sum_{n_x=1}^{n_{x_{\max}}} \sum_{n_y=1}^{n_{y_{\max}}} \sum_{n_z=1}^{n_{z_{\max}}} \frac{L_{29,\pm}}{M_{29,\pm}}\right]^{-1} \left[\sum_{n_x=1}^{n_{x_{\max}}} \sum_{n_y=1}^{n_{y_{\max}}} \sum_{n_z=1}^{n_{z_{\max}}} \frac{Q_{29,\pm}}{\left(M_{29,\pm}\right)^2}\right], \tag{1.91}$$

where $Q_{29,\pm} = A_{19,\pm} \left[\left(1 + A_{19,\pm}^2\right) \cos H_{19} + 2A_{19,\pm}\right]$.

1.2.11 Magnetothermopower in Quantum Dots of Gallium Antimonide

The conduction electrons of n-GaSb obey the three dispersion relations as provided by Seiler et al. [243], Mathur et al. [244], and Zhang [245], respectively. The TPSM in QDs of GaSb is being presented in accordance with the aforementioned models for the joint purpose of coherent presentation and relative assessment.

(a) In accordance with the model of Seiler et al. [243], the dispersion relation of the conduction electrons in n-GaSb assume the form

$$E = \left[-\frac{E_{g0}}{2} + \frac{E_{g0}}{2}\left[1 + \alpha_4 k^2\right]^{1/2} + \frac{\overline{\zeta_0}\hbar^2 k^2}{2m_0} + \frac{\overline{v_0}\theta_1(k)\hbar^2}{2m_0} \pm \frac{\overline{w_0}\theta_2(k)\hbar^2}{2m_0}\right], \tag{1.92}$$

where

$$\alpha_4 = 4P_0^2\left(E_{g_0} + \frac{2}{3}\Delta\right)\left[E_{g_0}^2\left(E_{g_0} + \Delta\right)\right]^{-1},$$

$\theta_1(k) = k^{-2}\left(k_x^2 k_y^2 + k_y^2 k_z^2 + k_z^2 k_x^2\right)$ represents the warping of the Fermi surface; the function $\theta_2(k) = \left[\left\{k^2\left(k_x^2 k_y^2 + k_y^2 k_z^2 + k_z^2 k_x^2\right) - 9k_x^2 k_y^2 k_z^2\right\}^{1/2} k^{-1}\right]$ indicates the inversion asymmetry splitting of the conduction band; and $\overline{\zeta_0}\,(= -2.1)$, $\overline{v_0}\,(= -1.49)$, and $\overline{w_0}\,(= 0.42)$ represents the constants of the spectrum.

The totally quantized energy $E_{\mathrm{QD20},\pm}$ is given by

$$E_{\mathrm{QD20},\pm} = \left[-\frac{E_{g0}}{2} + \frac{E_{g0}}{2}\left[1+\alpha_4\left[\left(\frac{\pi n_x}{d_x}\right)^2+\left(\frac{\pi n_y}{d_y}\right)^2+\left(\frac{\pi n_z}{d_z}\right)^2\right]\right]^{1/2} + \left(\frac{\overline{\zeta_0}\hbar^2}{2m_0}\right)\left[\left(\frac{\pi n_x}{d_x}\right)^2+\left(\frac{\pi n_y}{d_y}\right)^2+\left(\frac{\pi n_z}{d_z}\right)^2\right] + \left(\frac{\overline{v_0}\hbar^2}{2m_0}\right)S_6 \pm \left(\frac{\overline{w_0}S_7\hbar^2}{2m_0}\right)\right], \tag{1.93}$$

where

$$S_6 = \left[\left(\frac{\pi n_x}{d_x}\right)^2+\left(\frac{\pi n_y}{d_y}\right)^2+\left(\frac{\pi n_z}{d_z}\right)^2\right]^{-2} \times \left[\left(\frac{\pi^2 n_x n_y}{d_x d_y}\right)^2+\left(\frac{\pi^2 n_y n_z}{d_y d_z}\right)^2+\left(\frac{\pi^2 n_z n_x}{d_x d_z}\right)^2\right],$$

$$S_7 = \left[\left\{\left(\left(\frac{\pi n_x}{d_x}\right)^2+\left(\frac{\pi n_y}{d_y}\right)^2+\left(\frac{\pi n_z}{d_z}\right)^2\right) \times \left(\frac{n_x^2 n_y^2\pi^4}{d_x^2 d_y^2}+\frac{n_y^2 n_z^2\pi^4}{d_y^2 d_z^2}+\frac{n_z^2 n_x^2\pi^4}{d_z^2 d_x^2}\right) - \frac{9\pi^6 n_x^2 n_y^2 n_z^2}{d_x^2 d_y^2 d_z^2}\right\}^{1/2}\left[\left(\frac{\pi n_x}{d_x}\right)^2+\left(\frac{\pi n_y}{d_y}\right)^2+\left(\frac{\pi n_z}{d_z}\right)^2\right]^{-1/2}\right].$$

The hole concentration is given by

$$p_0 = \frac{g_v}{d_x d_y d_z}\sum_{n_x=1}^{n_{x_{\max}}}\sum_{n_y=1}^{n_{y_{\max}}}\sum_{n_z=1}^{n_{z_{\max}}}\frac{L_{30,\pm}}{M_{30,\pm}}, \tag{1.94}$$

where $L_{30,\pm} = \left[1 + A_{20,\pm}\cos H_{20}\right]$, $A_{20,\pm} = \exp[E_{\mathrm{QD20},\pm} - E_{\mathrm{FQD}}/k_B T]$, $H_{20} = \Gamma_{20}/k_B T$, Γ_{20} is the broadening parameter in this case, and $M_{30,\pm} = 1 + A_{20,\pm}^2 + 2A_{20,\pm}\cos H_{20}$.

The TPSM in this case, using (1.13) and (1.94), can be expressed as

$$G_0 = \frac{\pi^2 k_B}{3e}\left[\sum_{n_x=1}^{n_{x_{\max}}}\sum_{n_y=1}^{n_{y_{\max}}}\sum_{n_z=1}^{n_{z_{\max}}}\frac{L_{30,\pm}}{M_{30,\pm}}\right]^{-1}\left[\sum_{n_x=1}^{n_{x_{\max}}}\sum_{n_y=1}^{n_{y_{\max}}}\sum_{n_z=1}^{n_{z_{\max}}}\frac{Q_{30,\pm}}{(M_{30,\pm})^2}\right], \tag{1.95}$$

where $Q_{30,\pm} = A_{20,\pm}\left[\left(1 + A_{20,\pm}^2\right)\cos H_{20} + 2A_{20,\pm}\right]$.

(b) In accordance with the model of Mathur et al. [244], the electron dispersion law in n-GaSb can be written as

$$E = \alpha_9 k^2 + \frac{E_{g1}}{2}\left[\sqrt{1+\alpha_{10}k^2} - 1\right], \tag{1.96}$$

where

$$\alpha_9 = \frac{\hbar^2}{2m_0},$$
$$E_{g1} = E_{go} + \left[\frac{5\times 10^{-5}T^2}{2\,(112+T)}\right] eV$$

and

$$\alpha_{10} = \left(\frac{(2\hbar^2)}{(E_{g1})}\right)\left[\frac{1}{m^*} - \frac{1}{m_0}\right].$$

From (1.96), we get

$$k^2 = \frac{E}{\alpha_9} + \alpha_{11} - \left[\alpha_{12}E + \alpha_{13}\right]^{1/2}, \tag{1.97}$$

where

$$\alpha_{11} = \frac{E_{g1}^2}{8\alpha_9^2}\left[\alpha_{10} + \frac{4\alpha_9}{E_{g1}}\right],$$
$$\alpha_{12} = \left(\frac{E_{g1}^2}{\alpha_9^3}\right)$$

and

$$\alpha_{13} = \frac{E_{g1}^4}{64\alpha_9^4}\left[\alpha_{10}^2 + \frac{16\alpha_9^2}{E_{g1}^2} + \frac{8\alpha_9\alpha_{10}}{E_{g1}}\right].$$

The totally quantized energy E_{QD21} is given by

$$E_{QD21} = \alpha_9\left[\left(\frac{\pi n_x}{d_x}\right)^2 + \left(\frac{\pi n_y}{d_y}\right)^2 + \left(\frac{\pi n_z}{d_z}\right)^2\right] + \frac{E_{g1}}{2}\left[\sqrt{1 + \alpha_{10}\left[\left(\frac{\pi n_x}{d_x}\right)^2 + \left(\frac{\pi n_y}{d_y}\right)^2 + \left(\frac{\pi n_z}{d_z}\right)^2\right]} - 1\right] \tag{1.98}$$

The electron concentration is given by

$$n_0 = \frac{2g_v}{d_x d_y d_z} \sum_{n_x=1}^{n_{x_{\max}}} \sum_{n_y=1}^{n_{y_{\max}}} \sum_{n_z=1}^{n_{z_{\max}}} \frac{L_{31}}{M_{31}}, \tag{1.99}$$

where $L_{31} = [1 + A_{21} \cos H_{21}]$, $A_{21} = \exp[E_{QD21} - E_{FQD}/k_B T]$, $H_{21} = \Gamma_{21}/k_B T$, Γ_{21} is the broadening parameter in this case, and $M_{31} = 1 + A_{21}^2 + 2A_{21} \cos H_{21}$.

The TPSM in this case, using (1.13) and (1.99), can be expressed as

$$G_0 = \frac{\pi^2 k_B}{3e} \left[\sum_{n_x=1}^{n_{x_{\max}}} \sum_{n_y=1}^{n_{y_{\max}}} \sum_{n_z=1}^{n_{z_{\max}}} \frac{L_{31}}{M_{31}} \right]^{-1} \left[\sum_{n_x=1}^{n_{x_{\max}}} \sum_{n_y=1}^{n_{y_{\max}}} \sum_{n_z=1}^{n_{z_{\max}}} \frac{Q_{31}}{(M_{31})^2} \right], \tag{1.100}$$

where $Q_{31} = A_{21} \left[\left(1 + A_{21}^2\right) \cos H_{21} + 2A_{21} \right]$.

(c) The dispersion relation of the conduction electrons in n-GaSb can be expressed in accordance with Zhang [245] as

$$E = \left[\overline{E_1} + \overline{E_2} Z_1(k)\right] k^2 + \left[\overline{E_3} + \overline{E_4} Z_1(k)\right] k^4 + k^6 \left[\overline{E_5} + \overline{E_6} Z_1(k) + \overline{E_7} Z_2(k)\right], \tag{1.101}$$

where

$$Z_1(k) = \frac{5\sqrt{21}}{4} \left[\frac{k_x^4 + k_y^4 + k_z^4}{k^4} - \frac{3}{5} \right],$$

$$Z_2(k) = \left[\frac{639639}{32} \right]^{1/2} \left[\frac{k_x^2 k_y^2 k_z^2}{k^6} + \frac{1}{12} \left(\frac{k_x^4 + k_y^4 + k_z^4}{k^4} - \frac{3}{5} \right) - \frac{1}{105} \right],$$

the coefficients are in eV; the values of $k$are in $10(a/2\pi)$ times of k in atomic units (a is lattice constant); and $\overline{E_1}$, $\overline{E_2}$, $\overline{E_3}$, $\overline{E_4}$, $\overline{E_5}$, $\overline{E_6}$, and $\overline{E_7}$ are energy band constants.

The totally quantized energy E_{QD22} is given by

$$E_{QD22} = \left[\overline{E_1} + \overline{E_2} Z_3\right] \left[\left(\frac{\pi n_x}{d_x}\right)^2 + \left(\frac{\pi n_y}{d_y}\right)^2 + \left(\frac{\pi n_z}{d_z}\right)^2 \right] + \left[\left(\frac{\pi n_x}{d_x}\right)^2 + \left(\frac{\pi n_y}{d_y}\right)^2 + \left(\frac{\pi n_z}{d_z}\right)^2 \right]^2 . \left[\overline{E_3} + \overline{E_4} Z_3\right] + \left[\left(\frac{\pi n_x}{d_x}\right)^2 + \left(\frac{\pi n_y}{d_y}\right)^2 + \left(\frac{\pi n_z}{d_z}\right)^2 \right]^3 . \left[\overline{E_5} + \overline{E_6} Z_3 + \overline{E_7} Z_4\right], \tag{1.102}$$

where

$$Z_3 = \frac{5\sqrt{21}}{4}\left[\left[\left(\frac{n_x\pi}{d_x}\right)^4 + \left(\frac{n_y\pi}{d_y}\right)^4 + \left(\frac{n_z\pi}{d_z}\right)^4\right] \times \left[\left(\frac{n_x\pi}{d_x}\right)^2 + \left(\frac{n_y\pi}{d_y}\right)^2 + \left(\frac{n_z\pi}{d_z}\right)^2\right]^{-2} - \frac{3}{5}\right]$$

and

$$Z_4 = \left[\frac{639,639}{32}\right]^{1/2}\left[\left(\frac{n_x}{d_x}\frac{n_y}{d_y}\frac{n_z}{d_z}\right)^2\left[\left(\frac{n_x}{d_x}\right)^2 + \left(\frac{n_y}{d_y}\right)^2 + \left(\frac{n_z}{d_z}\right)^2\right]^{-3} + \frac{1}{(15)\sqrt{21}}Z_3 - \frac{1}{105}\right]$$

The electron concentration is given by

$$n_0 = \frac{2g_v}{d_x d_y d_z}\sum_{n_x=1}^{n_{x_{\max}}}\sum_{n_y=1}^{n_{y_{\max}}}\sum_{n_z=1}^{n_{z_{\max}}}\frac{L_{32}}{M_{32}} \tag{1.103}$$

where $L_{32} = [1 + A_{22}\cos H_{22}]$, $A_{22} = \exp[E_{QD22} - E_{FQD}/k_BT]$, $H_{22} = \Gamma_{22}/k_BT$, Γ_{22} is the broadening parameter in this case, and $M_{32} = 1 + A_{22}^2 + 2A_{22}\cos H_{22}$.

The TPSM in this case, using (1.13) and (1.103), can be expressed as

$$G_0 = \frac{\pi^2 k_B}{3e}\left[\sum_{n_x=1}^{n_{x_{\max}}}\sum_{n_y=1}^{n_{y_{\max}}}\sum_{n_z=1}^{n_{z_{\max}}}\frac{L_{32}}{M_{32}}\right]^{-1}\left[\sum_{n_x=1}^{n_{x_{\max}}}\sum_{n_y=1}^{n_{y_{\max}}}\sum_{n_z=1}^{n_{z_{\max}}}\frac{Q_{32}}{(M_{32})^2}\right], \tag{1.104}$$

where $Q_{32} = A_{22}\left[\left(1 + A_{22}^2\right)\cos H_{22} + 2A_{22}\right]$.

1.2.12 Magnetothermopower in Quantum Dots of Stressed Materials

The dispersion relation of the conduction electrons in bulk specimens of stressed materials can be written as [223]

$$\frac{k_x^2}{[a^*(E)]^2} + \frac{k_y^2}{[b^*(E)]^2} + \frac{k_z^2}{[c^*(E)]^2} = 1, \tag{1.105}$$

where

$$\left[a^*(E)\right]^2 = \frac{\bar{K}_0(E)}{\bar{A}_0(E) + \frac{1}{2}\bar{D}_0(E)},$$

$$\bar{K}_0(E) = \left[E - C_1\varepsilon - \frac{2C_2^2\varepsilon_{xy}^2}{3E'_{g_0}}\right]\left(\frac{3E'_{g_0}}{2B_2^2}\right),$$

C_1 is the conduction band deformation potential, ε is the trace of the strain tensor $\hat{\varepsilon}$ which can be written as

$$\hat{\varepsilon} = \begin{bmatrix} \varepsilon_{xx} & \varepsilon_{xy} & 0 \\ \varepsilon_{xy} & \varepsilon_{yy} & 0 \\ 0 & 0 & \varepsilon_{zz} \end{bmatrix},$$

C_2 is a constant which describes the strain interaction between the conduction and valance bands, $E'_{g_0} = E_{g_0} + E - C_1\varepsilon$, E_{g_0} is the band gap in the absence of stress, B_2 is the momentum-matrix element,

$$\bar{A}_0(E) = \left[1 - \frac{(\bar{a}_0 + C_1)}{E'_{g_0}} + \frac{3\bar{b}_0\varepsilon_{xx}}{2E'_{g_0}} - \frac{\bar{b}_0\varepsilon}{2E'_{g_0}}\right],$$

$$\bar{a}_0 = -\frac{1}{3}\left(\bar{b}_0 + 2\bar{m}\right),$$

$$\bar{b}_0 = \frac{1}{3}\left(\bar{l} - \bar{m}\right),$$

$$\bar{d}_0 = \frac{2\bar{n}}{\sqrt{3}},$$

$\bar{l}, \bar{m}, \bar{n}$ are the matrix elements of the strain perturbation operator,

$$\bar{D}_0(E) = \left(\bar{d}_0\sqrt{3}\right)\frac{\varepsilon_{xy}}{E'_{g_0}},$$

$$\left[b^*(E)\right]^2 = \frac{\bar{K}_0(E)}{\bar{A}_0(E) - \frac{1}{2}\bar{D}_0(E)},$$

$$\left[c^*(E)\right]^2 = \frac{\bar{K}_0(E)}{\bar{L}_0(E)},$$

and

$$\bar{L}_0(E) = \left[1 - \frac{(\bar{a}_0 + C_1)}{E'_{g_0}} + \frac{3\bar{b}_0\varepsilon_{zz}}{E'_{g_0}} - \frac{\bar{b}_0\varepsilon}{2E'_{g_0}}\right].$$

The totally quantized energy E_{QD23}in this case assumes the form

$$\left(\frac{\pi n_x}{d_x}\right)^2\left[a^*\left(E_{QD23}\right)\right]^{-2}+\left(\frac{\pi n_y}{d_y}\right)^2\left[b^*\left(E_{QD23}\right)\right]^{-2}$$
$$+\left(\frac{\pi n_z}{d_z}\right)^2\left[c^*\left(E_{QD23}\right)\right]^{-2}=1 \tag{1.106}$$

The electron concentration is given by

$$n_0=\frac{2g_v}{d_xd_yd_z}\sum_{n_x=1}^{n_{x_{\max}}}\sum_{n_y=1}^{n_{y_{\max}}}\sum_{n_z=1}^{n_{z_{\max}}}\frac{L_{33}}{M_{33}}, \tag{1.107}$$

where $L_{33}=[1+A_{23}\cos H_{23}]$, $A_{23}=\exp[E_{QD23}-E_{FQD}/k_BT]$, $H_{23}=\Gamma_{23}/k_BT$, Γ_{23} is the broadening parameter in this case, and $M_{33}=1+A_{23}^2+2A_{23}\cos H_{23}$.

The TPSM in this case, using (1.13) and (1.107), can be expressed as

$$G_0=\frac{\pi^2k_B}{3e}\left[\sum_{n_x=1}^{n_{x_{\max}}}\sum_{n_y=1}^{n_{y_{\max}}}\sum_{n_z=1}^{n_{z_{\max}}}\frac{L_{33}}{M_{33}}\right]^{-1}\left[\sum_{n_x=1}^{n_{x_{\max}}}\sum_{n_y=1}^{n_{y_{\max}}}\sum_{n_z=1}^{n_{z_{\max}}}\frac{Q_{33}}{(M_{33})^2}\right], \tag{1.108}$$

where $Q_{33}=A_{23}\left[\left(1+A_{23}^2\right)\cos H_{23}+2A_{23}\right]$.

1.2.13 Magnetothermopower in Quantum Dots of Bismuth

1.2.13.1 The McClure and Choi Model

The dispersion relation of the carriers in Bi can be written, following the McClure and Choi [174], as

$$E(1+\alpha E)=\frac{p_x^2}{2m_1}+\frac{p_y^2}{2m_2}+\frac{p_z^2}{2m_3}+\frac{p_y^2}{2m_2}\alpha E$$
$$\left\{1-\left(\frac{m_2}{m_2'}\right)\right\}+\frac{p_y^4\alpha}{4m_2m_2'}-\frac{\alpha p_x^2p_y^2}{4m_1m_2}-\frac{\alpha p_y^2p_z^2}{4m_2m_3} \tag{1.109}$$

where $p_i\equiv\hbar k_i$, $i=x,y,z$, m_1, m_2 and m_3 are the effective carrier masses at the band-edge along x, y and z directions, respectively, and m_2' is the effective-mass tensor component at the top of the valence band (for electrons) or at the bottom of the conduction band (for holes).

The totally quantized energy E_{QD24} is given by

$$E_{QD24}\left(1+\alpha E_{QD24}\right)=\frac{\hbar^2}{2m_1}\left(\frac{\pi n_x}{d_x}\right)^2+\frac{\hbar^2}{2m_2}\left(\frac{\pi n_y}{d_y}\right)^2+\frac{\hbar^2}{2m_3}\left(\frac{\pi n_z}{d_z}\right)^2$$
$$+\frac{\hbar^2\alpha E_{QD24}}{2m_2}\left(\frac{\pi n_y}{d_y}\right)^2\left(1-\frac{m_2}{m_2'}\right)+\frac{\alpha\hbar^4}{4m_2m_2'}\left(\frac{\pi n_y}{d_y}\right)^4$$
$$-\frac{\alpha\hbar^4}{4m_1m_2}\left(\frac{\pi^2n_x^2}{d_x^2}.\frac{\pi^2n_y^2}{d_y^2}\right)-\frac{\alpha\hbar^4\pi^4}{4m_2m_3}\left(\frac{n_yn_z}{d_yd_z}\right)^2 \quad (1.110)$$

The electron concentration can be written as

$$n_0=\frac{2g_v}{d_xd_yd_z}\sum_{n_x=1}^{n_{x_{max}}}\sum_{n_y=1}^{n_{y_{max}}}\sum_{n_z=1}^{n_{z_{max}}}\frac{L_{34}}{M_{34}}, \quad (1.111)$$

where $L_{34}=[1+A_{24}\cos H_{24}]$, $A_{24}=\exp[E_{QD24}-E_{FQD}/k_BT]$, $H_{24}=\Gamma_{24}/k_BT$, Γ_{24} is the broadening parameter in this case, and $M_{34}=1+A_{24}^2+2A_{24}\cos H_{24}$.

The TPSM in this case, using (1.13) and (1.111), can be expressed as

$$G_0=\frac{\pi^2k_B}{3e}\left[\sum_{n_x=1}^{n_{x_{max}}}\sum_{n_y=1}^{n_{y_{max}}}\sum_{n_z=1}^{n_{z_{max}}}\frac{L_{34}}{M_{34}}\right]^{-1}\left[\sum_{n_x=1}^{n_{x_{max}}}\sum_{n_y=1}^{n_{y_{max}}}\sum_{n_z=1}^{n_{z_{max}}}\frac{Q_{34}}{(M_{34})^2}\right], \quad (1.112)$$

where $Q_{34}=A_{24}[(1+A_{24}^2)\cos H_{24}+2A_{24}]$.

1.2.13.2 The Hybrid Model

The dispersion relation of the carriers in bulk specimens of Bi in accordance with the Hybrid model can be represented as [173]

$$E\left(1+\alpha E\right)=\frac{\theta_0\left(E\right)\left(\hbar k_y\right)^2}{2M_2}+\frac{\alpha\gamma_0\hbar^4k_y^4}{4M_2^2}+\frac{\hbar^2k_x^2}{2m_1}+\frac{\hbar^2k_z^2}{2m_3} \quad (1.113)$$

in which $\theta_0\left(E\right)\equiv\left[1+\alpha E\left(1-\gamma_0\right)+\bar{\delta}_0\right]$, $\gamma_0\equiv M_2/m_2$, $\bar{\delta}_0\equiv M_2/M_2'$, and the other notations are defined in [173].

The totally quantized energy E_{QD25} is given by

$$E_{QD25}\left(1+\alpha E_{QD25}\right)=\frac{\hbar^2}{2m_1}\left(\frac{\pi n_z}{d_z}\right)^2+\frac{\hbar^2}{2m_3}\left(\frac{\pi n_z}{d_z}\right)^2+\frac{\hbar^2}{2M_2}\left(\frac{\pi n_y}{d_y}\right)^2$$
$$\times\left[1+\bar{\delta}_0+\alpha E_{QD25}\left(1-\gamma_0\right)\right]+\frac{\gamma_0\hbar^4}{4M_2^2E_{g0}}\left(\frac{\pi n_y}{d_y}\right)^4 \quad (1.114)$$

The electron concentration can be written as

$$n_0 = \frac{2g_v}{d_x d_y d_z} \sum_{n_x=1}^{n_{x_{\max}}} \sum_{n_y=1}^{n_{y_{\max}}} \sum_{n_z=1}^{n_{z_{\max}}} \frac{L_{35}}{M_{35}}, \tag{1.115}$$

where $L_{35} = [1 + A_{25} \cos H_{25}]$, $A_{25} = \exp[E_{\mathrm{QD25}} - E_{\mathrm{FQD}}/k_B T]$, $H_{25} = \Gamma_{25}/k_B T$, Γ_{25} is the broadening parameter in this case, and $M_{35} = 1 + A_{25}^2 + 2A_{25} \cos H_{25}$.

The TPSM in this case, using (1.13) and (1.115), can be expressed as

$$G_0 = \frac{\pi^2 k_B}{3e} \left[\sum_{n_x=1}^{n_{x_{\max}}} \sum_{n_y=1}^{n_{y_{\max}}} \sum_{n_z=1}^{n_{z_{\max}}} \frac{L_{35}}{M_{35}} \right]^{-1} \left[\sum_{n_x=1}^{n_{x_{\max}}} \sum_{n_y=1}^{n_{y_{\max}}} \sum_{n_z=1}^{n_{z_{\max}}} \frac{Q_{35}}{(M_{35})^2} \right], \tag{1.116}$$

where $Q_{35} = A_{25}\left[\left(1 + A_{25}^2\right) \cos H_{25} + 2A_{25}\right]$.

1.2.13.3 The Cohen Model

In accordance with the Cohen model [172], the dispersion law of the carriers in Bi is given by

$$E\,(1 + \alpha E) = \frac{p_x^2}{2m_1} + \frac{p_z^2}{2m_3} - \frac{\alpha E p_y^2}{2m_2'} + \frac{p_y^2(1 + \alpha E)}{2m_2} + \frac{\alpha p_y^4}{4m_2 m_2'} \tag{1.117}$$

The totally quantized energy E_{QD26} assumes the form

$$E_{\mathrm{QD26}}\left(1 + \alpha E_{\mathrm{QD26}}\right) = \frac{\hbar^2}{2m_1}\left(\frac{\pi n_x}{d_x}\right)^2 + \frac{\hbar^2}{2m_3}\left(\frac{\pi n_z}{d_z}\right)^2 + \frac{\alpha \hbar^4}{4m_2 m_2'}\left(\frac{\pi n_y}{d_y}\right)^4 + \frac{\hbar^2}{2m_2}\left(\frac{\pi n_y}{d_y}\right)^2 \left[1 + \alpha E_{\mathrm{QD26}}\left(1 - \frac{m_2}{m_2'}\right)\right] \tag{1.118}$$

The electron concentration is given by

$$n_0 = \frac{2g_v}{d_x d_y d_z} \sum_{n_x=1}^{n_{x_{\max}}} \sum_{n_y=1}^{n_{y_{\max}}} \sum_{n_z=1}^{n_{z_{\max}}} \frac{L_{36}}{M_{36}}, \tag{1.119}$$

where $L_{36} = [1 + A_{26} \cos H_{26}]$, $A_{26} = \exp[E_{QD26} - E_{FQD}/k_B T]$, $H_{26} = \Gamma_{26}/k_B T$, Γ_{26} is the broadening parameter in this case, and $M_{36} = 1 + A_{26}^2 + 2A_{26} \cos H_{26}$.

The TPSM in this case, using (1.13) and (1.120), can be expressed as

$$G_0 = \frac{\pi^2 k_B}{3e} \left[\sum_{n_x=1}^{n_{x_{\max}}} \sum_{n_y=1}^{n_{y_{\max}}} \sum_{n_z=1}^{n_{z_{\max}}} \frac{L_{36}}{M_{36}} \right]^{-1} \left[\sum_{n_x=1}^{n_{x_{\max}}} \sum_{n_y=1}^{n_{y_{\max}}} \sum_{n_z=1}^{n_{z_{\max}}} \frac{Q_{36}}{(M_{36})^2} \right] \tag{1.120}$$

where $Q_{36} = A_{26}\left[\left(1 + A_{26}^2\right) \cos H_{26} + 2A_{26}\right]$.

1.2.13.4 The Lax Model

The electron energy spectra in bulk specimens of Bi in accordance with the Lax model can be written as [166]

$$E\,(1+\alpha E)=\frac{p_x^2}{2m_1}+\frac{p_y^2}{2m_2}+\frac{p_z^2}{2m_3} \tag{1.121}$$

The totally quantized energy E_{QD27} is given by

$$E_{\mathrm{QD27}}\left(1+\alpha E_{\mathrm{QD27}}\right)=\frac{\hbar^2}{2m_1}\left(\frac{\pi n_x}{d_x}\right)^2+\frac{\hbar^2}{2m_2}\left(\frac{\pi n_y}{d_y}\right)^2+\frac{\hbar^2}{2m_3}\left(\frac{\pi n_z}{d_z}\right)^2 \tag{1.122}$$

The electron concentration can be written as

$$n_0=\frac{2g_{\mathrm{v}}}{d_x d_y d_z}\sum_{n_x=1}^{n_{x_{\max}}}\sum_{n_y=1}^{n_{y_{\max}}}\sum_{n_z=1}^{n_{z_{\max}}}\frac{L_{37}}{M_{37}}, \tag{1.123}$$

where $L_{37}=[1+A_{27}\cos H_{27}]$, $A_{27}=\exp[E_{\mathrm{QD27}}-E_{\mathrm{FQD}}/k_{\mathrm{B}}T]$, $H_{27}=\Gamma_{27}/k_{\mathrm{B}}T$, Γ_{27} is the broadening parameter in this case, and $M_{37}=1+A_{27}^2+2A_{27}\cos H_{27}$.

The TPSM in this case, using (1.13) and (1.123), can be expressed as

$$G_0=\frac{\pi^2 k_{\mathrm{B}}}{3e}\left[\sum_{n_x=1}^{n_{x_{\max}}}\sum_{n_y=1}^{n_{y_{\max}}}\sum_{n_z=1}^{n_{z_{\max}}}\frac{L_{37}}{M_{37}}\right]^{-1}\left[\sum_{n_x=1}^{n_{x_{\max}}}\sum_{n_y=1}^{n_{y_{\max}}}\sum_{n_z=1}^{n_{z_{\max}}}\frac{Q_{37}}{(M_{37})^2}\right], \tag{1.124}$$

where $Q_{37}=A_{27}\left[\left(1+A_{27}^2\right)\cos H_{27}+2A_{27}\right]$.

1.2.13.5 Ellipsoidal Parabolic Model

The totally quantized energy $\overline{E_{\mathrm{QD27}}}$ for this model is given by

$$\overline{E_{\mathrm{QD27}}}=\frac{\hbar^2}{2m_1}\left(\frac{\pi n_x}{d_x}\right)^2+\frac{\hbar^2}{2m_2}\left(\frac{\pi n_y}{d_y}\right)^2+\frac{\hbar^2}{2m_3}\left(\frac{\pi n_z}{d_z}\right)^2 \tag{1.124a}$$

The electron concentration can be written as

$$n_0=\frac{2g_{\mathrm{v}}}{d_x d_y d_z}\sum_{n_x=1}^{n_{x_{\max}}}\sum_{n_y=1}^{n_{y_{\max}}}\sum_{n_z=1}^{n_{z_{\max}}}\frac{\overline{L_{37}}}{\overline{M_{37}}}, \tag{1.124b}$$

where $\overline{L_{37}} = \left[1 + \overline{A_{27}} \cos H_{27}\right]$, $\overline{A_{27}} = \exp[\overline{E_{QD27}} - E_{FQD}/k_B T]$, $H_{27} = \Gamma_{27}/k_B T$, Γ_{27} is the broadening parameter in this case, and $\overline{M_{37}} = 1 + \left(\overline{A_{27}}\right)^2 + 2\overline{A_{27}} \cos H_{27}$.

The TPSM in this case, using (1.13) and (1.124b), can be expressed as

$$G_0 = \frac{\pi^2 k_B}{3e}\left[\sum_{n_x=1}^{n_{x_{max}}}\sum_{n_y=1}^{n_{y_{max}}}\sum_{n_z=1}^{n_{z_{max}}}\frac{\overline{L_{37}}}{\overline{M_{37}}}\right]^{-1}\left[\sum_{n_x=1}^{n_{x_{max}}}\sum_{n_y=1}^{n_{y_{max}}}\sum_{n_z=1}^{n_{z_{max}}}\frac{\overline{Q_{37}}}{\left(\overline{M_{37}}\right)^2}\right], \tag{1.124c}$$

where $\overline{Q_{37}} = \overline{A_{27}}\left[\left(1 + \left(\overline{A_{27}}\right)^2\right) \cos H_{27} + 2\overline{A_{27}}\right]$.

1.2.14 Magnetothermopower in Quantum Dots of IV–VI Materials

In addition to Cohen and Lax models, the carriers of IV–VI materials are also described by the three more types of dispersion relations as provided by Dimmock [246], Bangert and Kastner [247] and Foley et al. [248, 249], respectively. The TPSM in QDs of IV–VI materials in accordance with the aforementioned models has been discussed for the purpose of complete presentation:

(a) The dispersion relation of the conduction electrons in IV–VI materials can be expressed in accordance with Dimmock [246] as

$$\left[\bar{\varepsilon} - \frac{E_{g_0}}{2} - \frac{\hbar^2 k_s^2}{2m_t^-} - \frac{\hbar^2 k_z^2}{2m_l^-}\right]\left[\bar{\varepsilon} + \frac{E_{g_0}}{2} + \frac{\hbar^2 k_s^2}{2m_t^-} + \frac{\hbar^2 k_z^2}{2m_l^-}\right] = P_\perp^2 k_s^2 + P_{||}^2 k_z^2, \tag{1.125}$$

where $\bar{\varepsilon}$ is the energy as measured from the center of the band gap E_{g_0} and $m_t^\pm$ and $m_l^\pm$ represent the contributions to the transverse and longitudinal effective masses of the external L_6^+ and L_6^- bands arising from the $\vec{k} \cdot \vec{p}$ perturbations with the other bands taken to the second order.

Using $\bar{\varepsilon} = E + \left(E_{g_0}/2\right)$, $P_\perp^2 = \hbar^2 E_{g_0}/2m_t^*$, and $P_{||}^2 = \hbar^2 E_{g_0}/2m_l^*$ (m_t^* and m_l^* are the transverse and longitudinal effective masses at $k = 0$) in (1.125), the totally quantized energy E_{QD28} in this case can be written as

$$\left[E_{QD28} - \frac{\hbar^2}{2m_t^-}\left\{\left(\frac{\pi n_x}{d_x}\right)^2 + \left(\frac{\pi n_y}{d_y}\right)^2\right\} - \frac{\hbar^2}{2m_l^-}\left(\frac{\pi n_z}{d_z}\right)^2\right]$$
$$\left[1 + \alpha E_{QD28} + \frac{\alpha\hbar^2}{2m_t^+}\left\{\left(\frac{\pi n_x}{d_x}\right)^2 + \left(\frac{\pi n_y}{d_y}\right)^2\right\} + \alpha\frac{\hbar^2}{2m_l^+}\left(\frac{\pi n_z}{d_z}\right)^2\right]$$

$$= \frac{\hbar^2}{2m_t^*}\left[\left(\frac{\pi n_x}{d_x}\right)^2 + \left(\frac{\pi n_y}{d_y}\right)^2\right] + \frac{\hbar^2}{2m_l^*}\left(\frac{\pi n_z}{d_z}\right)^2. \tag{1.126}$$

The electron concentration is given by

$$n_0 = \frac{2g_v}{d_x d_y d_z}\sum_{n_x=1}^{n_{x_{\max}}}\sum_{n_y=1}^{n_{y_{\max}}}\sum_{n_z=1}^{n_{z_{\max}}}\frac{L_{38}}{M_{38}}, \tag{1.127}$$

where $L_{38} = [1 + A_{28}\cos H_{28}]$, $A_{28} = \exp[E_{QD28} - E_{FQD}/k_B T]$, $H_{28} = \Gamma_{28}/k_B T$, Γ_{28} is the broadening parameter in this case, and $M_{38} = 1 + A_{28}^2 + 2A_{28}\cos H_{28}$.

The TPSM in this case, using (1.13) and (1.127), can be expressed as

$$G_0 = \frac{\pi^2 k_B}{3e}\left[\sum_{n_x=1}^{n_{x_{\max}}}\sum_{n_y=1}^{n_{y_{\max}}}\sum_{n_z=1}^{n_{z_{\max}}}\frac{L_{38}}{M_{38}}\right]^{-1}\left[\sum_{n_x=1}^{n_{x_{\max}}}\sum_{n_y=1}^{n_{y_{\max}}}\sum_{n_z=1}^{n_{z_{\max}}}\frac{Q_{38}}{(M_{38})^2}\right], \tag{1.128}$$

where $Q_{38} = A_{28}\left[\left(1 + A_{28}^2\right)\cos H_{28} + 2A_{28}\right]$.

(b) The electron dispersion law in IV–VI compounds can be written in accordance with Bangert and Kastner [247] as

$$\Gamma(E) = \bar{F}_1(E)k_s^2 + \bar{F}_2(E)k_z^2, \tag{1.129}$$

where $\Gamma(E) \equiv 2E$,

$$\bar{F}_1(E) \equiv \left[\frac{(\bar{R})^2}{E + E_{g_0}} + \frac{(\bar{s})^2}{E + \Delta_c'} + \frac{(\bar{Q})^2}{E + \Delta_c''}\right],$$

$$\bar{F}_2(E) \equiv \left[\frac{2(\bar{A})^2}{E + E_{g_0}} + \frac{(\bar{s} + \bar{Q})^2}{E + \Delta_c''}\right],$$

$\bar{R}$, $\bar{s}$, Δ_c', $\bar{Q}$, Δ_c'', and $\bar{A}$ are the spectrum constants.

The totally quantized energy E_{QD29} can be expressed as

$$\Gamma\left(E_{QD29}\right) = F_1\left(E_{QD29}\right)\left[\left(\frac{\pi n_x}{d_x}\right)^2 + \left(\frac{\pi n_y}{d_y}\right)^2\right] + F_2\left(E_{QD29}\right)\left(\frac{\pi n_z}{d_z}\right)^2, \tag{1.130}$$

The electron concentration is given by

$$n_0 = \frac{2g_v}{d_x d_y d_z}\sum_{n_x=1}^{n_{x_{\max}}}\sum_{n_y=1}^{n_{y_{\max}}}\sum_{n_z=1}^{n_{z_{\max}}}\frac{L_{39}}{M_{39}}, \tag{1.131}$$

where $L_{39} = [1 + A_{29} \cos H_{29}]$, $A_{29} = \exp[E_{QD29} - E_{FQD}/k_B T]$, $H_{29} = \Gamma_{29}/k_B T$, Γ_{29} is the broadening parameter in this case, and $M_{39} = 1 + A_{29}^2 + 2A_{29} \cos H_{29}$.

The TPSM in this case, using (1.13) and (1.131), can be expressed as

$$G_0 = \frac{\pi^2 k_B}{3e} \left[\sum_{n_x=1}^{n_{x_{max}}} \sum_{n_y=1}^{n_{y_{max}}} \sum_{n_z=1}^{n_{z_{max}}} \frac{L_{39}}{M_{39}} \right]^{-1} \left[\sum_{n_x=1}^{n_{x_{max}}} \sum_{n_y=1}^{n_{y_{max}}} \sum_{n_z=1}^{n_{z_{max}}} \frac{Q_{39}}{(M_{39})^2} \right], \tag{1.132}$$

where $Q_{39} = A_{29}\left[\left(1 + A_{29}^2\right) \cos H_{29} + 2A_{29}\right]$.

(c) In accordance with Foley et al. [248, 249], the electron dispersion relation assumes the form

$$E + \frac{E_{g0}}{2} = \frac{\hbar^2 k_s^2}{2m_\perp^-} + \frac{\hbar^2 k_z^2}{2m_{||}^-} + \left[\left[\frac{\hbar^2 k_s^2}{2m_\perp^+} + \frac{\hbar^2 k_z^2}{2m_{||}^+} + \frac{E_{g0}}{2} \right]^2 + P_{||}^2 k_z^2 + P_\perp^2 k_s^2 \right]^{1/2}, \tag{1.133}$$

where

$$\frac{1}{m_\perp^\pm} = \frac{1}{2} \left[\frac{1}{m_{tc}} \pm \frac{1}{m_{tv}} \right],$$

$$\frac{1}{m_{||}^\pm} = \frac{1}{2} \left[\frac{1}{m_{lc}} \pm \frac{1}{m_{lv}} \right],$$

m_{tc} and m_{lc} are the transverse and longitudinal effective electron masses of the conduction electrons at the edge of the conduction band and m_{tv} and m_{lv} are the transverse and longitudinal effective hole masses of the holes at the edge of the valence band.

The totally quantized energy E_{QD30} for this model is given by

$$\begin{aligned} E_{QD30} = & -\frac{E_{g0}}{2} + \frac{\hbar^2}{2m_\perp^-} \left[\left(\frac{n_x \pi}{d_x} \right)^2 + \left(\frac{n_y \pi}{d_y} \right)^2 \right] + \frac{\hbar^2}{2m_{||}^-} \left(\frac{n_z \pi}{d_z} \right)^2 \\ & + \left[\left[\frac{E_{g0}}{2} + \frac{\hbar}{2m_{||}^+} \left(\frac{n_z \pi}{d_z} \right)^2 + \frac{\hbar^2}{2m_\perp^+} \left\{ \left(\frac{n_x \pi}{d_x} \right)^2 + \left(\frac{n_y \pi}{d_y} \right)^2 \right\} \right]^2 \right. \\ & \left. + P_{||}^2 \left(\frac{n_z \pi}{d_z} \right)^2 + P_\perp^2 \left\{ \left(\frac{n_x \pi}{d_x} \right)^2 + \left(\frac{n_y \pi}{d_y} \right)^2 \right\} \right]^{1/2} \end{aligned} \tag{1.134}$$

The electron concentration can be expressed as

$$n_0 = \frac{2g_v}{d_x d_y d_z} \sum_{n_x=1}^{n_{x_{max}}} \sum_{n_y=1}^{n_{y_{max}}} \sum_{n_z=1}^{n_{z_{max}}} \frac{L_{40}}{M_{40}}, \tag{1.135}$$

where $L_{40} = [1 + A_{30} \cos H_{30}]$, $A_{30} = \exp[E_{\mathrm{QD30}} - E_{\mathrm{FQD}}/k_{\mathrm{B}}T]$, $H_{30} = \Gamma_{30}/k_{\mathrm{B}}T$, Γ_{30} is the broadening parameter in this case, and $M_{40} = 1 + A_{30}^2 + 2A_{30} \cos H_{30}$.

The TPSM in this case, using (1.13) and (1.135), can be written as

$$G_0 = \frac{\pi^2 k_{\mathrm{B}}}{3e}\left[\sum_{n_x=1}^{n_{x_{\max}}}\sum_{n_y=1}^{n_{y_{\max}}}\sum_{n_z=1}^{n_{z_{\max}}}\frac{L_{40}}{M_{40}}\right]^{-1}\left[\sum_{n_x=1}^{n_{x_{\max}}}\sum_{n_y=1}^{n_{y_{\max}}}\sum_{n_z=1}^{n_{z_{\max}}}\frac{Q_{40}}{(M_{40})^2}\right], \tag{1.136}$$

where $Q_{40} = A_{30}\left[\left(1 + A_{30}^2\right)\cos H_{30} + 2A_{30}\right]$.

1.2.15 *Magnetothermopower in Quantum Dots of Lead Germanium Telluride*

The dispersion law of n-type $Pb_{1-x}Ge_xTe$ with $x = 0.01$ can be expressed following Vassilev [250] as

$$\begin{aligned}&\left[E - 0.606k_s^2 - 0.722k_z^2\right]\left[E + \bar{E}_{g0} + 0.411k_s^2 + 0.377k_z^2\right]\\&\quad = 0.23k_s^2 + 0.02k_z^2 \pm \left[0.06\bar{E}_{g0} + 0.061k_s^2 + 0.0066k_z^2\right]k_s,\end{aligned} \tag{1.137}$$

where $\bar{E}_{g0} = 0.21$ eV, k_x and k_y and k_z are in the units of $10^9\ \mathrm{m}^{-1}$.

The totally quantized energy $\left(E_{QD31,\pm}\right)$is given by

$$\begin{aligned}&\left[E_{\mathrm{QD31},\pm} - 0.606\left(\left(\frac{\pi n_x}{d_x}\right)^2 + \left(\frac{\pi n_y}{d_y}\right)^2\right) - 0.0722\left(\frac{\pi n_z}{d_z}\right)^2\right]\\&\times\left[E_{\mathrm{QD31},\pm} + 0.471\left(\left(\frac{\pi n_x}{d_x}\right)^2 + \left(\frac{\pi n_y}{d_y}\right)^2\right) + \bar{E}_{g0} + 0.0377\left(\frac{\pi n_z}{d_z}\right)^2\right]\\&= 0.23\left(\left(\frac{\pi n_x}{d_x}\right)^2 + \left(\frac{\pi n_y}{d_y}\right)^2\right) + 0.02\left(\frac{\pi n_z}{d_z}\right)^2 \pm \left[\left(\frac{\pi n_x}{d_x}\right)^2 + \left(\frac{\pi n_y}{d_y}\right)^2\right]^{1/2}\\&\times\left[0.06\overline{E_{g0}} + 0.061\left(\left(\frac{\pi n_x}{d_x}\right)^2 + \left(\frac{\pi n_y}{d_y}\right)^2\right) + 0.0066\left(\frac{\pi n_z}{d_z}\right)^2\right]\end{aligned} \tag{1.138}$$

The electron concentration is given by

$$n_0 = \frac{g_{\mathrm{v}}}{d_x d_y d_z}\sum_{n_x=1}^{n_{x_{\max}}}\sum_{n_y=1}^{n_{y_{\max}}}\sum_{n_z=1}^{n_{z_{\max}}}\frac{L_{41,\pm}}{M_{41,\pm}}, \tag{1.139}$$

where $L_{41,\pm} = \left[1 + A_{31,\pm} \cos H_{31}\right]$, $A_{31,\pm} = \exp[E_{\mathrm{QD}31,\pm} - E_{\mathrm{FQD}}/k_{\mathrm{B}}T]$, $H_{31} = \Gamma_{31}/k_{\mathrm{B}}T$, Γ_{31} is the broadening parameter in this case, and $M_{41,\pm} = 1 + A_{31,\pm}^2 + 2A_{31,\pm} \cos H_{31}$.

The TPSM in this case, using (1.13) and (1.139), can be expressed as

$$G_0 = \frac{\pi^2 k_{\mathrm{B}}}{3e}\left[\sum_{n_x=1}^{n_{x_{\max}}}\sum_{n_y=1}^{n_{y_{\max}}}\sum_{n_z=1}^{n_{z_{\max}}}\frac{L_{41,\pm}}{M_{41,\pm}}\right]^{-1}\left[\sum_{n_x=1}^{n_{x_{\max}}}\sum_{n_y=1}^{n_{y_{\max}}}\sum_{n_z=1}^{n_{z_{\max}}}\frac{Q_{41,\pm}}{(M_{41,\pm})^2}\right], \quad (1.140)$$

where $Q_{41,\pm} = A_{31,\pm}\left[\left(1 + A_{31,\pm}^2\right)\cos H_{31} + 2A_{31,\pm}\right]$.

1.2.16 *Magnetothermopower in Quantum Dots of Zinc and Cadmium Diphosphides*

The dispersion relation of the holes of Cadmium and Zinc diphosphides can approximately be written following Chuiko [251] as

$$E = \left[\beta_1 + \frac{\beta_2\beta_3(k)}{8\beta_4}\right]k^2 \pm \left\{\left[\beta_4\beta_3(k)\left(\beta_5 - \frac{\beta_2\beta_3(k)}{8\beta_4}\right)k^2\right] + 8\beta_4^2\left(1 - \frac{\beta_3^2(k)}{4}\right) - \beta_2\left(1 - \frac{\beta_3^2(k)}{4}\right)k^2\right\}^{1/2}, \quad (1.141)$$

where β_1, β_2, β_4, and β_5 are system constants and $\beta_3(k) = k_x^2 + k_y^2 - 2k_z^2/k^2$.

The totally quantized energy $E_{\mathrm{QD}32,\pm}$ is given by

$$E_{\mathrm{QD}32,\pm} = \left[\beta_1 + \frac{\beta_2\beta_6}{8\beta_4}\right]\beta_7^2 \pm \left\{\left[\beta_4\beta_6\left(\beta_5 - \frac{\beta_2\beta_6}{8\beta_4}\right)\beta_7^2\right] + 8\beta_4^2\left(1 - \frac{\beta_6^2}{4}\right) - \beta_2\left(1 - \frac{\beta_6^2}{4}\right)\beta_7^2\right\}^{1/2}, \quad (1.142)$$

where

$$\beta_6 = \frac{1}{\beta_7^2}\left[\left(\frac{\pi n_x}{d_x}\right)^2 + \left(\frac{\pi n_y}{d_y}\right)^2 - 2\left(\frac{\pi n_z}{d_z}\right)^2\right]$$

and

$$\beta_7^2 = \left[\left(\frac{\pi n_x}{d_x}\right)^2 + \left(\frac{\pi n_y}{d_y}\right)^2 + \left(\frac{\pi n_z}{d_z}\right)^2\right].$$

The hole concentration can be written as

$$p_0 = \frac{g_{\mathrm{v}}}{d_x d_y d_z}\sum_{n_x=1}^{n_{x_{\max}}}\sum_{n_y=1}^{n_{y_{\max}}}\sum_{n_z=1}^{n_{z_{\max}}}\frac{L_{42,\pm}}{M_{42,\pm}}, \quad (1.143)$$

where $L_{42,\pm} = [1 + A_{32,\pm} \cos H_{32}]$, $A_{32,\pm} = \exp[E_{QD32,\pm} - E_{FQD}/k_B T]$, $H_{32} = \Gamma_{32}/k_B T$, Γ_{32} is the broadening parameter in this case, and $M_{42,\pm} = 1 + A_{32,\pm}^2 + 2A_{32,\pm} \cos H_{32}$.

The TPSM in this case, using (1.13) and (1.143), can be expressed as

$$G_0 = \frac{\pi^2 k_B}{3e} \left[\sum_{n_x=1}^{n_{x_{\max}}} \sum_{n_y=1}^{n_{y_{\max}}} \sum_{n_z=1}^{n_{z_{\max}}} \frac{L_{42,\pm}}{M_{42,\pm}} \right]^{-1} \left[\sum_{n_x=1}^{n_{x_{\max}}} \sum_{n_y=1}^{n_{y_{\max}}} \sum_{n_z=1}^{n_{z_{\max}}} \frac{Q_{42,\pm}}{(M_{42,\pm})^2} \right], \quad (1.144)$$

where $Q_{42,\pm} = A_{32,\pm} \left[\left(1 + A_{32,\pm}^2\right) \cos H_{32} + 2A_{32,\pm} \right]$.

1.2.17 Magnetothermopower in Quantum Dots of Bismuth Telluride

The dispersion relation of the holes in Bi_2Te_3 can be expressed as [252, 253]

$$E(1 + \alpha E) = \frac{\hbar^2}{2m_0} \left(\left(\overline{\overline{\alpha_{11}}}\right) k_x^2 + \alpha_{22} k_y^2 + \alpha_{33} k_z^2 + 2\alpha_{23} k_y k_z \right), \quad (1.145)$$

where $\overline{\overline{\alpha_{11}}}, \alpha_{22}, \alpha_{33}, \text{and} \alpha_{23}$ are spectrum constants.

The totally quantized energy E_{QD33} can be written as

$$E_{QD33}\left(1 + \alpha E_{QD33}\right) = \frac{\hbar^2}{2m_0} \left[\left(\overline{\overline{\alpha_{11}}}\right) \left(\frac{n_x \pi}{d_x}\right)^2 + \alpha_{22} \left(\frac{n_y \pi}{d_y}\right)^2 + \alpha_{33} \left(\frac{n_z \pi}{d_z}\right)^2 + 2\alpha_{23} \left(\frac{\pi^2 n_y n_z}{d_y d_z}\right) \right]. \quad (1.146)$$

The hole concentration is given by

$$p_0 = \frac{2g_v}{d_x d_y d_z} \sum_{n_x=1}^{n_{x_{\max}}} \sum_{n_y=1}^{n_{y_{\max}}} \sum_{n_z=1}^{n_{z_{\max}}} \frac{L_{43}}{M_{43}}, \quad (1.147)$$

where $L_{43} = [1 + A_{33} \cos H_{33}]$, $A_{33} = \exp[E_{QD33} - E_{FQD}/k_B T]$, $H_{33} = \Gamma_{33}/k_B T$, Γ_{33} is the broadening parameter in this case, and $M_{43} = 1 + A_{33}^2 + 2A_{33} \cos H_{33}$.

The TPSM in this case, using (1.13) and (1.147), can be expressed as

$$G_0 = \frac{\pi^2 k_B}{3e} \left[\sum_{n_x=1}^{n_{x_{\max}}} \sum_{n_y=1}^{n_{y_{\max}}} \sum_{n_z=1}^{n_{z_{\max}}} \frac{L_{43}}{M_{43}} \right]^{-1} \left[\sum_{n_x=1}^{n_{x_{\max}}} \sum_{n_y=1}^{n_{y_{\max}}} \sum_{n_z=1}^{n_{z_{\max}}} \frac{Q_{43}}{(M_{43})^2} \right], \quad (1.148)$$

where $Q_{43} = A_{33} \left[\left(1 + A_{33}^2\right) \cos H_{33} + 2A_{33} \right]$.

1.2.18 Magnetothermopower in Quantum Dots of Antimony

The dispersion relation of the conduction electrons in Antimony (Sb) can be written following Ketterson [254] as

$$\frac{2m_0E}{\hbar^2} = \overline{\alpha_{11}}k_x^2 + \overline{\alpha_{22}}k_y^2 + \overline{\alpha_{33}}k_z^2 + 2\overline{\alpha_{23}}k_yk_z, \tag{1.149}$$

$$\frac{2m_0E}{\hbar^2} = a_1k_x^2 + a_2k_y^2 + a_3k_z^2 + a_4k_yk_z + a_5k_zk_x + a_6k_xk_y, \tag{1.150}$$

$$\frac{2m_0E}{\hbar^2} = a_1k_x^2 + a_2k_y^2 + a_3k_z^2 + a_4k_yk_z - a_5k_zk_x - a_6k_xk_y, \tag{1.151}$$

where $a_1 = (4)^{-1}(\overline{\alpha_{11}} + 3\overline{\alpha_{22}})$, $a_2 = (4)^{-1}(\overline{\alpha_{22}} + 3\overline{\alpha_{11}})$, $a_3 = \overline{\alpha_{33}}$, $a_4 = \overline{\alpha_{33}}$, $a_5 = \sqrt{3}$, $a_6 = \sqrt{3}(\overline{\alpha_{22}} - \overline{\alpha_{11}})$, $\overline{\alpha_{11}}, \overline{\alpha_{22}}, \overline{\alpha_{33}}$, and $\overline{\alpha_{23}}$ are the system constants.

The totally quantized energy, E_{QD34}, whose bulk carrier dispersion law is described by (1.149) can be written as

$$\frac{2m_0E_{QD34}}{\hbar^2} = \bar{\alpha}_{11}\left(\frac{\pi n_{x1}}{d_x}\right)^2 + \bar{\alpha}_{22}\left(\frac{\pi n_{y1}}{d_y}\right)^2 + \bar{\alpha}_{33}\left(\frac{\pi n_{z1}}{d_y}\right)^2 + 2\bar{\alpha}_{23}\left(\frac{\pi^2 n_{y1}n_{z1}}{d_yd_z}\right). \tag{1.152}$$

Therefore, the electron concentration for the dispersion relation as described by (1.149) in this case assumes the form

$$n_{01} = \frac{2g_v}{d_xd_yd_z}\sum_{n_x=1}^{n_{x_{max}}}\sum_{n_y=1}^{n_{y_{max}}}\sum_{n_z=1}^{n_{z_{max}}}\frac{L_{44}}{M_{44}}, \tag{1.153}$$

where $L_{44} = [1 + A_{34}\cos H_{34}]$, $A_{34} = \exp[E_{QD34} - E_{FQD}/k_BT]$, $H_{34} = \Gamma_{34}/k_BT$, Γ_{34} is the broadening parameter in this case, and $M_{44} = 1 + A_{34}^2 + 2A_{34}\cos H_{34}$.

The totally quantized energy, E_{QD35}, whose bulk carrier dispersion law is described by (1.150) can be written as

$$E_{QD35} = \frac{\hbar^2}{2m_0}\left[a_1\left(\frac{n_{x2}\pi}{d_x}\right)^2 + a_2\left(\frac{n_{y2}\pi}{d_y}\right)^2 + a_3\left(\frac{n_{z2}\pi}{d_z}\right)^2 + a_4\frac{n_{y2}n_{z2}\pi^2}{d_yd_z} + a_5\frac{n_{x2}n_{z2}\pi^2}{d_xd_z} + a_6\frac{\pi^2 n_{x2}n_{y2}}{d_xd_y}\right] \tag{1.154}$$

Similarly for QDs, whose bulk carrier dispersion law is given by (1.150), the corresponding electron concentration can be expressed as

$$n_{02} = \frac{2g_{\mathrm{v}}}{d_x d_y d_z} \sum_{n_x=1}^{n_{x_{\max}}} \sum_{n_y=1}^{n_{y_{\max}}} \sum_{n_z=1}^{n_{z_{\max}}} \frac{L_{45}}{M_{45}}, \tag{1.155}$$

where $L_{45} = [1 + A_{35} \cos H_{35}]$, $A_{35} = \exp[E_{\mathrm{QD35}} - E_{\mathrm{FQD}}/k_{\mathrm{B}}T]$, $H_{35} = \Gamma_{35}/k_{\mathrm{B}}T$, Γ_{35} is the broadening parameter in this case, and $M_{45} = 1 + A_{35}^2 + 2A_{35} \cos H_{35}$.

The totally quantized energy, E_{QD36}, whose bulk carrier dispersion law is described by (1.151) can be written as

$$E_{\mathrm{QD36}} = \frac{\hbar^2}{2m_0}\left[a_1\left(\frac{n_{x3}\pi}{d_x}\right)^2 + a_2\left(\frac{n_{y3}\pi}{d_y}\right)^2 + a_3\left(\frac{n_{z3}\pi}{d_z}\right)^2 + a_4\frac{n_{y3}n_{z3}\pi^2}{d_y d_z} + a_5\frac{n_{x3}n_{z3}\pi^2}{d_x d_z} + a_6\frac{\pi^2 n_{x3}n_{y3}}{d_x d_y}\right] \tag{1.156}$$

Similarly for QDs, whose bulk carrier dispersion law is given by (1.151), the corresponding electron concentration can be written as

$$n_{03} = \frac{2g_{\mathrm{v}}}{d_x d_y d_z} \sum_{n_x=1}^{n_{x_{\max}}} \sum_{n_y=1}^{n_{y_{\max}}} \sum_{n_z=1}^{n_{z_{\max}}} \frac{L_{46}}{M_{46}}, \tag{1.157}$$

where $L_{46} = [1 + A_{36} \cos H_{36}]$, $A_{36} = \exp[E_{\mathrm{QD36}} - E_{\mathrm{FQD}}/k_{\mathrm{B}}T]$, $H_{36} = \Gamma_{36}/k_{\mathrm{B}}T$, Γ_{36} is the broadening parameter in this case, and $M_{46} = 1 + A_{36}^2 + 2A_{36} \cos H_{36}$.

Therefore, the total electron concentration in this case is given by

$$n_0 = \frac{2g_{\mathrm{v}}}{d_x d_y d_z}\left[\sum_{n_x=1}^{n_{x_{\max}}} \sum_{n_y=1}^{n_{y_{\max}}} \sum_{n_z=1}^{n_{z_{\max}}} \left[\frac{L_{44}}{M_{44}} + \frac{L_{45}}{M_{45}} + \frac{L_{46}}{M_{46}}\right]\right]. \tag{1.158}$$

The TPSM, in this case, using (1.13) and (1.158) can be expressed as

$$G_0 = \left(\frac{\pi^2 k_{\mathrm{B}}^2 T}{3eh_{11}}\right)\left[\sum_{n_x=1}^{n_{x_{\max}}} \sum_{n_y=1}^{n_{y_{\max}}} \sum_{n_z=1}^{n_{z_{\max}}} \left[\frac{L_{44}}{M_{44}} + \frac{L_{45}}{M_{45}} + \frac{L_{46}}{M_{46}}\right]\right]^{-1}, \tag{1.159}$$

where

$$(h_{11})^{-1} = \left[\sum_{n_x=1}^{n_{x_{\max}}} \sum_{n_y=1}^{n_{y_{\max}}} \sum_{n_z=1}^{n_{z_{\max}}} \left[\frac{Q_{44}}{(M_{44})^2} + \frac{Q_{45}}{(M_{45})^2} + \frac{Q_{46}}{(M_{46})^2}\right]\right],$$

$Q_{44} = A_{34}\left[\left(1 + A_{34}^2\right)\cos H_{34} + 2A_{34}\right]$, $Q_{45} = A_{35}[\left(1 + A_{35}^2\right)\cos H_{35}+2A_{35}]$, and $Q_{46} = A_{36}\left[\left(1 + A_{36}^2\right)\cos H_{36} + 2A_{36}\right]$.

1.3 Results and Discussion

Using (1.7) and (1.14) and the band constants from Table 1.1, the normalized TPSM in QDs of nonlinear optical materials (taking $CdGeAs_2$ as an example) has been plotted as a function of film thickness for the generalized band model [in accordance with (1.6)] as shown by curve (a) where the curves (b), (c), and (d) are valid for three (using (1.17) and two (using (1.21)) band models of Kane together with parabolic energy bands, respectively. The case for $\delta = 0$ has been plotted in the same figure and is represented by curve (e) for the purpose of assessing the influence of crystal field splitting on the TPSM in QDs of $CdGeAs_2$. It appears from the figure that the TPSM for QDs of $CdGeAs_2$ oscillates with film thickness and for specific values of the film thickness as determined by the energy band constants and the transition of Fermi energy from one allowed set of quantum numbers to another allowed set and the band structure of the material, the oscillatory dependence shows spikes. The influence of crystal field splitting effectively reduces the value of the TPSM for relatively large values of the thickness in the whole range of thickness as considered. From the numerical values for the three and two band models of Kane, it appears that the influence of spin–orbit splitting constant lessens the value of the TPSM for relatively large values of the nanothickness.

Figure 1.2 exhibits the variations of normalized TPSM in QDs of $CdGeAs_2$ as a function of electron concentration for all the cases mentioned as above for Fig. 1.1. From Fig. 1.2, we observe that the TPSM for the range of concentrations below the concentration zone $10^{23}\ \mathrm{m}^{-3}$, the influence of said band structures of nonlinear optical materials on the TPSM is not prominent, and they exhibit converging tendencies, whereas for higher values of the carrier degeneracy, the TPSM decreases with increasing carrier concentration. For generalized band model of QDs of $CdGeAs_2$, the numerical values of the TPSM are least, whereas, for the most approximate parabolic energy bands of the same, the TPSM exhibits the highest values with respect to concentration in this case. The quantum oscillations of the TPSM in QDs exhibit different numerical magnitudes as compared with the same in UFs and QWs. It may be noted that the QDs lead to the discrete energy levels, somewhat like atomic energy levels, which produce very large changes. This is in accordance to the inherent nature of the quantum confinement of the carrier gas as dealt with here. In QDs, there remain no free carrier states in between any two allowed sets of size-quantized levels unlike that found for UFs and QWs where the quantum confinements are one dimension and two dimension, respectively. Consequently, the crossing of the Fermi level by the size-quantized levels in QDs would have much greater impact on the redistribution of the carriers among the allowed levels, as compared with that found for UFs and QWs, respectively.

Table 1.1 The numerical values of the constants of the energy–wave vector dispersion relations of few materials

Materials	Numerical values of the energy band constants
1. The conduction electrons of n-Cadmium Germanium Arsenide can be described by three types of band models	1. The values of the energy band constants in accordance with the generalized electron dispersion relation of nonlinear optical materials are as follows $E_{g_0} = 0.57$ eV, $\Delta_{\|} = 0.30$ eV, $\Delta_{\perp} = 0.36$ eV, $m_{\|}^* = 0.034m_0$, $m_{\perp}^* = 0.039m_0$, $T = 4K$, $\delta = -0.21$ eV and $g_v = 1$ [223]. 2. In accordance with the three band model of Kane, the spectrum constants are given by $\Delta = \left(\Delta_{\|} + \Delta_{\perp}\right)/2 = 0.33$ eV, $E_{g_0} = 0.57$ eV, $m^* = \left(m_{\|}^* + m_{\perp}^*\right)/2 = 0.0365\,m_0$ and $\delta = 0$ eV. 3. In accordance with two band model of Kane, $E_{g_0} = 0.57$ eV and $m^* = 0.0365\,m_0$.
2. n-Indium Arsenide	The values $E_{g_0} = 0.36$ eV, $\Delta = 0.43$ eV, $m^* = 0.026m_0$ and $g_v = 1$ [224, 225] are valid for three band model of Kane.
3. n-Gallium Arsenide	The values $E_{g_0} = 1.55$ eV, $\Delta = 0.35$ eV, $m^* = 0.07m_0$ and $g_v = 1$ are valid for three band model of Kane [224, 225]. The values $\alpha_{13} = -1.97 \times 10^{-37}$ eVm^4 and $a_{15} = -2.3 \times 10^{-34}$ eVm^4 are valid for the model of Newson and Kurobe [227]. The values $\alpha_{11} = -2132 \times 10^{-40}$ eVm^4, $\alpha_{12} = 9030 \times 10^{-50}$ eVm^5, $\beta_{11} = -2493 \times 10^{-40}$ eVm^4, $\beta_{12} = 12594 \times 10^{-50}$ eVm^5, $\gamma_{11} = 30 \times 10^{-30}$ eVm^3, $\gamma_{12} = -154 \times 10^{-42}$ eVm^4 are valid for the model of Rossler [228].
4. n-Gallium Aluminum Arsenide	$E_{g_0} = \left(1.424 + 1.266x + 0.26x^2\right)$ eV, $\Delta = (0.34 - 0.5x)$ eV, $m^* = [0.066 + 0.088x]\,m_0$ and $g_v = 1$ [255, 256].
5. n-Mercury Cadmium Telluride	$E_{g_0} = (-0.302 + 1.93x + 5.35 \times 10^{-4}(1 - 2x)\,T - 0.810x^2 + 0.832x^3)$ eV, $\Delta = \left(0.63 + 0.24x - 0.27x^2\right)$ eV, $m^* = 0.1m_0 E_{g_0}(\text{eV})^{-1}$ and $g_v = 1$ [257].
6. n-Indium Gallium Arsenide Phosphide lattice matched to Indium Phosphide	$E_{g_0} = \left(1.337 - 0.73y + 0.13y^2\right)$ eV, $\Delta = \left(0.114 + 0.26y - 0.22y^2\right)$ eV, $m^* = (0.08 - 0.039y)\,m_0$, $y = (0.1896 - 0.4052x)/(0.1896 - 0.0123x)$ and $g_v = 1$ [258].
7. n-Indium Antimonide	$E_{g_0} = 0.2352$ eV, $\Delta = 0.81$ eV, $m^* = 0.01359m_0$ and $g_v = 1$ [224, 225].
8. n-Gallium Antimonide	The values of $E_{g_0} = 0.81$ eV, $\Delta = 0.80$ eV, $P_0 = 9.48 \times 10^{-10}$ eVm, $\bar{\varsigma}_0 = -2.1$, $\bar{v}_0 = -1.49$, $\bar{\omega}_0 = 0.42$ and $g_v = 1$ are valid for the model of Seiler et al. [243]. The values $\overline{E_1} = 1.024$ eV, $\overline{E_2} = 0$ eV, $\overline{E_3} = -1.132$ eV, $\overline{E_4} = 0.05$ eV, $\overline{E_5} = 1.107$ eV, $\overline{E_6} = -0.113$ eV and $\overline{E_7} = -0.0072$ eV are valid for the model of Zhang [245].

9. p-Cadmium Sulphide	$m^*_{\parallel} = 0.7m_0,\ m^*_{\perp} = 1.5m_0,\ C_0 = 1.4 \times 10^{-10}$ eVm and $g_v = 1$ [122].
10. n-Lead Telluride	The values $m_t^- = 0.070m_0$, $m_l^- = 0.54m_0$, $m_t^+ = 0.010m_0$, $m_l^+ = 1.4m_0$, $P_{\parallel} = 141$ me Vnm, $P_{\perp} = 486$ me Vnm, $E_{g_0} = 190$ meV and $g_v = 4$ [224, 225] are valid for the Dimmock model [243]. The values $(\bar{R})^2 = 2.3 \times 10^{-19}\,(\text{eVm})^2$, $E_{g_0} = 0.16$ eV, $(\bar{s})^2 = 4.6\,(\bar{R})^2$, $\Delta'_c = 3.07$ eV, $(\bar{Q})^2 = 1.3\,(\bar{R})^2$, $\Delta''_c = 3.28$ eV, $(\bar{A})^2 = 0.83 \times 10^{-19}\,(\text{eVm})^2$ are valid for the model of Bangert and Kastner [247]. The values $m_{tv} = 0.0965m_0$, $m_{lv} = 1.33m_0$, $m_{tc} = 0.088m_0$, $m_{lc} = 0.83m_0$ are valid for the model of Foley et al. [248, 249]. The values $m_1 = 0.0239m_0$, $m_2 = 0.024m_0$, $m'_2 = 0.31m_0$, $m_3 = 0.24m_0$ are valid for the Cohen model [172].
11. n-Lead Tin Telluride	$m_t^- = 0.070m_0,\ m_l^- = 0.54m_0,\ m_t^+ = 0.010m_0,\ m_l^+ = 1.4m_0,\ P_{\parallel} = 141$ meV nm, $P_{\perp} = 486$ meV nm, $E_{g_0} = 190$ meV, $g_v = 4$ [259] and $\varepsilon_{sc} = 33\varepsilon_0$ [223, 260–263].
12. n-Lead Tin Selenide	$m_t^- = 0.063m_0,\ m_l^- = 0.41m_0,\ m_t^+ = 0.089m_0,\ m_l^+ = 1.6m_0,\ P_{\parallel} = 137$ meV nm, $P_{\perp} = 464$ meV nm, $E_{g_0} = 90$ meV, $g_v = 4$ and $\varepsilon_{sc} = 60\varepsilon_0$ [223, 260–263].
13. Stressed n-Indium Antimonide	The values $m^* = 0.048m_o$, $E_{g_0} = 0.081$ eV, $B_2 = 9 \times 10^{-10}$ eVm, $C_1^c = 3$ eV, $C_2^c = 2$ eV, $\overline{a_0} = -10$ eV, $\overline{b_0} = -1.7$ eV, $\overline{d_0} = -4.4$ eV, $S_{xx} = 0.6 \times 10^{-3}(\text{kbar})^{-1}$, $S_{yy} = 0.42 \times 10^{-3}(\text{kbar})^{-1}$, $S_{zz} = 0.39 \times 10^{-3}(\text{kbar})^{-1}$, $S_{xy} = 0.5 \times 10^{-3}(\text{kbar})^{-1}$, $\varepsilon_{xx} = \sigma S_{xx}$, $\varepsilon_{yy} = \sigma S_{yy}$, $\varepsilon_{zz} = \sigma S_{zz}$, $\varepsilon_{xy} = \sigma S_{xy}$, σ is the stress in kilobar, and $g_v = 1$ are valid for the model of Seiler et al. [264–267].
14. Bismuth	$E_{g_0} = 0.0153$ eV, $m_1 = 0.00194m_0$, $m_2 = 0.313m_0$, $m_3 = 0.00246m_0$, $m'_2 = 0.36m_0$, $g_v = 3$, $g_s = 2$ [268], $M_2 = 0.128m_0$ and $M'_2 = 0.80m_0$ [173].
15. Mercury Telluride	$m_v^* = 0.028m_0$, $g_v = 1$ and $\varepsilon_\infty = 15.2\varepsilon_0$ [240].
16. Platinum Antimonide	For valence bands, along <111> direction, $\lambda_0 = 0.33$ eV, $l = 1.09$ eV, $\nu = 0.17$ eV, $\bar{\bar{n}} = 0.22$ eV, $\bar{\bar{a}} = 0.643$ nm, $I_0 = 0.30\,(\text{eV})^2$, $\delta_0 = 0.33$ eV and $g_v = 8$ [239].
17. n-Gallium Phosphide	$m^*_{\parallel} = 0.92m_0$, $m^*_{\perp} = 0.25m_0$, $k_0 = 1.7 \times 10^{19} m^{-1}$, $\lvert V_G \rvert = 0.21$ eV, $g_v = 6$ and $g_s = 2$ [232].
18. Germanium	$E_{g_0} = 0.785eV$, $m^*_{\parallel} = 1.57m_0$ and $m^*_{\perp} = 0.0807m_0$ [260, 261].

(continued)

Table 1.1 (Continued)

Materials	Numerical values of the energy band constants
19. Tellurium	The values $A_6 = 6.7 \times 10^{-16}$ meVm^2, $A_7 = 4.2 \times 10^{-16}$ meVm^2, $A_8 = (6 \times 10^{-8}\text{ meVm})^2$ and $A_9 = (3.6 \times 10^{-8}\text{ meVm})^2$ are valid for the model of Bouat et al. [236]. The values $t_1 = 0.06315$ eV, $t_2 = -10.0\hbar^2/2m_0$, $t_3 = -5.55\hbar^2/2m_0$, $t_4 = 0.3 \times 10^{-36}$ eVm^4, $t_5 = 0.3 \times 10^{-36} eVm^4$, $t_6 = -5.55\hbar^2/2m_0$ and $t_7 = 6.18 \times 10^{-20}$ (eVm)^2 are valid for the model of Ortenberg and Button [237].
20. Graphite	The values $\Delta_1 = -0.0002$ eV, $\gamma_1 = 0.392$ eV, $\gamma_5 = 0.194$ eV, $\bar{c} = 0.674$ nm, $\gamma_2 = -0.019$ eV, $\bar{a} = 0.246$ nm, $\gamma_0 = 3$ eV, $\gamma_4 = 0.193$ eV are valid for the model of Brandt et al. [269].
21. Lead Germanium Telluride	The values $\overline{E_{g0}} = 0.21$ eV and $g_v = 4$ [250, 270] are valid for the model of Vassilev.
22. Cadmium Antimonide	The values $A_{10} = -4.65 \times 10^{-19}$ eVm^2, $A_{11} = -2.035 \times 10^{-19}$ eVm^2, $A_{12} = -5.12 \times 10^{-19}$ eVm^2, $A_{13} = -0.25 \times 10^{-10}$ eVm, $A_{14} = 1.42 \times 10^{-19}$ eVm^2, $A_{15} = 0.405 \times 10^{-19}$ eVm^2, $A_{16} = -4.07 \times 10^{-19}$ eVm^2, $A_{17} = 3.22 \times 10^{-10}$ eVm, $A_{18} = 1.69 \times 10^{-20}$ (eVm)^2 and $A_{19} = 0.070$ eV [241, 242] are valid for the model of Yamada.
23. Cadmium Diphosphide	The values $\beta_1 = 8.6 \times 10^{-21}$ eVm^2, $\beta_2 = 1.8 \times 10^{-21}$ (eVm)^2, $\beta_4 = 0.0825$ eV and $\beta_5 = -1.9 \times 10^{-19}$ eVm^2 are valid for the model of Chuiko [251].
24. Zinc Diphosphide	The values $\beta_1 = 8.7 \times 10^{-21}$ eVm^2, $\beta_2 = 1.9 \times 10^{-21}$ (eVm)^2, $\beta_4 = 0.0875$ eV and $\beta_5 = -1.9 \times 10^{-19}$ eVm^2 are valid for the model of Chuiko [251].
25. Bismuth Telluride	The values $E_{g0} = 0.145$ eV, $\overline{\overline{\alpha_{11}}} = 3.25$, $\alpha_{22} = 4.81$, $\alpha_{33} = 9.02$, $\alpha_{23} = 4.15$, $g_s = 2$ and $g_v = 6$ are valid for the model of Stordeur et al. [252, 253].
26. Antimony	The values $\bar{\alpha}_{11} = 16.7$, $\bar{\alpha}_{22} = 5.98$, $\bar{\alpha}_{33} = 11.61$ and $\bar{\alpha}_{23} = 7.54$ are valid for the model of Ketterson [254].
27. Zinc Selenide	$m_2^* = 0.16m_0$, $\Delta_2 = 0.42$ eV and $E_{g02} = 2.82$ eV [260, 261].
28. Lead Selenide	$m_t^- = 0.23m_0$, $m_l^- = 0.32m_0$, $m_t^+ = 0.115m_0$, $m_l^+ = 0.303m_0$, $P_{\|\|} \approx 138$ meV nm, $P_\perp = 471$ meV nm and $E_{g0} = 0.28$ eV [271].
29. Carbon nanotubes	$t_c = 2.5$ eV, $a_c = 0.14$ nm and $r_0 = 0.7$ nm [272–275].

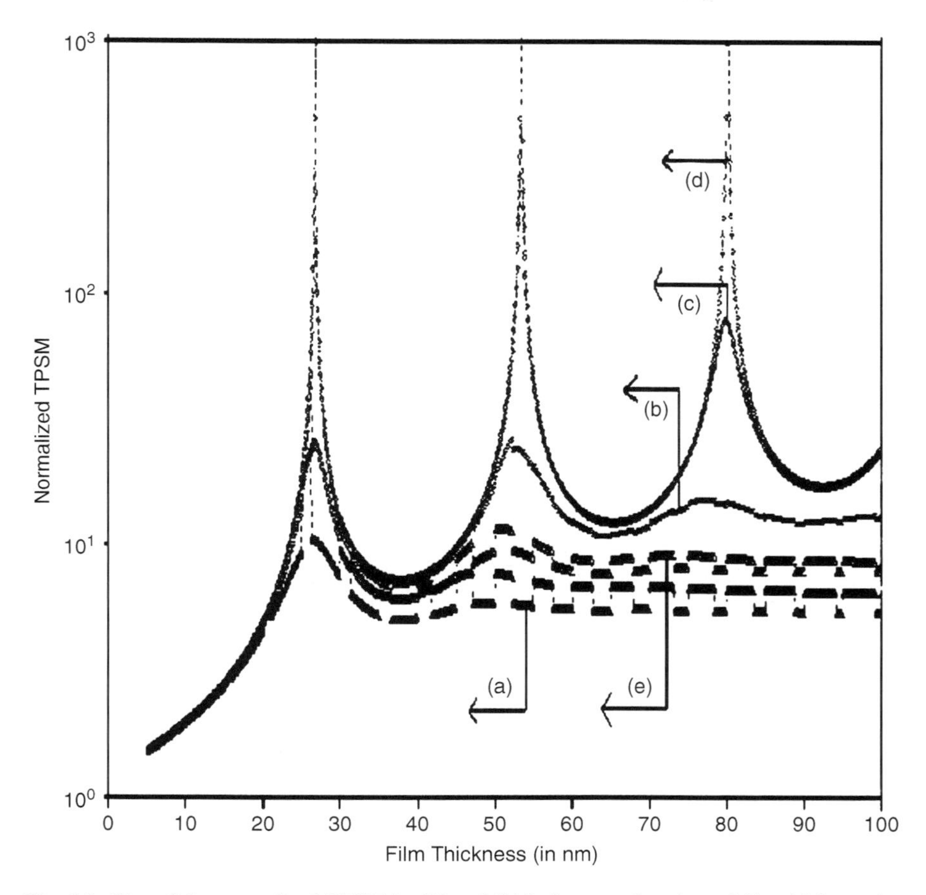

Fig. 1.1 Plot of the normalized TPSM in QDs of $CdGeAs_2$ as a function of film thickness has been shown in accordance with the (**a**) generalized band model ($\delta \neq 0$), (**b**) three and (**c**) two band models of Kane together with (**d**) parabolic energy bands. The special case for $\delta = 0$(**e**) has also been shown to assess the influence of crystal field splitting

In Fig. 1.3, the TPSM has been plotted for QDs of InAs as a function of film thickness in accordance with the three and two band models of Kane together with parabolic energy bands. It is observed from Fig. 1.3 that the TPSM shows oscillatory dependence with respect to film thickness, and the influence of the three band model of Kane is to change the numerical values of the TPSM as compared with the two band model of Kane and parabolic energy bands. Figure 1.4 exhibits the variation of TPSM with carrier concentration in QDs of InAs for all the cases of Fig. 1.3. It appears that the TPSM in accordance with all the band models exhibits wide separation from one another, and for three band model of Kane, although the TPSM is less as compared with the other two types of band models, it is very prominent and decreases rapidly with concentration after a particular value of carrier degeneracy.

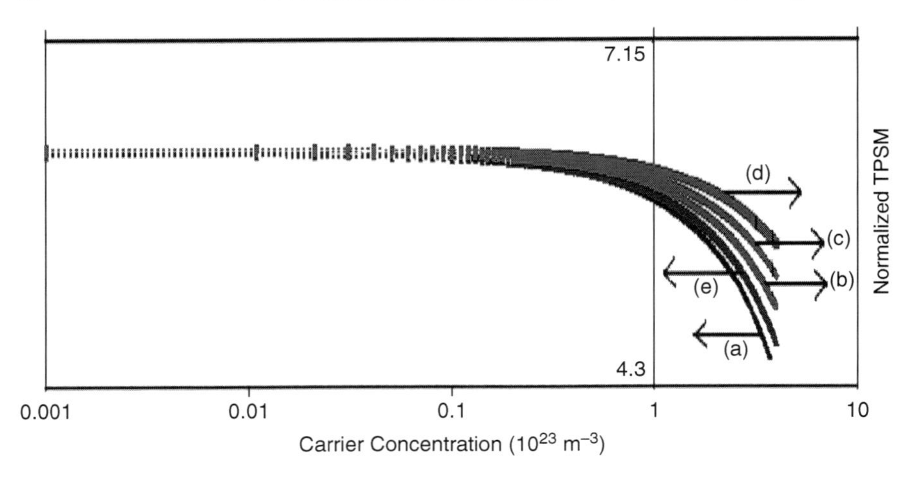

Fig. 1.2 Plot of the normalized TPSM in QDs of $CdGeAs_2$ as a function of carrier concentration for all cases of Fig. 1.1

In Fig. 1.5, the plot of the normalized TPSM in QDs of GaAs as a function of film thickness in accordance with the (a) three and (b) two band models of Kane together with (c) the parabolic energy bands has been shown. The influence of energy band models on the TPSM in QDs of GaAs is apparent from the plots of Fig. 1.5. Figure 1.6 exhibits the dependence of the TPSM in QDs of n-GaAs on n_0, and in this case, the shape of the curves for three band model of Kane and that of parabolic energy bands exhibit wide difference although their natures are same. From the same figure, it is apparent that the TPSM in QDs of n-GaAs in accordance with three band Kane model decreases rapidly with increasing electron concentration after the value of 10^{23} m^{-3}. The influence of the spin–orbit splitting constant is rather very large after a fixed value of carrier concentration. In Figs. 1.7 and 1.8, we have shown the normalized TPSM in QDs of InSb as a function of film thickness and concentration, respectively, in accordance with the three [using (1.18) and (1.19)] and two [using (1.22) and (1.23)] band models of Kane [224, 225] together with parabolic energy bands as shown by curves (a), (b), and (c) in the respective figures. The nature of variation of TPSM in QDs of InSb with respect to thickness and concentration remains same as compared with the TPSM in InAs and GaAs, although the numerical value changes precisely which is the manifestation of the energy band constants. Figures 1.9, 1.10 and 1.12 illustrate the film thickness dependence of the normalized TPSM in QDs of InAs, GaAs, and InSb in accordance with the models of Stillman et al. [226] [using (1.27) and (1.28)], Newson et al. [227] [using (1.32) and (1.31)], and Rossler et al. [228] [using (1.35) and (1.36)] as represented by curves (a), (b), and (c), respectively. From Fig. 1.9, it appears that the TPSM in QDs of InAs exhibits greatest value for the model of Stillman et al. and the lowest for the model of Rossler et al. The model of Newson et al. generates the values of

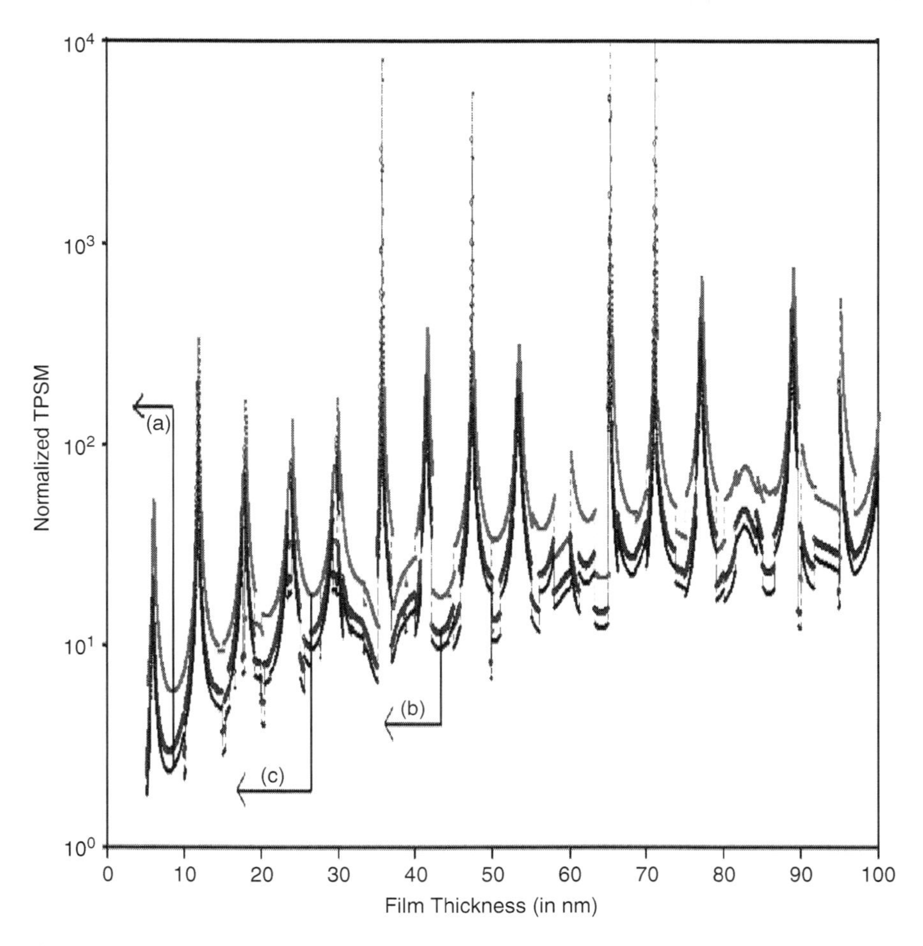

Fig. 1.3 Plot of the normalized TPSM in QDs of InAs as a function of film thickness in accordance with the (**a**) three and (**b**) two band models of Kane together with (**c**) parabolic energy bands

TPSM which lies in between. In Figs. 1.10 and 1.12, the TPSM has been plotted for QDs of GaAs and InSb as a function of film thickness for all the cases of Fig. 1.9, and the same observation of the aforementioned figure also holds true for Figs. 1.10 and 1.12, respectively. In Fig. 1.11, the TPSM in QDs of GaAs has been plotted with respect to carrier concentration in accordance with all the models of Fig. 1.9. It appears that the model of Stillman et al. exhibits the highest value, although for the whole range of carrier concentration, the TPSM in accordance with the models of Newson et al. and Rossler et al. maintains the same separation and the same shape. Besides, the TPSM in this case for the model of Rossler et al. decreases sharply with increase in carrier concentration for relatively large values of concentration, whereas for small values of electron degeneracy, the TPSM in accordance with the models of Newson et al. and Rossler et al. exhibits converging tendency. Figure 1.13

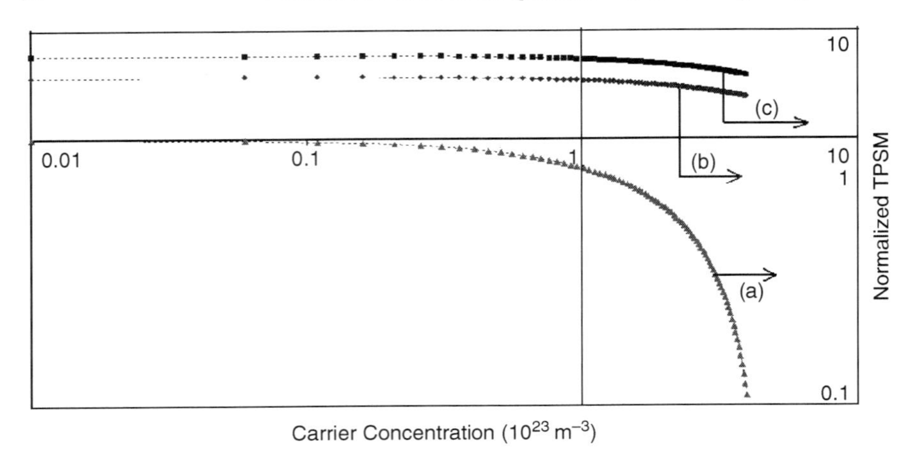

Fig. 1.4 Plot of the normalized TPSM in QDs of InAs as a function of carrier concentration for all the cases of Fig. 1.3

exhibits the influence of carrier concentration on the normalized TPSM in QDs of InSb for all the cases of Fig. 1.9. For small values of carrier concentration, the TPSM in QDs of InSb exhibits relative difference for all the models, and for higher values of carrier degeneracy, the decreasing nature of TPSM with increasing concentration is very prominent for all the models, and they exhibit converging nature.

Figure 1.14 exhibits the film thickness dependence of the normalized TPSM for QDs of InSb and InAs in accordance with the model of Agafonov et al. [231] [using (1.48) and (1.49)] as represented by the curves (a) and (b), respectively. The numerical values of the TPSM of InSb are greater than InAs in accordance with the said figure which is the direct influence of the energy band constants. Figure 1.15 illustrates the concentration dependence of normalized TPSM for all the cases of Fig. 1.14. For small values of the electron concentration, the TPSM in both cases of InSb and InAs shows wide difference, and with increasing values of the carrier degeneracy, the TPSM decreases and ultimately converges. Figure 1.16 demonstrates the film thickness dependence of the normalized TPSM for QDs of InSb and InAs in accordance with the model of Johnson et al. [230] [using (1.44) and (1.45)] as represented by the curves (a) and (b), respectively. The numerical values of the TPSM of InSb are greater than InAs in accordance with Fig. 1.16, which is the consequence of constants of the energy spectra. Figure 1.17 shows the concentration dependence of normalized TPSM for all the cases of Fig. 1.16.

For small values of the electron concentration, the TPSM in both cases of InSb and InAs shows converging tendency, and with increasing values of the carrier degeneracy, the TPSM decreases and shows appreciable difference with one another. Figure 1.18 depicts the film thickness dependence of the normalized TPSM for QDs of InSb and InAs in accordance with the model of Palik et al. [229] [using (1.40) and (1.41)] as represented by the curves (a) and (b) in this context. The numerical values of the TPSM of InSb are greater than InAs in accordance with Fig. 1.18,

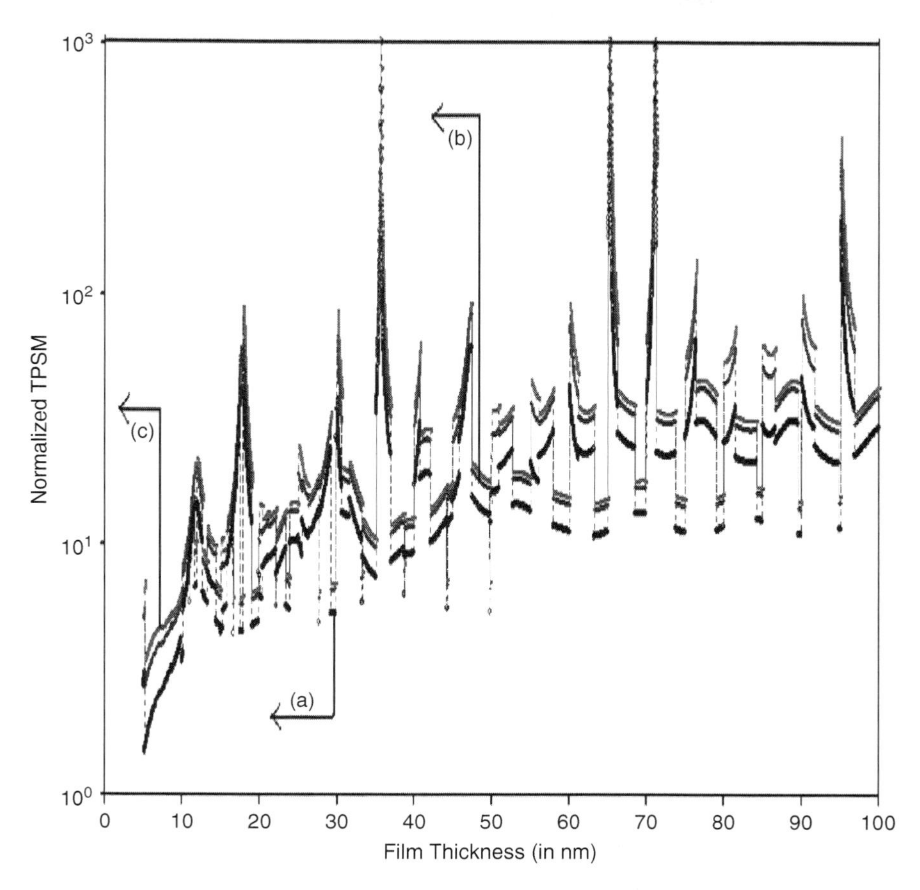

Fig. 1.5 Plot of the normalized TPSM in QDs of GaAs as a function of film thickness in accordance with the (**a**) three and (**b**) two band models of Kane together with (**c**) the parabolic energy bands

which is the direct influence of energy band constants. Figure 1.19 models the concentration dependence of normalized TPSM for all the cases of Fig. 1.18. For small values of the electron concentration, the influence of energy band constants of InSb and InAs on the TPSM becomes very small whereas, and with increasing values of the carrier degeneracy, the effect of the constants of the carrier energy spectrum of the two said materials on the TPSM is rather large. From Fig. 1.19, we infer that the TPSM decreases with increasing carrier concentration after the concentration value 10^{22} m^{-3}, and the TPSM for both the materials show substantial difference with one another. Figure 1.20 exhibits the variation of the normalized TPSM with the film thickness in QDs of II–VI materials in accordance with Hopfield model [using (1.52) and (1.53), taking p-CdS as an example and considering both the cases $C_0 = 0$ and $C_0 \neq 0$] and GaP [using (1.56) and (1.57) in accordance with the model of Rees] as shown by curves (a), (b), and (c), respectively. The TPSM is higher for

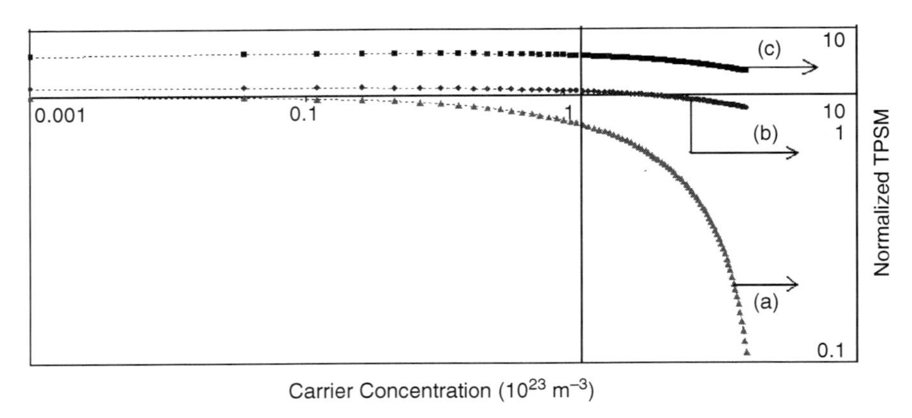

Fig. 1.6 Plot of the normalized TPSM in QDs of GaAs as a function of carrier concentration for all the cases of Fig. 1.5

QDs of CdS with $C_0 \neq 0$ as compared to the case for $C_0 = 0$. The TPSM is lowest for GaP as compared to the aforementioned two cases of CdS. Figure 1.21 demonstrates the concentration variation of the normalized TPSM for all cases of Fig. 1.20. For small values of the electron concentration, the TPSM in Fig. 1.21 for all cases of Fig. 1.20 preserves difference for small values of carrier concentration which is the direct signature of the energy band constants of CdS and GaP, respectively, and with increasing values of the carrier degeneracy, the TPSM decreases and shows significant difference with one another. The effect of spin splitting of the carriers in QDs of p-CdS on TPSM can be numerically investigated from Figs. 1.20 and 1.21. It appears that the absence of the spin splitting constant decreases the numerical value of the TPSM in CdS for a particular range of film thickness. In Fig. 1.22, the normalized TPSM has been plotted as a function of film thickness for QDs of Germanium for both the models of Wang et al. [235] [using (1.65) and (1.66)] and Cardona et al. [233,234] [using (1.60) and (1.61)] as shown by curves (a) and (b), respectively. The TPSM for the model of Cardona et al. shows sharp oscillatory peaks as compared to the model of Wang et al. Figure 1.23 exhibits the normalized TPSM as a function of film thickness in QDs of Tellurium [by using the models of Bouat et al. [236] (using (1.69) and (1.70)) and Ortenberg et al. [237] (using (1.73) and (1.74)] and stressed Kane type materials (taking n-InSb as an example) in accordance with the models of Seiler et al. [264–267] [using (1.107) and (1.108)] as shown by curves (a), (b), and (c), respectively. It appears from the figure that the TPSM for QDs of Te oscillates with increasing thickness, exhibiting oscillatory spikes in accordance with both the models of Te, and the TPSM in QDs of stressed InSb in accordance with the model of Seiler et al. exhibits the spikes in the whole range of thickness as considered. Figure 1.24 shows the dependence of the normalized TPSM on the carrier concentration for all the cases of Fig. 1.23. From Fig. 1.24, it appears that the TPSM in Te in accordance with both the aforementioned band models exhibits wide separation for relatively low values of carrier degeneracy, whereas for the relatively large

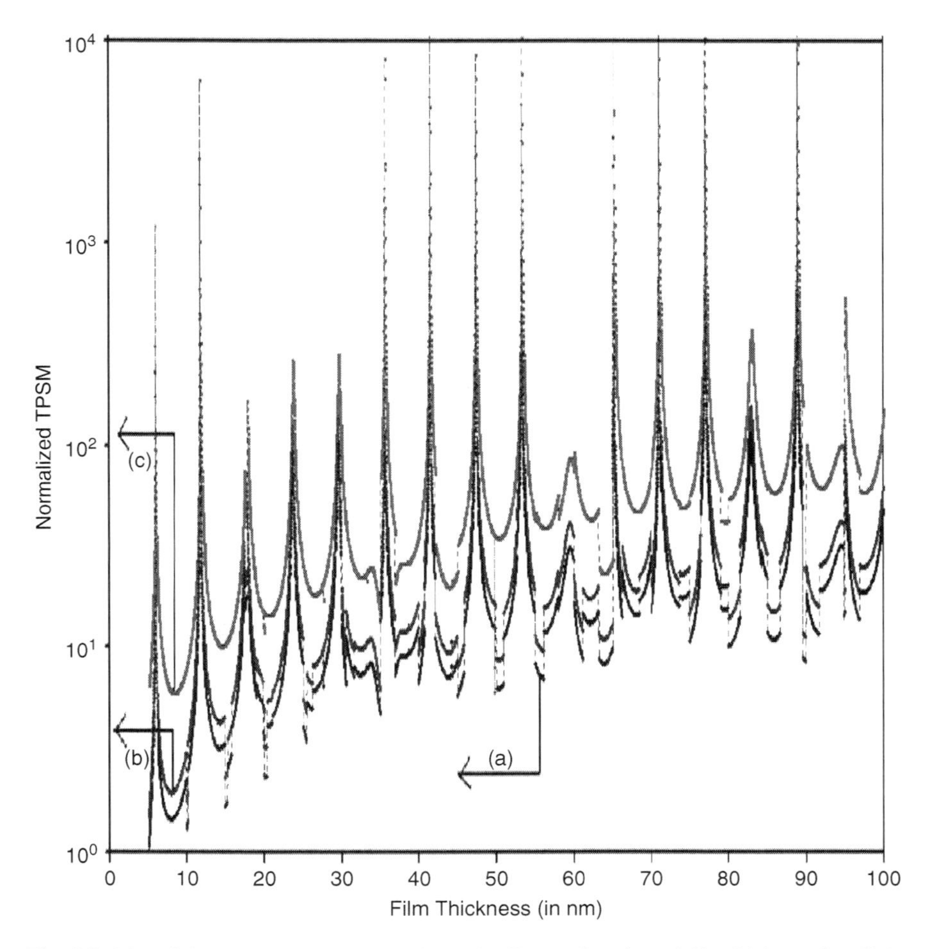

Fig. 1.7 Plot of the normalized TPSM in QDs of InSb as a function of film thickness for all the cases of Fig. 1.3

values of the carrier concentration, both of them decrease and cut at a point. For stressed InSb, the model of Seiler et al. predicts sharp fall of TPSM with increasing electron statistics. It appears that at extremely low and high film thicknesses, the TPSM in QDs of Te dominates over that of the corresponding stressed InSb, although, at mid zone thickness, the TPSM in stressed InSb exhibits a high peak together with the fact that the periods of oscillations of the TPSM are comparatively higher in Te than that of stressed compounds. In Fig. 1.25, the normalized TPSM has been plotted as a function of carrier concentration for QDs of Graphite [using (1.77) and (1.78)], Platinum Antimonide [using (1.82) and (1.83)], zero gap [using (1.86) and (1.87)], and Lead Germanium Telluride [using (1.90) and (1.91)] in accordance with the models of Ushio et al. [238], Emtage [239], Ivanov-Omskii et al. [240], and

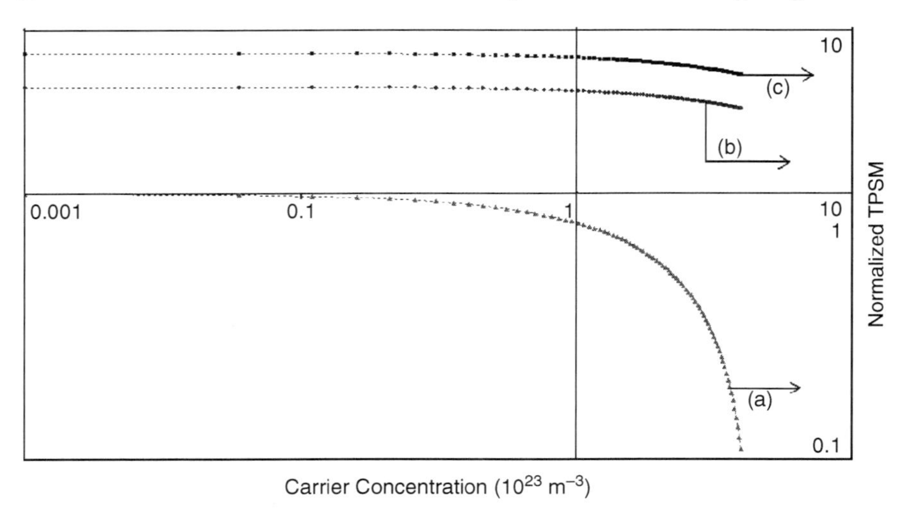

Fig. 1.8 Plot of the normalized TPSM in QDs of InSb as a function of carrier concentration for all the cases of Fig. 1.4

Vassilev [250] as shown by curves (a), (b), (c), and (d), respectively. The TPSM, in all the cases, decreases with increasing carrier degeneracy and is highest for QDs of $Pb_{1-x}Ge_xTe$ and lowest for QDs of Graphite with respect to concentration, and the TPSM in QDs of $Pb_{1-x}Ge_xTe$ is higher than that of the QDs of graphite, $PtSb_2$, and zerogap, respectively. Figure 1.26 depicts the plot of normalized TPSM as a function of carrier concentration in accordance with the models of Seiler et al. [243] [using (1.94) and (1.95)], Mathur et al. [244] [using (1.99) and (1.100)], and Zhang [245] [using (1.103) and (1.104)] as shown by curves (a), (b), and (c), respectively. It appears from the figure that the TPSM for QDs of Gallium Antimonide decreases with increasing carrier concentration after the concentration zone 10^{22} m^{-3} exhibiting highest numerical value for the model of Zhang and the lowest for the model of Seiler et al. Besides, in the concentration regime from 0.001×10^{23} m^{-3} to less than 0.1×10^{23} m^{-3}, the influence of band structure on the TPSM in this case is insignificant, and they exhibit the converging tendency toward a fixed value. Figure 1.27 exhibits the thickness dependence of the normalized TPSM for all the cases of Fig. 1.26, and Fig. 1.27 shows the oscillatory variation with respect to thickness as usual. The numerical value of the same is greatest in accordance with the model of Zhang and lowest for the band structure of GaSb as defined by Seiler et al. The numerical value of the TPSM as presented by the model of Mathur et al. appears to fall in the mid-zone.

Figure 1.28 illustrates the variation of the normalized TPSM with film thickness in QDs of Bismuth in accordance with the models of McClure et al. [174] [using (1.111) and (1.112)], Hybrid [173] [using (1.115) and (1.116)], Cohen [172] [using (1.119) and (1.120)] and Lax et al. [166] [using (1.123) and (1.124)] as shown by curves (a), (b), (c), and (d), respectively. It appears from the figure that TPSM

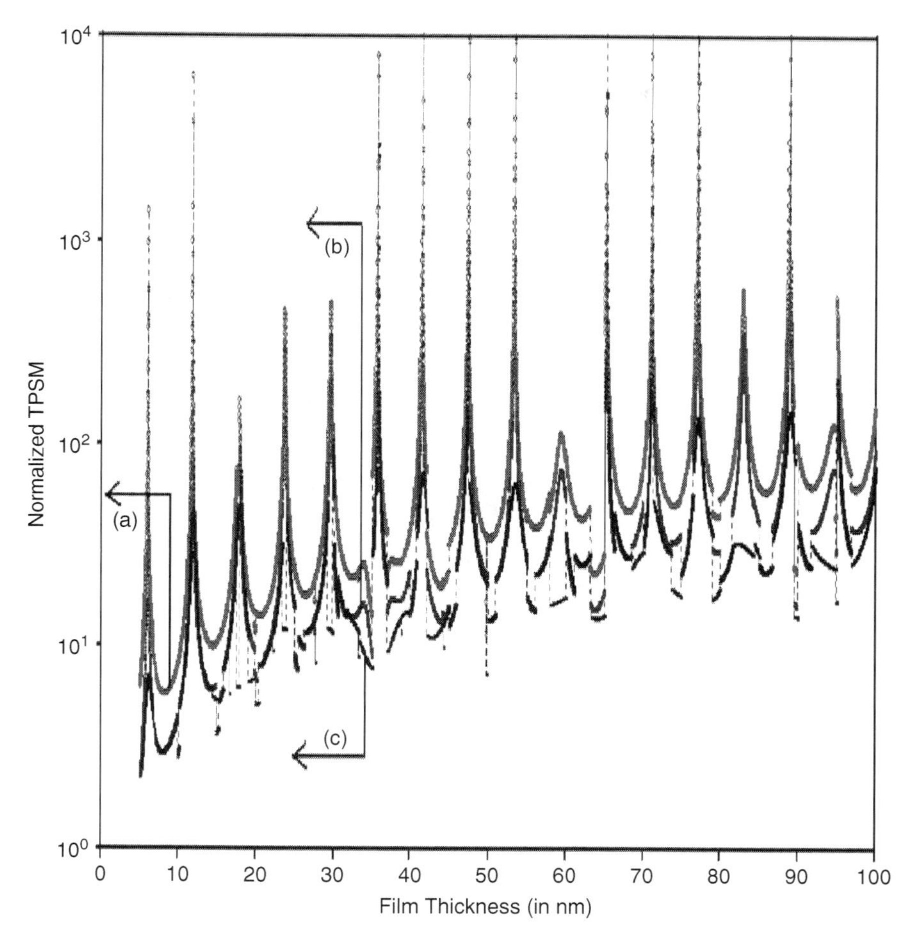

Fig. 1.9 Plot of the normalized TPSM in QDs of InAs as a function of film thickness in accordance with the models of (**a**) Stillman et al., (**b**) Newson et al. and (**c**) Rossler et al., respectively

for QDs of all the models of Bismuth exhibits regular oscillation with thickness, although the most prominent oscillatory lobe together with high sharp peak in the mid thickness zone is exhibited by the model of McClure and Choi. Figure 1.29 shows the dependence of normalized TPSM on the carrier concentration in QDs of Bismuth for all types of band models as stated in Fig. 1.28. From Fig. 1.29, it appears that the TPSM decreases with increasing electron concentration for all the models of Bismuth. Figure 1.30 shows the variation of TPSM on the carrier concentration for QDs of IV–VI materials (taking PbTe as an example) in accordance with the models of Dimmock [246] [using (1.127) and (1.128)], Bangert et al. [247] [using (1.131) and (1.132)] and Foley et al. [248, 249] [using (1.135) and (1.136)] together with Bismuth Telluride by using the model of Stordeur et al. [252,253] [using (1.147) and

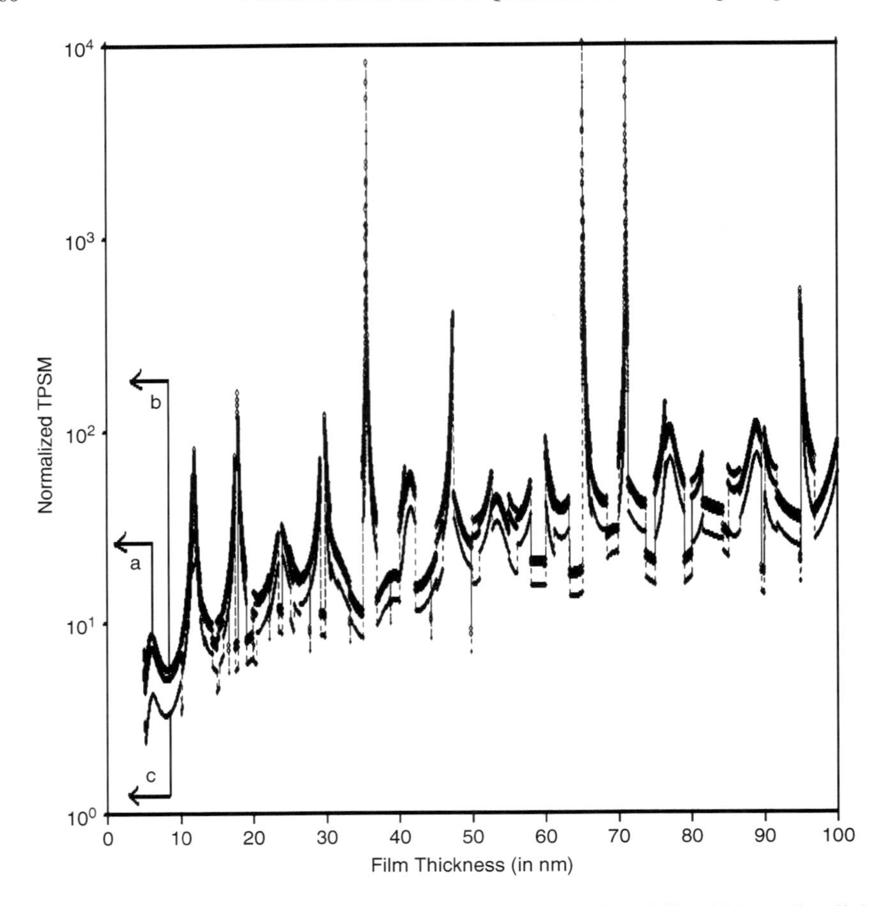

Fig. 1.10 Plot of the normalized TPSM in QDs of GaAs as a function of film thickness for all the cases of Fig. 1.9

(1.148)] as shown by the curves (a), (b), (c), and (d), respectively. For small values of the electron concentration, the TPSM in Fig. 1.30 shows substantial difference with each other, and with increasing values of the carrier degeneracy, the TPSM decreases and shows converging tendency. Figure 1.31 models the variation of the TPSM with the film thickness for QDs of IV–VI materials and Bismuth Telluride in accordance with the band models of Fig. 1.30. From the figure, it appears that the TPSM in PbTe (in accordance with the said three models) and Bi_2Te_3 in accordance with the model of Stordeur et al. exhibits dense oscillations with respect to film thickness, and it also appears that the numerical values of TPSM in QDs of PbTe and Bi_2Te_3 in accordance with all the band models are higher than all other materials.

Figure 1.32 shows the plot of the normalized TPSM as a function of film thickness in QDs of II–V materials (taking CdSb as an example) using the model of Yamada [241,242] [using (1.90) and (1.91)], Zinc and Cadmium diphosphides using

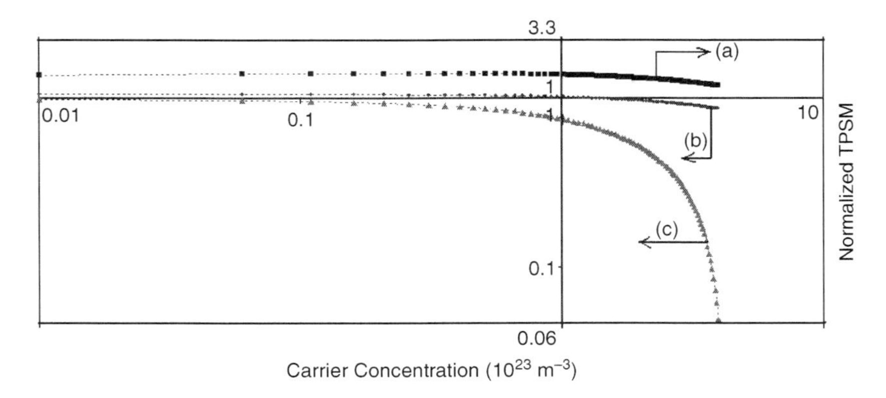

Fig. 1.11 Plot of the normalized TPSM in QDs of GaAs as a function of carrier concentration for all the cases of Fig. 1.9

the model of Chuiko [251] [using (1.143) and (1.144)], and Antimony in accordance with the model of Ketterson [254] [using (1.158) and (1.159)] as shown by the curves (a), (b), (c), and (d), respectively. It appears from the figure that the TPSM exhibits oscillatory spikes for fixed values of the film thickness which is again dependent on the specific band structure of II–V (Yamada model), zinc diphosphide and cadmium diphosphide (Chuiko model), and antimony (Ketterson model), respectively. From Fig. 1.32, it appears that the TPSM is largest for II–V compound and lowest for Antimony. From the aforementioned discussions, we realize that the signature of three-dimensional quantization is forthwith evident from Figs. 1.1, 1.3, 1.5, 1.7, 1.9, 1.10, 1.12, 1.14, 1.16, 1.18, 1.20, 1.22, 1.23, 1.27, 1.28, 1.31, and 1.32 for all materials as discussed having different band structures. It can be facilely discerned from the same that the normalized TPSM oscillates with film thickness exhibiting spikes for various values of film thickness which are totally band-structure dependent. The occurrence of peaks in the said figures originates from the totally quantized energy levels of the carriers of the concerned dots. The TPSM spectra are found bearing composite oscillations as function of the nanothickness. These are generally due to selection rules in the quantum numbers along the three confined directions. The dependence of the normalized TPSM on the carrier concentration is manifested by Figs. 1.2, 1.4, 1.6, 1.8, 1.11, 1.13, 1.15, 1.17, 1.19, 1.21, 1.24, 1.25, 1.26, 1.29, and 1.30 for the different materials as considered here. It can be ascertained from the same figures that the TPSM of all the corresponding materials decreases with increasing carrier concentration for relatively higher values of the carrier degeneracy. Although the TPSM varies in various manners with all the variables in all the limiting cases as evident from all the figures, the rate of variations in each case are totally band-structure dependent.

In formulating the generalized electron energy spectrum for nonlinear optical materials, we have considered the crystal-field splitting, the anisotropies in the momentum-matrix elements, and the spin–orbit splitting parameters, respectively. In the absence of the crystal field splitting constant together with the assumptions of

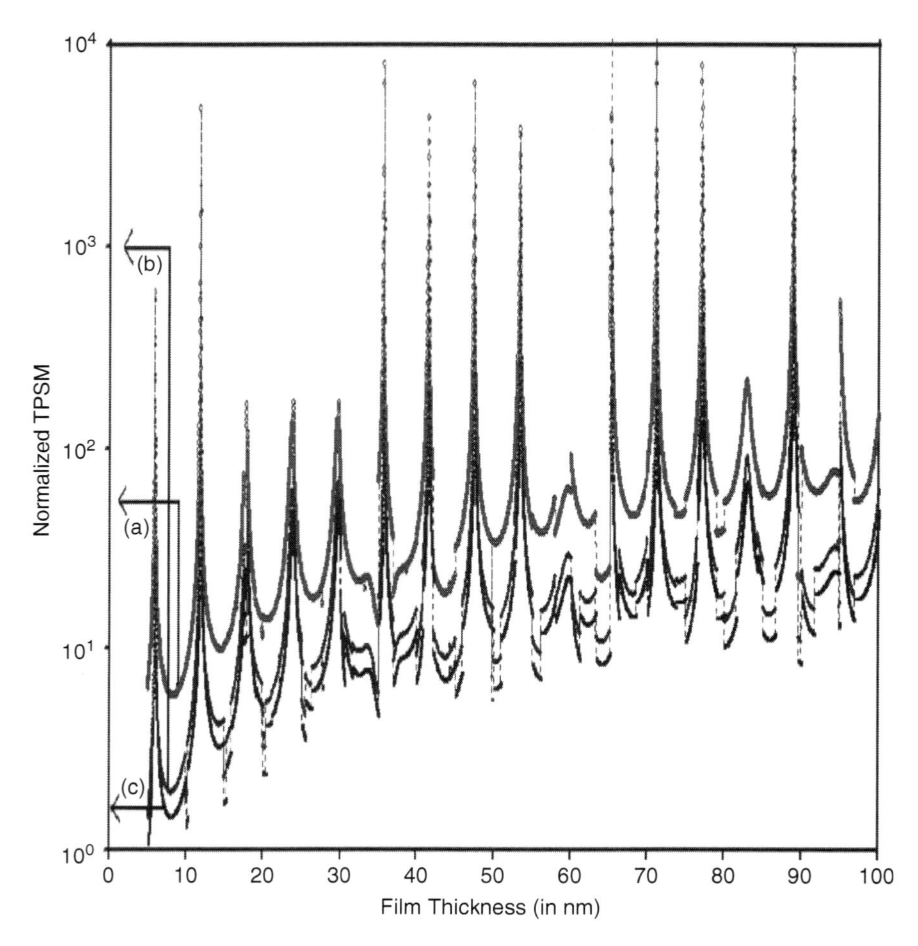

Fig. 1.12 Plot of the normalized TPSM in QDs of InSb as a function of film thickness for all the cases of Fig. 1.9

isotropic effective electron mass and isotropic spin–orbit splitting, our basic relation as given by (1.2) converts into (1.16). Equation (1.16) is the well-known three band Kane model [223] and is valid for III–V compounds, in general.

It should be used as such for studying the electronic properties of n-InAs, where the spin–orbit splitting parameter (Δ) is of the order of band gap (E_{g0}). For many important materials $\Delta \gg E_{g0}$ or $\Delta \ll E_{g0}$, and under such constraints, (1.16) assumes the form $E(1+EE_{g0}^{-1}) = \hbar^2k^2/2m^*$ which is (1.20) and is the well-known two band Kane model [224, 225]. Also under the condition, $E_{g0} \to \infty$, the above equation gets simplified to the well-known form of wide-gap parabolic energy bands as $E = \hbar^2k^2/2m^*$. Besides the three and two band models of Kane together with parabolic energy bands, the III–V materials are also being described in accordance

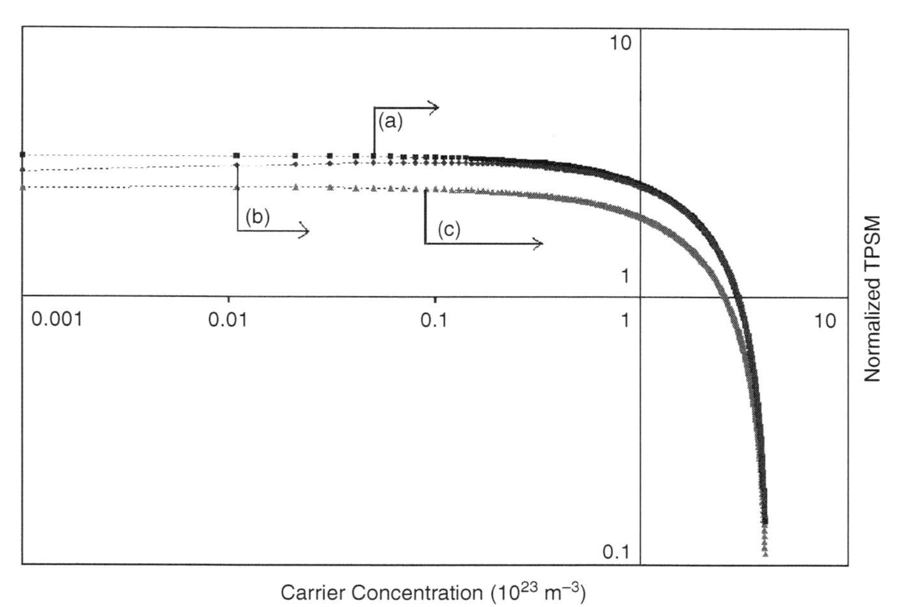

Fig. 1.13 Plot of the normalized TPSM in QDs of InSb as a function of carrier concentration for all the cases of Fig. 1.9

with the models of Stillman et al. [226], Newson and Kurobe [227], Rossler [228], Palik et al. [229], Johnson and Dickey [230], and Agafonov et al. [231], respectively. We have investigated the TPSM in QDs of III–V compounds in accordance with the all aforementioned band models for the purpose of relative assessment and thorough investigation. The TPSM in QDs of II–VI materials and GaP have been studied in accordance with the models of Hopfield and Rees, respectively.

Germanium is widely used since the inception of semiconductor science, and TPSM in QDs of Germanium has been presented in accordance with the models of Cardona et al. [233, 234] and Wang and Ressler [235]. Tellurium is an important elemental semiconductor, and the TPSM in QDs of Tellurium has been investigated in accordance with the models of Bouat et al. [236] and Ortenberg and Button [237], respectively. Graphite finds extensive use in materials science, and we have presented the TPSM in QDs of Graphite in accordance with the model of Ushio et al. [238]. In this chapter, investigations have been made of TPSM in QDs of $PtSb_2$, zerogap, and II–V materials in accordance with the models of Emtage [239], Ivanov-Omskii et al. [240], and Yamada [241, 242], respectively. The TPSM in QDs of GaSb has been studied by using the models of Seiler et al. [243], Mathur et al. [244], and Zhang [245], respectively. The stressed compounds find extensive applications in piezoelectric systems, and the TPSM in QDs of such materials has been studied by using the model of Seiler et al. [223]. The energy band models of Bismuth can be described in accordance with the models of McClure and Choi [174], Hybrid [173], Cohen [172], Lax [166], and ellipsoidal parabolic, respectively.

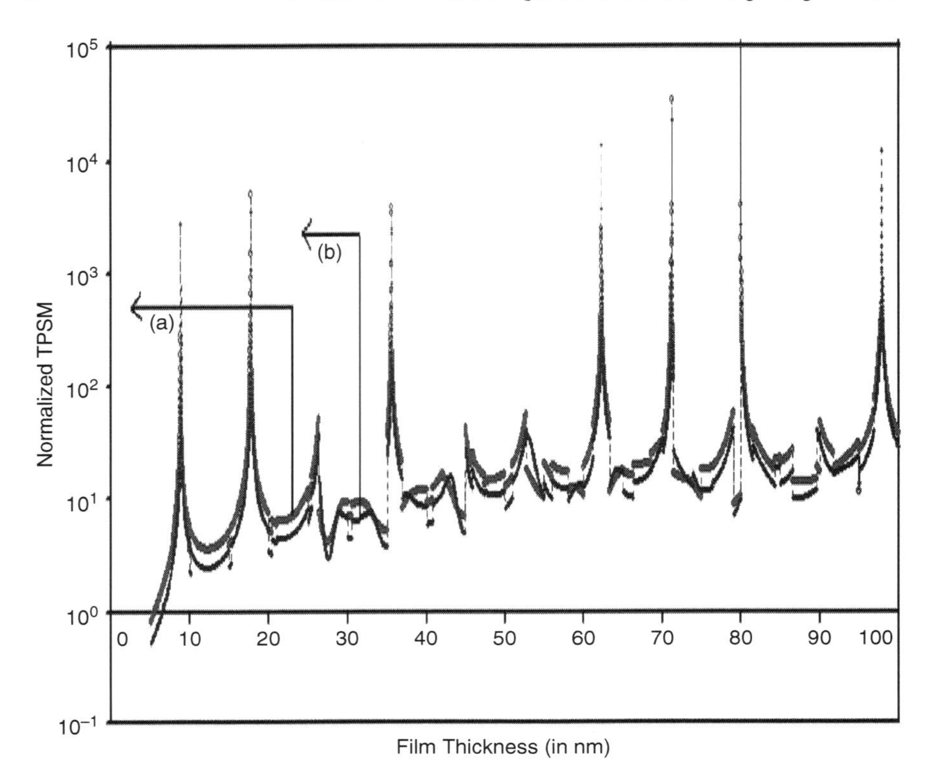

Fig. 1.14 Plot of the normalized TPSM in QDs of (**a**) InSb and (**b**) InAs as a function of film thickness in accordance with model of Agafonov et al.

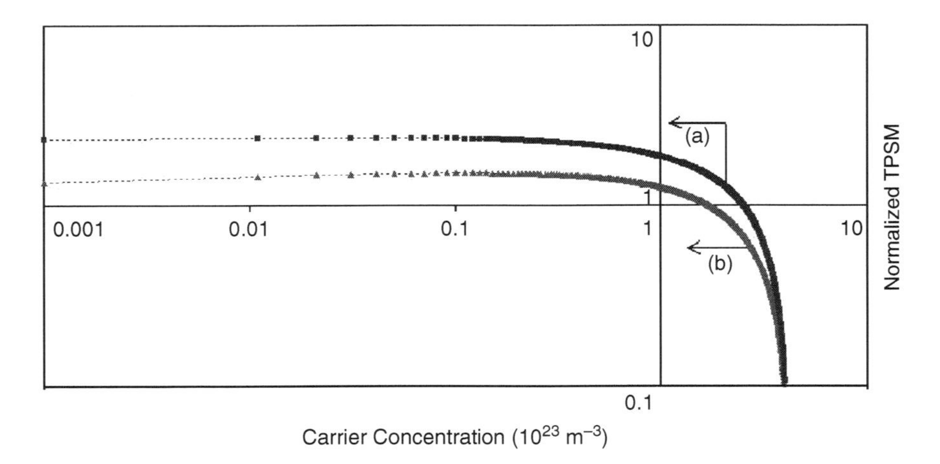

Fig. 1.15 Plot of the normalized TPSM in QDs of (**a**) InSb and (**b**) InAs as a function of carrier concentration for the case of Fig. 1.14

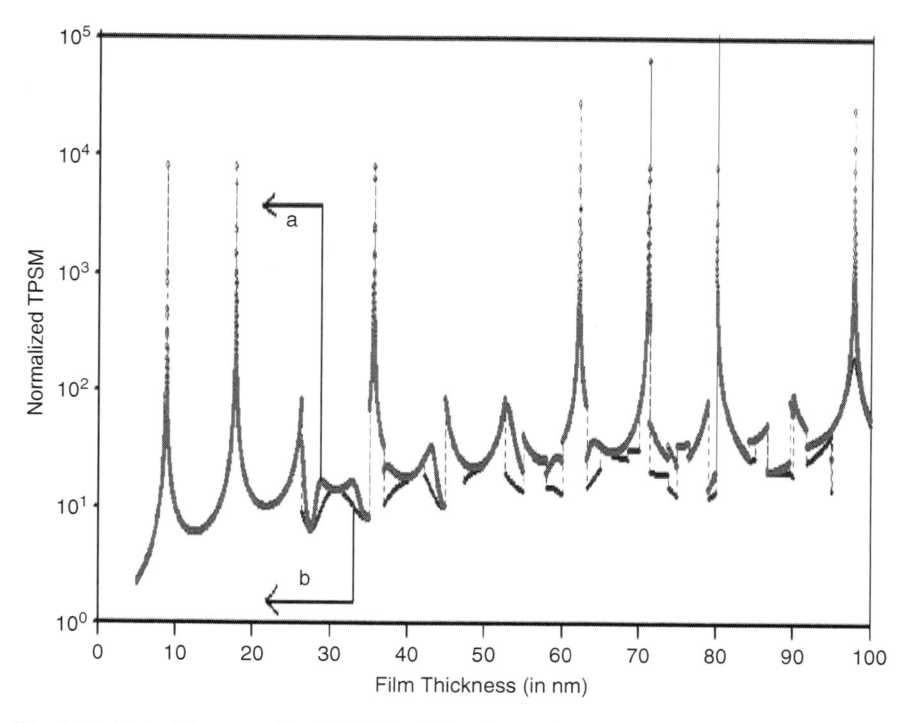

Fig. 1.16 Plot of the normalized TPSM in QDs of (**a**) InSb and (**b**) InAs as a function of film thickness in accordance with the model of Johnson et al.

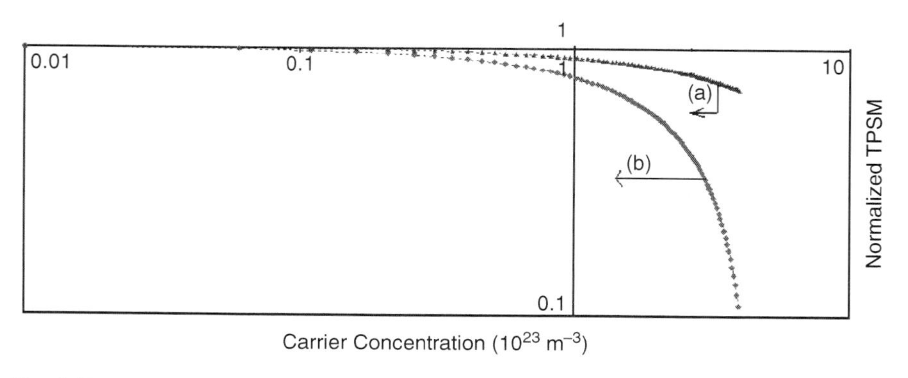

Fig. 1.17 Plot of the normalized TPSM in QDs of (**a**) InSb and (**b**) InAs as a function of carrier concentration for the case of Fig. 1.16

In Sect. 1.2.13, the TPSM in QDs of Bismuth in accordance with the aforementioned band models has been investigated. The TPSM in QDs of IV–VI materials has been presented by using the models of Dimmock [246], Bangert and Kastner [247], and Foley et al. [248, 249], respectively. Finally, the TPSM in QDs of Lead

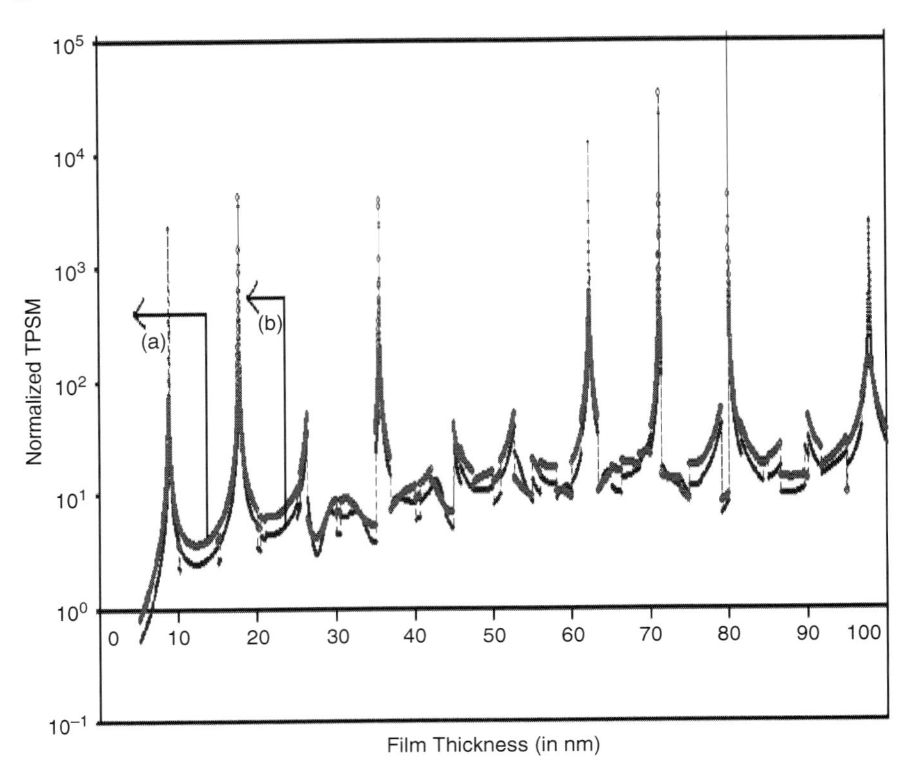

Fig. 1.18 Plot of the normalized TPSM in QDs of (**a**) InSb and (**b**) InAs as a function of film thickness in accordance with the model of Palik et al.

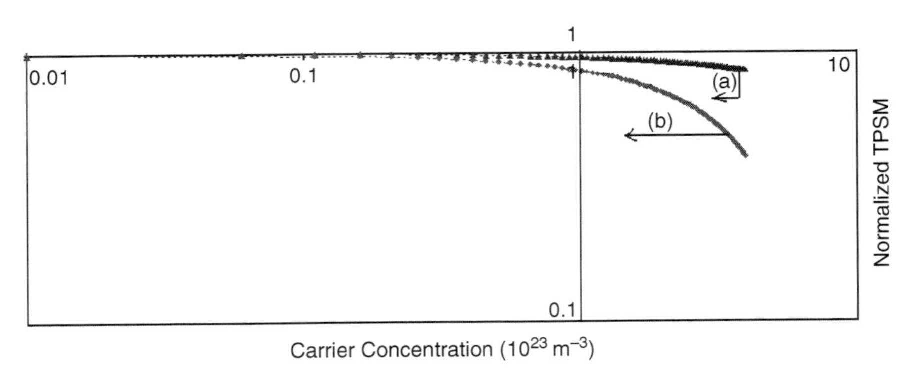

Fig. 1.19 Plot of the normalized TPSM in QDs of (**a**) InSb and (**b**) InAs as a function of carrier concentration for the case of Fig. 1.18

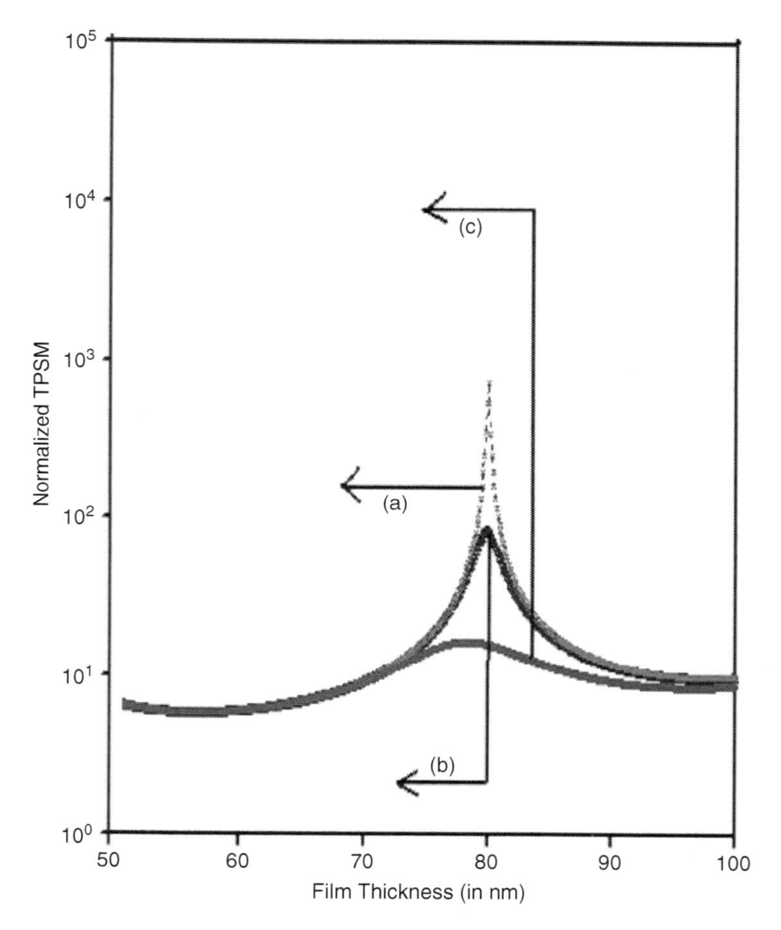

Fig. 1.20 Plot of the normalized TPSM in QDs of CdS with (**a**) $C_0 \neq 0$, (**b**) $C_0 = 0$ and (**c**) GaP as a function of film thickness in accordance with the models of Hopfield and Rees, respectively

Germanium Telluride, Zinc and Cadmium diphosphides, Bi_2Te_3, and Antimony in accordance with the appropriate carrier dispersion laws of Vassilev [250], Chuiko [251], Stordeur et al. [252, 253], and Ketterson [254] has been presented in Sects. 1.2.15–1.2.18, respectively.

It is imperative to state that our investigations exclude the many-body, hot electron, spin, and the allied quantum dot effects in this simplified theoretical formalism due to the absence of proper analytical techniques for including them for the generalized systems as considered here. For the purpose of simplified numerical computation, broadening has been neglected for obtaining all the plots. The inclusion of broadening will change the numerical magnitudes without altering the physics inside. Our simplified approach will be appropriate for the purpose of

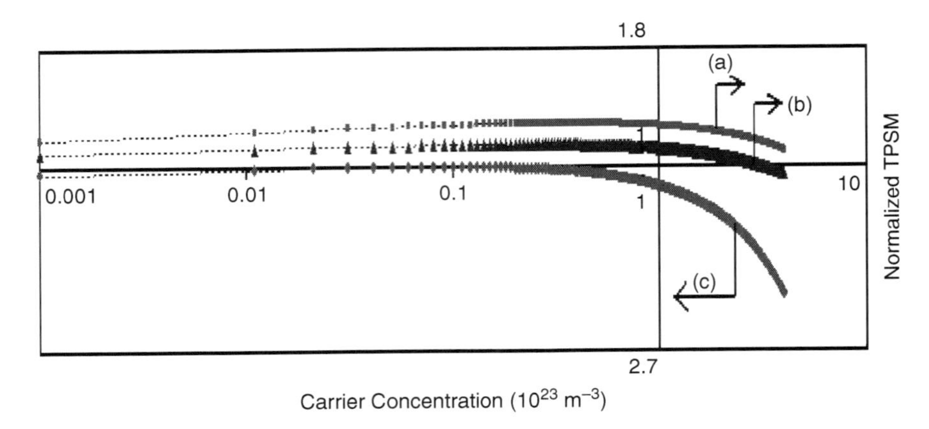

Fig. 1.21 Plot of the normalized TPSM in QDs of CdS with (**a**) $C_0 \neq 0$, (**b**) $C_0 = 0$ and (**c**) GaP as a function of carrier concentration for all the cases of Fig. 1.20

comparison when the methods of tackling the formidable problems after inclusion of the said effects for the generalized systems emerge. It is vital to elucidate that the results of our simple theory get transformed to the well-known formulation of TPSM for wide-gap materials having parabolic energy bands. This indirect test not only exhibits the mathematical compatibility of our formulation but also shows the fact that our simple analysis is a more generalized one, since one can obtain the corresponding results for materials having parabolic energy bands under certain limiting conditions from our present derivation. The experimental results for the verification of the theoretical analyses of this chapter are still not available in the literature. It is worth noting that our generalized formulation would be useful to analyze the experimental results when they materialize. The inclusion of the said effects would certainly increase the accuracy of the results, although the qualitative features of the TPSM would not change in the presence of the aforementioned effects. It is worth remarking that the influence of the broadening parameter has not been included in numerical computations for the purpose of simplified approach, although readers will enjoy the computer analysis in this context by including the said feature.

It can be noted that on the basis of the dispersion relations of the various quantized structures as discussed above, the heat capacity, the dia- and paramagnetic susceptibilities, and the various important dc/ac transport coefficients can be probed for all types of quantized structures as considered here. Thus, our theoretical formulation comprises the dispersion-relation-dependent properties of various technologically important quantum-confined materials having different band structures. We have not considered other types of compounds in order to keep the presentation concise and succinct. With different sets of energy band parameters, we shall get different numerical values of the TPSM.

The nature of variations of the TPSM as shown here would be similar for the other types of materials, and the simplified analysis of this chapter exhibits the basic

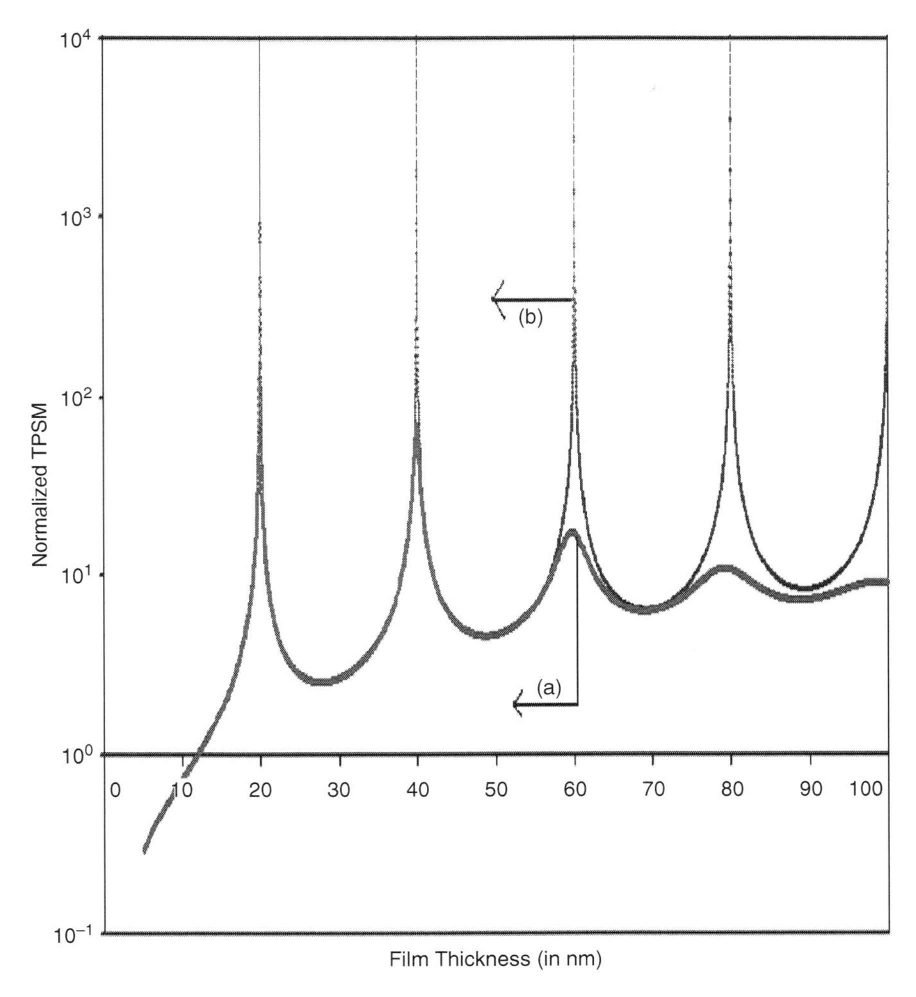

Fig. 1.22 Plot of the normalized TPSM in QDs of Germanium as a function of film thickness in accordance with the models of (**a**) Wang et al. and (**b**) Cardona et al., respectively

qualitative features of the TPSM in such quantized structures. It may be noted that the basic aim of this chapter is not solely to demonstrate the influence of quantum confinement on the TPSM for a wide class of quantized materials but also to formulate the appropriate carrier statistics in the most generalized form, since the transport and other phenomena in modern nanostructured devices having different band structures and the derivation of the expressions of many important carrier properties are based on the temperature-dependent carrier statistics in such materials. For the purpose of condensed presentation, the carrier statistics and the TPSM under large magnetic field for the QDs of the respective materials as discussed in this chapter have been presented in Table 1.2.

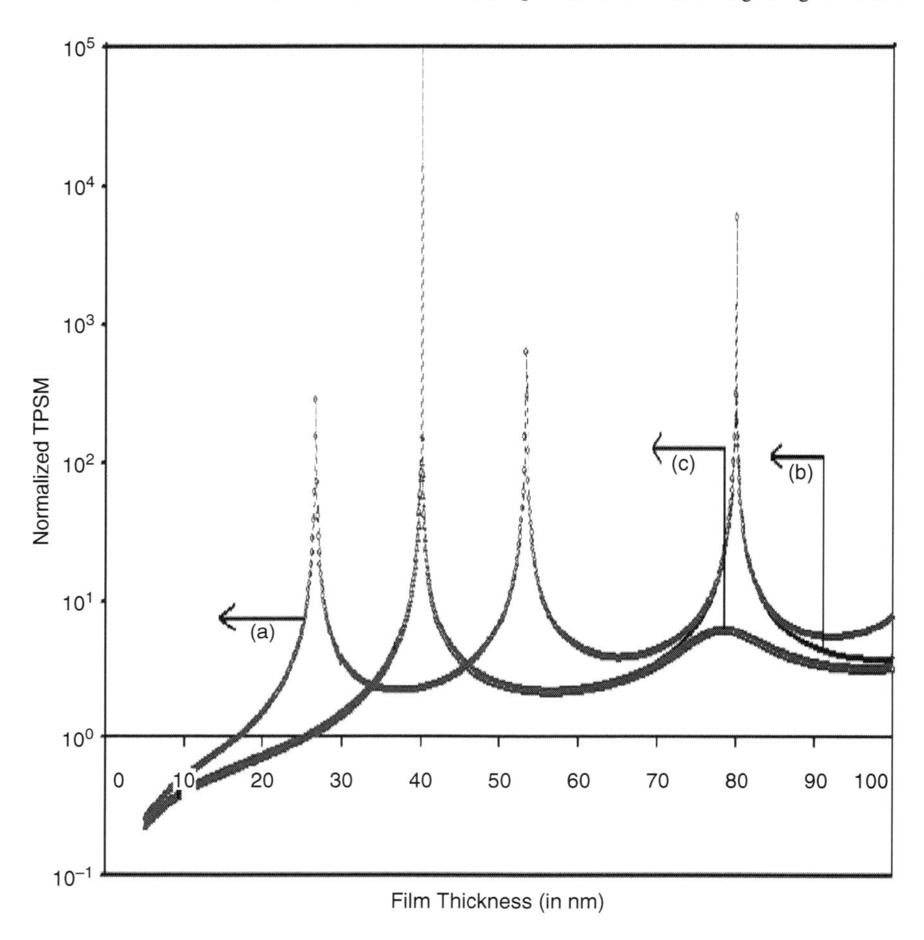

Fig. 1.23 Plot of the normalized TPSM in QDs of Tellurium and stressed Kane type material (n-InSb) as a function of film thickness in accordance with the models of (**a**) Bouat et al., (**b**) Ortenberg et al. and (**c**) Seiler et al., respectively

1.4 Open Research Problems

The problems under these sections which end with the end of this monograph form the integral part of the same and are intended for researchers together with advanced readers from a variety of disciplines as indicated in the Preface for rendering their own contribution in this pin-pointed research topic of thermopower in nanostructured materials. The numerical values of the energy band constants of the materials needed for appropriate numerical computations are given in Table 1.1.

(R1.1) Investigate the diffusion thermoelectric power (DTP), phonon-drag thermoelectric power (PTP), and thermoelectric figure-of-merit (Z) in the

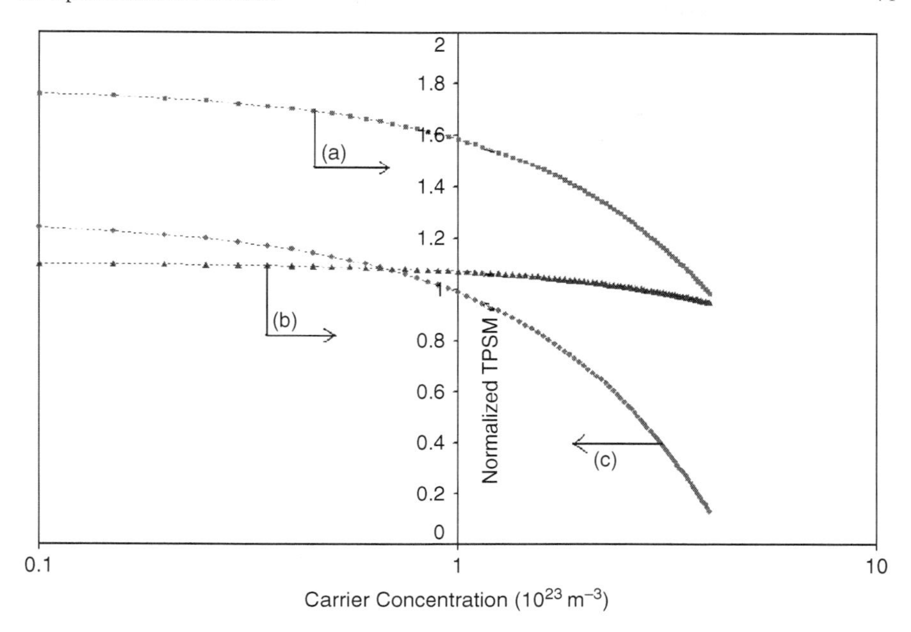

Fig. 1.24 Plot of the normalized TPSM in QDs of Tellurium and stressed Kane type material (n-InSb) as a function of carrier concentration for all the cases of Fig. 1.23

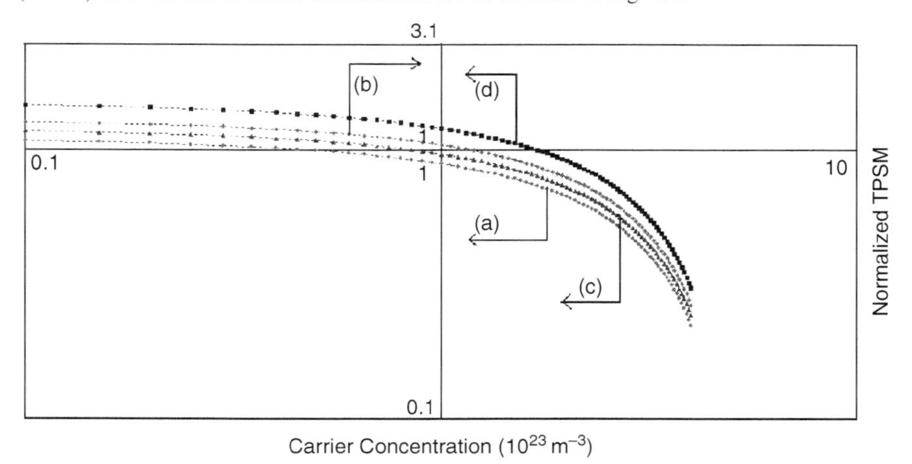

Fig. 1.25 Plot of the normalized TPSM in QDs of (**a**) Graphite, (**b**) Platinum antimonide, (**c**) zero gap (HgTe) and (**d**) $Pb_{1-x}Ge_xTe$ as a function of carrier concentration in accordance with the models of Ushio et al., Emtage, Ivanov-Omskii et al. and Vassilev, respectively

absence of magnetic field by considering all types of scattering mechanisms for QDs in the presence of three finite potential wells and three parabolic potential wells applied separately in the three orthogonal directions for all the materials whose unperturbed carrier energy spectra are defined in this chapter.

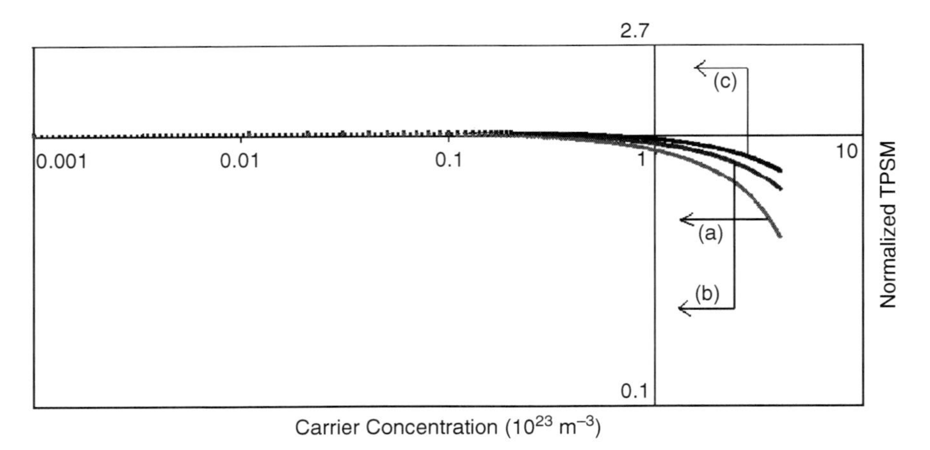

Fig. 1.26 Plot of the normalized TPSM in QDs of GaSb as a of carrier concentration in accordance with the models of (**a**) Seiler et al., (**b**) Mathur et al. and (**c**) Zhang, respectively

(R1.2) Investigate the DTP, PTP, and Z in the absence of magnetic field by considering all types of scattering mechanisms for QDs in the presence of three finite potential wells and three parabolic potential wells applied separately in the three orthogonal directions under an arbitrarily oriented (a) nonuniform electric field and (b) alternating electric field, respectively, for all the materials whose unperturbed carrier energy spectra are defined in this chapter.

(R1.3) Investigate the DTP, PTP, and Z for QDs by considering all types of scattering mechanisms in the presence of three finite potential wells and three parabolic potential wells applied separately in the three orthogonal directions under an arbitrarily oriented alternating magnetic field by including broadening and the electron spin for all the materials whose unperturbed carrier energy spectra are defined in this chapter.

(R1.4) Investigate the DTP, PTP, and Z for QDs by considering all types of scattering mechanisms by considering the presence of three finite potential wells and three parabolic potential wells applied separately in the three orthogonal directions under an arbitrarily oriented alternating magnetic field and crossed alternating electric field by including broadening and the electron spin for all the materials whose unperturbed carrier energy spectra are defined in this chapter.

(R1.5) Investigate the DTP, PTP, and Z for QDs by considering all types of scattering mechanisms in the presence of three finite potential wells and three parabolic potential wells applied separately in the three orthogonal directions under an arbitrarily oriented alternating magnetic field and crossed alternating nonuniform electric field by including broadening and the electron spin for all the materials whose unperturbed carrier energy spectra are defined in this chapter.

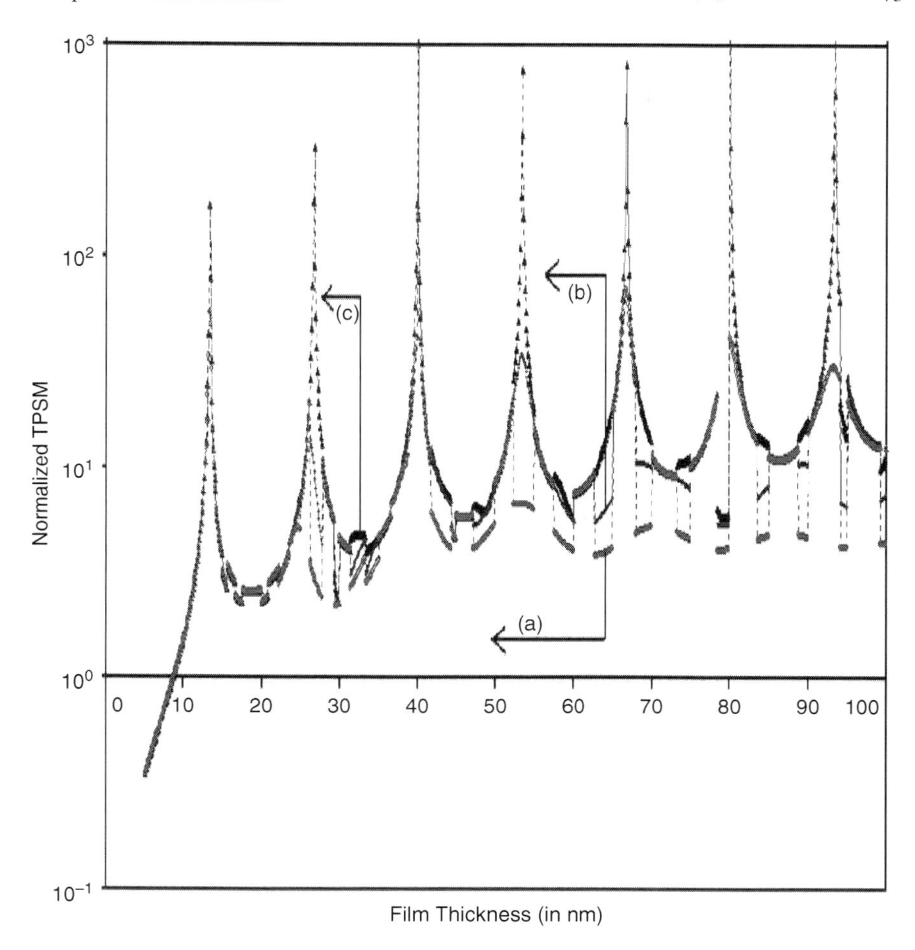

Fig. 1.27 Plot of the normalized TPSM in QDs of Gallium antimonide as a function of film thickness for all the cases of Fig. 1.26

(R1.6) Investigate the DTP, PTP, and Z in the absence of magnetic field for QDs by considering all types of scattering mechanisms in the presence of three finite potential wells and three parabolic potential wells applied separately in the three orthogonal directions under exponential, Kane, Halperin, Lax, and Bonch-Bruevich band tails [224, 225] for all the materials whose unperturbed carrier energy spectra are defined in this chapter.

(R1.7) Investigate the DTP, PTP, and Z in the absence of magnetic field for QDs by considering all types of scattering mechanisms in the presence of three finite potential wells and three parabolic potential wells applied separately in the three orthogonal directions for all the materials as defined in (R1.6) under an arbitrarily oriented (a) nonuniform electric field and

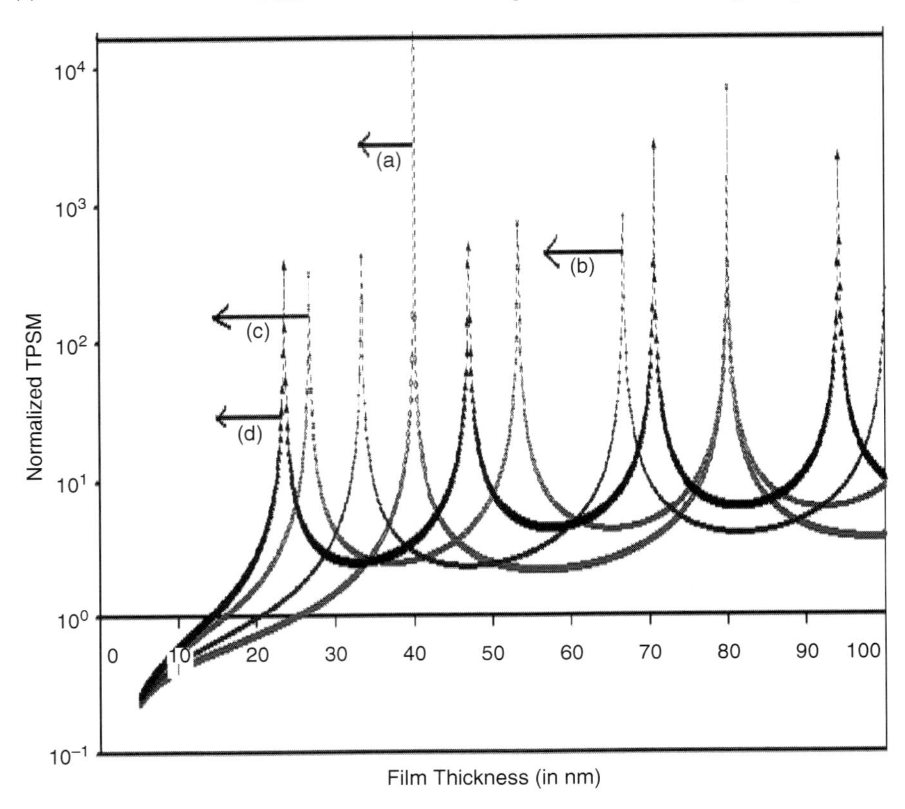

Fig. 1.28 Plot of the normalized TPSM in QDs of Bismuth as a function of film thickness in accordance with the models of (**a**) McClure et al, (**b**) Takaoka et al. (Hybrid model), (**c**) Cohen and (d) Lax et al., respectively

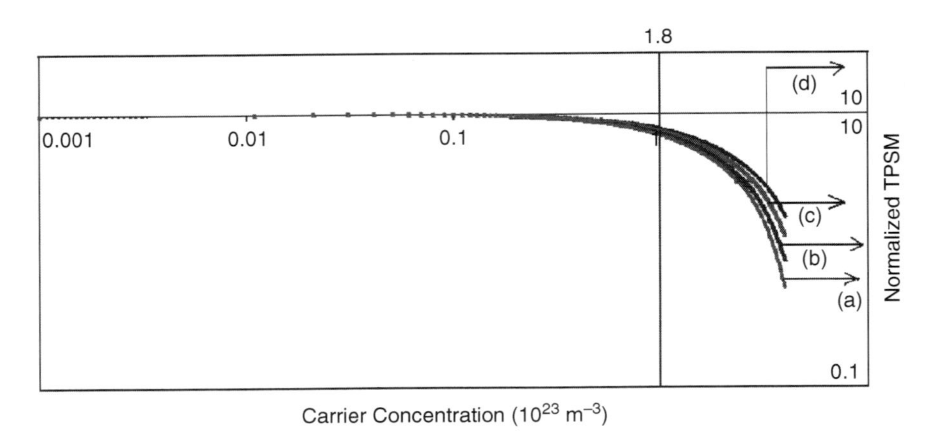

Fig. 1.29 Plot of the normalized TPSM in QDs of Bismuth as a function of carrier concentration for all the cases of Fig. 1.28

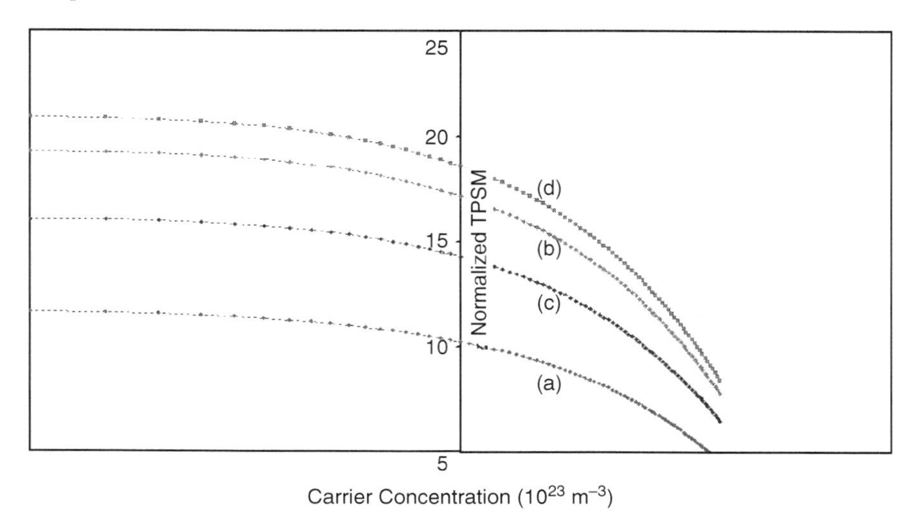

Fig. 1.30 Plot of the normalized TPSM in QDs of PbTe and Bi_2Te_3 as a function of carrier concentration in accordance with the models of (**a**) Dimmock, (**b**) Bangert et al. and (**c**) Foley et al. The plot (d) refers to Bi_2Te_3 in accordance with the model of Stordeur et al.

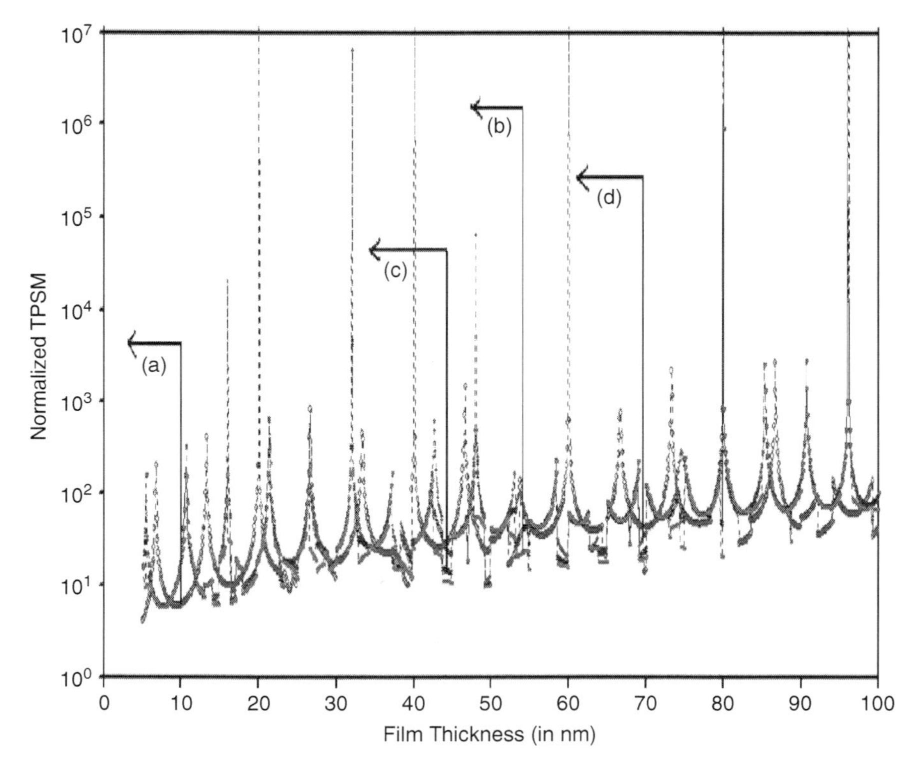

Fig. 1.31 Plot of the normalized TPSM in QDs of PbTe and Bi_2Te_3as a function of film thickness for all the cases of Fig. 1.30

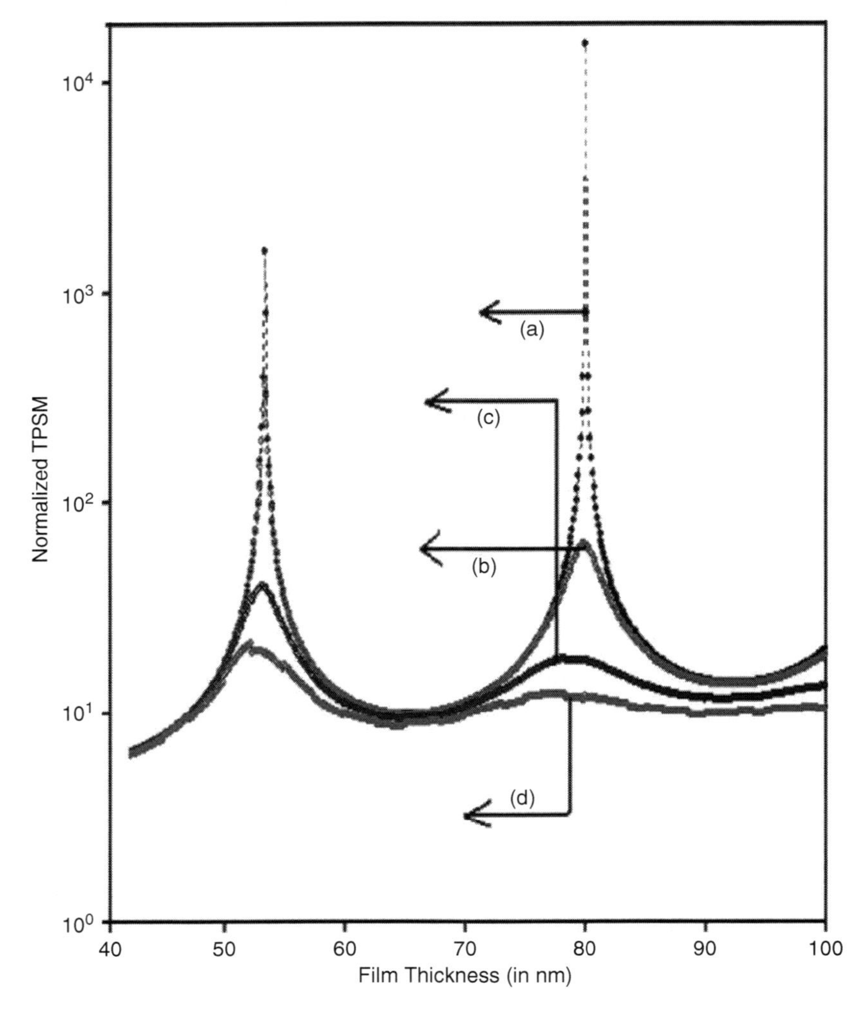

Fig. 1.32 Plot of the normalized TPSM in QDs of (**a**) II–V compound (CdSb), (**b**) zinc diphosphide, (**c**) cadmium diphosphide and (**d**) Antimony as a function of film thickness in accordance with the models of Yamada, Chuiko and Ketterson, respectively

(b) alternating electric field, respectively, whose unperturbed carrier energy spectra are defined in this chapter.

(R1.8) Investigate the DTP, PTP, and Z for QDs by considering all types of scattering mechanisms in the presence of three finite potential wells and three parabolic potential wells applied separately in the three orthogonal directions for all the materials as described in (R1.6) under an arbitrarily oriented alternating magnetic field by including broadening and the electron spin whose unperturbed carrier energy spectra are defined in this chapter.

(R1.9) Investigate the DTP, PTP, and Z for QDs by considering all types of scattering mechanisms in the presence of three finite potential wells and three parabolic potential wells applied separately in the three orthogonal directions as discussed in (R1.6) under an arbitrarily oriented alternating magnetic field and crossed alternating electric field by including broadening and the electron spin for all the materials whose unperturbed carrier energy spectra are defined in this chapter.

(R1.10) Investigate all the appropriate problems of this chapter for the following materials:

1. The energy spectrum of the valance bands of CuCl in accordance with Yekimov et al. [276] can be written as

$$E_h = (\gamma_6 - 2\gamma_7)\frac{\hbar^2 k^2}{2m_0} \tag{R1.1}$$

and

$$E_{l,s} = (\gamma_6 + \gamma_7)\frac{\hbar^2 k^2}{2m_0} - \frac{\Delta_1}{2} \pm \left[\frac{\Delta_1^2}{4} + \gamma_7 \Delta_1 \frac{\hbar^2 k^2}{2m_0} + 9\left(\frac{\gamma_7 \hbar^2 k^2}{2m_0}\right)^2\right]^{1/2}, \tag{R1.2}$$

where $\gamma_6 = 0.53$, $\gamma_7 = 0.07$, $\Delta_1 = 70\,\text{meV}$.

2. In the presence of stress, χ_6, along $\langle 001\rangle$ and $\langle 111\rangle$ directions, the energy spectra of the holes in semiconductors having diamond structure valance bands can be, respectively, expressed following Roman [277] et al. as

$$E = A_6 k^2 \pm \left[\bar{B}_7^2 k^4 + \delta_6^2 + B_7 \delta_6 \left(2k_z^2 - k_s^2\right)\right]^{1/2} \tag{R1.3}$$

and

$$E = A_6 k^2 \pm \left[\bar{B}_7^2 k^4 + \delta_7^2 + \frac{D_6}{\sqrt{3}}\delta_7 \left(2k_z^2 - k_s^2\right)\right]^{1/2}, \tag{R1.4}$$

where A_6, B_7, D_6, and C_6 are inverse mass band parameters in which $\delta_6 \equiv l_7\left(\bar{S}_{11} - \bar{S}_{12}\right)\chi_6$ and $\bar{S}_{ij}$ are the usual elastic compliance constants,

$$\bar{B}_7^2 \equiv \left(B_7^2 + \frac{c_6^2}{5}\right)$$

and

$$\delta_7 \equiv \left(\frac{d_8 S_{44}}{2\sqrt{3}}\right)\chi_6.$$

For gray tin, $d_8 = -4.1\,\text{eV}$, $l_7 = -2.3\,\text{eV}$, $A_6 = 19.2\frac{\hbar^2}{2m_0}$, $B_7 = 26.3\frac{\hbar^2}{2m_0}$,

Table 1.2 The carrier statistics and the thermoelectric power under large magnetic field in quantum dots of nonlinear optical, III–V, II–VI, n-GaP, n-Ge, Te, graphite, $PtSb_2$, zerogap, II–V, GaSb, stressed materials, Bismuth, IV–VI, $Pb_{1-x}Ge_xTe$, Zinc and Cadmium diphosphides, Bi_2Te_3, and antimony

Type of materials	Carrier statistics	TPSM
1. Nonlinear optical materials	In accordance with the generalized electron dispersion relation as given by (1.6), $$n_0 = \frac{2g_v}{d_x d_y d_z} \sum_{n_x=1}^{n_{x_{\max}}} \sum_{n_y=1}^{n_{y_{\max}}} \sum_{n_z=1}^{n_{z_{\max}}} \frac{L_{11}}{M_{11}} \quad (1.7)$$	On the basis of (1.7), $$G_0 = \frac{\pi^2 k_B}{3e} \left[\sum_{n_x=1}^{n_{x_{\max}}} \sum_{n_y=1}^{n_{y_{\max}}} \sum_{n_z=1}^{n_{z_{\max}}} \frac{L_{11}}{M_{11}} \right]^{-1} \left[\sum_{n_x=1}^{n_{x_{\max}}} \sum_{n_y=1}^{n_{y_{\max}}} \sum_{n_z=1}^{n_{z_{\max}}} \frac{Q_{11}}{(M_{11})^2} \right] \quad (1.14)$$
2. III–V materials, the conduction electrons of which can be defined by eight types of energy–wave vector dispersion relations as described in the column beside	(a) Three band model of Kane: In accordance with the three band model of Kane (1.17), which is the special case of (1.6) $$n_0 = \frac{2g_v}{d_x d_y d_z} \sum_{n_x=1}^{n_{x_{\max}}} \sum_{n_y=1}^{n_{y_{\max}}} \sum_{n_z=1}^{n_{z_{\max}}} \frac{L_{12}}{M_{12}} \quad (1.18)$$	On the basis of (1.18), $$G_0 = \frac{\pi^2 k_B}{3e} \left[\sum_{n_x=1}^{n_{x_{\max}}} \sum_{n_y=1}^{n_{y_{\max}}} \sum_{n_z=1}^{n_{z_{\max}}} \frac{L_{12}}{M_{12}} \right]^{-1} \left[\sum_{n_x=1}^{n_{x_{\max}}} \sum_{n_y=1}^{n_{y_{\max}}} \sum_{n_z=1}^{n_{z_{\max}}} \frac{Q_{12}}{(M_{12})^2} \right] \quad (1.19)$$
	(b) Two band model of Kane:	
	In accordance with the two band model of Kane (1.21), $$n_0 = \frac{2g_v}{d_x d_y d_z} \sum_{n_x=1}^{n_{x_{\max}}} \sum_{n_y=1}^{n_{y_{\max}}} \sum_{n_z=1}^{n_{z_{\max}}} \frac{L_{13}}{M_{13}} \quad (1.22)$$	On the basis of (1.22), $$G_0 = \frac{\pi^2 k_B}{3e} \left[\sum_{n_x=1}^{n_{x_{\max}}} \sum_{n_y=1}^{n_{y_{\max}}} \sum_{n_z=1}^{n_{z_{\max}}} \frac{L_{13}}{M_{13}} \right]^{-1} \left[\sum_{n_x=1}^{n_{x_{\max}}} \sum_{n_y=1}^{n_{y_{\max}}} \sum_{n_z=1}^{n_{z_{\max}}} \frac{Q_{13}}{(M_{13})^2} \right] \quad (1.23)$$
	(c) The model of Stillman et al.: In accordance with the model of Stillman et al. (1.26),	On the basis of (1.27),

$$n_0 = \frac{2g_v}{d_x d_y d_z} \sum_{n_x=1}^{n_{x_{\max}}} \sum_{n_y=1}^{n_{y_{\max}}} \sum_{n_z=1}^{n_{z_{\max}}} \frac{L_{14}}{M_{14}} \quad (1.27)$$

$$G_0 = \frac{\pi^2 k_B}{3e} \left[\sum_{n_x=1}^{n_{x_{\max}}} \sum_{n_y=1}^{n_{y_{\max}}} \sum_{n_z=1}^{n_{z_{\max}}} \frac{L_{14}}{M_{14}} \right]^{-1} \left[\sum_{n_x=1}^{n_{x_{\max}}} \sum_{n_y=1}^{n_{y_{\max}}} \sum_{n_z=1}^{n_{z_{\max}}} \frac{Q_{14}}{(M_{14})^2} \right] \quad (1.28)$$

(d) The model of Newson and Kurobe:

In accordance with the model of Newson and Kurobe (1.30),

$$n_0 = \frac{2g_v}{d_x d_y d_z} \sum_{n_x=1}^{n_{x_{\max}}} \sum_{n_y=1}^{n_{y_{\max}}} \sum_{n_z=1}^{n_{z_{\max}}} \frac{L_{15}}{M_{15}} \quad (1.31)$$

On the basis of (1.31),

$$G_0 = \frac{\pi^2 k_B}{3e} \left[\sum_{n_x=1}^{n_{x_{\max}}} \sum_{n_y=1}^{n_{y_{\max}}} \sum_{n_z=1}^{n_{z_{\max}}} \frac{L_{15}}{M_{15}} \right]^{-1} \left[\sum_{n_x=1}^{n_{x_{\max}}} \sum_{n_y=1}^{n_{y_{\max}}} \sum_{n_z=1}^{n_{z_{\max}}} \frac{Q_{15}}{(M_{15})^2} \right] \quad (1.32)$$

(e) The model of Rossler:

In accordance with the model of Rossler (1.34),

$$n_0 = \frac{g_v}{d_x d_y d_z} \sum_{n_x=1}^{n_{x_{\max}}} \sum_{n_y=1}^{n_{y_{\max}}} \sum_{n_z=1}^{n_{z_{\max}}} \frac{L_{16,\pm}}{M_{16,\pm}} \quad (1.35)$$

On the basis of (1.35),

$$G_0 = \frac{\pi^2 k_B}{3e} \left[\sum_{n_x=1}^{n_{x_{\max}}} \sum_{n_y=1}^{n_{y_{\max}}} \sum_{n_z=1}^{n_{z_{\max}}} \frac{L_{16,\pm}}{M_{16,\pm}} \right]^{-1} \left[\sum_{n_x=1}^{n_{x_{\max}}} \sum_{n_y=1}^{n_{y_{\max}}} \sum_{n_z=1}^{n_{z_{\max}}} \frac{Q_{16,\pm}}{(M_{16,\pm})^2} \right] \quad (1.36)$$

(f) The model of Palik et al.:

In accordance with the model of Palik et al. (1.39),

$$n_0 = \frac{2g_v}{d_x d_y d_z} \sum_{n_x=1}^{n_{x_{\max}}} \sum_{n_y=1}^{n_{y_{\max}}} \sum_{n_z=1}^{n_{z_{\max}}} \frac{L_{17}}{M_{17}} \quad (1.40)$$

On the basis of (1.40),

$$G_0 = \frac{\pi^2 k_B}{3e} \left[\sum_{n_x=1}^{n_{x_{\max}}} \sum_{n_y=1}^{n_{y_{\max}}} \sum_{n_z=1}^{n_{z_{\max}}} \frac{L_{17}}{M_{17}} \right]^{-1} \left[\sum_{n_x=1}^{n_{x_{\max}}} \sum_{n_y=1}^{n_{y_{\max}}} \sum_{n_z=1}^{n_{z_{\max}}} \frac{Q_{17}}{(M_{17})^2} \right] \quad (1.41)$$

(g) The model of Johnson and Dickey:

In accordance with the model of Johnson and Dickey (1.43),

$$n_0 = \frac{2g_v}{d_x d_y d_z} \sum_{n_x=1}^{n_{x_{\max}}} \sum_{n_y=1}^{n_{y_{\max}}} \sum_{n_z=1}^{n_{z_{\max}}} \frac{L_{18}}{M_{18}} \quad (1.44)$$

On the basis of (1.44),

$$G_0 = \frac{\pi^2 k_B}{3e} \left[\sum_{n_x=1}^{n_{x_{\max}}} \sum_{n_y=1}^{n_{y_{\max}}} \sum_{n_z=1}^{n_{z_{\max}}} \frac{L_{18}}{M_{18}} \right]^{-1} \left[\sum_{n_x=1}^{n_{x_{\max}}} \sum_{n_y=1}^{n_{y_{\max}}} \sum_{n_z=1}^{n_{z_{\max}}} \frac{Q_{18}}{(M_{18})^2} \right] \quad (1.45)$$

(continued)

Table 1.2 (Continued)

Type of materials	Carrier statistics	TPSM
	(h) The model of Agafonov et al.: In accordance with the model of Agafonov et al. (1.47), $n_0 = \frac{2g_v}{d_x d_y d_z} \sum_{n_x=1}^{n_{x_{max}}} \sum_{n_y=1}^{n_{y_{max}}} \sum_{n_z=1}^{n_{z_{max}}} \frac{L_{19}}{M_{19}}$ (1.48)	On the basis of (1.48), $G_0 = \frac{\pi^2 k_B}{3e} \left[\sum_{n_x=1}^{n_{x_{max}}} \sum_{n_y=1}^{n_{y_{max}}} \sum_{n_z=1}^{n_{z_{max}}} \frac{L_{19}}{M_{19}} \right]^{-1} \left[\sum_{n_x=1}^{n_{x_{max}}} \sum_{n_y=1}^{n_{y_{max}}} \sum_{n_z=1}^{n_{z_{max}}} \frac{Q_{19}}{(M_{19})^2} \right]$ (1.49)
3. II–VI materials as described by the Hopfield model	In accordance with (1.51), $n_0 = \frac{g_v}{d_x d_y d_z} \sum_{n_x=1}^{n_{x_{max}}} \sum_{n_y=1}^{n_{y_{max}}} \sum_{n_z=1}^{n_{z_{max}}} \frac{L_{20,\pm}}{M_{20,\pm}}$ (1.52)	On the basis of (1.52), $G_0 = \frac{\pi^2 k_B}{3e} \left[\sum_{n_x=1}^{n_{x_{max}}} \sum_{n_y=1}^{n_{y_{max}}} \sum_{n_z=1}^{n_{z_{max}}} \frac{L_{20,\pm}}{M_{20,\pm}} \right]^{-1} \left[\sum_{n_x=1}^{n_{x_{max}}} \sum_{n_y=1}^{n_{y_{max}}} \sum_{n_z=1}^{n_{z_{max}}} \frac{Q_{20,\pm}}{(M_{20,\pm})^2} \right]$ (1.53)
4. n-GaP as described by the Rees model	In accordance with (1.55), $n_0 = \frac{2g_v}{d_x d_y d_z} \sum_{n_x=1}^{n_{x_{max}}} \sum_{n_y=1}^{n_{y_{max}}} \sum_{n_z=1}^{n_{z_{max}}} \frac{L_{21}}{M_{21}}$ (1.56)	On the basis of (1.56), $G_0 = \frac{\pi^2 k_B}{3e} \left[\sum_{n_x=1}^{n_{x_{max}}} \sum_{n_y=1}^{n_{y_{max}}} \sum_{n_z=1}^{n_{z_{max}}} \frac{L_{21}}{M_{21}} \right]^{-1} \left[\sum_{n_x=1}^{n_{x_{max}}} \sum_{n_y=1}^{n_{y_{max}}} \sum_{n_z=1}^{n_{z_{max}}} \frac{Q_{21}}{(M_{21})^2} \right]$ (1.57)
5. n-Ge, the conduction electrons of which can be defined by two types of energy band models as described in the column beside	(a) In accordance with the model of Cardona et al. (1.59), $n_0 = \frac{2g_v}{d_x d_y d_z} \sum_{n_x=1}^{n_{x_{max}}} \sum_{n_y=1}^{n_{y_{max}}} \sum_{n_z=1}^{n_{z_{max}}} \frac{L_{22}}{M_{22}}$ (1.60)	On the basis of (1.60), $G_0 = \frac{\pi^2 k_B}{3e} \left[\sum_{n_x=1}^{n_{x_{max}}} \sum_{n_y=1}^{n_{y_{max}}} \sum_{n_z=1}^{n_{z_{max}}} \frac{L_{22}}{M_{22}} \right]^{-1} \left[\sum_{n_x=1}^{n_{x_{max}}} \sum_{n_y=1}^{n_{y_{max}}} \sum_{n_z=1}^{n_{z_{max}}} \frac{Q_{22}}{(M_{22})^2} \right]$ (1.61)
	(b) In accordance with the model of Wang and Ressler (1.64), $n_0 = \frac{2g_v}{d_x d_y d_z} \sum_{n_x=1}^{n_{x_{max}}} \sum_{n_y=1}^{n_{y_{max}}} \sum_{n_z=1}^{n_{z_{max}}} \frac{L_{23}}{M_{23}}$ (1.65)	On the basis of (1.65), $G_0 = \frac{\pi^2 k_B}{3e} \left[\sum_{n_x=1}^{n_{x_{max}}} \sum_{n_y=1}^{n_{y_{max}}} \sum_{n_z=1}^{n_{z_{max}}} \frac{L_{23}}{M_{23}} \right]^{-1} \left[\sum_{n_x=1}^{n_{x_{max}}} \sum_{n_y=1}^{n_{y_{max}}} \sum_{n_z=1}^{n_{z_{max}}} \frac{Q_{23}}{(M_{23})^2} \right]$ (1.66)

6. Tellurium, the conduction electrons of which can be defined by two types of energy band models as described in the column beside	In accordance with the model of Bouat et al. (1.68), $n_0 = \frac{g_v}{d_x d_y d_z} \sum_{n_x=1}^{n_{x_{max}}} \sum_{n_y=1}^{n_{y_{max}}} \sum_{n_z=1}^{n_{z_{max}}} \frac{L_{24,\pm}}{M_{24,\pm}}$ (1.69)	On the basis of (1.69), $G_0 = \frac{\pi^2 k_B}{3e} \left[\sum_{n_x=1}^{n_{x_{max}}} \sum_{n_y=1}^{n_{y_{max}}} \sum_{n_z=1}^{n_{z_{max}}} \frac{L_{24,\pm}}{M_{24,\pm}} \right]^{-1} \left[\sum_{n_x=1}^{n_{x_{max}}} \sum_{n_y=1}^{n_{y_{max}}} \sum_{n_z=1}^{n_{z_{max}}} \frac{Q_{24,\pm}}{(M_{24,\pm})^2} \right]$ (1.70)
	In accordance with the model of Ortenberg and Button (1.72), $n_0 = \frac{g_v}{d_x d_y d_z} \sum_{n_x=1}^{n_{x_{max}}} \sum_{n_y=1}^{n_{y_{max}}} \sum_{n_z=1}^{n_{z_{max}}} \frac{L_{25,\pm}}{M_{25,\pm}}$ (1.73)	On the basis of (1.73), $G_0 = \frac{\pi^2 k_B}{3e} \left[\sum_{n_x=1}^{n_{x_{max}}} \sum_{n_y=1}^{n_{y_{max}}} \sum_{n_z=1}^{n_{z_{max}}} \frac{L_{25,\pm}}{M_{25,\pm}} \right]^{-1} \left[\sum_{n_x=1}^{n_{x_{max}}} \sum_{n_y=1}^{n_{y_{max}}} \sum_{n_z=1}^{n_{z_{max}}} \frac{Q_{25,\pm}}{(M_{25,\pm})^2} \right]$ (1.74)
7. Graphite, as defined by the model of Ushio et al.	In accordance with (1.76), $n_0 = \frac{g_v}{d_x d_y d_z} \sum_{n_x=1}^{n_{x_{max}}} \sum_{n_y=1}^{n_{y_{max}}} \sum_{n_z=1}^{n_{z_{max}}} \frac{L_{26,\pm}}{M_{26,\pm}}$ (1.77)	On the basis of (1.77), $G_0 = \frac{\pi^2 k_B}{3e} \left[\sum_{n_x=1}^{n_{x_{max}}} \sum_{n_y=1}^{n_{y_{max}}} \sum_{n_z=1}^{n_{z_{max}}} \frac{L_{26,\pm}}{M_{26,\pm}} \right]^{-1} \left[\sum_{n_x=1}^{n_{x_{max}}} \sum_{n_y=1}^{n_{y_{max}}} \sum_{n_z=1}^{n_{z_{max}}} \frac{Q_{26,\pm}}{(M_{26,\pm})^2} \right]$ (1.78)
8. $PtSb_2$, as defined by the Emtage model	In accordance with (1.81), $p_0 = \frac{2g_v}{d_x d_y d_z} \sum_{n_x=1}^{n_{x_{max}}} \sum_{n_y=1}^{n_{y_{max}}} \sum_{n_z=1}^{n_{z_{max}}} \frac{L_{27}}{M_{27}}$ (1.82)	On the basis of (1.82), $G_0 = \frac{\pi^2 k_B}{3e} \left[\sum_{n_x=1}^{n_{x_{max}}} \sum_{n_y=1}^{n_{y_{max}}} \sum_{n_z=1}^{n_{z_{max}}} \frac{L_{27}}{M_{27}} \right]^{-1} \left[\sum_{n_x=1}^{n_{x_{max}}} \sum_{n_y=1}^{n_{y_{max}}} \sum_{n_z=1}^{n_{z_{max}}} \frac{Q_{27}}{(M_{27})^2} \right]$ (1.83)
9. Zerogap materials, as defined by Ivanov-Omskii et al. model	In accordance with (1.85), $p_0 = \frac{2g_v}{d_x d_y d_z} \sum_{n_x=1}^{n_{x_{max}}} \sum_{n_y=1}^{n_{y_{max}}} \sum_{n_z=1}^{n_{z_{max}}} \frac{L_{28}}{M_{28}}$ (1.86)	On the basis of (1.86), $G_0 = \frac{\pi^2 k_B}{3e} \left[\sum_{n_x=1}^{n_{x_{max}}} \sum_{n_y=1}^{n_{y_{max}}} \sum_{n_z=1}^{n_{z_{max}}} \frac{L_{28}}{M_{28}} \right]^{-1} \left[\sum_{n_x=1}^{n_{x_{max}}} \sum_{n_y=1}^{n_{y_{max}}} \sum_{n_z=1}^{n_{z_{max}}} \frac{Q_{28}}{(M_{28})^2} \right]$ (1.87)

(continued)

Table 1.2 (Continued)

Type of materials	Carrier statistics	TPSM
10. II–V materials, as defined by the model of Yamada	In accordance with (1.89), $$p_0 = \frac{g_v}{d_x d_y d_z} \sum_{n_x=1}^{n_{x_{\max}}} \sum_{n_y=1}^{n_{y_{\max}}} \sum_{n_z=1}^{n_{z_{\max}}} \frac{L_{29,\pm}}{M_{29,\pm}} \quad (1.90)$$	On the basis of (1.90), $$G_0 = \frac{\pi^2 k_B}{3e} \left[\sum_{n_x=1}^{n_{x_{\max}}} \sum_{n_y=1}^{n_{y_{\max}}} \sum_{n_z=1}^{n_{z_{\max}}} \frac{L_{29,\pm}}{M_{29,\pm}} \right]^{-1} \left[\sum_{n_x=1}^{n_{x_{\max}}} \sum_{n_y=1}^{n_{y_{\max}}} \sum_{n_z=1}^{n_{z_{\max}}} \frac{Q_{29,\pm}}{(M_{29,\pm})^2} \right] \quad (1.91)$$
11. Gallium Antimonide, the carriers of which can be defined by three types of energy band models as described in the column beside	In accordance with the model of Seiler et al. (1.93), $$p_0 = \frac{g_v}{d_x d_y d_z} \sum_{n_x=1}^{n_{x_{\max}}} \sum_{n_y=1}^{n_{y_{\max}}} \sum_{n_z=1}^{n_{z_{\max}}} \frac{L_{30,\pm}}{M_{30,\pm}} \quad (1.94)$$	On the basis of (1.94), $$G_0 = \frac{\pi^2 k_B}{3e} \left[\sum_{n_x=1}^{n_{x_{\max}}} \sum_{n_y=1}^{n_{y_{\max}}} \sum_{n_z=1}^{n_{z_{\max}}} \frac{L_{30}}{M_{30}} \right]^{-1} \left[\sum_{n_x=1}^{n_{x_{\max}}} \sum_{n_y=1}^{n_{y_{\max}}} \sum_{n_z=1}^{n_{z_{\max}}} \frac{Q_{30}}{(M_{30})^2} \right] \quad (1.95)$$
	In accordance with the model of Mathur et al. (1.98), $$n_0 = \frac{2g_v}{d_x d_y d_z} \sum_{n_x=1}^{n_{x_{\max}}} \sum_{n_y=1}^{n_{y_{\max}}} \sum_{n_z=1}^{n_{z_{\max}}} \frac{L_{31}}{M_{31}} \quad (1.99)$$	On the basis of (1.99), $$G_0 = \frac{\pi^2 k_B}{3e} \left[\sum_{n_x=1}^{n_{x_{\max}}} \sum_{n_y=1}^{n_{y_{\max}}} \sum_{n_z=1}^{n_{z_{\max}}} \frac{L_{31}}{M_{31}} \right]^{-1} \left[\sum_{n_x=1}^{n_{x_{\max}}} \sum_{n_y=1}^{n_{y_{\max}}} \sum_{n_z=1}^{n_{z_{\max}}} \frac{Q_{31}}{(M_{31})^2} \right] \quad (1.100)$$
	In accordance with the model of Zhang (1.102), $$n_0 = \frac{2g_v}{d_x d_y d_z} \sum_{n_x=1}^{n_{x_{\max}}} \sum_{n_y=1}^{n_{y_{\max}}} \sum_{n_z=1}^{n_{z_{\max}}} \frac{L_{32}}{M_{32}} \quad (1.103)$$	On the basis of (1.103), $$G_0 = \frac{\pi^2 k_B}{3e} \left[\sum_{n_x=1}^{n_{x_{\max}}} \sum_{n_y=1}^{n_{y_{\max}}} \sum_{n_z=1}^{n_{z_{\max}}} \frac{L_{32}}{M_{32}} \right]^{-1} \left[\sum_{n_x=1}^{n_{x_{\max}}} \sum_{n_y=1}^{n_{y_{\max}}} \sum_{n_z=1}^{n_{z_{\max}}} \frac{Q_{32}}{(M_{32})^2} \right] \quad (1.104)$$
12. Stressed materials, as defined by the model of Seiler et al.	In accordance with (1.106), $$n_0 = \frac{2g_v}{d_x d_y d_z} \sum_{n_x=1}^{n_{x_{\max}}} \sum_{n_y=1}^{n_{y_{\max}}} \sum_{n_z=1}^{n_{z_{\max}}} \frac{L_{33}}{M_{33}} \quad (1.107)$$	On the basis of (1.107), $$G_0 = \frac{\pi^2 k_B}{3e} \left[\sum_{n_x=1}^{n_{x_{\max}}} \sum_{n_y=1}^{n_{y_{\max}}} \sum_{n_z=1}^{n_{z_{\max}}} \frac{L_{33}}{M_{33}} \right]^{-1} \left[\sum_{n_x=1}^{n_{x_{\max}}} \sum_{n_y=1}^{n_{y_{\max}}} \sum_{n_z=1}^{n_{z_{\max}}} \frac{Q_{33}}{(M_{33})^2} \right] \quad (1.108)$$

13. Bismuth, the carriers of which can be defined by five types of energy band models as described in the column beside	The McClure and Choi model: In accordance with (1.110), $n_0 = \frac{2g_v}{d_x d_y d_z} \sum_{n_x=1}^{n_{x_{\max}}} \sum_{n_y=1}^{n_{y_{\max}}} \sum_{n_z=1}^{n_{z_{\max}}} \frac{L_{34}}{M_{34}}$ (1.111)	On the basis of (1.111), $G_0 = \frac{\pi^2 k_B}{3e} \left[\sum_{n_x=1}^{n_{x_{\max}}} \sum_{n_y=1}^{n_{y_{\max}}} \sum_{n_z=1}^{n_{z_{\max}}} \frac{L_{34}}{M_{34}} \right]^{-1} \left[\sum_{n_x=1}^{n_{x_{\max}}} \sum_{n_y=1}^{n_{y_{\max}}} \sum_{n_z=1}^{n_{z_{\max}}} \frac{Q_{34}}{(M_{34})^2} \right]$ (1.112)
	The Hybrid Model: In accordance with (1.114), $n_0 = \frac{2g_v}{d_x d_y d_z} \sum_{n_x=1}^{n_{x_{\max}}} \sum_{n_y=1}^{n_{y_{\max}}} \sum_{n_z=1}^{n_{z_{\max}}} \frac{L_{35}}{M_{35}}$ (1.115)	On the basis of (1.115), $G_0 = \frac{\pi^2 k_B}{3e} \left[\sum_{n_x=1}^{n_{x_{\max}}} \sum_{n_y=1}^{n_{y_{\max}}} \sum_{n_z=1}^{n_{z_{\max}}} \frac{L_{35}}{M_{35}} \right]^{-1} \left[\sum_{n_x=1}^{n_{x_{\max}}} \sum_{n_y=1}^{n_{y_{\max}}} \sum_{n_z=1}^{n_{z_{\max}}} \frac{Q_{35}}{(M_{35})^2} \right]$ (1.116)
	The Cohen Model: In accordance with (1.118), $n_0 = \frac{2g_v}{d_x d_y d_z} \sum_{n_x=1}^{n_{x_{\max}}} \sum_{n_y=1}^{n_{y_{\max}}} \sum_{n_z=1}^{n_{z_{\max}}} \frac{L_{36}}{M_{36}}$ (1.119)	On the basis of (1.119), $G_0 = \frac{\pi^2 k_B}{3e} \left[\sum_{n_x=1}^{n_{x_{\max}}} \sum_{n_y=1}^{n_{y_{\max}}} \sum_{n_z=1}^{n_{z_{\max}}} \frac{L_{36}}{M_{36}} \right]^{-1} \left[\sum_{n_x=1}^{n_{x_{\max}}} \sum_{n_y=1}^{n_{y_{\max}}} \sum_{n_z=1}^{n_{z_{\max}}} \frac{Q_{36}}{(M_{36})^2} \right]$ (1.120)
	The Lax Model: In accordance with (1.122), $n_0 = \frac{2g_v}{d_x d_y d_z} \sum_{n_x=1}^{n_{x_{\max}}} \sum_{n_y=1}^{n_{y_{\max}}} \sum_{n_z=1}^{n_{z_{\max}}} \frac{L_{37}}{M_{37}}$ (1.123)	On the basis of (1.123), $G_0 = \frac{\pi^2 k_B}{3e} \left[\sum_{n_x=1}^{n_{x_{\max}}} \sum_{n_y=1}^{n_{y_{\max}}} \sum_{n_z=1}^{n_{z_{\max}}} \frac{L_{37}}{M_{37}} \right]^{-1} \left[\sum_{n_x=1}^{n_{x_{\max}}} \sum_{n_y=1}^{n_{y_{\max}}} \sum_{n_z=1}^{n_{z_{\max}}} \frac{Q_{37}}{(M_{37})^2} \right]$ (1.124)
	Ellipsoidal Parabolic Model:In accordance with (1.124a), $n_0 = \frac{2g_v}{d_x d_y d_z} \sum_{n_x=1}^{n_{x_{\max}}} \sum_{n_y=1}^{n_{y_{\max}}} \sum_{n_z=1}^{n_{z_{\max}}} \frac{L_{37}}{M_{37}}$ (1.124b)	On the basis of (1.124b), $G_0 = \frac{\pi^2 k_B}{3e} \left[\sum_{n_x=1}^{n_{x_{\max}}} \sum_{n_y=1}^{n_{y_{\max}}} \sum_{n_z=1}^{n_{z_{\max}}} \frac{\overline{L_{37}}}{\overline{\overline{M_{37}}}} \right]^{-1} \left[\sum_{n_x=1}^{n_{x_{\max}}} \sum_{n_y=1}^{n_{y_{\max}}} \sum_{n_z=1}^{n_{z_{\max}}} \frac{\overline{Q_{37}}}{\left(\overline{M_{37}}\right)^2} \right]$ (1.124c)

(continued)

Table 1.2 (Continued)

Type of materials	Carrier statistics	TPSM
14. IV–VI materials, the carriers of which can be defined by three types of energy band models as described in the column beside	In accordance with the model of Dimmock (1.126), $$n_0 = \frac{2g_v}{d_x d_y d_z} \sum_{n_x=1}^{n_{x_{\max}}} \sum_{n_y=1}^{n_{y_{\max}}} \sum_{n_z=1}^{n_{z_{\max}}} \frac{L_{38}}{M_{38}} \quad (1.127)$$	On the basis of (1.127), $$G_0 = \frac{\pi^2 k_B}{3e} \left[\sum_{n_x=1}^{n_{x_{\max}}} \sum_{n_y=1}^{n_{y_{\max}}} \sum_{n_z=1}^{n_{z_{\max}}} \frac{L_{38}}{M_{38}} \right]^{-1} \left[\sum_{n_x=1}^{n_{x_{\max}}} \sum_{n_y=1}^{n_{y_{\max}}} \sum_{n_z=1}^{n_{z_{\max}}} \frac{Q_{38}}{(M_{38})^2} \right] \quad (1.128)$$
	In accordance with the model of Bangert and Kastner (1.130), $$n_0 = \frac{2g_v}{d_x d_y d_z} \sum_{n_x=1}^{n_{x_{\max}}} \sum_{n_y=1}^{n_{y_{\max}}} \sum_{n_z=1}^{n_{z_{\max}}} \frac{L_{39}}{M_{39}} \quad (1.131)$$	On the basis of (1.131), $$G_0 = \frac{\pi^2 k_B}{3e} \left[\sum_{n_x=1}^{n_{x_{\max}}} \sum_{n_y=1}^{n_{y_{\max}}} \sum_{n_z=1}^{n_{z_{\max}}} \frac{L_{39}}{M_{39}} \right]^{-1} \left[\sum_{n_x=1}^{n_{x_{\max}}} \sum_{n_y=1}^{n_{y_{\max}}} \sum_{n_z=1}^{n_{z_{\max}}} \frac{Q_{39}}{(M_{39})^2} \right] \quad (1.132)$$
	In accordance with the model of Foley et al. (1.134), $$n_0 = \frac{2g_v}{d_x d_y d_z} \sum_{n_x=1}^{n_{x_{\max}}} \sum_{n_y=1}^{n_{y_{\max}}} \sum_{n_z=1}^{n_{z_{\max}}} \frac{L_{40}}{M_{40}} \quad (1.135)$$	On the basis of (1.135) $$G_0 = \frac{\pi^2 k_B}{3e} \left[\sum_{n_x=1}^{n_{x_{\max}}} \sum_{n_y=1}^{n_{y_{\max}}} \sum_{n_z=1}^{n_{z_{\max}}} \frac{L_{40}}{M_{40}} \right]^{-1} \left[\sum_{n_x=1}^{n_{x_{\max}}} \sum_{n_y=1}^{n_{y_{\max}}} \sum_{n_z=1}^{n_{z_{\max}}} \frac{Q_{40}}{(M_{40})^2} \right] \quad (1.136)$$
15. $Pb_{1-x}Ge_xTe$, which follows the model of Vassilev	In accordance with (1.138), $$n_0 = \frac{g_v}{d_x d_y d_z} \sum_{n_x=1}^{n_{x_{\max}}} \sum_{n_y=1}^{n_{y_{\max}}} \sum_{n_z=1}^{n_{z_{\max}}} \frac{L_{41,\pm}}{M_{41,\pm}} \quad (1.139)$$	On the basis of (1.139), $$G_0 = \frac{\pi^2 k_B}{3e} \left[\sum_{n_x=1}^{n_{x_{\max}}} \sum_{n_y=1}^{n_{y_{\max}}} \sum_{n_z=1}^{n_{z_{\max}}} \frac{L_{41,\pm}}{M_{41,\pm}} \right]^{-1} \left[\sum_{n_x=1}^{n_{x_{\max}}} \sum_{n_y=1}^{n_{y_{\max}}} \sum_{n_z=1}^{n_{z_{\max}}} \frac{Q_{41,\pm}}{(M_{41,\pm})^2} \right] \quad (1.140)$$

16. Zinc and Cadmium diphosphides as defined by Chuiko model	In accordance with (1.142), $p_0 = \frac{g_v}{d_x d_y d_z} \sum_{n_x=1}^{n_{x\max}} \sum_{n_y=1}^{n_{y\max}} \sum_{n_z=1}^{n_{z\max}} \frac{L_{42,\pm}}{M_{42,\pm}}$ (1.143)	On the basis of (1.143), $G_0 = \frac{\pi^2 k_B}{3e} \left[\sum_{n_x=1}^{n_{x\max}} \sum_{n_y=1}^{n_{y\max}} \sum_{n_z=1}^{n_{z\max}} \frac{L_{42,\pm}}{M_{42,\pm}}\right]^{-1} \left[\sum_{n_x=1}^{n_{x\max}} \sum_{n_y=1}^{n_{y\max}} \sum_{n_z=1}^{n_{z\max}} \frac{Q_{42,\pm}}{\left(M_{42,\pm}\right)^2}\right]$ (1.144)
17. Bi_2Te_3, which follows the model of Stordeur et al.	In accordance with (1.146), $p_0 = \frac{2g_v}{d_x d_y d_z} \sum_{n_x=1}^{n_{x\max}} \sum_{n_y=1}^{n_{y\max}} \sum_{n_z=1}^{n_{z\max}} \frac{L_{43}}{M_{43}}$ (1.147)	On the basis of (1.147), $G_0 = \frac{\pi^2 k_B}{3e} \left[\sum_{n_x=1}^{n_{x\max}} \sum_{n_y=1}^{n_{y\max}} \sum_{n_z=1}^{n_{z\max}} \frac{L_{43}}{M_{43}}\right]^{-1} \left[\sum_{n_x=1}^{n_{x\max}} \sum_{n_y=1}^{n_{y\max}} \sum_{n_z=1}^{n_{z\max}} \frac{Q_{43}}{(M_{43})^2}\right]$ (1.148)
18. Sb, which follows Ketterson model	In accordance with (1.152), (1.154) and (1.156), $n_0 = \frac{2g_v}{d_x d_y d_z} \left[\sum_{n_x=1}^{n_{x\max}} \sum_{n_y=1}^{n_{y\max}} \sum_{n_z=1}^{n_{z\max}} \left[\frac{L_{44}}{M_{44}} + \frac{L_{45}}{M_{45}} + \frac{L_{46}}{M_{46}}\right]\right]$ (1.158)	On the basis of (1.158), $G_0 = \left(\frac{\pi^2 k_B^2 T}{3eh_{11}}\right) \left[\sum_{n_x=1}^{n_{x\max}} \sum_{n_y=1}^{n_{y\max}} \sum_{n_z=1}^{n_{z\max}} \left[\frac{L_{44}}{M_{44}} + \frac{L_{45}}{M_{45}} + \frac{L_{46}}{M_{46}}\right]\right]^{-1}$ (1.159)

$$D_6 = 31 \frac{\hbar^2}{2m_0},$$

and

$$c_6^2 = -1,112 \frac{\hbar^2}{2m_0}.$$

3. The dispersion relation of the holes in p-InSb is given by [278]

$$\bar{E} = c_4 (1 + \gamma_4 f_4) k^2 \pm \frac{1}{3} \left[2\sqrt{2}\sqrt{c_4}\sqrt{16 + 5\gamma_4}\sqrt{E_4 g_4 k} \right], \qquad \text{(R1.5)}$$

where $\bar{E}$ is the energy of the hole as measured from the top of the valance and within it,

$$c_4 \equiv \frac{\hbar^2}{2m_0} + \theta_4,$$
$$\theta_4 \equiv 4.7 \frac{\hbar^2}{2m_0},$$
$$\gamma_4 \equiv \frac{b_4}{c_4},$$
$$b_4 \equiv \frac{3}{2} b_5 + 2\theta_4,$$
$$b_5 \equiv 2.4 \frac{\hbar^2}{2m_0},$$
$$f_4 \equiv \frac{1}{4} \left[\sin^2 2\theta + \sin^4 \theta \, \sin^2 2\phi \right],$$

θ is measured from the positive Z-axis, ϕ is measured from positive X-axis,

$$g_4 \equiv \sin\theta \left[\cos^2\theta + \frac{1}{4} \sin^4\theta \sin^2 2\phi \right],$$

and $E_4 = 5 \times 10^{-4}$ eV.

(R1.11) Investigate all the appropriate problems of this chapter after proper modifications for wedge shaped, cylindrical, ellipsoidal, conical, triangular, colloidal, pyramidal, circular, lateral parabolic rotational, parabolic cylindrical, and position-controlled QDs, respectively.

(R1.12) (a) Investigate all the problems of (R1.11) in the presence of strain after proper modifications. (b) Investigate all the appropriate problems of this chapter for quantum rods.

(R1.13) Investigate the influence of deep traps and surface states separately for all the appropriate problems of this chapter after proper modifications.

(R1.14) Investigate the influence of the localization of carriers for all the appropriate problems of this chapter after proper modifications.

References

1. K.V. Klitzing, G. Dorda, M. Pepper, Phys. Rev. Lett. **45**, 494 (1980)
2. K.V. Klitzing, Rev. Mod. Phys. **58**, 519 (1986)
3. J. Hajdu, G. Landwehr, in *Strong and Ultrastrong Magnetic Fields and Their Applications*, ed. by F. Herlach (Springer, Berlin, 1985), p. 17
4. I.M. Tsidilkovskii, *Band Structure of Semiconductors* (Pergamon Press, Oxford, 1982)
5. E.A. Arushanov, A.F. Knyazev, A.N. Natepov, S.I. Radautsan, Sov. Phys. Semiconduct. **15**, 828 (1981)
6. S.P. Zelenim, A.S. Kondratev, A.E. Kuchma, Sov. Phys. Semiconduct. **16**, 355 (1982)
7. F.M. Peeters, P. Vasilopoulos, Phys. Rev. B **46**, 4667 (1992)
8. W. Zawadzki, in *Two-Dimensional Systems, Heterostructures and Superlattices*, Springer Series in Solid State Sciences, vol. 53, ed. by G. Bauer, F. Kuchar, H. Heinrich (Springer, Berlin, Heidelberg, 1984)
9. B.M. Askerov, N.F. Gashimzcede, M.M. Panakhov, Sov. Phys. Sol. State **29**, 465 (1987)
10. G.P. Chuiko, Sov. Phys. Semiconduct. **19**, 1279 (1985)
11. S. Pahari, S. Bhattacharya, K.P. Ghatak, J. Comput. Theor. Nanosci. **6**, 2088 (2009)
12. K.P. Ghatak, S. Bhattacharya, S. Bhowmik, R. Benedictus, S. Choudhury, J. Appl. Phys. **103**, 034303 (2008)
13. K.P. Ghatak, S. Bhattacharya, S. Pahari, D. De, S. Ghosh, M. Mitra, Ann. Phys. **17**, 195 (2008)
14. K.P. Ghatak, S.N. Biswas, J. Appl. Phys. **70**, 299 (1991)
15. K.P. Ghatak, S.N. Biswas, J. Low Temp. Phys. **78**, 219 (1990)
16. K.P. Ghatak, M. Mondal, J. Appl. Phys. **65**, 3480 (1989)
17. K.P. Ghatak, B. Nag, Nanostruct. Mater. **5**, 769 (1995)
18. K.P. Ghatak, M. Mondal, Phys. Stat. Sol. (b) **185**, K5 (1994)
19. K.P. Ghatak, B. Mitra, Il Nuovo Cimento D **15**, 97 (1993)
20. K.P. Ghatak, S.N. Biswas, Phys. Stat. Sol (b) **140**, K107 (1987)
21. K.P. Ghatak, Il Nuovo Cimento D **13**, 1321 (1991)
22. K.P. Ghatak, A. Ghoshal, Phys. Stat. Sol. (b) **170**, K27 (1992)
23. K.P. Ghatak, B. De, B. Nag, P.K. Chakraborty, Mol. Cryst. Liq. Cryst. Sci. Technol. Sect. B Nonlinear Opt. **16**, 221 (1996)
24. K.P. Ghatak, M. Mitra, B. Goswami, B. Nag, Mol. Cryst. Liq. Cryst. Sci. Technol. Sect. B Nonlinear Opt. **16**, 167 (1996)
25. K.P. Ghatak, D.K. Basu, D. Basu, B. Nag, Nuovo Cimento della Societa Italiana di Fisica D – Condensed Matter, Atomic, Molecular and Chemical Physics, Biophysics **18**, 947 (1996).
26. B. Mitra, K.P. Ghatak, Phys. Lett. A **141**, 81 (1989)
27. S.K. Biswas, A.R. Ghatak, A. Neogi, A. Sharma, S. Bhattacharya, K.P. Ghatak, Physica E: Low-Dimensional Systems and Nanostructures **36**, 163 (2007)
28. M. Mondal, A. Ghoshal, K.P. Ghatak, Il Nuovo Cimento D **14**, 63 (1992)
29. K.P. Ghatak, M. Mondal Phys. Stat. Sol. (b), **135**, 819 (1986)
30. L.J. Singh, S. Choudhury, D. Baruah, S.K. Biswas, S. Pahari, K.P. Ghatak, Physica B **368**, 188 (2005)
31. K.P. Ghatak, B. De, M. Mondal, S.N. Biswas, Mater. Res. Soc. Symp. Proc. **198**, 327 (1990)
32. K.P. Ghatak, Proc. SPIE Int. Soc. Opt. Eng. **1584**, 435 (1992)
33. K.P. Ghatak, B. De, Mater. Res. Soc. Symp. Proc. **234**, 55 and 59 (1991)
34. K.P. Ghatak, B. De, M. Mondal, S.N. Biswas, Mater. Res. Soc. Symp. Proc. **184**, 261 (1990)
35. K.P. Ghatak, *Influence of Band Structure on Some Quantum Processes in Tetragonal Semiconductors*, D. Eng. Thesis (Jadavpur University, Kolkata, India, 1991)
36. K.P. Ghatak, S.N. Biswas, Mater. Res. Soc. Symp. Proc. **216**, 465 (1990)
37. T.M. Tritt, M. Kanatzidis, G. Mahan, H.B. Lyon Jr. (eds.), New Materials for Small Scale Thermoelectric Refrigeration and Power Generation Applications, Vol. 545, *Proceedings of the 1998 Materials Research Society*
38. C. Yu, Y.S. Kim, D. Kim, J.C. Grunlan, Nano Lett. **8**, 4428 (2008)

39. J. Cibert, P.M. Petroff, G.J. Dolan, S.J. Pearton, A.C. Gossard, J.H. English, Appl. Phys. Lett. **49**, 1275 (1988)
40. P.M. Petroff, A.C. Gossard, W. Wiegmann, Appl. Phys. Lett. **45**, 620 (1984)
41. J.M. Gaines, P.M. Petroff, H. Kroemer, R.J. Simes, R.S. Geels, J.H. English, J. Vac. Sci. Technol. B **6**, 1378 (1988)
42. T. Fukui, H. Saito, Appl. Phys. Lett. **50**, 824 (1987)
43. H. Sakaki, Jpn. J. Appl. Phys. **19**, L735 (1980)
44. P.M. Petroff, A.C. Gossard, R.A. Logan, W. Wiegmann, Appl. Phys. Lett. **41**, 635 (1982)
45. S. Bhattacharya, R. Sarkar, D. De, S. Mukherjee, S. Pahari, A. Saha, S. Roy, N.C. Paul, S. Ghosh, K.P. Ghatak, J. Comput. Theor. Nanosci. **6**, 112 (2009)
46. K.P. Ghatak, S. Bhattacharya, D. De, R. Sarkar, S. Pahari, A. Dey, A.K. Dasgupta, S.N. Biswas, J. Comput. Theor. Nanosci. **5**, 1345 (2008)
47. K.P. Ghatak, S. Dutta, D.K. Basu, B. Nag, Nuovo Cimento della Societa Italiana di Fisica D – Condensed matter, atomic, molecular and chemical physics, biophysics **20**, 227 (1998)
48. B. Nag, K.P. Ghatak, J. Phys. Chem. Solids **59**, 713 (1998)
49. K.P. Ghatak, B. Nag, D. Bhattacharyya, J. Low Temp. Phys. **101**, 983 (1995)
50. K.P. Ghatak, D. Bhattacharyya, Phys. Lett. A **184**, 366 (1994)
51. K.P. Ghatak, S.N. Biswas, Solid-State Electron. **37**, 1437 (1994)
52. K.P. Ghatak, S.N. Biswas, Mol. Cryst. Liq. Cryst. Sci. Technol. Sect. B Nonlinear Opt. **4**, 39 (1993)
53. K.P. Ghatak, B. Mitra, Int. J. Electron. **72**, 541 (1992)
54. K.P. Ghatak, B. De, M. Mondal, Phys. Stat. Sol. (b) **105**, K53 (1991)
55. K.P. Ghatak, B. Mitra, M. Mondal, Ann. Phys. **48**, 283 (1993)
56. D. Bimberg, M. Grundmann, N.N. Ledentsov, *Quantum Dot Heterostructures* (Wiley, 1999)
57. G. Konstantatos, I. Howard, A. Fischer, S. Howland, J. Clifford, E. Klem, L. Levina, E.H. Sargent, Nature **442**, 180 (2006)
58. J.K. Jaiswal, H. Mattoussi, J.M. Mauro, S.M. Simon, Nat. Biotechnol. **21**, 47 (2003)
59. A. Watson, X. Wu, M. Bruchez, Biotechniques **34**, 296 (2003)
60. J. Nakanishi, Y. Kikuchi, T. Takarada, H. Nakayama, K. Yamaguchi, M. Maeda, J. Am. Chem. Soc. **126**, 16314 (2004)
61. X. Michalet, F.F. Pinaud, L.A. Bentolila, J.M. Tsay, S. Doose, J.J. Li, G. Sundaresan, A.M. Wu, S.S. Gambhir, S. Weiss, Science **307**, 538 (2005)
62. W.G.J.H.M. van Sark, P.L.T.M. Frederix, D.J. Van den Heuvel, H.C.G.A. Bol, J.N.J. van Lingen, C. de Mello Donegá, A. Meijerink, J. Phys. Chem. B **105**, 8281 (2001)
63. E.J. Sánchez, L. Novotny, X.S. Xie, Phys. Rev. Lett. **82**, 4014 **(1999)**
64. B. Bailey, D. L. Farkas, D.L. Taylor, F. Lanni, Nature **366**, 44 (1993)
65. L.V. Asryan, R.A. Suris in *Selected Topics in Electronics and Systems*, 25, ed. by E. Borovitskaya, M.S. Shur, (World Scientific, Singapore, 2002)
66. L.V. Asryan, R.A. Suris, Int. J. High Speed Electron. Syst., Special Issue on "Quantum Dot Heterostructures - Fabrication, Application, Theory," **12**(1), 111 (2002)
67. L.V. Asryan, S. Luryi, in *Future Trends in Microelectronics: The Nano Millennium*, ed. by S. Luryi, J.M. Xu, A. Zaslavsky (Wiley Interscience, New York, 2002), p. 219
68. R.A. Freitas Jr., J. Comput. Theor. Nanosci. **2**, 1 (2005)
69. A. Ferreira, C. Mavroidis, IEEE Robot. Autom. Mag. **13**, 78 (2006)
70. A. Dubey, G. Sharma, C. Mavroidis, S.M. Tomassone, K. Nikitczuk, M.L. Yarmush, J. Comput. Theor. Nanosci. **1**, 18 (2004)
71. C. Mavroidis, A. Dubey, M.L. Yarmush, Annu. Rev. Biomed. Eng. **6**, 363 (2004)
72. Y. Liu, J. A. Starzyk, Z. Zhu, IEEE Trans. Neural Netw. (2008)
73. J.A. Starzyk, H. He, IEEE Trans. Neural Netw., **18**(2), 344 (2007)
74. J.A. Starzyk, H. He, IEEE Trans. Circuits Syst. II **54**(2), 176 (2007)
75. E.S. Hasaneen, E. Heller, R. Bansal, W. Huang, F. Jain, Solid State Electron. **48**, 2055 (2004)
76. T. Kawazoe, S. Tanaka, M. Ohtsu, J. Nanophoton. **2**, 029502 (2008)
77. H.J. Krenner, S. Stufler, M. Sabathil, E.C Clark, P. Ester, M. Bichler, G. Abstreiter, J.J Finley, A. Zrenner, New J. Phys. **7**, 184 (2005)
78. A. E. Zhukov, A. R. Kovsh, Quantum Electron. **38**, 409 (2008)

79. M Sugawara, T Akiyama, N Hatori, Y Nakata, H Ebe, H Ishikawa, Meas. Sci. Technol. **13**, 1683 (2002)
80. M. van der Poel, D. Birkedal, J. Hvam, M. Laemmlin, D. Bimberg, Conference on Lasers and Electro-Optics (CLEO) **1**, 16 (2004)
81. J.M. Costa-Fernandez, Anal. Bioanal. Chem. **384**, 37 (2006)
82. H.S. Djie, C.E. Dimas, D.N. Wang, B.S. Ooi, J.C.M. Hwang, G.T. Dang, W.H. Chang, IEEE Sens. J. **7**, 251 (2007)
83. X.-X. Zhu, Y.-C. Cao, X. Jin, J. Yang, X.-F. Hua, H.-Q. Wang, B. Liu, Z. Wang, J.-H. Wang, L. Yang, Y.-D. Zhao, Nanotechnology **19**, 025708 (2008)
84. X. Gao, W.C.W. Chan, S. Nie, J. Biomed. Opt. **7**, 532 (2002)
85. X. Michalet, F.F. Pinaud, L.A. Bentolila, J.M. Tsay, S. Doose, J.J. Li, G. Sundaresan, A.M. Wu, S.S. Gambhir, S. Weiss, Science **307**, 538 (2005)
86. J.K. Jaiswal, E.R. Goldman, H. Mattoussi, S.M. Simon, Nat. Methods **1**, 73 (2004)
87. H. Matsueda, Int. J. Circ. Theor. Appl. **31**, 23 (2003)
88. X. Hu, S. Das Sarma, Phys. Stat. Sol. (b) **238**, 360 (2003)
89. G.-L. Chen, D.M.T. Kuo, W.-T. Lai, P.-W. Li, Nanotechnology **18**, 475402 (2007)
90. A.G. Pogosov, M.V. Budantsev, A.A. Shevyrin, A.E. Plotnikov, A.K. Bakarov, A.I. Toropov, JETP Lett. **87**, 150 (2008)
91. K.W. Johnston, A.G. Pattantyus-Abraham, J.P. Clifford, S.H. Myrskog, D.D. MacNeil, L. Levina, E.H. Sargent, Appl. Phys. Lett. **92**, 151115 (2008)
92. K.S. Leschkies, R. Divakar, J. Basu, E.E. Pommer, J.E. Boercker, C.B. Carter, U.R. Kortshagen, D.J. Norris, E.S. Aydil, Nano Lett. **7**, 1793 (2007)
93. I.-S. Liu, H.-H. Lo, C.-T. Chien, Y.-Y. Lin, C.-W. Chen, Y.-F. Chen, W.-F. Su, S.-C. Liou, J. Mater. Chem. **18**, 675 (2008)
94. N. Hitoshi, Y. Sugimoto, K. Nanamoto, N. Ikeda, Y. Tanaka, Y. Nakamura, S. Ohkouchi, Y. Watanabe, K. Inoue, H. Ishikawa, K. Asakawa, Opt. Express **12**, 6606 (2004)
95. N. Yamamoto, T. Matsuno, H. Takai, N. Ohtani, Jpn. J. Appl. Phys. **44**, 4749 (2005)
96. T. Yamada, Y. Kinoshita, S. Kasai, H. Hasegawa, Y. Amemiya, Jpn. J. Appl. Phys. **40**, 4485 (2001)
97. K. Asakawa, Y. Sugimoto, Y. Watanabe, N. Ozaki, A. Mizutani, Y. Takata, Y. Kitagawa, H. Ishikawa, N. Ikeda, K. Awazu, X. Wang, A. Watanabe, S. Nakamura, S. Ohkouchi, K. Inoue, M. Kristensen, O. Sigmund, P.I. Borel, R. Baets, New J. Phys. **8**, 208 (2006)
98. A.R. Clapp, I.L. Medintz, B.R. Fisher, G.P. Anderson, H. Mattoussi, J. Am. Chem. Soc. **127**, 1242 (2005)
99. L. Shi, B. Hernandez, M. Selke, J. Am. Chem. Soc. **128**, 6278 (2006)
100. C. Wu, J. Zheng, C. Huang, J. Lai, S. Li, C. Chen, Y. Zhao, Angew. Chem. Int. Ed. **46**, 5393 (2007)
101. J.L. Shay, J.W. Wernick, *Ternary Chalcopyrite Semiconductors-Growth, Electronic Properties and Applications* (Pergamon Press, London, 1975)
102. J.E. Rowe, J.L. Shay, Phys. Rev. B **3**, 451 (1973)
103. H. Kildal, Phys. Rev. B **10**, 5082 (1974)
104. J. Bodnar, in *Proceeding of the International Conference of the Physics of Narrow-gap Semiconductors* (Polish Science Publishers, Warsaw, 1978)
105. G.P. Chuiko, N.N. Chuiko, Sov. Phys. Semiconduct. **15**, 739 (1981)
106. K.P. Ghatak and S.N. Biswas, Proc. SPIE **1484,** 149 (1991)
107. A. Rogalski, J. Alloys Compd. **371**, 53 (2004)
108. A. Baumgartner, A. Chaggar, A. Patanè, L. Eaves, M. Henini, Appl. Phys. Lett. **92**, 091121 (2008)
109. J. Devenson, R. Teissier, O. Cathabard, A. N. Baranov, Proc. SPIE **6909**, 69090U (2008)
110. B. S. Passmore, J. Wu, M. O. Manasreh, G. J. Salamo, Appl. Phys. Lett. **91**, 233508 (2007)
111. M. Mikhailova, N. Stoyanov, I. Andreev, B. Zhurtanov, S. Kizhaev, E. Kunitsyna, K. Salikhov, Y. Yakovlev, Proc. SPIE **6585**, 658526 (2007)
112. W. Kruppa, J.B. Boos, B.R. Bennett, N.A. Papanicolaou, D. Park, R. Bass, Electron. Lett. **42**, 688 (2006)

113. E.O. Kane, J. Phys. Chem. Sol. **1**, 249 (1957)
114. J.A. Zapien, Y.K. Liu, Y.Y. Shan, H. Tang, C.S. Lee, S.T. Lee, Appl. Phys. Lett. **90**, 213114 (2007)
115. R.M. Park, Proc. SPIE **2524**, 142 (1995)
116. S.-G. Hur, E.-T. Kim, J.-H. Lee, G.-H. Kim, S.-G. Yoon, Electrochem. Solid-State Lett. **11**, H176 (2008)
117. H. Kroemer, Rev. Mod. Phys. **73**, 783 (2001)
118. T. Nguyen Duy, J. Meslage, G. Pichard, J. Cryst. Growth **72**, 490 (1985)
119. T. Aramoto, F. Adurodija, Y. Nishiyama, T. Arita, A. Hanafusa, K. Omura, A. Morita, Solar Energy Mat. Solar Cells **75**, 211 (2003)
120. H.B. Barber, J. Electron. Mater. **25**, 1232 (1996)
121. S. Taniguchi, T.Hino, S. Itoh, K. Nakano, N. Nakayama, A. Ishibashi, M. Ikeda, Electron. Lett. **32**, 552 (1996)
122. J.J. Hopfield, J. Appl. Phys. **32**, 2277 (1961)
123. F. Hatami, V. Lordi, J.S. Harris, H. Kostial, W.T. Masselink, J. Appl. Phys. **97**, 096106 (2005)
124. B.W. Wessels, J. Electrochem. Soc. **122**, 402 (1975)
125. D.W.L. Tolfree, J. Sci. Instrum. **41,** 788 (1964)
126. P.B. Hart, Proc. IEEE **61**, 880 (1973)
127. H. Choi, M. Chang, M. Jo, S.J. Jung, H. Hwang, Electrochem. Solid-State Lett. **11**, H154 (2008)
128. S. Cova, M. Ghioni, A. Lacaita, C. Samori, F. Zappa, Appl. Opt. **35**, 1956 (1996)
129. H.W.H. Lee, B.R. Taylor, S.M. Kauzlarich, *Nonlinear Optics: Materials, Fundamentals, and Applications*, (Technical Digest, 12, 2000)
130. E. Brundermann, U. Heugen, A. Bergner, R. Schiwon, G.W. Schwaab, S. Ebbinghaus, D.R. Chamberlin, E.E. Haller, M. Havenith, *29th International Conference on Infrared and Millimeter Waves and 12th International Conference on Terahertz Electronics,* 283 2004
131. P.K. Weimer, Proc. IEEE **52**, 608 (1964)
132. G. Ribakovs, A.A. Gundjian, IEEE J. Quantum Electron. **QE-14**, 42 (1978)
133. S.K. Dey, J. Vac. Sci. Technol. **10**, 227 (1973)
134. S.J. Lynch, Thin Solid Films **102**, 47 (1983)
135. V.V. Kudzin, V.S. Kulakov, D.R. Pape', S.V. Kulakov, V.V. Molotok, IEEE Ultrasonics Symp., **1**, 749 (1997)
136. M.S. Dresselhaus, J.G. Mavroides, IBM J. Res. Develop. **8**, 262 (1964)
137. P.R. Schroeder, M.S. Dresselhaus, A. Javan, Phys. Rev. Lett.**20**, 1292 (1968)
138. M.S. Dresselhaus in *Proceedings of the Conference on the Physic of Semimetals and Narrow Gap Semiconductors*, ed. by D.L. Carter, R.T. Bate (Pergamon Press, New York,1970)
139. M.S. Dresselhaus, G. Dresselhaus Adv. Phys. **51**, 1 (2002)
140. M.A. Hines, G.D. Scholes, Adv. Mater. **15**, 1844 (2003)
141. C.A. Wang, R.K. Huang, D.A. Shiau, M.K. Connors, P.G. Murphy, P.W. O'Brien, A.C. Anderson, D.M. DePoy, G. Nichols, M.N. Palmisiano, Appl. Phys. Lett **83,** 1286 (2003)
142. C.W. Hitchcock, R.J. Gutmann, J.M. Borrego, I.B. Bhat, G.W. Charache, IEEE Trans. Electron. Devices**46**, 2154 (1999)
143. S. Matsushita, K. Kawai, H. Uchida, IEEE J. Lightwave Technol. **LT-3**, 533 (1985)
144. N.K. Ailawadi, R.C. Alferness, G.D. Bergland, R.A. Thompson, IEEE Mag. Lightwave Telecommun. Syst. **2**, 38 (1991)
145. K. Kubota, J. Noda, O. Mikami, IEEE J. Quantum Electron. **QE-16**, 754 (1980)
146. R.C. Alferness, C.H. Joyner, L.L. Buhl, S.K. Korotky, IEEE J. Quantum Electron. **QE-19**, 1339 (1983)
147. R.G. Walker, IEEE J. Quantum Electron. **27**, 654 (1991)
148. G.C. Young, W.W. Anderson, L.B. Anderson, IEEE Trans. Electron Dev. **24**, 492 (1977)
149. R.L. Gordon, V.I. Neeley, H.R. Curtin, Proc. IEEE **54**, 2014 (1966)
150. P.K. Weimer, Proc. IRE **50**, 1462 (1962)
151. M.J. Lee, S.W. Wright, C.P. Judge, P.Y. Cheung, Display Research Conference, International Conference Record, 211 (1991)

152. A.N. Baranov, T.I. Voronina, N.S. Zimogorova, L.M. Kauskaya, Y.P. Yakoviev, Sov. Phys. Semiconduct. **19**, 1676 (1985)
153. M. Yano, Y. Suzuki, T. Ishii, Y. Matsushima, M. Kimata, Jpn. J. Appl. Phys. **17**, 2091 (1978)
154. F.S. Yuang, Y.K. Su, N.Y. Li, Jpn. J. Appl. Phys. **30**, 207 (1991)
155. F.S. Yuang, Y.K. Su, N.Y. Li, K.J. Gan, J. Appl. Phys. **68**, 6383 (1990)
156. Y.K. Su, S.M. Chen, J. Appl. Phys. **73**, 8349 (1993)
157. S.K. Haywood, A.B. Henriques, N.J. Mason, R.J. Nicholas, P.J. Walker, Semiconduct. Sci. Technol. **3**, 315 (1988)
158. F. Hüe, M. Hÿtch, H. Bender, F. Houdellier, A. Claverie, Phys. Rev. Lett. **100**, 156602 (2008)
159. S. Banerjee, K.A. Shore, C.J. Mitchell, J.L. Sly, M. Missous, IEE Proc. Circ. Dev. Syst. **152**, 497 (2005)
160. M. Razeghi, A. Evans, S. Slivken, J.S. Yu, J.G. Zheng, V.P. Dravid, Proc. SPIE **5840**, 54 (2005)
161. R.A. Stradling, Semiconduct. Sci. Technol.**6**, C52 (1991)
162. R.V. Belosludov, A.A. Farajian, H. Mizuseki, K. Miki, Y. Kawazoe, Phys. Rev. B **75**, 113411 (2007)
163. J. Heremans, C.M. Thrush, Y.-M Lin, S. Cronin, Z. Zhang, M.S. Dresselhaus, J.F. Mansfield, Phys. Rev. B. **61**, 2921 (2000)
164. D. Shoenberg, Proc. Roy. Soc. (London) **170**, 341 (1939)
165. B. Abeles, S. Meiboom, Phys. Rev. **101**, 544 (1956)
166. B. Lax, J.G. Mavroides, H.J. Zieger, R.J. Keyes, Phys. Rev. Lett. **5**, 241 (1960)
167. M. Maltz, M.S. Dresselhaus, Phys. Rev. B **2**, 2877 (1970)
168. M. Cankurtaran, H. Celik, T. Alper, J. Phys. F Metal Phys. **16**, 853 (1986)
169. Y.-H. Kao, Phys. Rev. **129**, 1122 (1963)
170. R.J. Dinger, A.W. Lawson, Phys. Rev. B **3**, 253 (1971)
171. J.F. Koch, J.D. Jensen, Phys. Rev. **184**, 643 (1969)
172. M.H. Cohen, Phys. Rev. **121**, 387 (1961)
173. S. Takaoka, H. Kawamura, K. Murase, S. Takano, Phys. Rev. B **13**, 1428 (1976)
174. J.W. McClure, K.H. Choi, Solid State Comm. **21**, 1015 (1977)
175. G.P. Agrawal, N.K. Dutta, *Semiconductor Lasers*, (Van Nostrand Reinhold, New York, 1993)
176. S. Chatterjee, U. Pal, Opt. Eng. **32**, 2923 (1993)
177. T.K. Chaudhuri, Int. J. Energy Res. **16**, 481 (1992)
178. J.H. Dughaish, Physica B **322**, 205 (2002)
179. C. Wood, Rep. Prog. Phys. **51**, 459 (1988)
180. K.-F. Hsu, S. Loo, F. Guo, W. Chen, J.S. Dyck, C. Uher, T. Hogan, E.K. Polychroniadis, M.G. Kanatzidis, Science **303**, 818 (2004)
181. J. Androulakis, K.F. Hsu, R. Pcionek, H. Kong, C. Uher, J.J. D'Angelo, A. Downey, T. Hogan, M.G. Kanatzidis, Adv. Mater. **18**, 1170 (2006)
182. P.F.P. Poudeu, J. D'Angelo, A.D. Downey, J.L. Short, T.P. Hogan, M.G. Kanatzidis, Angew. Chem. Int. Ed. **45**, 3835 (2006)
183. P.F. Poudeu, J. D'Angelo, H. Kong, A. Downey, J.L. Short, R. Pcionek, T. P. Hogan, C. Uher, M.G. Kanatzidis, J. Am. Chem. Soc. **128**, 14347 (2006)
184. J.R. Sootsman, R.J. Pcionek, H. Kong, C. Uher, M.G. Kanatzidis, Chem. Mater. **18**, 4993 (2006)
185. A.J. Mountvala, G. Abowitz, J. Am. Ceram. Soc. **48**, 651 (1965)
186. E.I. Rogacheva, I.M. Krivulkin, O.N. Nashchekina, A.Yu. Sipatov, V.A. Volobuev, M.S. Dresselhaus, Appl. Phys. Lett. **78**, 3238 (2001)
187. H.S. Lee, B. Cheong, T.S. Lee, K.S. Lee, W.M. Kim, J.W. Lee, S.H. Cho, J.Y. Huh, Appl. Phys. Lett. **85**, 2782 (2004)
188. K. Kishimoto, M. Tsukamoto, T. Koyanagi, J. Appl. Phys. **92**, 5331 (2002)
189. E.I. Rogacheva, O.N. Nashchekina, S.N. Grigorov, M.A. Us, M.S. Dresselhaus, S.B. Cronin, Nanotechnology **14**, 53 (2003)
190. E.I. Rogacheva, O.N. Nashchekina, A.V. Meriuts, S.G. Lyubchenko, M.S. Dresselhaus, G. Dresselhaus, Appl. Phys. Lett. **86**, 063103 (2005)

191. E.I. Rogacheva, S.N. Grigorov, O.N. Nashchekina, T.V. Tavrina, S.G. Lyubchenko, A.Yu. Sipatov, V.V. Volobuev, A.G. Fedorov, M.S. Dresselhaus, Thin Solid Films **493**, 41 (2005)
192. X. Qiu, Y. Lou, A.C.S. Samia, A. Devadoss, J.D. Burgess, S. Dayal, C. Burda, Angew. Chem. Int. Ed. **44**, 5855 (2005)
193. C. Wang, G. Zhang, S. Fan, Y. Li, J. Phys. Chem. Solids **62**, 1957 (2001)
194. B. Poudel, W.Z. Wang, D.Z. Wang, J.Y. Huang, Z.F. Ren, J. Nanosci. Nanotechnol. **6**, 1050 (2006)
195. B. Zhang, J. He, T.M. Tritt, Appl. Phys. Lett. **88**, 043119 (2006)
196. W. Heiss, H. Groiss, E. Kaufmann, G. Hesser, M. Böberl, G. Springholz, F. Schäffler, K. Koike, H. Harada, M. Yano, Appl. Phys. Lett. **88**, 192109 (2006)
197. C. Wang, G. Zhang, S. Fan, Y. Li, J. Phys. Chem. Solids **62**, 1957 (2001)
198. B. Poudel, W.Z. Wang, D.Z. Wang, J.Y. Huang, Z.F. Ren, J. Nanosci. Nanotechnol. **6**, 1050 (2006)
199. B. Zhang, J. He, T.M. Tritt, Appl. Phys. Lett. **88**, 043119 (2006)
200. W. Heiss, H. Groiss, E. Kaufmann, G. Hesser, M. Böberl, G. Springholz, F. Schäffler, K. Koike, H. Harada, M. Yano, Appl. Phys. Lett. **88**, 192109 (2006)
201. B.A. Volkov, L.I. Ryabova, D.R. Khokhlov, Phys. Usp. **45**, 819 (2002), and references therein.
202. J.M. Thornton, J. Mol. Biol. **151**, 261 (1981);.
203. T.E. Creighton, Methods Enzymol. **131**, 83 (1986)
204. T.E. Creighton, Bio Essays **8**, 57 (1988)
205. V.J. Hruby, Life Sci. **31**, 189 (1982)
206. V.J. Hruby, F. Al-Obeidi, W. Kazmierski, Biochem. J. **268**, 249 (1990)
207. R. Wetzel, Trends Biochem. Sci. **12**, 478 (1987)
208. T. Kimura, R. Matsueda, Y. Nakagawa, E.T. Kaiser, Anal. Biochem. **122**, 274 (1982)
209. D. Andreu, F. Albericio, N.A. Sole, M.C. Munson, M. Ferrer, G. Barany, in *Methods in Molecular Biology, 35 Peptide Synthesis Protocols*, ed. by M.W. Pennington, B.M. Dunn (Humana Press Inc., New Jersey, 1994)
210. H.J. Goldsmid, R.W. Douglas, Br. J. Appl. Phys. **5**, 386 (1954)
211. F.D. Rosi, B. Abeles, R.V. Jensen, J. Phys. Chem. Sol. **10**, 191 (1959)
212. T.M. Tritt (ed.), *Semiconductors and Semimetals, Vols. 69, 70 and 71: Recent Trends in Thermoelectric Materials Research I, II and III* (Academic Press, USA, 2000)
213. D.M. Rowe (ed.), *CRC Handbook of Thermoelectrics* (CRC Press, USA, 1995)
214. D.M. Rowe, C. M. Bhandari, *Modern Thermoelectrics* (Reston Publishing Company, Virginia, 1983)
215. D.M. Rowe (ed.), *Thermoelectrics Handbook: Macro to Nano* (CRC Press, USA, 2006)
216. F. Smektala, C. Quemard, V. Couderc, A. Barthélémy, J. Non-Cryst. Sol. **274**, 232 (2000)
217. G. Boudebs, F. Sanchez, J. Troles, F. Smektala, Opt. Comm. **199**, 425 (2001)
218. J.M. Harbold, F.O. Ilday, F.W. Wise, J.S. Sanghera, V.Q. Nguyen, L.B. Shaw, I.D. Aggarwal, Opt. Lett. **27**, 119 (2002)
219. J.M. Harbold, F.O. Ilday, F.W. Wise, B. Aitken, IEEE Photon. Technol. Lett. **14**, 822 (2002)
220. V.T. Plaksiy, V.M. Svetlichniy, *Symposium on Physics and Engineering of Millimeter and Sub-Millimeter Waves, 2001. The Fourth International Kharkov* **1**, 331 (2001)
221. Z. Helin, D.M. Rowe, S. G. K. Williams, 2001. Proceedings of XX International Conference on Thermoelectrics, 314 (2001)
222. J. Bodnar, *Physics of Narrow-gap Semiconductors,Proceedings of International Conference, Warozawa*, Ed. by J. Rautuszkiewicz, M. Gorska, E. Kaczmarek, (PWN-Polish Scientific Publisher, Warszwa, Polland, 1978)
223. K.P. Ghatak, S. Bhattacharya, D. De, *Einstein Relation in Compound Semiconductors and Their Nanostructures*, Springer Series in Materials Science, vol. **116** (Springer-Verlag, Germany, 2008)
224. B.R. Nag, *Electron Transport in Compound Semiconductors* (Springer-Verlag, Germany, 1980)
225. E.O. Kane, in *Semiconductors and Semi metals*, Vol. 1, ed. by R.K. Willardson, A.C. Beer (Academic Press, New York, 1966)

226. G.E. Stillman, C.M. Wolfe, J.O. Dimmock in *Semiconductors and Semi metals*, ed. by R.K. Willardon, A.C. Beer **12**, (Academic Press, IV,V, USA, 1977)
227. D.J. Newson, A. Kurobe, Semiconduct. Sci. Technol. **3**, 786 (1988)
228. U. Rossler, Solid State Comm. **49**, 943 (1984)
229. E.D. Palik, G.S. Picus, S. Teither, R.E. Wallis, Phys. Rev. **122**, 475 (1961)
230. E.J. Johnson, D.H. Dickey, Phys. Rev. **1**, 2676 (1970)
231. V.G. Agafonov, P.M. Valov, B.S. Ryvkin, I.D. Yarashetskin, Sov. Phys. Semiconduct. **12**, 1182 (1978)
232. G. J. Rees, *Physics of Compounds, Proceedings of the 13th International Conference.* ed. by F.G. Fumi (North Holland Company, The Netherlands, 1976), p. 116
233. M. Cardona, W. Paul, H. Brooks Helv, Acta Phys. **33**, 329 (1960)
234. A.F.Gibson in *Proceeding of International school of physics "ENRICO FERMI"* course XIII, ed. by R.A Smith, (Academic Press, 1963), p. 171
235. C.C. Wang, N.W. Ressler, Phys. Rev. **2**, 1827 (1970)
236. J. Bouat, J.C. Thuillier, Surface Sci. **73**, 528 (1978)
237. M.V. Ortenberg, K.J. Button, Phys. Rev. B **16**, 2618 (1977)
238. H. Ushio, T. Dau, Y. Uemura, J. Phys. Soc. Jpn. **33**, 1551 (1972)
239. P.R. Emtage, Phys. Rev. **138**, A246 (1965)
240. V.I. Ivanov-Omskii, A.Sh. Mekhtisev, S.A. Rustambekova, E.N. Ukraintsev, Phys. Stat. Sol. (b) **119**, 159 (1983)
241. Y. Yamada, J. Phys. Soc. Jpn. **35**, 1600 (1973)
242. M. Singh, P.R. Wallace, S.D. Jog, E. Arushanov, J. Phys. Chem. Sol. **45**, 409 (1984)
243. D.G. Seiler, W.M. Beeker, K.M. Roth, Phys. Rev. **1**, 764 (1970)
244. P.C. Mathur, S. Jain, Phys. Rev. **19**, 1359 (1979)
245. H.I. Zhang, Phys. Rev. B **1**, 3450 (1970)
246. J.O. Dimmock in *The Physics of Semimetals and Narrowgap Semiconductors* ed. by D.L. Carter, R.T. Bates (Pergamon Press, Oxford, 1971)
247. E. Bangert, P. Kastner, Phys. Stat. Sol. (b) **61**, 503 (1974)
248. G.M.T. Foley, P.N. Langenberg, Phys. Rev. B **15**, 4850 (1977)
249. W.E. Spicer, G.J. Lapeyre, Phys. Rev. **139**, A565 (1965)
250. L.A. Vassilev, Phys. Stat. Sol. (b) **121**, 203 (1984)
251. G.P. Chuiko, Sov. Phys. Semiconduct. **19**, 1381 (1985)
252. M. Stordeur, W. Kuhnberger, Phys. Stat. Sol. (b) **69**, 377 (1975)
253. H. Köhler, Phys. Stat. Sol. (b), **74**, 591 (1976)
254. J.B. Ketterson, Phys. Rev. **129**, 18 (1963)
255. S. Adachi, J. Appl. Phys. **58**, R1 (1985)
256. S. Adachi, *GaAs and Related Materials: Bulk Semiconductors and Superlattice Properties*, (World Scientific, USA, 1994)
257. G.L. Hansen, J.L. Schmit, T.N. Casselman, J. Appl. Phys. **63**, 7079 (1982)
258. S. Adachi, J. Appl. Phys. **53**, 8775 (1982)
259. M. Meltz, M.S. Dresselhaus, Phys. Rev. **2B**, 2877 (1970)
260. O. Madelung, *Semiconductors: Data Handbook*, 3rd edn. (Springer-Verlag, Germany, 2003)
261. S. Adachi, *Properties of Group-IV, III–V and II–VI Semiconductors* (John Wiley and Sons, USA, 2005)
262. M. Krieehbaum, P. Kocevar, H. Pascher, G. Bauer, IEEE QE **24**, 1727 (1988)
263. J.R. Lowney, S.D. Senturia, J. Appl. Phys. **47**, 1771 (1976)
264. D.G. Seiler, B.D. Bajaj, A.E. Stephens, Phys. Rev. B **16**, 2822 (1977)
265. A.V. Germaneko, G.M. Minkov, Phys. Stat. Sol. (b) **184**, 9 (1994)
266. G.L. Bir, G.E. Pikus, *Symmetry and Strain-Induced effects in Semiconductors* (Nauka, Russia, 1972) (in Russian)
267. M. Mondal, K.P. Ghatak, Phys. Stat. Sol. (b) **135**, K21 (1986)
268. C.C. Wu, C.J. Lin, J. Low. Temp. Phys. **57**, 469 (1984)
269. N.B. Brandt, V.N. Davydov, V.A. Kulbachinskii, O.M. Nikitina, Sov. Phys. Sol. Stat. **29**, 1014 (1987)
270. S. Takaoka, K. Murase, Phys. Rev. B **20**, 2823 (1979)

271. I. Kang, F.W. Wise, J. Opt. Soc. Am. B **14**, 1632 (1997)
272. M.S. Lundstrom, J. Guo, *Nanoscale Transistors, Device Physics, Modeling and Simulation* (Springer, 2006, USA)
273. R. Saito, G. Dresselhaus, M.S. Dresselhaus, *Physical Properties of Carbon Nanotubes* (Imperial College Press, London, **1998**)
274. X. Yang, J. Ni, Phys. Rev. B **72**, 195426 (2005)
275. W. Mintmire, C.T. White, Phys. Rev. Lett. **81**, 2506 (1998)
276. A.I. Yekimov, A.A. Onushchenko, A.G. Plyukhin, Al.L. Efros, J. Exp. Theor. Phys. **88**, 1490 (1985)
277. B.J. Roman, A.W. Ewald, Phys. Rev. B **5**, 3914 (1972)
278. R.W. Cunningham, Phys. Rev. **167**, 761 (1968)
279. K.P. Ghatak, S. Karmakar, D. De, S. Pahari, S.K. Chakraborty, S.K. Biswas, S. Chowdhury, J. Comput. Theor. Nanoscience **3**, 153 (2006)
280. D.R. Lovett, *Semimetals and Narrow-Bandgap Semiconductors* (Pion Limited, London, 1977)

Chapter 2
Thermoelectric Power in Ultrathin Films and Quantum Wires Under Large Magnetic Field

2.1 Introduction

The asymmetry of the wave-vector space of the charge carriers in semiconductors indicates the fact that in ultrathin films (UFs), the restriction of the motion of the carriers in the direction normal to the film (say, the z−direction) may be viewed as carrier confinement in an infinitely deep 1D rectangular potential well, leading to quantization [known as quantum size effect (QSE)] of the wave vector of the carrier along the direction of the potential well, allowing 2D carrier transport parallel to the surface of the film epitomizing new physical features not exhibited in bulk semiconductors [1–4]. The low-dimensional heterostructures based on various materials are widely explored because of the enhancement of carrier mobility [5]. These properties make such structures befitting for applications in quantum well lasers [6], heterojunction FETs [7, 8], high-speed digital networks [9–12], high-frequency microwave circuits [13], optical modulators [14], optical switching systems [15], and other devices.

Therefore, it can be readily fathomed that the constant energy 3D wave-vector space of bulk semiconductors becomes 2D wave-vector surface in UFs or one dimensional in quantum wires (QWs) is owing to two-dimensional quantization. Thus, the consequence of the concept of reduction of symmetry of the wave-vector space as noted already can unlock the physics of low-dimensional structures. In QWs, the motions of the carriers along any two orthogonal directions are being prohibited and the carriers are forced to move along the rest free direction in the wave-vector space [16–19]. With the onset of modern fabricational techniques, such one-dimensional structures have been experimentally realized and have profusely resulted in plethora of important applications in the realm of nanoscience in the quantum regime. They have spawned much interest and excitement in the analysis of nanostructured devices for investigating their electronic, optical, and allied properties [20–27]. Examples of such new applications are based on the different transport properties of ballistic charge carriers which include quantum resistors [28, 29], resonant tunneling diodes and band filters [30, 31], quantum sensors [32–34], quantum switches [35], quantum logic gates [36, 37], quantum transistors and subtuners [38–40], and other nanostructured devices. In this chapter, the TPSM is investigated in UFs and QWs of nonlinear optical, Kane type III–V, II–VI, bismuth, IV–VI, stressed materials, and carbon nanotubes (CNs) from Sects. 2.2.1 to 2.2.7, respectively.

With the discovery of CNs in 1991 by Iijima [41], the CNs have been recognized as fascinating materials with nanometer dimensions uncovering new phenomena in the sphere of low-dimensional science and technology. The significant physical properties of these quantum materials make them ideal candidates to reveal new phenomena in nanoelectronics. The CNs find wide applications in conductive [42, 43] and high-strength composites [44], chemical sensors [45], field emission displays [46, 47], hydrogen storage media [48, 49], nanotweezers [50], nanogears [51], nanocantilever devices [52], nanomotors [53, 54], and nanoelectronic devices [55,56]. Single walled carbon nanotubes (SWCNs) emerge to be excellent materials for single molecule electronics [57–61] such as nanotube-based diodes [62,63], single electron transistors [64, 65], random access memory cells [66], logic circuits [67], gigahertz oscillators [68–73], data storage nanodevices [74–79], nanorelay [80–85], and other low-dimensional devices. The CNs can be bespoke into a metal or a semiconductor based on the diameter and the chiral index numbers (m, n), where the integers m and n denote the number of unit vectors along two directions in the honeycomb crystal lattice of graphene [86]. For armchair and zigzag nanotubes, the chiral indices are given as $m = n$ and $m = 0$, respectively [86]. Another class of CN called as chiral CN has distinct integers m and n. Besides, a CN can be a metallic if $m - n = 3q$, where $q = 1, 2, 3, \ldots$; otherwise, it is a semiconductor. Metallic SWCNs have received substantial attention as potential substitutions for traditional interconnect materials such as Cu due to their excellent inherent electrical and thermal properties. Since the carriers are confined, in a metallic SWCN, the inclusion of the subband energy owing to Born–Von Karman (BVK) boundary conditions [87] for their unique band structure becomes prominent. The quantization of the motion of the carriers in such structures leads to the discontinuity in the DOS function due to van Hove singularity [88] (VHS) of the wave vectors. In Sect. 2.2.7 we have explored the TPSM in CNs. Section 2.3 contains the results and discussion pertaining to this chapter. Section 2.4 contains the open research problems pertinent to this chapter.

2.2 Theoretical Background

2.2.1 *Magnetothermopower in Quantum-Confined Nonlinear Optical Materials*

The 2D electron energy spectrum in QWs in UFs of nonlinear optical materials in the presence of size quantization along x-direction can be expressed following (1.2) as

$$\gamma(E) = f_1(E)\left(\frac{\pi n_x}{d_x}\right)^2 + f_1(E)k_y^2 + f_2(E)k_z^2. \qquad (2.1)$$

It appears that the formulation of TPSM requires an expression of the two-dimensional density-of-states function per subband ($N_{2D}(E)$) which can, in turn, be written in this case using (2.1) as

$$N_{2D}(E) = \frac{g_v}{(2\pi)} \frac{\partial}{\partial E} \{\phi_1(E, n_x)\}, \tag{2.2}$$

where

$$\phi_1(E, n_x) \equiv [f_1(E) f_2(E)]^{-1/2} \left[\gamma(E) - f_1(E) \left(\frac{\pi n_x}{d_x}\right)^2\right].$$

The evaluation of TPSM in this case requires the expression of 2D electron statistics per unit area, which can be expressed in this case using (2.2) as

$$n_0 = \frac{g_v}{2\pi} \sum_{n_x=1}^{n_{x\max}} [\phi_1(E_{F2D}, n_x) + \phi_2(E_{F2D}, n_x)], \tag{2.3}$$

where

$$\phi_2(E_{F2D}, n_x) \equiv \sum_{r=1}^{s_0} Z_{r2}[\phi_1(E_{F2D}, n_x)],$$

$$Z_{r,Y} \equiv \left[2(k_B T)^{2r} \left(1 - 2^{1-2r}\right) \zeta(2r) \frac{\partial^{2r}}{\partial E_{FYD}^{2r}}\right],$$

$Y = 2, \zeta(2r)$ is the Zeta function of order $2r$, and E_{F2D} is the Fermi energy in the presence of size quantization as measured from the edge of the conduction band in the vertically upward direction in the absence of any quantization.

Therefore, using (1.13) and (2.3), the TPSM in this case is given by

$$G_0 = \left(\pi^2 k_B^2 T / (3e)\right) \left[\sum_{n_x=1}^{n_{x\max}} [\phi_1(E_{F2D}, n_x) + \phi_2(E_{F2D}, n_x)]\right]^{-1} \left[\sum_{n_x=1}^{n_{x\max}} [\{\phi_1(E_{F2D}, n_x)\}' + \{\phi_2(E_{F2D}, n_x)\}']\right], \tag{2.4}$$

where the primes indicate the differentiation of the differentiable functions with respect to the appropriate Fermi energy.

In this context, the TPSM from QWs of nonlinear optical materials can be formulated in the following way.

For electron motion along x-direction, the 1D electron dispersion law in this case can be written following (1.2) as

$$\gamma(E) = f_1(E)k_x^2 + f_1(E)\left(\pi n_y/d_y\right)^2 + f_2(E)\left(\pi n_z/d_z\right)^2 . \tag{2.5}$$

The subband energy (E') is defined through the equation

$$\gamma(E') = f_1(E')\left(\pi n_y/d_y\right)^2 + f_2(E')\left(\pi n_z/d_z\right)^2 . \tag{2.6}$$

The 1D DOS function per subband is given by

$$N_{1D}(E) = \frac{2g_v}{\pi}\frac{\partial k_x}{\partial E}. \tag{2.7}$$

Thus, it appears that the evaluation of TPSM requires an expression of 1D carrier statistics per unit length which can, in turn, be written in this case combining (2.5), (2.7), and the Fermi–Dirac occupation probability factor as

$$n_0 = \left(\frac{2g_v}{\pi}\right)\sum_{n_y=1}^{n_{y\max}}\sum_{n_z=1}^{n_{z\max}}\left[t_1\left(E_{F1D}, n_y, n_z\right) + t_2\left(E_{F1D}, n_y, n_z\right)\right], \tag{2.8}$$

where

$$t_1\left(E_{F1D}, n_y, n_z\right) \equiv \left[\gamma\left(E_{F1D}\right) - f_1\left(E_{F1D}\right)\left(\pi n_y/d_y\right)^2 - f_2\left(E_{F1D}\right)\left(\pi n_z/d_z\right)^2\right]^{1/2}\left[f_1\left(E_{F1D}\right)\right]^{-1/2},$$

E_{F1D} is the Fermi energy in the presence of two-dimensional quantization as measured from the edge of the conduction band in the vertically upward direction in the absence of any quantization,
$t_2\left(E_{F1D}, n_y, n_z\right) \equiv \sum_{r=1}^{s_0} Z_{r,Y}\left[t_1\left(E_{F1D}, n_y, n_z\right)\right]$ and $Y = 1$.

Therefore, using (1.13) and (2.8), the TPSM in this case is given by

$$G_0 = \left(\pi^2 k_B^2 T/(3e)\right)\left[\sum_{n_y=1}^{n_{y\max}}\sum_{n_z=1}^{n_{z\max}}\left[t_1\left(E_{F1D}, n_y, n_z\right) + t_2\left(E_{F1D}, n_y, n_z\right)\right]\right]^{-1} \times\left[\sum_{n_y=1}^{n_{y\max}}\sum_{n_z=1}^{n_{z\max}}\left[\left\{t_1\left(E_{F1D}, n_y, n_z\right)\right\}' + \left\{t_2\left(E_{F1D}, n_y, n_z\right)\right\}'\right]\right]. \tag{2.9}$$

2.2.2 Magnetothermopower in Quantum-Confined Kane Type III–V Materials

2.2.2.1 The Three Band Model of Kane

The dispersion relation of the 2D electrons in this case is given by

$$\frac{\hbar^2 k_y^2}{2m^*} + \frac{\hbar^2 k_z^2}{2m^*} + \frac{\hbar^2}{2m^*}(\pi n_x/d_x)^2 = I(E). \tag{2.10}$$

The subband energy (E_{n_x}) can be written as

$$I(E_{n_x}) = \frac{\hbar^2}{2m^*}(\pi n_x/d_x)^2 . \tag{2.11}$$

The electron statistics per unit area in this case can be expressed as

$$n_0 = \frac{m^* g_v}{\pi\hbar^2}\sum_{n_x=1}^{n_{x\max}}[T_3(E_{F2D}, n_x) + T_4(E_{F2D}, n_x)], \tag{2.12}$$

where

$$T_3(E_{F2D}, n_x) \equiv \left[I(E_{F2D}) - \frac{\hbar^2}{2m^*}\left(\frac{\pi n_x}{d_x}\right)^2\right],$$

$$T_4(E_{F2D}, n_x) \equiv \sum_{r=1}^{s_v} Z_{r,Y}[T_3(E_{F2D}, n_x)] \text{ and } Y = 2.$$

The TPSM in this case assumes the form

$$G_0 = \left(\pi^2 k_B^2 T/(3e)\right)\left[\sum_{n_x=1}^{n_{x\max}}[T_3(E_{F2D}, n_x) + T_4(E_{F2D}, n_x)]\right]^{-1} \times\left[\sum_{n_x=1}^{n_{x\max}}\left[\{T_3(E_{F2D}, n_x)\}' + \{T_4(E_{F2D}, n_x)\}'\right]\right]. \tag{2.13}$$

The one-dimensional electron dispersion law is given by

$$\frac{\hbar^2 k_x^2}{2m^*} + G_2(n_y, n_z) = I(E), \tag{2.14}$$

where

$$G_2(n_y, n_z) \equiv \left(\hbar^2\pi^2/2m^*\right)\left[(n_y/d_y)^2 + (n_z/d_z)^2\right].$$

The subband energy E' is defined through the equation

$$G_2\left(n_y,n_z\right) = I(E'). \tag{2.15}$$

The electron statistics per unit length in this case is given by

$$n_0 = \frac{2g_v\sqrt{2m^*}}{\pi\hbar}\sum_{n_y=1}^{n_{y\max}}\sum_{n_z=1}^{n_{z\max}}\left[T_5(E_{F1D},n_y,n_z) + T_6(E_{F1D},n_y,n_z)\right], \tag{2.16}$$

where $T_5\left(E_{F1D},n_y,n_z\right) \equiv \left[I(E_{F1D}) - G_2\left(n_y,n_z\right)\right]^{1/2}$, $T_6\left(E_{F1D},n_y,n_z\right) \equiv \sum_{r=1}^{S_0} Z_{r,Y}\left[T_5\left(E_{F1D},n_y,n_z\right)\right]$, and $Y = 1$.

Combining (2.16) and (1.13), the TPSM in this case assumes the form

$$G_0 = \left(\pi^2k_B^2T/\left(3e\right)\right)\left[\sum_{n_y=1}^{n_{y\max}}\sum_{n_z=1}^{n_{z\max}}\left[T_5(E_{F1D},n_y,n_z) + T_6(E_{F1D},n_y,n_z)\right]\right]^{-1}$$
$$\times\left[\sum_{n_y=1}^{n_{y\max}}\sum_{n_z=1}^{n_{z\max}}\left[\left\{T_5(E_{F1D},n_y,n_z)\right\}' + \left\{T_6(E_{F1D},n_y,n_z)\right\}'\right]\right]. \tag{2.17}$$

2.2.2.2 The Two Band Model of Kane

For UFs of III–V materials whose bulk energy band structures obey the two band model of Kane, the 2D electron dispersion law and the density-of-states per subband assume the forms

$$E(1+\alpha E) = \frac{\hbar^2k_y^2}{2m^*} + \frac{\hbar^2k_z^2}{2m^*} + \frac{\hbar^2}{2m^*}\left(\frac{n_x\pi}{d_x}\right)^2, \tag{2.18}$$

$$N_{2D}(E) = \frac{m^*g_v}{\pi\hbar^2}\left(1+2\alpha E\right). \tag{2.19}$$

Combining (2.19) with the Fermi–Dirac occupation probability factor, the electron statistics per unit area in this case can be written as

$$n_0 = \frac{m^*g_vk_BT}{\pi\hbar^2}\sum_{n_x=1}^{n_{x\max}}\left[(1+2\alpha E_{n_x})F_0(\eta_n) + 2\alpha k_BTF_1(\eta_n)\right], \tag{2.20}$$

where $\eta_n \equiv ((E_{F2D} - E_{n_x})/k_BT)$ and E_{n_x} is given by

$$E_{n_x} = [2\alpha]^{-1}\left[-1 + \sqrt{1 + (2\alpha\hbar^2/m^*)\left(\pi n_x/d_x\right)^2}\right]. \tag{2.21}$$

The symbols $F_0(\eta_{n_z})$ and $F_1(\eta_{n_z})$ are the special cases of the one parameter Fermi–Dirac integral of order j which assumes the form [89],

$$F_j(\eta) = \left(\frac{1}{\Gamma(j+1)}\right)\int_0^{\infty}\frac{x^j\,dx}{1+\exp(x-\eta)},\ j > -1 \tag{2.22}$$

or for all j, analytically continued as a complex contour integral around the negative x-axis

$$F_j(\eta) = \left(\frac{\Gamma(-j)}{2\pi\sqrt{-1}}\right)\int_{-\infty}^{+0}\frac{x^j\,dx}{1+\exp(-x-\eta)}, \tag{2.23}$$

where η is the dimensionless x independent variable, $\Gamma(j+1) = j\Gamma(j)$, $\Gamma(1/2) = \sqrt{\pi}$, and $\Gamma(0) = 1$.

Combining (2.20) with (1.13), the TPSM in this case assumes the form

$$G_0 = \left(\pi^2 k_B/(3e)\right)\left[\sum_{n_x=1}^{n_{x\max}}\left[(1+2\alpha E_{n_x})F_0(\eta_n) + 2\alpha k_B T F_1(\eta_n)\right]\right]^{-1} \times\left[\sum_{n_x=1}^{n_{x\max}}\left[(1+2\alpha E_{n_x})F_{-1}(\eta_n) + 2\alpha k_B T F_0(\eta_n)\right]\right], \tag{2.24}$$

where we have used the well-known formula [90]

$$\frac{\partial}{\partial\eta}F_j(\eta) = F_{j-1}(\eta). \tag{2.25}$$

The expression of 1D dispersion relation for QWs of III–V materials whose energy band structures are defined by the two band model of Kane assumes the form

$$E(1+\alpha E) = \frac{\hbar^2 k_x^2}{2m^*} + G_2(n_y, n_z). \tag{2.26}$$

In this case, the quantized energy E' is given by

$$E' = (2\alpha)^{-1}\left[-1+\sqrt{1+4\alpha G_2(n_y, n_z)}\right]. \tag{2.27}$$

The electron statistics per unit length in this case is given by

$$n_0 = \frac{2g_v}{\pi}\frac{\sqrt{2m^*}}{\hbar}\sum_{n_y=1}^{n_{y\max}}\sum_{n_z=1}^{n_{z\max}}\left[T_7(E_{F1D}, n_y, n_z) + T_8(E_{F1D}, n_y, n_z)\right], \tag{2.28}$$

where $T_7(E_{F1D}, n_y, n_z) \equiv \left[E_{F1D}(1+\alpha E_{F1D}) - G_2(n_y, n_z)\right]^{1/2}$, $T_8(E_{F1D}, n_y, n_z) \equiv \sum_{r=1}^{S_0} Z_{r,Y}\left[T_7(E_{F1D}, n_y, n_z)\right]$, and $Y = 1$.

Combining (2.28) and (1.13), the TPSM in this case assumes the form

$$G_0 = \left(\pi^2 k_B^2 T/(3e)\right) \left[\sum_{n_y=1}^{n_{y\max}} \sum_{n_z=1}^{n_{z\max}} \left[T_7(E_{F1D}, n_y, n_z) + T_8(E_{F1D}, n_y, n_z) \right] \right]^{-1} \times \left[\sum_{n_y=1}^{n_{y\max}} \sum_{n_z=1}^{n_{z\max}} \left[\left\{ T_7(E_{F1D}, n_y, n_z) \right\}' + \left\{ T_8(E_{F1D}, n_y, n_z) \right\}' \right] \right]. \quad (2.29)$$

Under the condition, $\alpha E_{F1D} \ll 1$, the expressions of the 1D electron statistics per unit length and the corresponding TPSM can, respectively, be written as

$$n_0 = \frac{2g_v\sqrt{2m^*\pi k_B T}}{h} \sum_{n_y=1}^{n_{y\max}} \sum_{n_z=1}^{n_{z\max}} \frac{1}{\sqrt{\overline{i_1}}} \left[\left(1 + \frac{3}{2}\alpha.\overline{i_2}\right) F_{-1/2}\left(\overline{\eta_2}\right) + \frac{3}{4}\alpha k_B T F_{1/2}\left(\overline{\eta_2}\right) \right], \quad (2.30)$$

where $\overline{i_1} \equiv \left[1 + \alpha G_2\left(n_y, n_z\right)\right]$, $\overline{i_2} \equiv G_2\left(n_y, n_z\right)\left(\overline{i_1}\right)^{-1}$, and $\overline{\eta_2} \equiv (E_{F1D} - \overline{i_2})/k_B T$.

$$G_0 = \left(\frac{\pi^2 k_B}{3e}\right) \frac{\sum_{n_y=1}^{n_{y\max}} \sum_{n_z=1}^{n_{z\max}} \frac{1}{\sqrt{\overline{i_1}}} \left[\left(1 + \frac{3}{2}\alpha.\overline{i_2}\right) F_{-3/2}\left(\overline{\eta_2}\right) + \frac{3}{4}\alpha k_B T F_{-1/2}\left(\overline{\eta_2}\right) \right]}{\sum_{n_y=1}^{n_{y\max}} \sum_{n_z=1}^{n_{z\max}} \frac{1}{\sqrt{\overline{i_1}}} \left[\left(1 + \frac{3}{2}\alpha.\overline{i_2}\right) F_{-1/2}\left(\overline{\eta_2}\right) + \frac{3}{4}\alpha k_B T F_{1/2}\left(\overline{\eta_2}\right) \right]}. \quad (2.31)$$

For parabolic energy bands, $\alpha \to 0$ and the expressions of the electron concentration per unit area and per unit length and the TPSM for UFs and QWs in this case can, respectively, be written as

$$n_0 = \frac{m^* g_v k_B T}{\pi \hbar^2} \sum_{n_x=1}^{n_{x\max}} \left[F_0(\eta_1) \right], \quad (2.32)$$

$$G_0 = \left(\pi^2 k_B/(3e)\right) \left[\sum_{n_x=1}^{n_{x\max}} \left[F_0(\eta_1) \right] \right]^{-1} \left[\sum_{n_x=1}^{n_{x\max}} \left[F_{-1}(\eta_1) \right] \right], \quad (2.33)$$

$$n_0 = \frac{2g_v\sqrt{2\pi m^* k_B T}}{h} \sum_{n_y=1}^{n_{y\max}} \sum_{n_z=1}^{n_{z\max}} \left[F_{-1/2}\left(\eta_2\right) \right], \quad (2.34)$$

$$G_0 = \left(\pi^2 k_B/(3e)\right) \left[\sum_{n_x=1}^{n_{x\max}} \left[F_{-1/2}(\eta_2) \right] \right]^{-1} \left[\sum_{n_x=1}^{n_{x\max}} \left[F_{-3/2}(\eta_2) \right] \right], \quad (2.35)$$

where

$$\eta_1 = (1/(k_B T))\left[E_{F2D} - \frac{\hbar^2}{2m^*}\left(\frac{n_x \pi}{d_x}\right)^2\right]$$

and

$$\eta_2 = (1/(k_B T))\left[E_{F1D} - G_2(n_y, n_z)\right].$$

Under the condition of nondegeneracy,

$$F_j(\eta) \approx \exp(\eta) \quad \text{for } \eta < 0 \text{ for all } j \ [90] \tag{2.36}$$

and (2.24), (2.31), (2.33), and (2.35) get simplified into the form as [91]

$$G_0 = \left(\pi^2 k_B/(3e)\right). \tag{2.37}$$

Equation (2.37) is the well-known classical equation for TPSM as mentioned in the preface.

Thus, (2.37) indicates that it is not only a function of basic three constants, but also the signature of any material is being totally devoid in the expression of classical TPSM equation a fact as stated in the preface, but proved mathematically in this context.

2.2.3 *Magnetothermopower in Quantum-Confined II–VI Materials*

In the presence of the size quantization along z-direction, the 2D carrier dispersion law of II–VI materials can be expressed following (1.50) as

$$E = A_0 k_s^2 + B_0\left(\frac{\pi n_z}{d_z}\right)^2 \pm C_0 k_s. \tag{2.38}$$

The expressions of the subband energy E_{n_z}, the 2D density-of-states function per subband, the total density-of-states function $N_{2DT}(E)$, and the electron concentration per unit area can, respectively, be given by

$$E_{n_z} = B_0(\pi n_z/d_z)^2, \tag{2.39}$$

$$N_{2D}(E) = \frac{g_v m_\perp^*}{\pi\hbar^2}\left[1 - \frac{(C_0/2\sqrt{A_0})}{\sqrt{E + \delta_{51}(n_z)}}\right], \tag{2.40}$$

$$N_{2DT}(E) = \frac{g_v m_\perp^*}{\pi\hbar^2}\sum_{n_z=1}^{n_{z\max}}\left[1 - \frac{(C_0/2\sqrt{A_0})}{\sqrt{E + \delta_{51}(n_z)}}\right] H\left(E - E_{n_z}\right), \tag{2.41}$$

$$n_0 = \frac{g_v m_\perp^* k_B T}{\pi\hbar^2}\sum_{n_z=1}^{n_{z\max}}\left[F_0\left(\eta_{n_{z3}}\right) - \frac{C_0 f_s(E_{F2D}, n_z)}{2\sqrt{A_0 k_B T}}\right], \tag{2.42}$$

where $H(E - E_{n_z})$ is the Heaviside step function,

$$\delta_{51}(n_z) \equiv \frac{1}{4A_0}\left[(C_0)^2 - 4A_0 B_0\left(\frac{\pi n_z}{d_z}\right)^2\right],$$

$$\eta_{n_{z3}} \equiv (k_B T)^{-1}[E_{F2D} - B_0(\pi n_z/d_z)^2],$$

$$f_s(E_{F2D}, n_z) \equiv \left[2\left(\sqrt{\eta_{n_{z3}} + \delta_{52}(n_z)} - \sqrt{\delta_{52}(n_z)}\right) + \sum_{r=1}^{s}\left[2(1-2^{1-2r})\zeta(2r)\frac{(-1)^{2r-1}(2r-1)!}{(\eta_{n_{z3}} + \delta_{52}(n_z))^{2r}}\right]\right],$$

and $\delta_{52}(n_z) \equiv \left[\frac{B_0(\pi n_z/d_z)^2 + \delta_{51}(n_z)}{k_B T}\right]$.

Combining (2.42) and (1.13), the TPSM in this case can be expressed as

$$G_0 = \left(\pi^2 k_B/(3e)\right)\left[\sum_{n_z=1}^{n_{z\max}}\left[F_0\left(\eta_{n_{z3}}\right) - \frac{C_0 f_s\left(E_{F2D}, n_z\right)}{2\sqrt{A_0 k_B T}}\right]\right]^{-1} \times\left[\sum_{n_z=1}^{n_{z\max}}\left[F_{-1}\left(\eta_{n_{z3}}\right) - \frac{C_0 f_s'\left(E_{F2D}, n_z\right)}{2\sqrt{A_0 k_B T}}\right]\right]. \tag{2.43}$$

The 1D carrier energy spectrum for QWs of II–VI materials can be written as

$$E = B_0 k_z^2 + G_{3,\pm}\left(n_x, n_y\right), \tag{2.44}$$

where

$$G_{3,\pm}\left(n_x, n_y\right) \equiv \left[A_0\left\{\left(\frac{\pi n_x}{d_x}\right)^2 + \left(\frac{\pi n_y}{d_y}\right)^2\right\} \pm C_0\left\{\left(\frac{\pi n_x}{d_x}\right)^2 + \left(\frac{\pi n_y}{d_y}\right)^2\right\}^{1/2}\right].$$

The 1D electron statistics per unit length assumes the form

$$n_0 = \frac{g_v}{\pi\sqrt{B_0}}\sum_{n_x=1}^{n_{x\max}}\sum_{n_y=1}^{n_{y\max}}\left[t_7\left(E_{F1D}, n_x, n_y\right) + t_8\left(E_{F1D}, n_x, n_y\right)\right], \tag{2.45}$$

where $t_7(E_{F1D}, n_x, n_y) \equiv [E_{F1D} - [G_{3,+}(n_x, n_y)]]^{1/2} + [E_{F1D} - [G_{3,-}(n_x, n_y)]]^{1/2}$, $t_8(E_{F1D}, n_x, n_y) \equiv \sum_{r=1}^{S_0} Z_{r,Y}[t_7(E_{F1D}, n_x, n_y)]$, and $Y = 1$.

Combining (2.45) and (1.13), the TPSM in this case assumes the form

$$G_0 = \left(\pi^2 k_B^2 T/(3e)\right)\left[\sum_{n_x=1}^{n_{x_{\max}}}\sum_{n_y=1}^{n_{y_{\max}}}\left[t_7\left(E_{F1D}, n_x, n_y\right) + t_8\left(E_{F1D}, n_x, n_y\right)\right]\right]^{-1}$$
$$\times\left[\sum_{n_x=1}^{n_{x_{\max}}}\sum_{n_y=1}^{n_{y_{\max}}}\left[\left\{t_7\left(E_{F1D}, n_x, n_y\right)\right\}' + \left\{t_8\left(E_{F1D}, n_x, n_y\right)\right\}'\right]\right]. \quad (2.46)$$

2.2.4 Magnetothermopower in Quantum-Confined Bismuth

2.2.4.1 The McClure and Choi Model

In the presence of size quantization along y-direction, the 2D dispersion relation of the carriers in bismuth in this case can be written as

$$k_x^2\left[\frac{\hbar^2}{2m_1} - \left\{\frac{\alpha\hbar^2}{4m_1m_2}\left(\frac{\pi n_y}{d_y}\right)^2\right\}\right] + k_z^2\left[\frac{\hbar^2}{2m_3} - \frac{\alpha\hbar^2}{4m_2m_3}\left(\frac{\pi n_y}{d_y}\right)^2\right]$$
$$= E\left(1+\alpha E\right) - \left\{\frac{\alpha}{4m_2m_2'}\left(\frac{\pi\hbar n_y}{d_y}\right)^4\right\} - \alpha E\left\{1 - \left(\frac{m_2}{m_2'}\right)\right\}\left(\frac{\pi\hbar n_y}{d_y}\right)^2. \quad (2.47)$$

The 2D area assumes the form

$$A\left(E\right) = \frac{2\pi\sqrt{m_1m_3}}{\hbar^2}t_{25}\left(E, n_y\right), \quad (2.48)$$

where

$$t_{25}\left(E, n_y\right) \equiv \left[1 - \frac{\alpha\hbar^2}{2m_2}\left(\frac{\pi\hbar n_y}{d_y}\right)^2\right]^{-1}\left[E\left(1+\alpha E\right) - \frac{\alpha}{4m_2m_2'}\left(\frac{\pi\hbar n_y}{d_y}\right)^4\right.$$
$$\left. - \alpha E\left\{1 - \left(\frac{m_2}{m_2'}\right)\right\}\left(\frac{\pi\hbar n_y}{d_y}\right)^2\right].$$

The subband energy E_{n_y} can be expressed as

$$E_{n_y}\left(1+\alpha E_{n_y}\right) - \frac{\alpha}{4m_2m_2'}\left(\frac{\pi\hbar n_y}{d_y}\right)^4 - \alpha E_{n_y}\left\{1 - \left(\frac{m_2}{m_2'}\right)\right\}\left(\frac{\pi\hbar n_y}{d_y}\right)^2 = 0. \quad (2.49)$$

The DOS function per subband is given by

$$N_{\mathrm{2D}}(E) = \frac{g_{\mathrm{v}}\sqrt{m_1 m_3}}{\pi\hbar^2}\left\{t_{25}\left(E, n_y\right)\right\}'. \tag{2.50}$$

The total 2D DOS function can be written as

$$N_{\mathrm{2DT}}(E) = \frac{g_{\mathrm{v}}\sqrt{m_1 m_3}}{\pi\hbar^2}\sum_{n_y=1}^{n_{y\max}}\left\{t_{25}\left(E, n_y\right)\right\}' H\left(E - E_{n_y}\right). \tag{2.51}$$

Using (2.51) and the Fermi–Dirac occupation probability factor, the electron statistics per unit area can be expressed as

$$n_0 = \frac{g_{\mathrm{v}}\sqrt{m_1 m_3}}{\pi\hbar^2}\sum_{n_y=1}^{n_{y\max}}\left[t_{25}\left(E_{\mathrm{F2D}}, n_y\right) + t_{26}\left(E_{\mathrm{F2D}}, n_y\right)\right], \tag{2.52}$$

where

$t_{26}\left(E_{\mathrm{F2D}}, n_y\right) \equiv \sum_{r=1}^{s_0} Z_{r,Y}\left[t_{25}\left(E_{\mathrm{F2D}}, n_y\right)\right]$ and $Y = 2$.

Combining (2.52) and (1.13), the TPSM in this case assumes the form

$$G_0 = \left(\pi^2 k_{\mathrm{B}}^2 T/(3e)\right)\left[\sum_{n_y=1}^{n_{y\max}}\left[t_{25}\left(E_{\mathrm{F2D}}, n_y\right) + t_{26}\left(E_{\mathrm{F2D}}, n_y\right)\right]\right]^{-1} \times\left[\sum_{n_y=1}^{n_{y\max}}\left[\left\{t_{25}\left(E_{\mathrm{F2D}}, n_y\right)\right\}' + \left\{t_{26}\left(E_{\mathrm{F2D}}, n_y\right)\right\}'\right]\right]. \tag{2.53}$$

The 1D dispersion relation of the carriers in Bi in this case can be written as

$$E(1+\alpha E) = \left\{\frac{\hbar^2 k_x^2}{2m_1}\left[1 - \frac{\alpha\hbar^2}{2m_2}\left(\frac{\pi n_y}{d_y}\right)^2\right] + G_{12} + \frac{\hbar^2}{2m_2}\alpha E\left\{1 - \left(\frac{m_2}{m_2'}\right)\right\}\left(\frac{\pi n_y}{d_y}\right)^2\right\}, \tag{2.54}$$

where

$$G_{12} \equiv \left\{\frac{\hbar^2}{2m_2}\left(\frac{\pi n_y}{d_y}\right)^2 + \frac{\hbar^2}{2m_3}\left(\frac{\pi n_z}{d_z}\right)^2 + \frac{\alpha\hbar^4}{4m_2 m_2'}\left(\frac{\pi n_y}{d_y}\right)^4 - \frac{\alpha}{4m_2 m_3}\left(\frac{\hbar^2 n_y n_z \pi^2}{d_y d_z}\right)^2\right\}.$$

Using (2.54), the 1D electron statistics per unit length can be expressed as

$$n_0 = \frac{2g_v}{\pi}\frac{\sqrt{2m_1}}{\hbar}\sum_{n_y=1}^{n_{y\max}}\sum_{n_z=1}^{n_{z\max}}\left[t_{27}\left(E_{F1D},n_y,n_z\right)+t_{28}\left(E_{F1D},n_y,n_z\right)\right], \quad (2.55)$$

$t_{27}(E_{F1D},n_y,n_z) \equiv \{[1-\frac{\alpha\hbar^2}{2m_2}(\frac{\pi n_y}{d_y})^2]^{-1/2}[E_{F1D}(1+\alpha E_{F1D})-G_{12}-\frac{\hbar^2}{2m_2}\alpha E_{F1D}$ $\{1-(\frac{m_2}{m_2'})\}(\frac{\pi\hbar n_y}{d_y})^2]^{1/2}\}$, $t_{28}(E_{F1D},n_y,n_z) \equiv \sum_{r=1}^{s_0} Z_{r,Y}[t_{27}(E_{F1D},n_y,n_z)]$, and $Y=1$.

Combining (2.55) and (1.13), the TPSM in this case assumes the form

$$G_0 = \left(\pi^2 k_B^2 T/(3e)\right)\left[\sum_{n_y=1}^{n_{y\max}}\sum_{n_z=1}^{n_{z\max}}\left[t_{27}\left(E_{F1D},n_y,n_z\right)+t_{28}\left(E_{F1D},n_y,n_z\right)\right]\right]^{-1}$$
$$\times\left[\sum_{n_y=1}^{n_{y\max}}\sum_{n_z=1}^{n_{z\max}}\left[\left\{t_{27}\left(E_{F1D},n_y,n_z\right)\right\}'+\left\{t_{28}\left(E_{F1D},n_y,n_z\right)\right\}'\right]\right]. \quad (2.56)$$

2.2.4.2 The Hybrid Model

In the presence of size quantization along y-direction, the 2D electron dispersion relation can be written as

$$\frac{\hbar^2 k_x^2}{2m_1}+\frac{\hbar^2 k_z^2}{2m_3} = E\left(1+\alpha E\right)-\frac{\theta_0\left(E\right)\hbar^2}{2M_2}\left(\frac{\pi n_y}{d_y}\right)^2-\frac{\alpha\gamma_0\hbar^4}{4M_2^2}\left(\frac{\pi n_y}{d_y}\right)^4. \quad (2.56a)$$

The 2D area is given by

$$A\left(E,n_y\right) = \frac{2\pi\sqrt{m_1 m_3}}{\hbar^2}t_{29}\left(E,n_y\right), \quad (2.56b)$$

$$t_{29}\left(E,n_y\right) = \left[E\left(1+\alpha E\right)-\frac{\theta_0\left(E\right)\hbar^2}{2M_2}\left(\frac{\pi n_y}{d_y}\right)^2-\frac{\alpha\gamma_0\hbar^4}{4M_2^4}\left(\frac{\pi n_y}{d_y}\right)^4\right].$$

The subband energy (E_{n_y}) is given as

$$E_{n_y}\left(1+\alpha E_{n_y}\right)-\frac{\theta_0\left(E_{n_y}\right)\hbar^2}{2M_2}\left(\frac{\pi n_y}{d_y}\right)^2-\frac{\alpha\gamma_0\hbar^4}{4M_2^2}\left(\frac{\pi n_y}{d_y}\right)^4 = 0. \quad (2.56c)$$

The total DOS function in this case can be written as

$$N_{2DT}\left(E\right) = \frac{g_v\sqrt{m_1 m_3}}{\pi\hbar^2}\sum_{n_y=1}^{n_{y\max}}\left\{t_{29}\left(E,n_y\right)\right\}' H\left(E-E_{n_y}\right). \quad (2.56d)$$

The use of (2.56d) leads to the 2D electron statistics per unit area in QWs in UFs of Bi in this case as

$$n_0 = \frac{g_v\sqrt{m_1 m_3}}{\pi\hbar^2}\sum_{n_y=1}^{n_{y\max}}\left[t_{29}\left(E_{F2D}, n_y\right) + t_{30}\left(E_{F2D}, n_y\right)\right] \tag{2.56e}$$

in which $t_{30}\left(E_{F2D}, n_y\right) = \sum_{r=1}^{s_0} Z_{r,Y}\left[t_{29}\left(E_{F2D}, n_y\right)\right]$ and $Y = 2$.

Combining (1.13) and (2.56e), the TPSM in this case assumes the form

$$G_0 = \left(\pi^2 k_B^2 T/(3e)\right)\left[\sum_{n_y=1}^{n_{y\max}}\left[t_{29}\left(E_{F2D}, n_y\right) + t_{30}\left(E_{F2D}, n_y\right)\right]\right]^{-1} \times\left[\sum_{n_y=1}^{n_{y\max}}\left[\left\{t_{29}\left(E_{F2D}, n_y\right)\right\}' + \left\{t_{30}\left(E_{F2D}, n_y\right)\right\}'\right]\right]. \tag{2.56f}$$

The 1D dispersion relation in this case can be expressed as

$$E\left(1+\alpha E\right) = \frac{\hbar^2 k_x^2}{2m_1} + G_{14} + \frac{\hbar^2}{2M_2}\left(\frac{\pi n_y}{d_y}\right)^2 \alpha E\left(1-\gamma_0\right), \tag{2.56g}$$

where

$$G_{14} = \left[\frac{\hbar^2}{2m_3}\left(\frac{\pi n_z}{d_z}\right)^2 + \frac{\hbar^2}{2M_2}\left(\frac{\pi n_y}{d_y}\right)^2\left(1+\bar{\delta}_0\right) + \frac{\alpha\gamma_0\hbar^4}{4M_2^2}\left(\frac{\pi n_y}{d_y}\right)^4\right].$$

The use of (2.56g) leads to the expression for the electron concentration per unit length as

$$n_0 = \frac{2g_v}{\pi}\frac{\sqrt{2m_1}}{\hbar}\sum_{n_y=1}^{n_{y\max}}\sum_{n_z=1}^{n_{z\max}}\left[t_{31}\left(E_{F1D}, n_y, n_z\right) + t_{32}\left(E_{F1D}, n_y, n_z\right)\right], \tag{2.56h}$$

where $t_{31}(E_{F1D}, n_y, n_z) \equiv [E_{F1D}(1+\alpha E_{1DF}) - G_{14} - \frac{\hbar^2}{2M_2}(\frac{\pi n_y}{d_y})^2 \alpha E_{F1D}(1-\gamma_0)]^{1/2}$, $t_{32}(E_{F1D}, n_y, n_z) \equiv \sum_{r=1}^{s_0} Z_{r,Y}[t_{31}(E_{F1D}, n_y, n_z)]$, and $Y = 1$.

Combining (1.13) and (2.56h), the TPSM in this case assumes the form

$$G_0 = \left(\pi^2 k_B^2 T/(3e)\right)\left[\sum_{n_y=1}^{n_{y\max}}\sum_{n_z=1}^{n_{z\max}}\left[t_{31}\left(E_{F1D}, n_y, n_z\right) + t_{32}\left(E_{F1D}, n_y, n_z\right)\right]\right]^{-1}$$
$$\times\left[\sum_{n_y=1}^{n_{y\max}}\sum_{n_z=1}^{n_{z\max}}\left[\left\{t_{31}\left(E_{F1D}, n_y, n_z\right)\right\}' + \left\{t_{32}\left(E_{F1D}, n_y, n_z\right)\right\}'\right]\right]. \quad (2.56i)$$

2.2.4.3 The Cohen Model

In the presence of size quantization along y-direction, the 2D electron dispersion law in this case is given by

$$E(1+\alpha E) + \frac{\alpha E\hbar^2}{2m_2'}\left(\frac{\pi n_y}{d_y}\right)^2 - \frac{(1+\alpha E)\hbar^2}{2m_2}\left(\frac{\pi n_y}{d_y}\right)^2 - \frac{\alpha\hbar^4}{4m_2 m_2'}\left(\frac{\pi n_y}{d_y}\right)^4$$
$$= \frac{\hbar^2 k_x^2}{2m_1} + \frac{\hbar^2 k_z^2}{2m_3}. \quad (2.57)$$

The subband energy can be written as

$$E_{n_y}\left(1+\alpha E_{n_y}\right) + \frac{\alpha E_{n_y}\hbar^2}{2m_2'}\left(\frac{\pi n_y}{d_y}\right)^2 - \frac{(1+\alpha E_{n_y})\hbar^2}{2m_2}\left(\frac{\pi n_y}{d_y}\right)^2$$
$$-\frac{\alpha\hbar^4}{4m_2 m_2'}\left(\frac{\pi n_y}{d_y}\right)^4 = 0. \quad (2.58)$$

The 2D area can be expressed as

$$A\left(E, n_y\right) = \frac{2\pi\sqrt{m_1 m_3}}{\hbar^2} t_{33}\left(E, n_y\right), \quad (2.59)$$

where

$$t_{33}\left(E, n_y\right) = E(1+\alpha E) + \frac{\alpha E\hbar^2}{2m_2'}\left(\frac{\pi n_y}{d_y}\right)^2 - \frac{(1+\alpha E)\hbar^2}{2m_2}\left(\frac{\pi n_y}{d_y}\right)^2$$
$$-\frac{\alpha\hbar^4}{4m_2 m_2'}\left(\frac{\pi n_y}{d_y}\right)^4.$$

The total DOS function in this case assumes the form

$$N_{2DT}(E) = \frac{g_v\sqrt{m_1 m_3}}{\pi\hbar^2}\sum_{n_y=1}^{n_{y\max}}\left\{t_{33}\left(E, n_y\right)\right\}' H\left(E - E_{n_y}\right). \quad (2.60)$$

The electron statistics per unit area in this case can be written as

$$n_0 = \frac{g_{\mathrm{v}}\sqrt{m_1 m_3}}{\pi \hbar^2} \sum_{n_y=1}^{n_{y\max}} \left[t_{33}\left(E_{\mathrm{F2D}}, n_y\right) + t_{34}\left(E_{\mathrm{F2D}}, n_y\right)\right], \tag{2.61}$$

where
$t_{34}\left(E_{\mathrm{F2D}}, n_y\right) = \sum_{r=1}^{s_0} Z_{r,Y}\left[t_{33}\left(E_{\mathrm{F2D}}, n_y\right)\right]$ and $Y = 2$.

Combining (2.61) and (1.13), the TPSM in this case assumes the form

$$G_0 = \left(\pi^2 k_{\mathrm{B}}^2 T/(3e)\right)\left[\sum_{n_y=1}^{n_{y\max}}\left[t_{33}\left(E_{\mathrm{F2D}}, n_y\right) + t_{34}\left(E_{\mathrm{F2D}}, n_y\right)\right]\right]^{-1} \times \left[\sum_{n_y=1}^{n_{y\max}}\left[\left\{t_{33}\left(E_{\mathrm{F2D}}, n_y\right)\right\}' + \left\{t_{34}\left(E_{\mathrm{F2D}}, n_y\right)\right\}'\right]\right]. \tag{2.62}$$

The 1D carrier dispersion law in this case can be written as

$$\alpha E^2 + E l_7 - G_{15} = \frac{\hbar^2 k_x^2}{2m_1}, \tag{2.63}$$

where $l_7 = [1 - \frac{\alpha\hbar^2}{2m_2}(\frac{\pi n_y}{d_y})^2 + \frac{\alpha\hbar^2}{2m_2'}(\frac{\pi n_y}{d_y})^2]$ and $G_{15} = [\frac{\hbar^2}{2m_3}(\frac{\pi n_z}{d_z})^2 + \frac{\hbar^2}{2m_2}(\frac{\pi n_y}{d_y})^2 + \frac{\alpha\hbar^4}{4m_2 m_2'}(\frac{\pi n_y}{d_y})^4]$.

The 1D electron concentration per unit length is given by

$$n_0 = \frac{2g_{\mathrm{v}}}{\pi}\frac{\sqrt{2m_1}}{\hbar} \sum_{n_y=1}^{n_{y\max}} \sum_{n_z=1}^{n_{z\max}} \left[t_{35}\left(E_{\mathrm{F1D}}, n_y, n_z\right) + t_{36}\left(E_{\mathrm{F1D}}, n_y, n_z\right)\right], \tag{2.64}$$

where $t_{35}\left(E_{\mathrm{F1D}}, n_y, n_z\right) = \left[\alpha E_{\mathrm{F1D}}^2 + E_{\mathrm{F1D}} l_7 - G_{15}\right]^{1/2}$, $t_{36}\left(E_{\mathrm{F1D}}, n_y, n_z\right) = \sum_{r=1}^{s_0} Z_{r,Y}\left[t_{35}\left(E_{\mathrm{F1D}}, n_y, n_z\right)\right]$, and $Y = 1$.

Combining (2.64) and (1.13), the TPSM in this case assumes the form

$$G_0 = \left(\pi^2 k_{\mathrm{B}}^2 T/(3e)\right)\left[\sum_{n_y=1}^{n_{y\max}} \sum_{n_z=1}^{n_{z\max}}\left[t_{35}\left(E_{\mathrm{F1D}}, n_y, n_z\right) + t_{36}\left(E_{\mathrm{F1D}}, n_y, n_z\right)\right]\right]^{-1} \times \left[\sum_{n_y=1}^{n_{y\max}} \sum_{n_z=1}^{n_{z\max}}\left[\left\{t_{35}\left(E_{\mathrm{F1D}}, n_y, n_z\right)\right\}' + \left\{t_{36}\left(E_{\mathrm{F1D}}, n_y, n_z\right)\right\}'\right]\right]. \tag{2.64a}$$

2.2.4.4 The Lax Model

The 2D electron dispersion relation in this case can be written as

$$E(1+\alpha E) - \frac{\hbar^2}{2m_2}\left(\frac{\pi n_y}{d_y}\right)^2 = \frac{\hbar^2 k_x^2}{2m_1} + \frac{\hbar^2 k_z^2}{2m_3}. \tag{2.65}$$

The subband energy (E_{n_y}) is given by

$$E_{n_y}\left(1+\alpha E_{n_y}\right) = \frac{\hbar^2}{2m_2}\left(\pi n_y/d_y\right)^2. \tag{2.66}$$

The 2D area can be written as

$$A\left(E,n_y\right) = \frac{2\pi\sqrt{m_1 m_3}}{\hbar^2} t_{37}\left(E,n_y\right), \tag{2.67}$$

where

$$t_{37}\left(E,n_y\right) = \left[E(1+\alpha E) - \frac{\hbar^2}{2m_2}\left(\frac{\pi n_y}{d_y}\right)^2\right].$$

The total DOS function in this case assumes the form

$$N_{\mathrm{2DT}}(E) = \frac{g_{\mathrm{v}}\sqrt{m_1 m_3}}{\pi\hbar^2}\sum_{n_y=1}^{n_{y\max}}(1+2\alpha E)\,H\left(E-E_{n_y}\right). \tag{2.68}$$

The use of (2.68) and Fermi–Dirac factor leads to the expression of the electron statistics per unit area as

$$n_0 = \frac{(k_{\mathrm{B}}Tg_{\mathrm{v}})\sqrt{m_1 m_3}}{\pi\hbar^2}\sum_{n_y=1}^{n_{y\max}}\left[\left[\left(1+2\alpha E_{n_y}\right)F_0\left(\eta_{n_y}\right)+2\alpha k_{\mathrm{B}}TF_1\left(\eta_{n_y}\right)\right]\right], \tag{2.69}$$

where

$$\eta_{n_y} = \frac{1}{k_{\mathrm{B}}T}\left[E_{\mathrm{F2D}}-E_{n_y}\right] \text{ and } E_{n_y} = (2\alpha)^{-1}\left[-1+\left[1+\frac{2\alpha\hbar^2}{m_2}\left(\frac{\pi n_y}{d_y}\right)^2\right]^{1/2}\right].$$

Combining (2.69) and (1.13), the TPSM in this case assumes the form

$$G_0 = \left(\pi^2 k_{\mathrm{B}}/(3e)\right)\left[\sum_{n_y=1}^{n_{y\max}}\left[\left[\left(1+2\alpha E_{n_y}\right)F_0\left(\eta_{n_y}\right)+2\alpha k_{\mathrm{B}}TF_1\left(\eta_{n_y}\right)\right]\right]\right]^{-1}$$
$$\times\left[\sum_{n_y=1}^{n_{y\max}}\left[\left[\left(1+2\alpha E_{n_y}\right)F_{-1}\left(\eta_{n_y}\right)+2\alpha k_{\mathrm{B}}TF_0\left(\eta_{n_y}\right)\right]\right]\right]. \tag{2.70}$$

The 1D dispersion relation in this case can be written as

$$E(1+\alpha E)=\frac{\hbar^2k_x^2}{2m_1}+G_{16}, \tag{2.70a}$$

where

$$G_{16}=\frac{\hbar^2}{2m_2}\left(\frac{\pi n_y}{d_y}\right)^2+\frac{\hbar^2}{2m_3}\left(\frac{\pi n_z}{d_z}\right)^2.$$

The 1D electron statistics per unit length is given by

$$n_0=\frac{2g_v}{\pi}\frac{\sqrt{2m_1}}{\hbar}\sum_{n_y=1}^{n_{y\max}}\sum_{n_z=1}^{n_{z\max}}\left[t_{37}\left(E_{F1D},n_y,n_z\right)+t_{38}\left(E_{F1D},n_y,n_z\right)\right], \tag{2.71}$$

where

$$t_{37}\left(E_{F1D},n_y,n_z\right)=\left[E_{F1D}\left(1+\alpha E_{F1D}\right)-G_{16}\right]^{1/2},$$

$$t_{38}\left(E_{F1D},n_y,n_z\right)=\sum_{r=1}^{s_0}Z_{r,Y}\left[t_{37}\left(E_{F1D},n_y,n_z\right)\right],$$

and $Y=1$.

Combining (2.71) and (1.13), the TPSM in this case assumes the form

$$G_0=\left(\pi^2k_B^2T/(3e)\right)\left[\sum_{n_y=1}^{n_{y\max}}\sum_{n_z=1}^{n_{z\max}}\left[t_{37}\left(E_{F1D},n_y,n_z\right)+t_{38}\left(E_{F1D},n_y,n_z\right)\right]\right]^{-1}$$
$$\times\left[\sum_{n_y=1}^{n_{y\max}}\sum_{n_z=1}^{n_{z\max}}\left[\left\{t_{37}\left(E_{F1D},n_y,n_z\right)\right\}'+\left\{t_{38}\left(E_{F1D},n_y,n_z\right)\right\}'\right]\right]. \tag{2.72}$$

It may be noted that under the conditions $\alpha\to 0$ and isotropic effective electron mass at the edge of the conduction band together with the conversion of the summation over the quantum numbers to the corresponding integrations, leads to the well-known expression of the TPSM for nondegenerate wide-gap materials as given by (2.37).

2.2.5 Magnetothermopower in Quantum-Confined IV–VI Materials

Using

$$\varepsilon=E+\left(E_{g0}/2\right),\ P_{\perp}^2=\frac{\hbar^2E_{g0}}{2m_t^*},\ P_{||}^2=\frac{\hbar^2E_{g0}}{2m_l^*}$$

(m_t^* and m_l^* are the transverse and longitudinal effective electron masses at $k = 0$) in (1.125), we can write

$$\left[E - \frac{\hbar^2 k_s^2}{2m_t^-} - \frac{\hbar^2 k_z^2}{2m_l^-}\right]\left[1 + \alpha E + \alpha\frac{\hbar^2 k_s^2}{2m_t^+} + \alpha\frac{\hbar^2 k_z^2}{2m_l^+}\right] = \frac{\hbar^2 k_s^2}{2m_t^*} + \frac{\hbar^2 k_z^2}{2m_l^*}. \quad (2.73)$$

Therefore, the surface electron concentration per unit area in UFs of IV–VI materials in accordance with the Dimmock model can be written as [89]

$$n_0 = \frac{g_v}{2\pi}\sum_{n_z=1}^{n_{z\max}}\left[\overline{T_{55}}\left(E_{F2D}, n_z\right) + \overline{T_{56}}\left(E_{F2D}, n_z\right)\right], \quad (2.74)$$

where

$$\overline{T_{55}}\left(E_{F2D}, n_z\right) \equiv \frac{\overline{A}\left(E_{F2D}, n_z\right)}{\pi},$$

$$\overline{A}\left(E_{F2D}, n_z\right) = \frac{\pi \overline{t_3}\left(E_{F2D}, n_z\right)}{\sqrt{\overline{A_1}\left(E_{F2D}, n_z\right)\overline{B_1}\left(E_{F2D}, n_z\right)}} \times\left[1 - \frac{1}{\overline{x_5}}\left(\frac{1}{\overline{x_1}} + \frac{3}{\overline{x_2}}\right)\frac{\alpha \overline{t_3}\left(E_{F2D}, n_z\right)\hbar^4}{8\left\{\overline{B_1}\left(E_{F2D}, n_z\right)\right\}^2}\right],$$

$$\overline{t_3}\left(E_{F2D}, n_z\right) \equiv \left[E_{F2D}\left(1 + \alpha E_{F2D}\right) + \alpha E_{F2D}\left(\frac{\hbar^2}{2\left(\overline{x_6}\right)}\right)\left(\frac{n_z\pi}{d_z}\right)^2 - \left(1+\alpha E_{F2D}\right)\left(\frac{\hbar^2}{2\left(\overline{x_3}\right)}\right)\left(\frac{n_z\pi}{d_z}\right)^2 - \alpha\left(\frac{\hbar^4}{4\left(\overline{x_3}\right)\left(\overline{x_6}\right)}\right)\left(\frac{n_z\pi}{d_z}\right)^4\right],$$

$$\overline{x_6} = \frac{3m_t^+ m_l^+}{2m_l^+ + m_t^+}, \quad \overline{x_3} = \frac{3m_t^- m_l^-}{2m_l^- + m_t^-}, \quad \overline{x_5} = \frac{m_t^+ + 2m_l^+}{3},$$

$$\overline{x_1} = m_t^-, \quad \overline{x_2} = \frac{m_t^- + 2m_l^-}{3},$$

$\overline{B_1}(E_{F2D}, n_z) \equiv \frac{\hbar^2}{2m_2}\left[1 + m_2\left[\frac{\alpha\hbar^2}{2(\overline{x_3})(\overline{x_5})}\left(\frac{n_z\pi}{d_z}\right)^2 + \frac{\alpha\hbar^2}{2(\overline{x_2})(\overline{x_6})}\left(\frac{n_z\pi}{d_z}\right)^2 + \frac{1+\alpha E_{F2D}}{(\overline{x_2})} - \frac{\alpha E_{F2D}}{(\overline{x_5})}\right]\right]$, $\overline{T_{56}}(E_{F2D}, n_z) \equiv \sum_{r=1}^{s} Z_{r,Y}[\overline{T_{55}}(E_{F2D}, n_z)]$, and $Y = 2$.

Combining (2.74) and (1.13), the TPSM in this case assumes the form

$$G_0 = \left(\pi^2 k_B^2 T/(3e)\right) \left[\sum_{n_z=1}^{n_{z\max}} \left[\overline{T_{55}}(E_{F2D}, n_z) + \overline{T_{56}}(E_{F2D}, n_z) \right] \right]^{-1} \times \left[\sum_{n_z=1}^{n_{z\max}} \left[\left\{ \overline{T_{55}}(E_{F2D}, n_z) \right\}' + \left\{ \overline{T_{56}}(E_{F2D}, n_z) \right\}' \right] \right]. \tag{2.75}$$

The 1D dispersion relation of the conduction electrons in QWs of IV–VI materials for the two-dimensional quantizations along y- and z-direction can be expressed as

$$\begin{aligned} &E(1+\alpha E) + \alpha E \left(\frac{\overline{x}}{\overline{x_4}} + \frac{\hbar^2}{2\overline{x_5}} \left(\frac{n_y \pi}{d_y} \right)^2 \right) + \alpha E \frac{\hbar^2}{2\overline{x_6}} \left(\frac{n_z \pi}{d_z} \right)^2 \\ &- (1+\alpha E) \left(\frac{\overline{x}}{\overline{x_1}} + \frac{\hbar^2}{2\overline{x_2}} \left(\frac{n_y \pi}{d_y} \right)^2 \right) \\ &- \alpha \left(\frac{\overline{x}}{\overline{x_1}} + \frac{\hbar^2}{2\overline{x_2}} \left(\frac{n_y \pi}{d_y} \right)^2 \right) \left(\frac{\overline{x}}{\overline{x_4}} + \frac{\hbar^2}{2\overline{x_5}} \left(\frac{n_y \pi}{d_y} \right)^2 \right) \\ &- \alpha \left(\frac{\overline{x}}{\overline{x_1}} + \frac{\hbar^2}{2\overline{x_2}} \left(\frac{n_y \pi}{d_y} \right)^2 \right) \frac{\hbar^2}{2\overline{x_6}} \left(\frac{n_z \pi}{d_z} \right)^2 \\ &- (1+\alpha E) \frac{\hbar^2}{2\overline{x_3}} \left(\frac{n_z \pi}{d_z} \right)^2 - \alpha \frac{\hbar^2}{2\overline{x_3}} \left(\frac{n_z \pi}{d_z} \right)^2 \left(\frac{\overline{x}}{\overline{x_4}} + \frac{\hbar^2}{2\overline{x_5}} \left(\frac{n_y \pi}{d_y} \right)^2 \right) \\ &- \alpha \frac{\hbar^2}{2\overline{x_3}} \left(\frac{n_z \pi}{d_z} \right)^2 \frac{\hbar^2}{2\overline{x_6}} \left(\frac{n_z \pi}{d_z} \right)^2 = \frac{\overline{x}}{m_1} + \frac{\hbar^2}{2m_2} \left(\frac{n_y \pi}{d_y} \right)^2 \\ &+ \frac{\hbar^2}{2m_3} \left(\frac{n_z \pi}{d_z} \right)^2, \end{aligned} \tag{2.75a}$$

where

$$\overline{x} \equiv \frac{\hbar^2 k_x^2}{2}.$$

The use of (2.75a) leads to the expression of 1D electron statistics per unit length as

$$n_0 = \frac{2g_v}{\pi \hbar \sqrt{\overline{g_1}}} \sum_{n_y=1}^{n_{y\max}} \sum_{n_z=1}^{n_{z\max}} \left[T_{613}(E_{F1D}, n_y, n_z) + T_{614}(E_{F1D}, n_y, n_z) \right], \tag{2.75b}$$

where $T_{614}(E_{F1D}, n_y, n_z) \equiv \sum_{r=1}^{s} Z_{r,Y}\left[T_{613}(E_{F1D}, n_y, n_z)\right]$, $Y = 1$,

$$T_{613}\left(E_{F1D}, n_y, n_z\right) \equiv \left[\sqrt{\bar{g}_2^2\left(E_{F1D}, n_y, n_z\right) + 4\bar{g}_1 c_1\left(E_{F1D}, n_y, n_z\right)} - \bar{g}_2\left(E_{F1D}, n_y, n_z\right)\right]^{1/2}, \bar{g}_1 \equiv \frac{\alpha}{(\overline{x_1})\overline{x_4}},$$

$$\bar{g}_2\left(E_{F1D}, n_y, n_z\right) \equiv \left[\frac{-\alpha E_{F1D}}{\overline{x_4}} + \frac{1 + \alpha E_{F1D}}{\overline{x_1}} + \alpha\left[\frac{\hbar^2}{2(\overline{x_2})\overline{x_4}}\left(\frac{n_y \pi}{d_y}\right)^2 + \frac{\hbar^2}{2(\overline{x_1})\overline{x_5}}\left(\frac{n_y \pi}{d_y}\right)^2\right] + \frac{\alpha\hbar^2}{2(\overline{x_1})\overline{x_6}}\left(\frac{n_z \pi}{d_z}\right)^2 + \frac{\alpha\hbar^2}{2(\overline{x_3})\overline{x_4}}\left(\frac{n_z \pi}{d_z}\right)^2 + \frac{1}{m_1}\right],$$

$$c_1\left(E_{F1D}, n_y, n_z\right) \equiv \left[E_{F1D}\left(1 + \alpha E_{F1D}\right) + \alpha E_{F1D}\frac{\hbar^2}{2\overline{x_5}}\left(\frac{n_y \pi}{d_y}\right)^2 + \alpha E_{F1D}\frac{\hbar^2}{2\overline{x_6}}\left(\frac{n_z \pi}{d_z}\right)^2 - \left(1 + \alpha E_{F1D}\right)\left(\frac{\hbar^2}{2\overline{x_2}}\left(\frac{n_y \pi}{d_y}\right)^2\right) - \alpha\frac{\hbar^4}{4(\overline{x_2})\overline{x_5}}\left(\frac{n_y \pi}{d_y}\right)^4 - \alpha\frac{\hbar^4}{4(\overline{x_2})\overline{x_6}}\left(\frac{n_y \pi}{d_y}\right)^2\left(\frac{n_z \pi}{d_z}\right)^2 - \left(1 + \alpha E_{F1D}\right)\frac{\hbar^2}{2\overline{x_3}}\left(\frac{n_z \pi}{d_z}\right)^2 - \alpha\frac{\hbar^4}{4(\overline{x_3})\overline{x_5}}\left(\frac{n_y \pi}{d_y}\right)^2\left(\frac{n_z \pi}{d_z}\right)^2 - \alpha\frac{\hbar^4}{4(\overline{x_3})\overline{x_6}}\left(\frac{n_y \pi}{d_y}\right)^2\left(\frac{n_z \pi}{d_z}\right)^2 - \frac{\hbar^2}{2m_2}\left(\frac{n_y \pi}{d_y}\right)^2 - \frac{\hbar^2}{2m_3}\left(\frac{n_z \pi}{d_z}\right)^2\right].$$

Combining (2.75b) and (1.13), the TPSM in this case assumes the form

$$G_0 = \left(\pi^2 k_B^2 T / (3e)\right)\left[\sum_{n_y=1}^{n_{y\max}}\sum_{n_z=1}^{n_{z\max}}\left[T_{613}\left(E_{F1D}, n_y, n_z\right) + T_{614}\left(E_{F1D}, n_y, n_z\right)\right]\right]^{-1} \times \left[\sum_{n_y=1}^{n_{y\max}}\sum_{n_z=1}^{n_{z\max}}\left[\left\{T_{613}\left(E_{F1D}, n_y, n_z\right)\right\}' + \left\{T_{614}\left(E_{F1D}, n_y, n_z\right)\right\}'\right]\right]. \quad (2.76)$$

2.2.6 Magnetothermopower in Quantum-Confined Stressed Materials

In the presence of size quantization along x-direction, the 2D electron energy spectrum in stressed materials can be expressed using (1.105) as

$$\frac{k_y^2}{[b^*(E)]^2}+\frac{k_z^2}{[c^*(E)]^2}=1-\frac{1}{[a^*(E)]^2}\left(\pi n_x/d_x\right)^2. \quad (2.77)$$

The subband energy (E_{n_x}) is given by

$$a^*(E_{n_x})=n_x\pi/d_x. \quad (2.78)$$

The area of 2D wave-vector space enclosed by (2.77) can be written as

$$A=\pi b^*(E)c^*(E)\left[1-\left(\frac{\pi n_x}{d_x a^*(E)}\right)^2\right]. \quad (2.79)$$

The electron statistics per unit area in this case can be expressed as

$$n_0=\frac{g_v}{2\pi}\sum_{n_x=1}^{n_{x\max}}[t_{23}(E_{F2D},n_x)+t_{24}(E_{F2D},n_x)], \quad (2.80)$$

where $t_{23}(E_{F2D},n_x)\equiv\{b^*(E_{F2D})c^*(E_{F2D})[1-(\frac{\pi n_x}{d_x a^*(E_{F2D})})^2]\}$, $t_{24}(E_{F2D},n_x)\equiv\sum_{r=1}^{s_0}Z_{r,Y}[t_{23}(E_{F2D},n_x)]$, and $Y=2$.

Combining (2.80) and (1.13), the TPSM in this case can be expressed as

$$G_0=\left(\pi^2 k_B T/(3e)\right)\left[\sum_{n_x=1}^{n_{x\max}}[t_{23}(E_{F2D},n_x)+t_{24}(E_{F2D},n_x)]\right]^{-1} \times\left[\sum_{n_x=1}^{n_{x\max}}[\{t_{23}(E_{F2D},n_x)\}'+\{t_{24}(E_{F2D},n_x)\}']\right]. \quad (2.81)$$

The 1D dispersion relation in stressed Kane type materials can be written extending (2.77) as

$$k_x^2=[a^*(E)]^2\left\{1-\frac{1}{[b^*(E)]^2}\left(\frac{\pi n_y}{d_y}\right)^2-\frac{1}{[c^*(E)]^2}\left(\frac{\pi n_z}{d_z}\right)^2\right\}. \quad (2.82)$$

The quantized energy level $E\prime$ in this case is given by

$$\frac{1}{[b^*(E')]^2}\left(\frac{\pi n_y}{d_y}\right)^2+\frac{1}{[c^*(E')]^2}\left(\frac{\pi n_z}{d_z}\right)^2=1. \tag{2.83}$$

Using (2.82), the 1D electron statistics per unit length can be written as

$$n_0=\frac{2g_v}{\pi}\sum_{n_y=1}^{n_{y\max}}\sum_{n_z=1}^{n_{z\max}}\left[p_1(E_{\mathrm{F1D}},n_y,n_z)+p_2(E_{\mathrm{F1D}},n_y,n_z)\right], \tag{2.84}$$

where
$p_1(E_{\mathrm{F1D}},n_y,n_z)\equiv\left[[a^*(E_{\mathrm{F1D}})]^2\left\{1-\frac{1}{[b^*(E_{\mathrm{F1D}})]^2}\left(\frac{\pi n_y}{d_y}\right)^2-\frac{1}{[c^*(E_{\mathrm{F1D}})]^2}\left(\frac{\pi n_z}{d_z}\right)^2\right\}\right]^{1/2}$,
$p_2(E_{\mathrm{F1D}},n_y,n_z)\equiv\sum_{r=1}^{s_0}Z_{r,Y}\left[p_1(E_{\mathrm{F1D}},n_y,n_z)\right]$, and $Y=1$.

Combining (2.84) and (1.13), the TPSM in this case can be expressed as

$$G_0=\left(\pi^2k_B^2T/(3e)\right)\left[\sum_{n_y=1}^{n_{y\max}}\sum_{n_z=1}^{n_{z\max}}\left[p_1(E_{\mathrm{F1D}},n_y,n_z)+p_2(E_{\mathrm{F1D}},n_y,n_z)\right]\right]^{-1}$$
$$\left[\sum_{n_y=1}^{n_{y\max}}\sum_{n_z=1}^{n_{z\max}}\left[\{p_1(E_{\mathrm{F1D}},n_y,n_z)\}'+\{p_2(E_{\mathrm{F1D}},n_y,n_z)\}'\right]\right]. \tag{2.85}$$

2.2.7 *Magnetothermopower in Carbon Nanotubes*

For armchair and zigzag CNs, the energy dispersion relations are given by [89]

$$E=t_c\left[1+4\cos\left(\frac{\nu\pi}{n}\right)\cos\left(\frac{k_ya_c\sqrt{3}}{2}\right)+4\cos^2\left(\frac{k_ya_c\sqrt{3}}{2}\right)\right]^{1/2},$$
$$\left(-\pi/\sqrt{3}a_c<k_y<\pi/\sqrt{3}a_c\right), \tag{2.86}$$

$$E=t_c\left[1+4\cos\left(\frac{3k_ya_c}{2}\right)\cos\left(\frac{\nu\pi}{n}\right)+4\cos^2\left(\frac{\nu\pi}{n}\right)\right]^{1/2},$$
$$\left(-\pi/3a_c<k_y<\pi/3a_c\right), \tag{2.87}$$

where t_c is the tight binding parameter, $\nu=1,2,\ldots,2n$, a_c is the nearest neighbor C–C bonding distance.

Using (2.86) and (2.87), the electron statistics for both the cases can, respectively, be written as [89],

$$n_0 = \frac{8}{\pi a_c \sqrt{3}} \sum_{i=1}^{i_{\max}} \left[A_{c_1}\left(E_{F_1}, i\right) + B_{c_1}\left(E_{F_1}, i\right) \right], \tag{2.88}$$

$$n_0 = \frac{8}{3\pi a_c} \sum_{i=1}^{i_{\max}} \left[A_{c_2}\left(E_{F_1}, i\right) + B_{c_2}\left(E_{F_1}, i\right) \right], \tag{2.89}$$

where

$$A_{c_1}\left(E_{F_1}, i\right) = \cos^{-1}\left[\frac{1}{8}\left[-\left(\frac{E_i^2}{t_c^2} - 5\right) + \left[\left(\frac{E_i^2}{t_c^2} - 5\right)^2 + 16\left(\frac{E_{F_1}^2}{t_c^2} - 1\right) \right]^{1/2} \right] \right],$$

$$E_i = \frac{|3i - m + n|}{2} |t_c| \frac{a_c}{r_0},$$

r_0 is the radius of the nanotube, E_{F_1} is the Fermi energy as measured from the middle of the band gap in the vertically upward direction $i = 1, 2, 3, \ldots, i_{\max}$, $B_{c_1}\left(E_{F_1}, i\right) = \sum_{r=1}^{s} Z_{r,Y}\left[A_{c_1}\left(E_{F_1}, i\right)\right]$, $Y = 1$,

$$A_{c_2}\left(E_{F_1}, i\right) = \cos^{-1}\left[\left[\frac{E_{F_1}^2}{t_c^2} - 1 - \left(\frac{E_i}{t_c} - 1\right)^2 \right] \left[\frac{2E_i}{t_c} - 1 \right]^{-1} \right],$$

and

$$B_{c_2}\left(E_{F_1}, i\right) = \sum_{r=1}^{s} Z_{r,Y}\left[B_{c_1}\left(E_{F_1}, i\right)\right].$$

Combining (2.88) and (2.89) separately with (1.13), the TPSM for armchair and zigzag nanotubes can, respectively, be expressed as

$$G_0 = \left(\pi^2 k_B^2 T / (3e)\right) \left[\sum_{i=1}^{i_{\max}} \left[A_{c_1}\left(E_{F_1}, i\right) + B_{c_1}\left(E_{F_1}, i\right) \right] \right]^{-1} \left[\sum_{i=1}^{i_{\max}} \left[\left\{A_{c_1}\left(E_{F_1}, i\right)\right\}' + \left\{B_{c_1}\left(E_{F_1}, i\right)\right\}' \right] \right] \tag{2.90}$$

$$G_0 = \left(\pi^2 k_B^2 T / (3e)\right) \left(\overline{\partial_{12}}\right) \cdot \left(\overline{\partial_{11}}\right), \tag{2.91}$$

where

$$\overline{\partial_{12}} = \left[\sum_{i=1}^{i_{\max}} [A_{c_2}(E_{F_1}, i) + B_{c_2}(E_{F_1}, i)]\right]^{-1}$$

and

$$\overline{\partial_{11}} = \left[\sum_{i=1}^{i_{\max}} [\{A_{c_2}(E_{F_1}, i)\}' + \{B_{c_2}(E_{F_1}, i)\}']\right].$$

2.3 Results and Discussion

Using (2.8) and (2.9) and taking the energy band constants as given in Table 1.1, the normalized TPSM in UFs of $CdGeAs_2$ (an example of nonlinear optical materials) has been plotted as a function of film thickness as shown in Fig. 2.1 in accordance with the generalized band model ($\delta \neq 0$), three (using (2.12) and (2.13)) and two (using (2.20) and (2.24)) band models of Kane together with parabolic (using (2.32) and (2.33)) energy bands as shown by curves (a), (c), (d), and (e), respectively. The special case for $\delta = 0$ has also been shown in plot (b) in the same figure to assess the influence of crystal field splitting.

The figure exhibits that the TPSM in UFs of $CdGeAs_2$ increases with increasing film thickness and shows nonideal quantum step behavior due to finite temperature. The TPSM is highest for the parabolic energy bands and lowest for the generalized band model ($\delta \neq 0$). Figure 2.2 exhibits the plots of the normalized TPSM in UFs of $CdGeAs_2$ as a function of the surface electron concentration per unit area

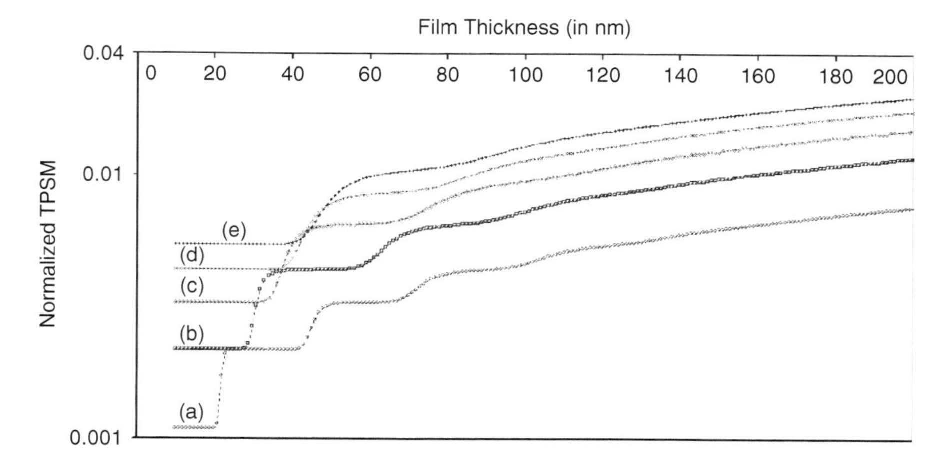

Fig. 2.1 Plot of the normalized TPSM in UFs of $CdGeAs_2$ as a function of film thickness in accordance with (**a**) the generalized band model ($\delta \neq 0$), (**b**) $\delta = 0$, (**c**) the three and (**d**) the two band models of Kane together with (**e**) the parabolic energy bands

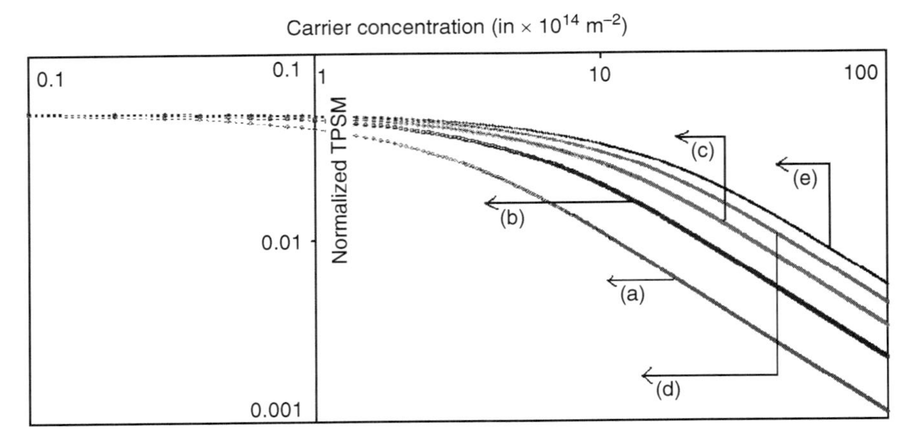

Fig. 2.2 Plot of the normalized TPSM in UFs of $CdGeAs_2$ as a function of carrier concentration for all cases of Fig. 2.1

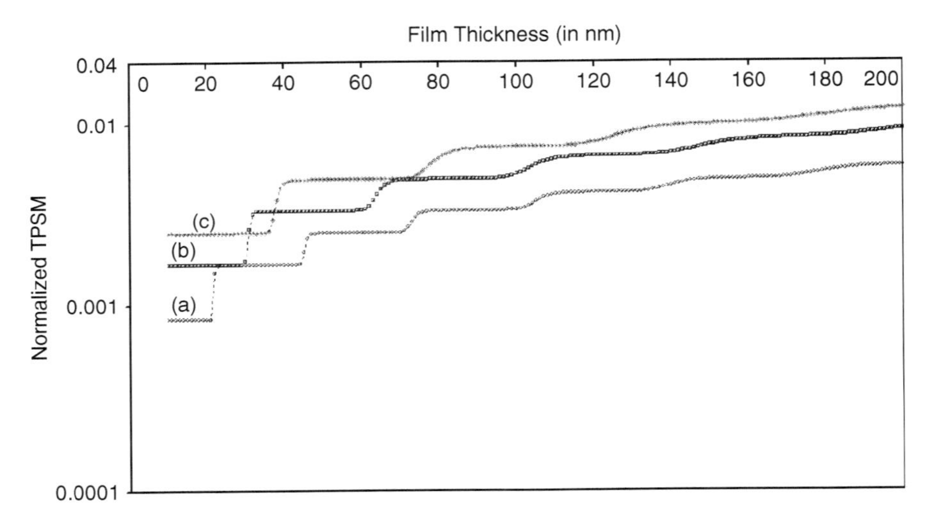

Fig. 2.3 Plot of the normalized TPSM in UFs of InAs as a function of film thickness in accordance with the (**a**) three and (**b**) two band models of Kane together with (**c**) parabolic energy bands

for all cases of Fig. 2.1. For low values of electron concentration, the TPSM in UFs of $CdGeAs_2$ for all cases of Fig. 2.1 shows converging tendencies, whereas with increasing electron concentration, the TPSM for all cases shows substantial difference with each other. The TPSM is highest for the parabolic energy bands and lowest for the generalized band model ($\delta \neq 0$).

Figure 2.3 exhibits the normalized TPSM for UFs of InAs as a function of film thickness for three and two band models of Kane together with parabolic energy bands as shown by curves (a), (b), and (c), respectively, in both the figures. The

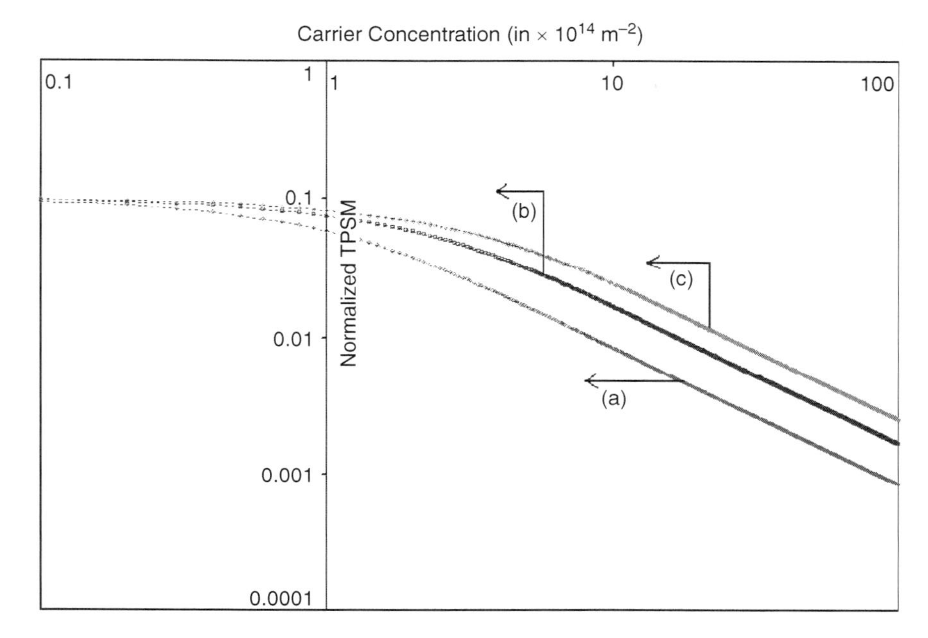

Fig. 2.4 Plot of the normalized TPSM in UFs of InAs as a function of carrier concentration for all cases of Fig. 2.3

TPSM increases with increasing film thickness in an oscillatory way and highest for parabolic energy bands and lowest for the three band model of Kane. The TPSM is highest for the parabolic energy bands and lowest for the three band model of Kane.

Figure 2.4 exhibits the plots of the normalized TPSM in UFs of InAs as a function of the surface electron concentration per unit area for all cases of Fig. 2.3. For low values of electron concentration, the TPSM in UFs of InAs for all cases of Fig. 2.3 shows converging tendencies, whereas with increasing electron concentration, the TPSM for all cases shows substantial difference with each other. The TPSM is highest for the parabolic energy bands and lowest for the three band model of Kane.

Figure 2.5 exhibits the normalized TPSM for UFs of InSb as a function of film thickness for three and two band models of Kane together with parabolic energy bands as shown by curves (a), (b), and (c), respectively, in both the figures. The TPSM increases with increasing film thickness in an oscillatory way and highest for parabolic energy bands and lowest for the three band model of Kane. The TPSM is highest for the parabolic energy bands and lowest for the three band model of Kane.

Figure 2.6 exhibits the plots of the normalized TPSM in UFs of InSb as a function of the surface electron concentration per unit area for all cases of Fig. 2.3. For low values of electron concentration, the TPSM in UFs of InSb for all cases of Fig. 2.6 shows converging tendencies, whereas with increasing electron concentration, the TPSM for all cases shows substantial difference with each other. The

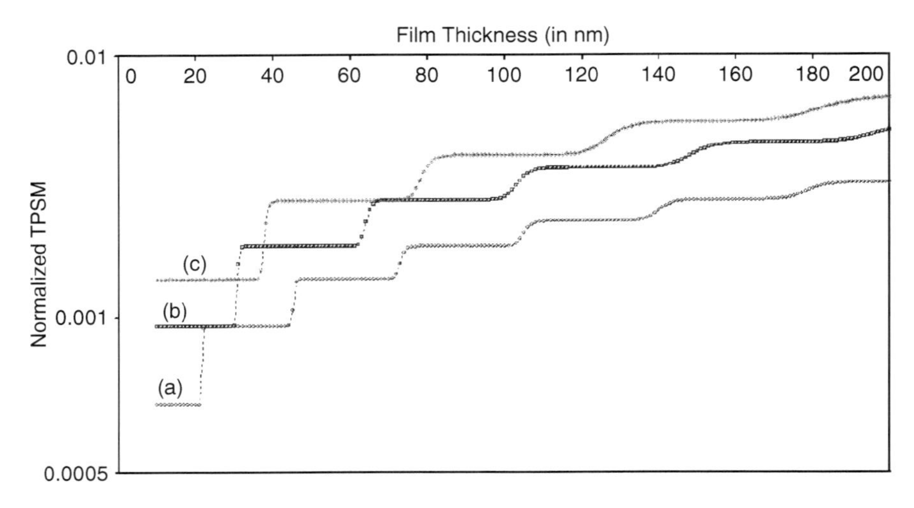

Fig. 2.5 Plot of the normalized TPSM of UFs of InSb as a function of film thickness for all cases of Fig. 2.3

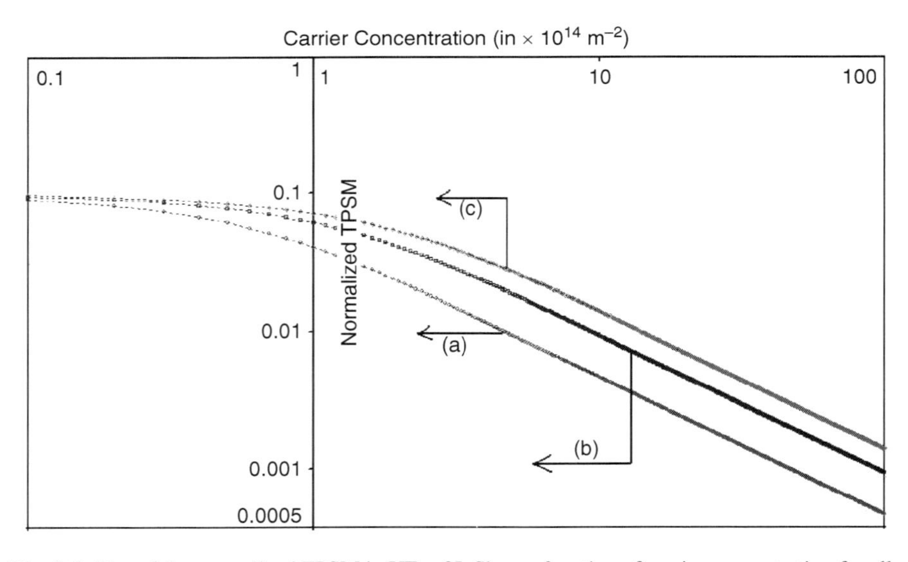

Fig. 2.6 Plot of the normalized TPSM in UFs of InSb as a function of carrier concentration for all cases of Fig. 2.3

TPSM is highest for the parabolic energy bands and lowest for the three band model of Kane.

In Fig. 2.7, the normalized TPSM has been plotted in UFs of CdS (using (2.42) and (2.43)) as a function of film thickness for both $C_0 = 0$ and $C_0 \neq 0$ as shown by curves (b) and (a), respectively, for the purpose of assessing the splitting of the two spin states by the spin orbit coupling and the crystalline field. It appears from the figure that for low values of film thickness, the numerical magnitude of the TPSM in the presence of C_0 is greater than that in the absence of the same and for relatively higher values of film thickness, the influence of the term C_0 decreases. Figure 2.8 shows the corresponding carrier statistics dependence of the TPSM for all cases of Fig. 2.7. It appears from Fig. 2.8 that for relatively large values of carrier concentration, the influence of the term C_0 increases. In Fig. 2.9, the normalized TPSM has been plotted for UFs of PbTe, PbSnTe (using (2.74) and (2.75)), and stressed InSb (using (2.80) and (2.81)) as a function of film thickness in accordance with the appropriate band models as shown by curves (a), (b), and (c), respectively. It appears that normalized TPSM changes with changing film thickness and for

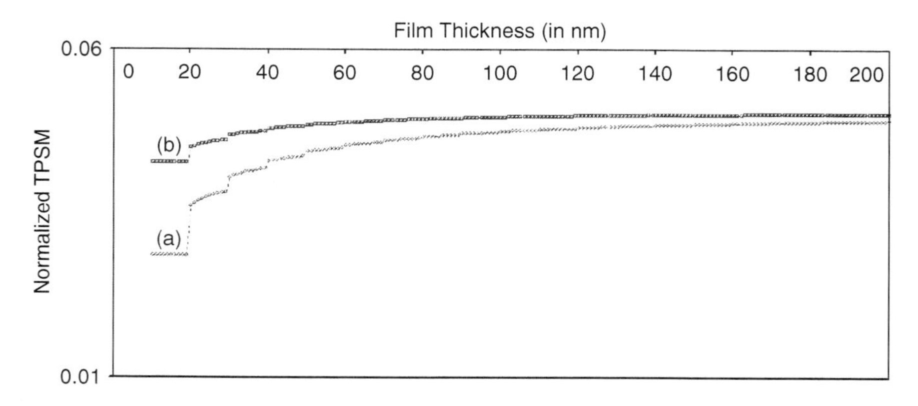

Fig. 2.7 Plot of the normalized TPSM in UFs of CdS as a function of film thickness for (**a**) $C_0 \neq 0$ and (**b**) $C_0 = 0$ in accordance with the model of Hopfield

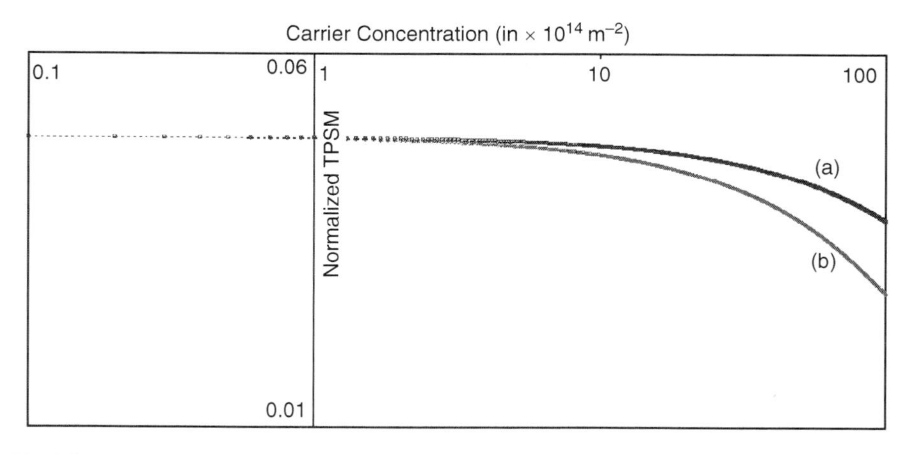

Fig. 2.8 Plot of the normalized TPSM of UFs of CdS as a function of carrier concentration for the cases of Fig. 2.7

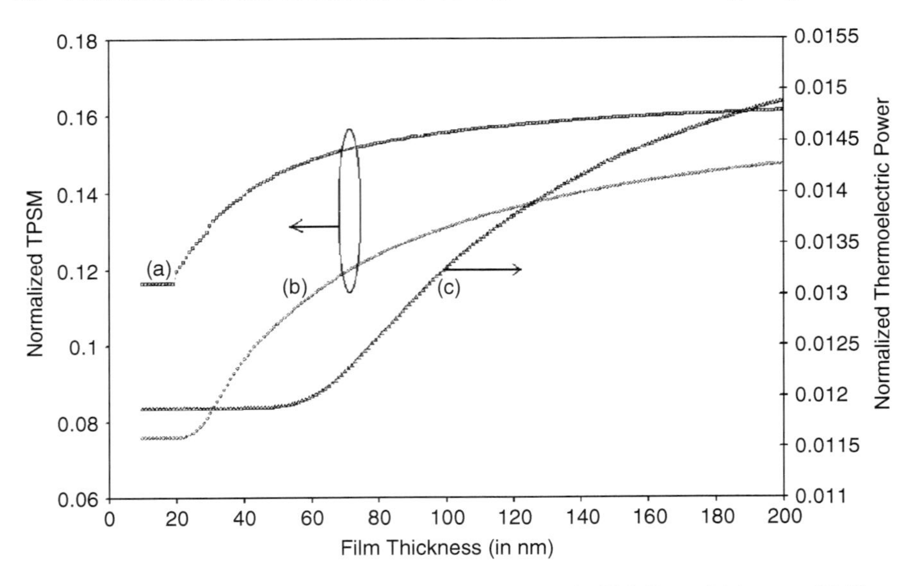

Fig. 2.9 Plot of the normalized TPSM for UFs of (**a**) PbTe, (**b**) PbSnTe, and (**c**) stressed InSb as a function of film thickness

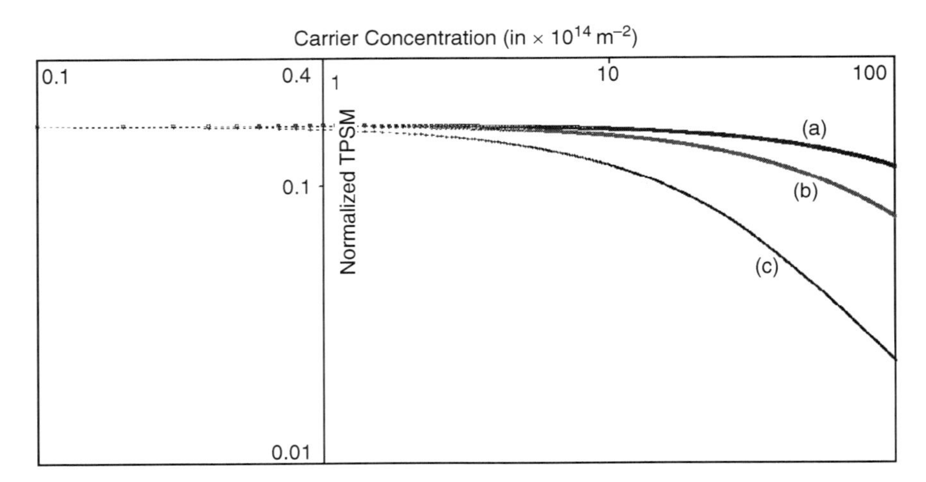

Fig. 2.10 Plot of the normalized TPSM for UFs of (**a**) PbTe, (**b**) PbSnTe, and (**c**) stressed InSb as a function of carrier concentration

relatively low values of film thickness, the numerical values of TPSM for PbTe, PbSnTe, and stressed InSb in accordance with the appropriate band models differ widely. Figure 2.10 exhibits the corresponding dependence on the surface electron concentration per unit area. From Fig. 2.10, we can write that for relatively large values of carrier concentration, the TPSM for stressed InSb exhibits the least value, whereas for PbTe exhibits the highest. Figure 2.11 demonstrates the plots of the

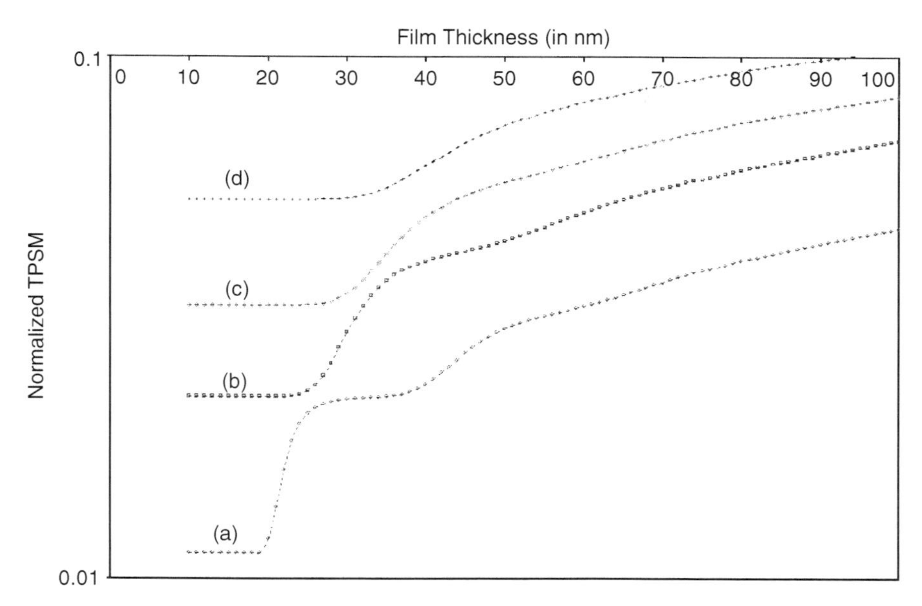

Fig. 2.11 Plot of the normalized TPSM in UFs of bismuth as a function of film thickness in accordance with the models of (**a**) McClure et al., (**b**) Takaoka et al. (Hybrid model), (**c**) Cohen, and (**d**) Lax et al., respectively

normalized TPSM in UFs of bismuth as a function of film thickness in accordance with the models of McClure and Choi (using (2.52) and (2.53)), Hybrid (using (2.56e) and (2.56f)), Cohen (using (2.61) and (2.62)) and Lax ellipsoidal (using (2.69) and (2.70)), respectively. It is apparent from Fig. 2.11 that the value of the TPSM in UFs of bismuth is least for McClure and Choi model and highest for the model of Lax et al. Besides, all the models of bismuth exhibit wide variation with respect to each other. In Fig. 2.12, the normalized TPSM has been plotted as a function of carrier concentration for all cases of Fig. 2.11. From Fig. 2.12, it is apparent that for relatively large values of electron concentration, the TPSM in UFs of bismuth exhibit wide variation for all the models of bismuth as considered here and they decrease with increasing carrier concentration.

Figures 2.13–2.24 exhibit the corresponding dependences of the normalized TPSM for QWs of all the materials in accordance with all the band models and obtained by using the appropriate equations as formulated in this chapter. Figure 2.25 demonstrates the plots of the normalized TPSM of (13, 6) chiral semiconductor CN, (16, 0) zigzag semiconductor CN, (10, 10) metallic armchair CN, and (22, 19) chiral metallic CNs as a function of electron statistics (using (2.88); (2.89) and (2.90); (2.91)) as shown by plots (a), (b), (c), and (d), respectively.

From Fig. 2.13, it appears that the TPSM in QWs of $CdGeAs_2$ for all the models of the same material exhibits quantized variations with increasing film thickness. For a range of film thickness, the dependence exhibits trapezoidal variations and

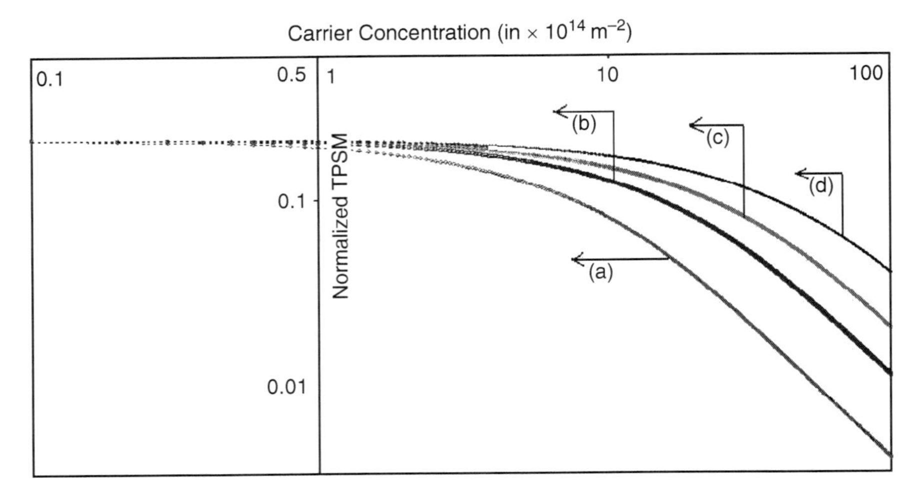

Fig. 2.12 Plot of the normalized TPSM for UFs of bismuth as a function of carrier concentration for all cases of Fig. 2.11

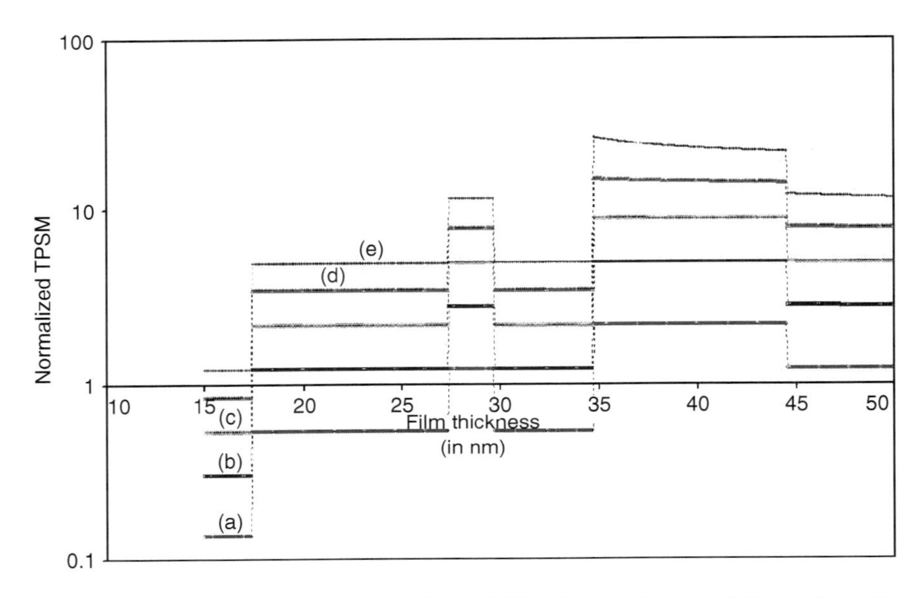

Fig. 2.13 Plot of the normalized TPSM for QWs of $CdGeAs_2$ as a function of film thickness for all cases of Fig. 2.1

for higher values of film thickness, the length and width of the trapezoid increase. From Fig. 2.14, it appears that the TPSM decreases with increasing carrier concentration per unit length and the value of the TPSM is least for the generalized band model and greatest for the parabolic band model of the same. From Fig. 2.15, we

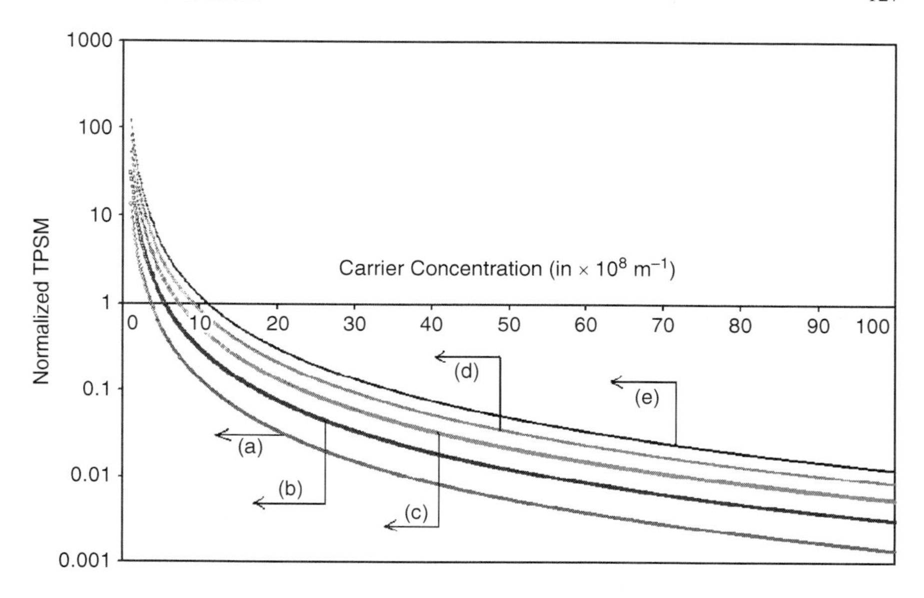

Fig. 2.14 Plot of the normalized TPSM for QWs of quantum wires of $CdGeAs_2$ as a function of carrier concentration for all cases of Fig. 2.2

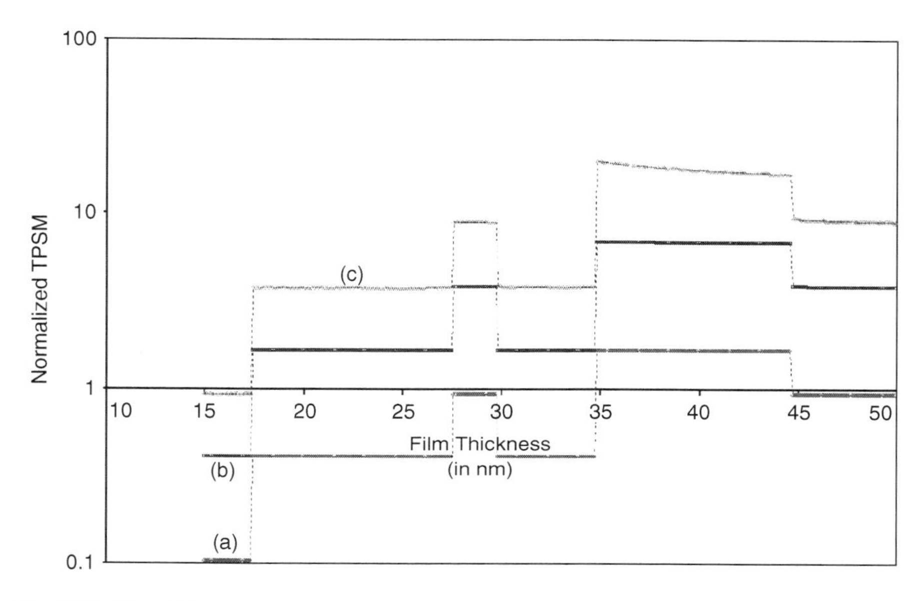

Fig. 2.15 Plot of the normalized TPSM for QWs of InAs as a function of film thickness for all cases of Fig. 2.3

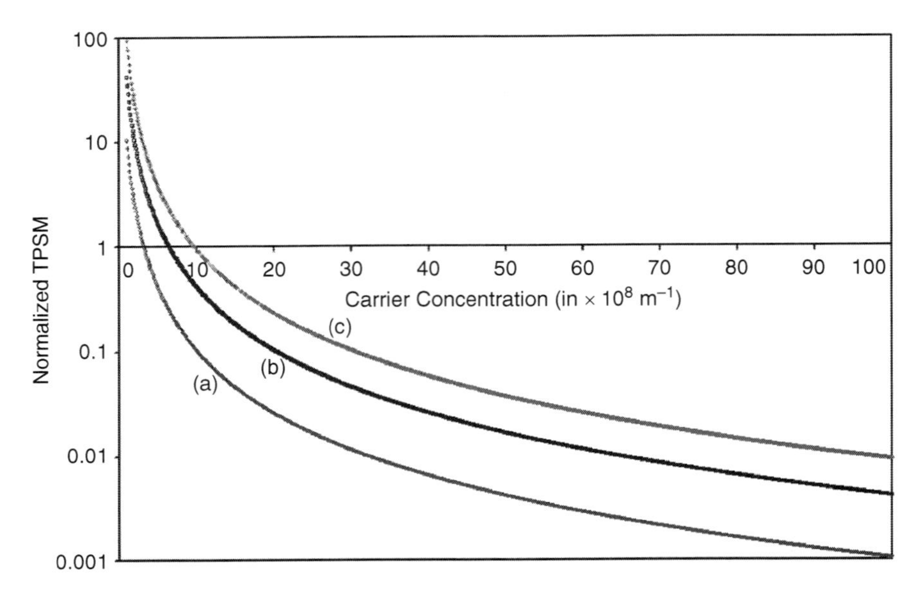

Fig. 2.16 Plot of the normalized TPSM for QWs of InAs as a function of carrier concentration for all cases of Fig. 2.4

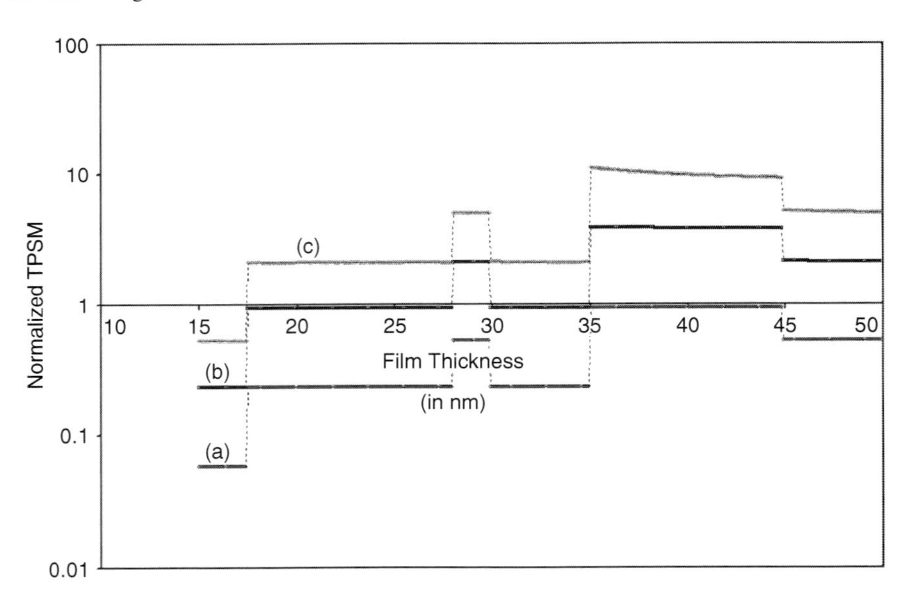

Fig. 2.17 Plot of the normalized TPSM for QWs of InSb as a function of film thickness for all cases of Fig. 2.5

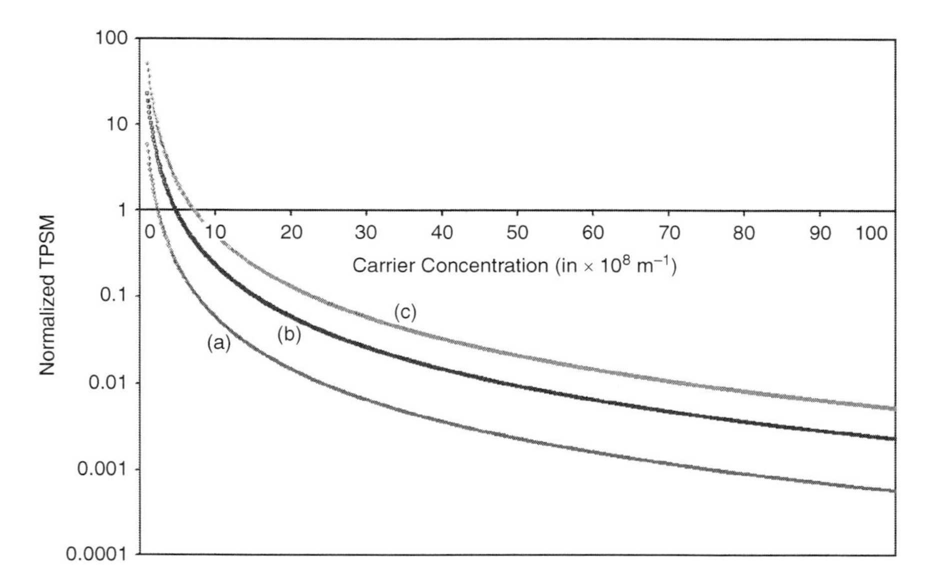

Fig. 2.18 Plot of the normalized TPSM for QWs of InSb as a function of carrier concentration for all cases of Fig. 2.6

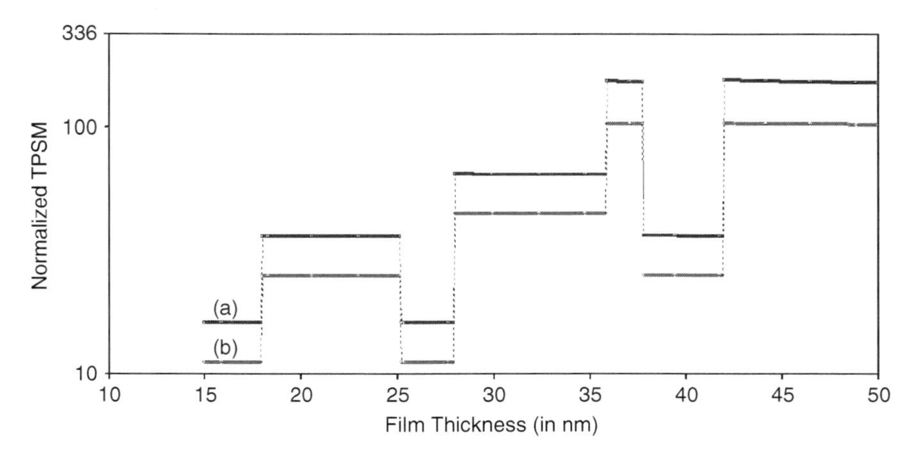

Fig. 2.19 Plot of the normalized TPSM for QWs of CdS as a function of film thickness for all cases of Fig. 2.7

observe that the TPSM for QWs of InAs exhibits the lowest value in accordance with the three band model of Kane model of the same, whereas for parabolic energy bands it exhibits the highest value. It is apparent from plot (a) of Fig. 2.15 that the influence of the energy band gap is to reduce the value of the TPSM as compared with parabolic energy bands and the influence of spin orbit splitting constant is to reduce the TPSM further in the whole range of thickness as compared with two band

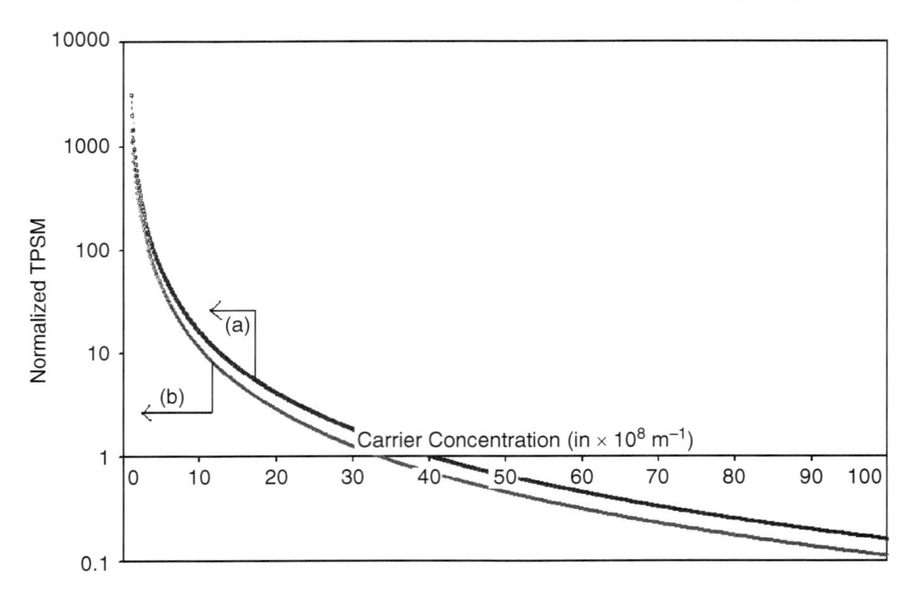

Fig. 2.20 Plot of the normalized TPSM for QWs of CdS as a function of carrier concentration for all cases of Fig. 2.8

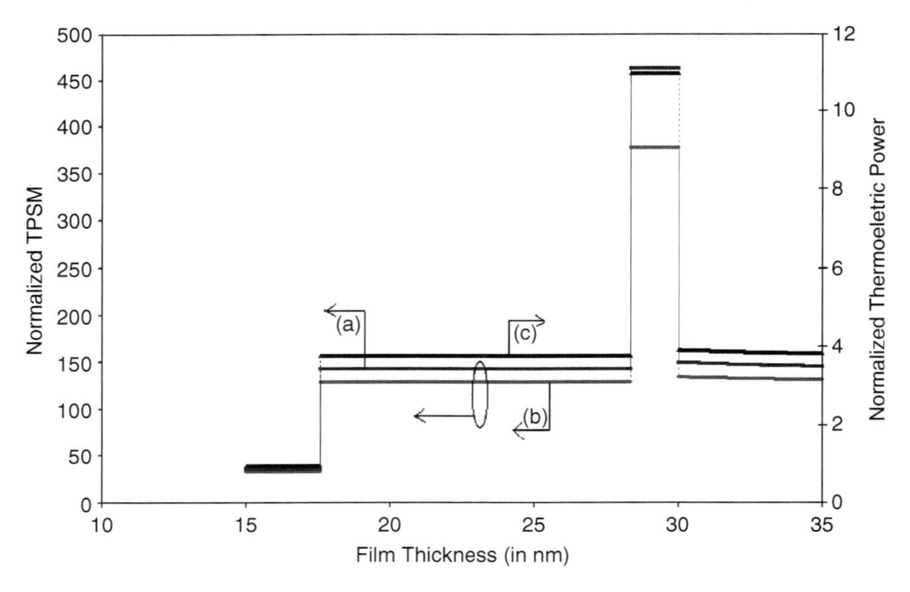

Fig. 2.21 Plot of the normalized TPSM for QWs of (a) PbTe, (b) PbSnTe, and (c) stressed InSb as a function of film thickness

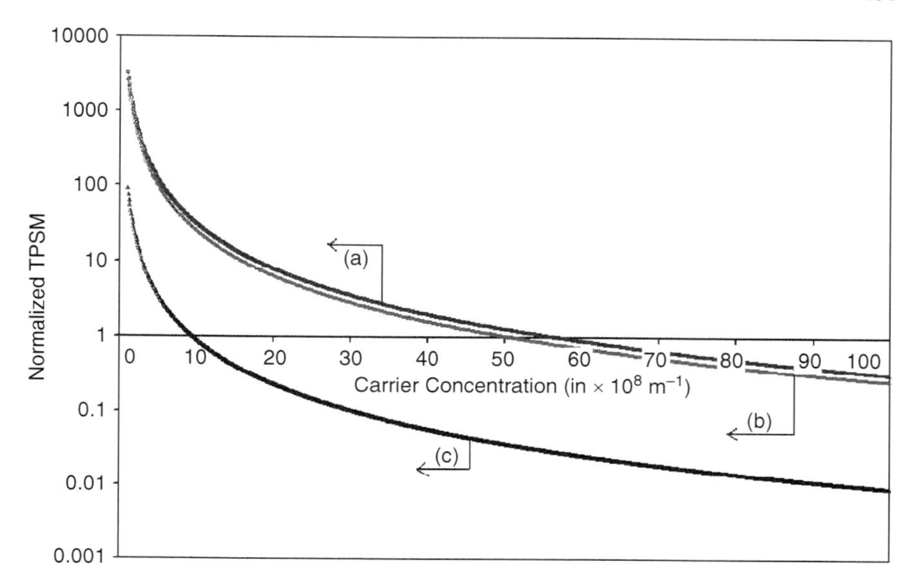

Fig. 2.22 Plot of the normalized TPSM for QWs of (**a**) PbTe, (**b**) PbSnTe, and (**c**) stressed InSb as a function of carrier concentration

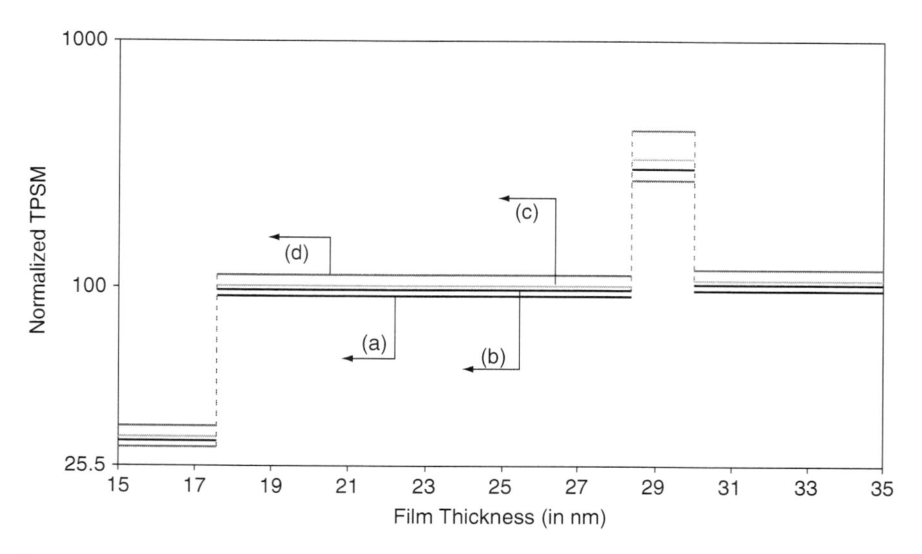

Fig. 2.23 Plot of the normalized TPSM for QWs of bismuth as a function of film thickness for all cases of Fig. 2.11

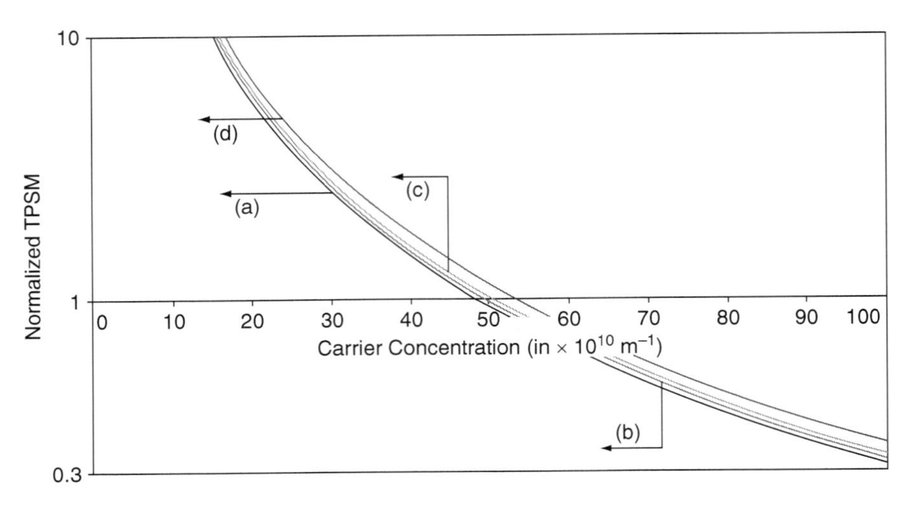

Fig. 2.24 Plot of the normalized TPSM for QWs of bismuth as a function of carrier concentration for all cases of Fig. 2.12

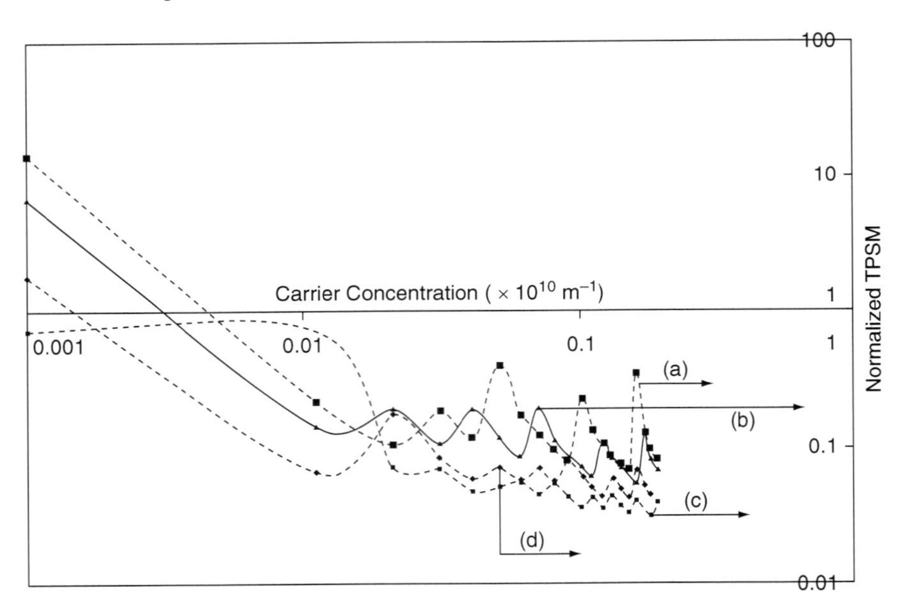

Fig. 2.25 Plot of the normalized TPSM of (**a**) (13, 6) chiral semiconductor CN, (**b**) (16, 0) zigzag semiconductor CN, (**c**) (10, 10) metallic armchair CN, and (**d**) (22, 19) chiral metallic CNs as a function of carrier concentration

model of Kane. From Fig. 2.16, we observe that TPSM decreases with increasing concentration and the value of TPSM in accordance with all the band models differs widely as concentration increases. Figures 2.17 and 2.18 exhibit the plot of the TPSM for QWs of InSb as functions of thickness and concentration, respectively. Nature of these two figures does not differ as compared with the plot of TPSM for QWs of InAs as given in Figs. 2.15 and 2.16, respectively. Important point to note is that although the nature of the plots is same, but the exact numerical values of the TPSM are determined by the numerical values of the energy band constants of InSb and InAs, respectively. From Fig. 2.19, we observe that the influence of the splitting of the two spin states by the spin orbit coupling and the crystalline field enhances the numerical values of the TPSM in QWs of CdS as compared with $C_0 = 0$. Besides, trapezoidal variations of TPSM in QWs of CdS with respect to thickness as appears from Fig. 2.19 are perfect. From Fig. 2.20, we observe that TPSM decreases with increasing carrier concentration per unit length and by comparing it with Fig. 2.8 of the corresponding plot for UFs of CdS, we can write that although TPSM decreases with increasing carrier degeneracy in the latter case but the nature and rate of decrement with increasing concentration are totally different in the QWs of CdS. From Fig. 2.21, we observe that the TPSM for QWs of PbTe, PbSnTe, and stressed InSb in accordance with the appropriate band models exhibits quantum steps and trapezoidal variations in the whole range of thickness as considered with widely different numerical values as apparent from the said figure. From Fig. 2.22, the TPSM decreases with increasing carrier degeneracy for QWs of PbTe, PbSnTe, and stressed InSb, respectively. From Fig. 2.23, we can write that the TPSM exhibits quantum step and quantum trapezoid variations with respect to film thickness for QWs of bismuth and from Fig. 2.24, it is apparent that TPSM decreases with increasing carrier degeneracy for all the models of the same in the present context.

Figure 2.25 exhibits the variation of the TPSM for metallic and semiconductor CNs, respectively, and it appears that the TPSM decreases with the increasing electron degeneracy. It appears from Fig. 2.25 that the TPSM in CNs exhibits oscillatory dependence with increasing carrier degeneracy in a completely different manner as compared with that of UFs and QWs, respectively. The numerical values of the TPSM in all (m, n) cases vary widely, and are determined thoroughly by the chiral indices and diameter of the CNs. From Fig. 2.25, the influence of chiral index numbers on the TPSM in CNs can be assessed. The oscillatory dependence is due to the crossing over of the Fermi level by the quantized level due to van Hove singularities. It should be noted that the rate of increment are totally dependent on the band structure and the spectrum constants of the CNs. This oscillatory dependence will be less and less prominent with increasing nanotube radius and carrier degeneracy, respectively. Ultimately, for larger diameters, the TPSM will be found to be less prominent resulting in monotonic decreasing variation.

The influence of 1D quantum confinement is immediately apparent from Figs. 2.1, 2.3, 2.5, 2.7, 2.9, and 2.11 and the same is also the case for 2D quantum confinement as inferred from Figs. 2.13, 2.15, 2.17, 2.19, 2.21, and 2.23, respectively, since the TPSM depends strongly on the thickness of the quantum-confined materials which is in direct contrast with bulk specimens. The TPSM increases with increasing film thickness in an oscillatory way with different numerical magnitudes for UFs and QWs, respectively. It appears from the aforementioned figures that the TPSM in QWs exhibits spikes for particular values of film thickness which, in turn, is not only the signature of the asymmetry of the wave-vector space but also the particular band structure of the specific material. Moreover, the TPSM in UFs and QWs of different compounds can become several orders of magnitude larger than that of the bulk specimens of the same materials, which is also a direct signature of quantum confinement. This oscillatory dependence will be less and less prominent with increasing film thickness.

It appears from Figs. 2.2, 2.4, 2.6, 2.8, 2.10, and 2.12 that the TPSM decreases with increasing carrier degeneracy for 1D quantum confinement as considered for the said figures. For relatively high values of carrier degeneracy, the influence of band structure of a specific 2D material is large and the plots of TPSM differ widely from one another, whereas for low values of the carrier degeneracy, they exhibit the converging tendency. For bulk specimens of the same material, the TPSM will be found to decrease continuously with increasing electron degeneracy in a nonoscillatory manner in an altogether different way.

For QWs, the TPSM increases with increasing film thickness in a step-like manner for all the appropriate figures. The appearance of the discrete jumps in the figures for QWs is due to the redistribution of the electrons among the quantized energy levels when the size quantum number corresponding to the highest occupied level changes from one fixed value to the others. With varying thickness, a change is reflected in the TPSM through the redistribution of the electrons among the size-quantized levels. It should be noted that although the TPSM varies in various manners with all the variables in all the cases as evident from all the figures, the rates of variations are totally band structure dependent. The two different signatures of 2D and 1D quantization of the carriers of UFs and QWs of all the materials as considered here are apparent from all the appropriate plots. Values of TPSM for QWs differ as compared with UFs and the nature of variations of the TPSM also changes accordingly.

For the purpose of condensed presentation, the carrier statistics and the TPSM for UFs and QWs for all the materials as considered in this chapter have been presented in Table 2.1.

2.4 Open Research Problems

(R2.1) Investigate the DTP, PTP, and Z in the absence of magnetic field by considering all types of scattering mechanisms for UFs by considering the presence of finite, symmetric infinite, asymmetric infinite, parabolic, finite circular, infinite circular, annular infinite and elliptic potential wells applied separately for all the materials whose unperturbed carrier energy spectra are defined in Chap. 1.

Table 2.1 The carrier statistics and the thermoelectric power under large magnetic field in ultrathin films and quantum wires of nonlinear optical, Kane type III–V, II–VI, bismuth, IV–VI, stressed materials, and carbon nanotubes

Type of materials	Carrier statistics	TPSM
1. Nonlinear optical materials	For ultrathin films $n_0 = \frac{g_v}{2\pi} \sum_{n_x=1}^{n_{x\max}} [\phi_1(E_{F2D}, n_x) + \phi_2(E_{F2D}, n_x)]$ (2.3)	For ultrathin films on the basis of (2.3), $G_0 = (\pi^2 k_B^2 T/(3e)) \left[\sum_{n_x=1}^{n_{x\max}} [\phi_1(E_{F2D}, n_x) + \phi_2(E_{F2D}, n_x)]\right]^{-1} \left[\sum_{n_x=1}^{n_{x\max}} \left[\{\phi_1(E_{F2D}, n_x)\}' + \{\phi_2(E_{F2D}, n_x)\}'\right]\right]$ (2.4)
	For quantum wires $n_0 = \left(\frac{2g_v}{\pi}\right) \sum_{n_y=1}^{n_{y\max}} \sum_{n_z=1}^{n_{z\max}} \left[t_1(E_{F1D}, n_y, n_z) . + t_2(E_{F1D}, n_y, n_z)\right]$ (2.8)	For quantum wires on the basis of (2.8), $G_0 = (\pi^2 k_B^2 T/(3e)) \left[\sum_{n_y=1}^{n_{y\max}} \sum_{n_z=1}^{n_{z\max}} \left[t_1(E_{F1D}, n_y, n_z) + t_2(E_{F1D}, n_y, n_z)\right]\right]^{-1} \left[\sum_{n_y=1}^{n_{y\max}} \sum_{n_z=1}^{n_{z\max}} \left[\{t_1(E_{F1D}, n_y, n_z)\}' + \{t_2(E_{F1D, n_y, n_z})\}'\right]\right]$ (2.9)
2. Kane type III–V materials	(a) The three band model of Kane	For ultrathin films on the basis of (2.12),
	For ultrathin films $n_0 = \frac{m^* g_v}{\pi \hbar^2} \sum_{n_x=1}^{n_{x\max}} \left[T_3(E_{F2D}, n_x) + T_4(E_{F2D}, n_x)\right]$ (2.12)	$G_0 = (\pi^2 k_B^2 T/(3e)) \left[\sum_{n_x=1}^{n_{x\max}} [T_3(E_{F2D}, n_x) + T_4(E_{F2D}, n_x)]\right]^{-1} \left[\sum_{n_x=1}^{n_{x\max}} \left[\{T_3(E_{F2D}, n_x)\}' + \{T_4(E_{F2D}, n_x)\}'\right]\right]$ (2.13)
	For quantum wires $n_0 = \frac{2g_v\sqrt{2m^*}}{\pi\hbar} \sum_{n_y=1}^{n_{y\max}} \sum_{n_z=1}^{n_{z\max}} \left[T_5(E_{F1D}, n_y, n_z) + T_6(E_{F1D}, n_y, n_z)\right]$ (2.16)	On the basis of (2.16), $G_0 = (\pi^2 k_B^2 T/(3e)) \left[\sum_{n_y=1}^{n_{y\max}} \sum_{n_z=1}^{n_{z\max}} [T_5(E_{F1D}, n_y, n_z) + T_6(E_{F1D}, n_y, n_z)]\right]^{-1} \left[\sum_{n_y=1}^{n_{y\max}} \sum_{n_z=1}^{n_{z\max}} \left[\{T_5(E_{F1D}, n_y, n_z)\}' + \{T_6(E_{F1D}, n_y, n_z)\}'\right]\right]$ (2.17)

(continued)

Table 2.1 (Continued)

Type of materials	Carrier statistics	TPSM
	(b) The two band model of Kane For ultrathin films $n_0 = \frac{m^* g_v k_B T}{\pi \hbar^2} \sum_{n_x=1}^{n_{x\max}} [(1+2\alpha E_{n_x}) F_0(\eta_n) + 2\alpha k_B T F_1(\eta_n)]$ (2.20)	On the basis of (2.20), $G_0 = (\pi^2 k_B/(3e)) \left[\sum_{n_x=1}^{n_{x\max}} [(1+2\alpha E_{n_x}) F_0(\eta_n) + 2\alpha k_B T F_1(\eta_n)]\right]^{-1} \left[\sum_{n_x=1}^{n_{x\max}} [(1+2\alpha E_{n_x}) F_{-1}(\eta_n) + 2\alpha k_B T F_0(\eta_n)]\right]$ (2.24)
	For quantum wires $n_0 = \frac{2g_v}{\pi} \frac{\sqrt{2m^*}}{\hbar} \sum_{n_y=1}^{n_{y\max}} \sum_{n_z=1}^{n_{z\max}} [T_7(E_{F1D}, n_y, n_z) + T_8(E_{F1D}, n_y, n_z)]$ (2.28)	On the basis of (2.28), $G_0 = (\pi^2 k_B^2 T/(3e)) \left[\sum_{n_y=1}^{n_{y\max}} \sum_{n_z=1}^{n_{z\max}} [T_7(E_{F1D}, n_y, n_z) + T_8(E_{F1D}, n_y, n_z)]\right]^{-1} \left[\sum_{n_y=1}^{n_{y\max}} \sum_{n_z=1}^{n_{z\max}} \left[\{T_7(E_{F1D}, n_y, n_z)\}' + \{T_8(E_{F1D}, n_y, n_z)\}'\right]\right]$ (2.29)
	The (2.28) which is valid for quantum wires whose bulk conduction electrons obey thetwo band model of Kane, under the constraint $\alpha E_{F1D} \ll 1$ assumes the form $n_0 = \left(\frac{2g_v \sqrt{2m^* \pi k_B T}}{h}\right) \sum_{n_y=1}^{n_{y\max}} \sum_{n_z=1}^{n_{z\max}} \frac{1}{\sqrt{i_1}} \left[\left(1+\frac{3}{2}\alpha(\bar{i_2})\right) F_{-1/2}(\overline{\eta_2}) + \frac{3}{4}\alpha k_B T F_{1/2}(\overline{\eta_2})\right]$ (2.30)	On the basis of (2.30), $G_0 = \frac{\pi^2 k_B}{3e} \frac{\sum_{n_y=1}^{n_{y\max}} \sum_{n_z=1}^{n_{z\max}} \frac{1}{\sqrt{i_1}} [(1+\frac{3}{2}\alpha(\bar{i_2})) F_{-3/2}(\overline{\eta_2}) + \frac{3}{4}\alpha k_B T F_{-1/2}(\overline{\eta_2})]}{\sum_{n_y=1}^{n_{y\max}} \sum_{n_z=1}^{n_{z\max}} \frac{1}{\sqrt{i_1}} [(1+\frac{3}{2}\alpha(\bar{i_2})) F_{-1/2}(\overline{\eta_2}) + \frac{3}{4}\alpha k_B T F_{1/2}(\overline{\eta_2})]}$ (2.31)
	For ultrathin films having parabolic energy bands, $\alpha \to 0$ $n_0 = \frac{m^* g_v k_B T}{\pi \hbar^2} \sum_{n_x=1}^{n_{x\max}} [F_0(\eta_1)]$ (2.32)	On the basis of (2.32), $G_0 = (\pi^2 k_B/(3e)) \left[\sum_{n_x=1}^{n_{x\max}} [F_0(\eta_1)]\right]^{-1} \left[\sum_{n_x=1}^{n_{x\max}} [F_{-1}(\eta_1)]\right]$ (2.33) On the basis of (2.34), $G_0 = (\pi^2 k_B/(3e)) \left[\sum_{n_y=1}^{n_{y\max}} \sum_{n_z=1}^{n_{z\max}} [F_{-1/2}(\eta_2)]\right]^{-1} \left[\sum_{n_y=1}^{n_{y\max}} \sum_{n_z=1}^{n_{z\max}} [F_{-3/2}(\eta_2)]\right]$ (2.35)

	For quantum wires having parabolic energy bands, $\alpha \to 0$ $n_0 = \frac{2g_v\sqrt{2\pi m^* k_B T}}{h}\sum_{n_y=1}^{n_{y\max}}\sum_{n_z=1}^{n_{z\max}}\left[F_{-1/2}(\eta_2)\right]$ (2.34)	On the basis of (2.42), $G_0 = (\pi^2 k_B/(3e))\left[\sum_{n_z=1}^{n_{z\max}}\left[F_0\left(\eta_{n_{z3}}\right) - \frac{C_0 f_s(E_{F2D}, n_z)}{2\sqrt{A_0 k_B T}}\right]\right]^{-1} \left[\sum_{n_z=1}^{n_{z\max}}\left[F_{-1}\left(\eta_{n_{z3}}\right) - \frac{C_0 f_s'(E_{F2D}, n_z)}{2\sqrt{A_0 k_B T}}\right]\right]$ (2.43)
3. II–VI materials	For ultrathin films $n_0 = \frac{g_v m_\perp^* k_B T}{\pi \hbar^2}\sum_{n_z=1}^{n_{z\max}}\left[F_0\left(\eta_{n_{z3}}\right) - \frac{C_0 f_s(E_{F2D}, n_z)}{2\sqrt{A_0 k_B T}}\right]$ (2.42) For quantum wires $n_0 = \frac{g_v}{\pi\sqrt{B_0}}\sum_{n_x=1}^{n_{x\max}}\sum_{n_y=1}^{n_{y\max}}\left[t_7(E_{F1D}, n_x, n_y) + t_8(E_{F1D}, n_x, n_y)\right]$ (2.45)	On the basis of (2.45) $G_0 = (\pi^2 k_B^2 T/(3e))\left[\sum_{n_x=1}^{n_{x\max}}\sum_{n_y=1}^{n_{y\max}}\left[t_7(E_{F1D}, n_x, n_y) + t_8(E_{F1D}, n_x, n_y)\right]\right]^{-1}\left[\sum_{n_x=1}^{n_{x\max}}\sum_{n_y=1}^{n_{y\max}}\left[\{t_7(E_{F1D}, n_x, n_y)\}' + \{t_8(E_{F1D}, n_x, n_y)\}'\right]\right]$ (2.46)
4. Bismuth	(a) The McClure and Choi model: For ultrathin films $n_0 = \frac{g_v\sqrt{m_1 m_3}}{\pi\hbar^2}\sum_{n_y=1}^{n_{y\max}}\left[t_{25}(E_{F2D}, n_y) + t_{26}(E_{F2D}, n_y)\right]$ (2.52)	On the basis of (2.52) $G_0 = (\pi^2 k_B^2 T/(3e))\left[\sum_{n_y=1}^{n_{y\max}}\left[t_{25}(E_{F2D}, n_y) + t_{26}(E_{F2D}, n_y)\right]\right]^{-1}\left[\sum_{n_y=1}^{n_{y\max}}\left[\{t_{25}(E_{F2D}, n_y)\}' + \{t_{26}(E_{F2D}, n_y)\}'\right]\right]$ (2.53)
	For quantum wires $n_0 = \frac{2g_v}{\pi}\frac{\sqrt{2m_1}}{\hbar}\sum_{n_y=1}^{n_{y\max}}\sum_{n_z=1}^{n_{z\max}}\left[t_{27}(E_{F1D}, n_y, n_z) + t_{28}(E_{F1D}, n_y, n_z)\right]$ (2.55)	On the basis of (2.55) $G_0 = (\pi^2 k_B^2 T/(3e))\left[\sum_{n_y=1}^{n_{y\max}}\sum_{n_z=1}^{n_{z\max}}\left[t_{27}(E_{F1D}, n_y, n_z) + t_{28}(E_{F1D}, n_y, n_z)\right]\right]^{-1}\left[\sum_{n_y=1}^{n_{y\max}}\sum_{n_z=1}^{n_{z\max}}\left[\{t_{27}(E_{F1D}, n_y, n_z)\}' + \{t_{28}(E_{F1D}, n_y, n_z)\}'\right]\right]$ (2.56)

(continued)

Table 2.1 (Continued)

Type of materials	Carrier statistics	TPSM
	(b) The Hybrid Model:	
	For ultrathin films	On the basis of (2.56e),
	$n_0 = \frac{g_v \sqrt{m_1 m_3}}{\pi \hbar^2} \sum_{n_y=1}^{n_{y\max}} \left[t_{29}\left(E_{F2D}, n_y\right) + t_{30}\left(E_{F2D}, n_y\right) \right]$ (2.56e)	$G_0 = \left(\pi^2 k_B^2 T/(3e)\right) \left[\sum_{n_y=1}^{n_{y\max}} \left[t_{29}\left(E_{F2D}, n_y\right) + t_{30}\left(E_{F2D}, n_y\right) \right] \right]^{-1} \left[\sum_{n_y=1}^{n_{y\max}} \left[\left\{ t_{29}\left(E_{F2D}, n_y\right) \right\}' + \left\{ t_{30}\left(E_{F2D}, n_y\right) \right\}' \right] \right]$ (2.56f)
	For quantum wires	On the basis of (2.56h),
	$n_0 = \frac{2g_v}{\pi} \frac{\sqrt{2m_1}}{\hbar} \sum_{n_y=1}^{n_{y\max}} \sum_{n_z=1}^{n_{z\max}} \left[t_{31}\left(E_{F1D}, n_y, n_z\right) + t_{32}\left(E_{F1D}, n_y, n_z\right) \right]$ (2.56*h*)	$G_0 = \left(\pi^2 k_B^2 T/(3e)\right) \left[\sum_{n_y=1}^{n_{y\max}} \sum_{n_z=1}^{n_{z\max}} \left[t_{31}\left(E_{F1D}, n_y, n_z\right) + t_{32}\left(E_{F1D}, n_y, n_z\right) \right] \right]^{-1} \left[\sum_{n_y=1}^{n_{y\max}} \sum_{n_z=1}^{n_{z\max}} \left[\left\{ t_{31}\left(E_{F1D}, n_y, n_z\right) \right\}' + \left\{ t_{32}\left(E_{F1D}, n_y, n_z\right) \right\}' \right] \right]$ (2.56i)
	(c) The Cohen Model:	On the basis of (2.61)
	For ultrathin flims	
	$n_0 = \frac{g_v \sqrt{m_1 m_3}}{\pi \hbar^2} \sum_{n_y=1}^{n_{y\max}} \left[t_{33}\left(E_{F2D}, n_y\right) + t_{34}\left(E_{F2D}, n_y\right) \right]$ (2.61)	$G_0 = \left(\pi^2 k_B^2 T/(3e)\right) \left[\sum_{n_y=1}^{n_{y\max}} \left[t_{33}\left(E_{F2D}, n_y\right) + t_{34}\left(E_{F2D}, n_y\right) \right] \right]^{-1} \left[\sum_{n_y=1}^{n_{y\max}} \left[\left\{ t_{33}\left(E_{F2D}, n_y\right) \right\}' + \left\{ t_{34}\left(E_{F2D}, n_y\right) \right\}' \right] \right]$ (2.62)
	For quantum wires	On the basis of (2.64)
	$n_0 = \frac{2g_v}{\pi} \frac{\sqrt{2m_1}}{\hbar} \sum_{n_y=1}^{n_{y\max}} \sum_{n_z=1}^{n_{z\max}} \left[t_{35}\left(E_{F1D}, n_y, n_z\right) + t_{36}\left(E_{F1D}, n_y, n_z\right) \right]$ (2.64)	$G_0 = \left(\pi^2 k_B^2 T/(3e)\right) \left[\sum_{n_y=1}^{n_{y\max}} \sum_{n_z=1}^{n_{z\max}} \left[t_{35}\left(E_{F1D}, n_y, n_z\right) + t_{36}\left(E_{F1D}, n_y, n_z\right) \right] \right]^{-1} \left[\sum_{n_y=1}^{n_{y\max}} \sum_{n_z=1}^{n_{z\max}} \left[\left\{ t_{35}\left(E_{F1D}, n_y, n_z\right) \right\}' + \left\{ t_{36}\left(E_{F1D}, n_y, n_z\right) \right\}' \right] \right]$ (2.64a)

	(d) The Lax model:	
	For ultrathin films $$n_0 = \frac{(k_B T g_v)\sqrt{m_1 m_3}}{\pi \hbar^2} \sum_{n_y=1}^{n_{y\max}} [[(1+2\alpha E_{n_y}) F_0(\eta_{n_y}) + 2\alpha k_B T F_1(\eta_{n_y})]] \quad (2.69)$$	On the basis of (2.69) $$G_0 = (\pi^2 k_B/(3e)) \left[\sum_{n_y=1}^{n_{y\max}} [[(1+2\alpha E_{n_y}) F_0(\eta_{n_y}) + 2\alpha k_B T F_1(\eta_{n_y})]]\right]^{-1} \left[\sum_{n_y=1}^{n_{y\max}} [[(1+2\alpha E_{n_y}) F_{-1}(\eta_{n_y}) + 2\alpha k_B T F_0(\eta_{n_y})]]\right] \quad (2.70)$$
	For quantum wires $$n_0 = \frac{2g_v}{\pi} \frac{\sqrt{2m_1}}{\hbar} \sum_{n_y=1}^{n_{y\max}} \sum_{n_z=1}^{n_{z\max}} [t_{37}(E_{F1D}, n_y, n_z) + t_{38}(E_{F1D}, n_y, n_z)] \quad (2.71)$$	On the basis of (2.71) $$G_0 = (\pi^2 k_B^2 T/(3e)) \left[\sum_{n_y=1}^{n_{y\max}} \sum_{n_z=1}^{n_{z\max}} [t_{37}(E_{F1D}, n_y, n_z) + t_{38}(E_{F1D}, n_y, n_z)]\right]^{-1} \left[\sum_{n_y=1}^{n_{y\max}} \sum_{n_z=1}^{n_{z\max}} \left[\{t_{37}(E_{F1D}, n_y, n_z)\}' + \{t_{38}(E_{F1D}, n_y, n_z)\}'\right]\right] \quad (2.72)$$
5. IV–VI materials	For ultrathin films $$n_0 = \frac{g_v}{2\pi} \sum_{n_z=1}^{n_{z\max}} [\overline{T_{55}}(E_{F2D}, n_z) + \overline{T_{56}}(E_{F2D}, n_z)] \quad (2.74)$$	On the basis of (2.74), $$G_0 = (\pi^2 k_B^2 T/(3e)) \left[\sum_{n_z=1}^{n_{z\max}} [\overline{T_{55}}(E_{F2D}, n_z) + \overline{T_{56}}(E_{F2D}, n_z)]\right]^{-1} \left[\sum_{n_z=1}^{n_{z\max}} \left[\{\overline{T_{55}}(E_{F2D}, n_z)\}' + \{\overline{T_{56}}(E_{F2D}, n_z)\}'\right]\right] \quad (2.75)$$
	For quantum wires $$n_0 = \frac{2g_v}{\pi \hbar \sqrt{\bar{g}_1}} \sum_{n_y=1}^{n_{y\max}} \sum_{n_z=1}^{n_{z\max}} \left[T_{613}(E_{F1D}, n_y, n_z) + T_{614}(E_{F1D}, n_y, n_z)\right]]^{-1} \quad (2.75b)$$	On the basis of (2.75b), $$G_0 = (\pi^2 k_B^2 T/(3e)) \left[\sum_{n_y=1}^{n_{y\max}} \sum_{n_z=1}^{n_{z\max}} [T_{613}(E_{F1D}, n_y, n_z)\right. \left[\sum_{n_y=1}^{n_{y\max}} \sum_{n_z=1}^{n_{z\max}} \left[\{T_{613}(E_{F1D}, n_y, n_z)\}' + \{T_{614}(E_{F1D}, n_y, n_z)\}'\right]\right] \quad (2.76)$$

(continued)

Table 2.1 (Continued)

Type of materials	Carrier statistics	TPSM
6. Stressed materials	For ultrathin films	On the basis of (2.80)
	$n_0 = \frac{g_v}{2\pi} \sum_{n_x=1}^{n_{x\max}} [t_{23}(E_{F2D}, n_x) + t_{24}(E_{F2D}, n_x)]$ (2.80)	$G_0 = (\pi^2 k_B T/(3e)) \left[\sum_{n_x=1}^{n_{x\max}} [t_{23}(E_{F2D}, n_x) + t_{24}(E_{F2D}, n_x)] \right]^{-1} \left[\sum_{n_x=1}^{n_{x\max}} [\{t_{23}(E_{F2D}, n_x)\}' + \{t_{24}(E_{F2D}, n_x)\}'] \right]$ (2.81)
	For quantum wires	On the basis of (2.84)
	$n_0 = \frac{2g_v}{\pi\hbar\sqrt{g_1}} \sum_{n_y=1}^{n_{y\max}} \sum_{n_z=1}^{n_{z\max}} \left[p_1 \left(E_{F1D}, n_y, n_z\right) + p_2 \left(E_{F1D}, n_y, n_z\right) \right]$ (2.84)	$G_0 = (\pi^2 k_B^2 T/(3e)) \left[\sum_{n_y=1}^{n_{y\max}} \sum_{n_z=1}^{n_{z\max}} [p_1(E_{F1D}, n_y, n_z) + p_2(E_{F1D}, n_y, n_z)] \right]^{-1} \left[\sum_{n_y=1}^{n_{y\max}} \sum_{n_z=1}^{n_{z\max}} \left[\{p_1(E_{F1D}, n_y, n_z)\}' + \{p_2(E_{F1D}, n_y, n_z)\}' \right] \right]$ (2.85)
7. Carbon nanotubes	For armchair nanotubes	On the basis of (2.88),
	$n_0 = \frac{8}{\pi a_c \sqrt{3}} \sum_{i=1}^{i_{\max}} \left[A_{c_1} \left(E_{F_1}, i\right) + B_{c_1} \left(E_{F_1}, i\right) \right]$ (2.88)	$G_0 = (\pi^2 k_B^2 T/(3e)) \left[\sum_{i=1}^{i_{\max}} \left[A_{c_1} \left(E_{F_1}, i\right) + B_{c_1} \left(E_{F_1}, i\right) \right] \right]^{-1} \left[\sum_{i=1}^{i_{\max}} \left[\{A_{c_1} \left(E_{F_1}, i\right)\}' + \{B_{c_1} \left(E_{F_1}, i\right)\}' \right] \right]$ (2.90)
	For zigzag nanotubes	On the basis of (2.89),
	$n_0 = \frac{8}{3\pi a_c} \sum_{i=1}^{i_{\max}} \left[A_{c_2} \left(E_{F_1}, i\right) + B_{c_2} \left(E_{F_1}, i\right) \right]$ (2.89)	$G_0 = (\pi^2 k_B^2 T/(3e)) \left(\overline{\partial_{12}}\right) \left(\overline{\partial_{11}}\right)$ (2.91)

(R2.2) Investigate the DTP, PTP, and Z in the absence of magnetic field by considering all types of scattering mechanisms for nipi structures, accumulation and inversion layers of all the materials whose unperturbed carrier energy spectra are defined in Chap. 1.

(R2.3) Investigate the DTP, PTP, and Z in the absence of magnetic field by considering all types of scattering mechanisms for QWs by considering the presence of finite and parabolic potential wells applied separately in the two different orthogonal directions of all the materials whose unperturbed carrier energy spectra are defined in Chap. 1.

(R2.4) Investigate the DTP, PTP, and Z in the absence of magnetic field by considering all types of scattering mechanisms for an elliptic hill and quantum square rings of all the materials whose unperturbed carrier energy spectra are defined in Chap. 1.

(R2.5) Investigate (R2.1)–(R2.5) under an arbitrarily oriented (a) nonuniform electric field and (b) alternating electric field, respectively, for all the materials whose unperturbed carrier energy spectra are defined in Chap. 1.

(R2.6) Investigate all the appropriate problems of this chapter under an arbitrarily oriented alternating magnetic field by including broadening and the electron spin for all the materials whose unperturbed carrier energy spectra are defined in Chap. 1.

(R2.7) Investigate the DTP, PTP, and Z for the appropriate problems of this chapter under an arbitrarily oriented alternating magnetic field and crossed alternating electric field by including broadening and the electron spin for all the materials whose unperturbed carrier energy spectra are defined in Chap. 1.

(R2.8) Investigate the DTP, PTP, and Z for the appropriate problems of this chapter under an arbitrarily oriented alternating magnetic field and crossed alternating nonuniform electric field by including broadening and the electron spin for all the materials whose unperturbed carrier energy spectra are defined in Chap. 1.

(R2.9) Investigate the DTP, PTP, and Z in the absence of magnetic field for all the appropriate problems of this chapter under exponential, Kane, Halperin, Lax, and Bonch–Bruevich band tails [92] for all the materials whose unperturbed carrier energy spectra are defined in Chap. 1.

(R2.10) Investigate the DTP, PTP, and Z in the absence of magnetic field for all the appropriate problems of this chapter for all the materials as defined in (R2.9) under an arbitrarily oriented (a) nonuniform electric field and (b) alternating electric field, respectively, whose unperturbed carrier energy spectra are defined in Chap. 1.

(R2.11) Investigate the DTP, PTP, and Z for all the appropriate problems of this chapter for all the materials as described in (R2.9) under an arbitrarily oriented alternating magnetic field by including broadening and the electron spin whose unperturbed carrier energy spectra are defined in Chap. 1.

(R2.12) Investigate the DTP, PTP, and Z for all the appropriate problems of this chapter for all the materials as discussed in (R2.9) under an arbitrarily oriented alternating magnetic field and crossed alternating electric field by including broadening and the electron spin for all the materials whose unperturbed carrier energy spectra are defined in Chap. 1.

(R2.13) Investigate all the appropriate problems for all types of systems as discussed in this chapter for p-InSb, p-CuCl, and stressed semiconductors having diamond structure valence bands whose dispersion relations of the carriers in bulk materials are given by Cunningham [93], Yekimov et al. [94] and Roman et al. [95], respectively.

(R2.14) Investigate the influence of deep traps and surface states separately for all the appropriate problems of this chapter after proper modifications.

(R2.15) Investigate the DTP, PTP, and Z for all the appropriate problems of this chapter for multiple quantum wells and wires of all the heavily doped materials as described in (R2.9).

References

1. L.L. Chang, H. Esaki, C.A. Chang, L. Esaki, Phys. Rev. Lett. **38**, 1489 (1977)
2. K. Less, M.S. Shur, J.J. Drunnond, H. Morkoc, IEEE Trans. Electron. Devices **ED-30**, 07 (1983)
3. M.J. Kelly, *Low Dimensional Semiconductors: Materials, Physics, Technology, Devices* (Oxford University Press, Oxford 1995)
4. C. Weisbuch, B. Vinter, *Quantum Semiconductor Structures* (Boston Academic Press, Boston, 1991)
5. N.T. Linch, Festkorperprobleme **23**, 27 (1985)
6. D.R. Sciferes, C. Lindstrom, R.D. Burnham, W. Streifer, T.L. Paoli, Electron. Lett. **19**, 170 (1983)
7. P.M. Solomon, Proc. IEEE. **70**, 489 (1982)
8. T.E. Schlesinger, T. Kuech, Appl. Phys. Lett. **49**, 519 (1986)
9. D. Kasemet, C.S. Hong, N. B. Patel, P.D. Dapkus, Appl. Phys. Lett. **41**, 912 (1982)
10. K. Woodbridge, P. Blood, E.D. Pletcher, P.J. Hulyer, Appl. Phys. Lett. **45**, 16 (1984)
11. S. Tarucha, H.O. Okamoto, Appl. Phys. Lett. **45**, 16 (1984)
12. H. Heiblum, D.C. Thomas, C.M. Knoedler, M.I Nathan, Appl. Phys. Lett. **47**, 1105 (1985)
13. O. Aina, M. Mattingly, F.Y. Juan, P.K. Bhattacharyya, Appl. Phys. Lett. **50**, 43 (1987)
14. I. Suemune, L.A. Coldren, IEEE J. Quant. Electron. **24**, 1178 (1988)
15. D.A.B. Miller, D.S. Chemla, T.C. Damen, J.H. Wood, A.C. Burrus, A.C. Gossard, W. Weigmann, IEEE J. Quant. Electron. **21**, 1462 (1985)
16. P. Harrison, *Quantum Wells, Wires and Dots* (Wiley, USA, 2002)
17. B.K. Ridley, *Electrons and Phonons in Semiconductors Multilayers* (Cambridge University Press, London, 1997)
18. G. Bastard, *Wave Mechanics Applied to Semiconductor Heterostructures* (Halsted, Les Ulis, Les Editions de Physique, New York, 1988)
19. V.V. Martin, A.A. Kochelap, M.A. Stroscio, *Quantum Heterostructures* (Cambridge University Press, London, 1999)
20. C.S. Lent, D.J. Kirkner, J. Appl. Phys. **67**, 6353 (1990)
21. F. Sols, M. Macucci, U. Ravaioli, K. Hess, Appl. Phys. Lett. **54**, 350 (1980)
22. C.S. Kim, A.M. Satanin, Y.S. Joe, R.M. Cosby, Phys. Rev. B **60**, 10962 (1999)
23. S. Midgley, J.B. Wang, Phys. Rev. B **64**, 153304 (2001)
24. T. Sugaya, J.P. Bird, M. Ogura, Y. Sugiyama, D.K. Ferry, K.Y. Jang, Appl. Phys. Lett. **80**, 434 (2002)

25. B. Kane, G. Facer, A. Dzurak, N. Lumpkin, R. Clark, L. PfeiKer, K. West, Appl. Phys. Lett. **72**, 3506 (1998)
26. C. Dekker, Phys Today **52**, 22 (1999)
27. A. Yacoby, H.L. Stormer, N.S. Wingreen, L.N. Pfeiffer, K.W. Baldwin, K.W. West, Phys. Rev. Lett. **77**, 4612 (1996)
28. Y. Hayamizu, M. Yoshita, S. Watanabe, H.A.L. PfeiKer, K. West, Appl. Phys. Lett. **81**, 4937 (2002)
29. S. Frank, P. Poncharal, Z.L. Wang, W.A. Heer, Science **280**, 1744 (1998)
30. I. Kamiya, I. I. Tanaka, K. Tanaka, F. Yamada, Y. Shinozuka, H. Sakaki, Physica E **13**, 131 (2002)
31. A.K. Geim, P.C. Main, N. LaScala, L. Eaves, T.J. Foster, P.H. Beton, J.W. Sakai, F.W. Sheard, M. Henini, G. Hill et al., Phys. Rev. Lett. **72**, 2061 (1994)
32. K. Schwab, E.A. henriksen, J. M. Worlock, M.L. Roukes, Nature **404**, 974 (2000)
33. L. Kouwenhoven, Nature **403**, 374 (2000)
34. S. Komiyama, O. Astafiev, V. Antonov, H. Hirai, Nature **403**, 405 (2000)
35. A.S. Melinkov, V.M. Vinokur, Nature **415**, 60 (2002)
36. E. Paspalakis, Z. Kis, E. Voutsinas, A.F. Terziz, Phys. Rev. B **69**, 155316 (2004)
37. J.H. Jefferson, M. Fearn, D.L.J. Tipton, T.P. Spiller, Phys. Rev. A **66**, 042328 (2002)
38. J. Appenzeller, C. Schroer, T. Schapers, A. Hart, A. Froster, B. Lengeler, H. Luth, Phys. Rev. B **53**, 9959 (1996)
39. J. Appenzeller, C. Schroer, J. Appl. Phys. **87**, 31659 (2002)
40. P. Debray, O.E. Raichev, M. Rahman, R. Akis, W.C. Mitchel, Appl. Phys. Lett. **74**, 768 (1999)
41. S. Iijima, Nature **354**, 56 (1991)
42. V.N. Popov, Mater. Sci. Eng. R. **43**, 61 (2004)
43. J. Sandler, M.S.P. Shaffer, T. Prasse, W. Bauhofer, K. Schulte, A.H. Windle, Polymer **40**, 5967 (1999)
44. D. Qian, E.C. Dickey, R. Andrews, T. Rantell, Appl. Phys. Lett. **76**, 2868 (2000)
45. J. Kong, N.R. Franklin, C.W. Zhou, M.G. Chapline, S. Peng, K.J. Cho, H.J. Dai, Science **287**, 622 (2000)
46. W.A. Deheer, A. Chatelain, D. Ugarte, Science **270**, 1179 (1995)
47. A.G. Rinzler, J.H. Hafner, P. Nikolaev, L Lou, S.G. Kim, D. Tomanek, P. Nordlander, D.T. Olbert, R.E. Smalley, Science **269**, 1550 (1995)
48. A.C. Dillon, K.M. Jones, T.A. Bekkedahl, C.H. Kiang, D.S. Bethune, M.J. Heben, Nature **386**, 377 (1997)
49. C. Liu, Y.Y. Fan, M. Liu, H.T. Cong, H.M. Cheng, M.S. Dresselhaus, Science **286**, 1127 (1999)
50. P. Kim, C.M. Lieber, Science **286**, 2148 (1999)
51. D. Srivastava, Nanotechnology **8**, 186 (1997)
52. C. Ke, H.D. Espinosa, Appl. Phys. Lett. **85**, 681 (2004)
53. J. W. Kang, H.J. Hwang, Nanotechnology **15**, 1633 (2004)
54. J. Cumings, A. Zettl, Science **289**, 602 (2000)
55. S.J. Tans, M.H. Devoret, H.J. Dai, A. Thess, R.E. Smalley, LJ. Geerligs, C. Dekker, Nature **386**, 474 (1997)
56. S.J. Tans, A.R.M. Verschueren, C. Dekker, Nature **393**, 49 (1998)
57. P. Avouris, Acc. Chem. Res. **35**, 1026 (2002)
58. P.G. Collins, A. Zettl, H. Bando, A. Thess, R.E. Smalley, Science **278**,100 (1997)
59. S. Saito, Science **278**, 77 (1997)
60. J.C. Charlier, Acc. Chem. Res. **35**, 1063 (2002)
61. R.H. Baughman, A.A. Zakhidov, W.A. de Heer, Science **297**, 787 (2002)
62. Z. Yao, H.W.C. Postma, L. Balents, C. Dekker, Nature **402**, 273 (1999)
63. C.W. Zhou, J. Kong, E. Yenilmez, H. Dai, Science **290**, 1552 (2000)
64. S.J. tans, A.R.M. Verschueren, C. Dekker, Nature **393**, 49 (1998)
65. H.W.Ch. Postma, T. Teepen, Z. Yao, M. Grifoni, C. Dekker, Science **293**, 76 (2001)
66. T. Rueckes, K. Kim, E. Joselevich, G.Y. Tseng, C.L. Cheung, C.M. Lieber, Science **289**, 94 (2000)
67. A. Bacthtold, P. Hadley, T. Nakanish, C. Dekker, Science **294**, 1317 (2001)

68. Q. Zheng, Q. Jiang, Phys. Rev. Lett. **88**, 045503 (2002)
69. Q. Zheng, J.S. Liu, Q. Jiang, Phys. Rev. B **65**, 245409 (2002)
70. Y. Zhao, C.-C. Ma, G.H. Chen, Q. Jiang, Phys. Rev. Lett. **91**, 175504 (2003)
71. S.B. Legoas, V.R. Coluci, S.F. Braga, P.Z. Coura, S.O. Dantus, D.S. Galvao, Phys. Rev. Lett. **90**, 055504 (2003)
72. S.B. Legaos, V.R. Coluci, S.F. Braga, P.Z. Coura, S.O. Dantus, D.S. Galvao, Nanotechnology **15**, S184 (2004)
73. J.W. Kang, H.W. Hwang, J. Appl. Phys. **96**, 3900 (2004)
74. W.Y. Choi, J.W. Kang, H.W. Hwang, Physica E **23**, 125 (2004)
75. H.J. Hwang, K.R. Byun, J.W. Kang, Physica E **23**, 208 (2004)
76. J.W. Kang, H.J. Hwang, Physica E **23**, 36 (2004)
77. J.W. Kang, H.J. Hwang, J. Appl. Phys. **73**, 4447 (2004)
78. J.W. Kang, H.J. Hwang, J. Phys. Soc. Jpn. **73**, 1077 (2004)
79. J.W. Kang, Y.W. Choi, H.J. Hwang, J. Comp. Theor. Nanosci. **1**, 199 (2004)
80. M. Dequesnes, S.V. Rotkin, N.R. Aluru, Nanotechnology **13**, 120 (2002)
81. J.M. Kinaret, T. Nord, S. Viefers, Appl. Phys. Lett. **82**, 1287 (2003)
82. C. Ke, H.D. Espinosa, Appl. Phys. Lett. **85**, 681 (2004)
83. L.M. Jonsson, T. Nord, J.M. Kinaret, S. Viefers, J. Appl. Phys. **96**, 629 (2004)
84. L.M. Jonsson, S. Axelsson, T. Nord, S. Viefers, J.M. Kinaret, Nanotechnology **15**, 1497 (2004)
85. S.W. Lee, D.S. Lee, R.E. Morjan, S.H. Jhang, M. Sveningsson, O.A. Nerushev, Y.W. Park, E.E.B. Campbell, Nano. Lett. **4**, 2027 (2004)
86. R. Saito, G. Dresselhaus, M.S. Dresselhaus, *Physical Properties of Carbon Nanotubes*, (Imperial College Press, London, 1998)
87. R. Heyd, A. Charlier, E. McRae, Phys. Rev. B **55**, 6820 (1997)
88. J.W. Mintmire, C.T. White, Phys. Rev. Lett. **81**, 2506 (1998)
89. K.P. Ghatak, S. Bhattacharya, D. De, *Einstein Relation in Compound Semiconductors and Their Nanostructures*, Springer Series in Materials Science, vol. 116 (Springer, Germany, 2008)
90. J.S. Blakemore, *Semiconductor Statistics* (Dover, USA, 1987)
91. K.P. Ghatak, S. Bhattacharya, S. Bhowmick, R. Benedictus, S. Chowdhury, J. Appl. Phys. **103**, 034303 (2008)
92. B.R. Nag, *Electron Transport in Compound Semiconductors* (Springer, Germany, 1980)
93. R.W. Cunningham, Phys. Rev. **167**, 761 (1968)
94. A.I. Yekimov, A.A. Onushchenko, A.G. Plyukhin, Al.L. Efros, J. Exp. Theor. Phys. **88**, 1490 (1985)
95. B.J. Roman, A.W. Ewald, Phys. Rev. B **5**, 3914 (1972)

Chapter 3
Thermoelectric Power in Quantum Dot Superlattices Under Large Magnetic Field

3.1 Introduction

In recent years, modern fabrication techniques have generated altogether a new dimension in the arena of quantum effect devices through the experimental realization of an important artificial structure known as semiconductor superlattice (SL) by growing two similar but different semiconducting compounds in alternate layers with finite thicknesses. The materials forming the alternate layers have the same kind of band structure but different energy gaps. The concept of SL was developed for the first time by Keldysh [1] and was successfully fabricated by Esaki and Tsu [2–5]. The SLs are being extensively used in thermal sensors [6,7], quantum cascade lasers [8–10], photodetectors [11, 12], light emitting diodes [13–16], multiplication [17], frequency multiplication [18], photocathodes [19,20], thin film transistor [21], solar cells [22,23], infrared imaging [24], thermal imaging [25,26], infrared sensing [27], and also in other microelectronic devices.

The most extensively studied III–V SL is the one consisting of alternate layers of GaAs and $Ga_{1-x}Al_xAs$ owing to the relative easiness of fabrication. The GaAs and $Ga_{1-x}Al_xAs$ layers form the quantum wells and the potential barriers, respectively. The III–V SL's are attractive for the realization of high-speed electronic and optoelectronic devices [28]. In addition to SLs with usual structure, other types of SLs such as II–VI [29], IV–VI [30], and HgTe/CdTe [31] SL's have also been investigated in the literature. The IV–VI SLs exhibit quite different properties as compared to the III–V SL due to the specific band structure of the constituent materials [32]. The epitaxial growth of II–VI SL is a relatively recent development and the primary motivation for studying the mentioned SLs made of materials with the large band gap is in their potential for optoelectronic operation in the blue [32]. HgTe/CdTe SL's have raised a great deal of attention since 1979, when as a promising new materials for long wavelength infrared detectors and other electro-optical applications [33]. Interest in Hg-based SL's has been further increased as new properties with potential device applications were revealed [33, 34]. These features arise from the unique zero band gap material HgTe [35] and the direct band gap semiconductor CdTe, which can be described by the three band mode of Kane [36]. The combination of the aforementioned materials with specified dispersion relation makes

HgTe/CdTe SL very attractive, especially because of the tailoring of the material properties for various applications by varying the energy band constants of the SLs.

We note that all the aforementioned SLs have been proposed with the assumption that the interfaces between the layers are sharply defined, of zero thickness, i.e., devoid of any interface effects. The SL potential distribution may be then considered as a one-dimensional array of rectangular potential wells. The aforementioned advanced experimental techniques may produce SLs with physical interfaces between the two materials that are crystallographically abrupt; adjoining their interface will change at least on an atomic scale. As the potential form changes from a well (barrier) to a barrier (well), an intermediate potential region exists for the electrons. The influence of finite thickness of the interfaces on the electron dispersion law is very important, since the electron energy spectrum governs the electron transport in SLs. In addition to it, for effective mass SLs, the electronic subbands appear continually in real space [37].

In Sects. 3.2.1–3.2.8, the TPSM in the quantum dots of the aforementioned SLs have been investigated. Section 3.3 contains the results and discussion. Section 3.4 contains the open research problems pertinent to this chapter.

3.2 Theoretical Background

3.2.1 Magnetothermopower in III–V Quantum Dot Superlattices with Graded Interfaces

The electron dispersion law in bulk specimens of the constituent materials of III–V SLs whose energy band structures are defined by three band model of Kane can be expressed following (1.16) as

$$(\hbar^2 k^2)/(2m_i^*) = EG(E, E_{g0_i}, \Delta_i), \tag{3.1}$$

$i = 1, 2, \ldots$ and

$$G(E, E_{g0_i}, \Delta_i) \equiv \frac{\left(E_{g0_i} + \frac{2}{3}\Delta_i\right)\left(E + E_{g0_i} + \Delta_i\right)\left(E + E_{g0_i}\right)}{\left[E_{g0_i}\left(E_{g0_i} + \Delta_i\right)\left(E + E_{g0_i} + \frac{2}{3}\Delta_i\right)\right]}.$$

Therefore, the dispersion law of the electrons of III–V SLs with graded interfaces can be expressed, following Jiang and Lin [38], as

$$\cos(L_0 k) = \frac{1}{2}\Phi(E, k_s), \tag{3.2}$$

where L_0 ($\equiv a_0 + b_0$) is the period length, a_0 and b_0 are the widths of the barrier and the well, respectively,

$$\Phi(E,k_s) \equiv \Big[2\cosh\{\beta(E,k_s)\}\cos\{\gamma(E,k_s)\} + \varepsilon(E,k_s)\sinh\{\beta(E,k_s)\}\sin\{\gamma(E,k_s)\}$$

$$+\Delta_0\Bigg[\left(\frac{K_1^2(E,k_s)}{K_2(E,k_s)} - 3K_2(E,k_s)\right)\cosh\{\beta(E,k_s)\}\ \sin\{\gamma(E,k_s)\}$$

$$+\left(3K_1(E,k_s) - \frac{\{K_2(E,k_s)\}^2}{K_1(E,k_s)}\right)\sinh\{\beta(E,k_s)\}\cos\{\gamma(E,k_s)\}\Bigg]$$

$$+\Delta_0\Big[2\left(\{K_1(E,k_s)\}^2 - \{K_2(E,k_s)\}^2\right)\cosh\{\beta(E,k_s)\}\cos\{\gamma(E,k_s)\}$$

$$+\frac{1}{12}\Big[\frac{5\{K_2(E,k_s)\}^3}{K_1(E,k_s)} + \frac{5\{K_1(E,k_s)\}^3}{K_2(E,k_s)}$$

$$-34K_2(E,k_s)K_1(E,k_s)\Big]\sin\{\beta(E,k_s)\}\sin\{\gamma(E,k_s)\}\Big]\Big],$$

$\beta(E,k_s) \equiv K_1(E,k_s)[a_0-\Delta_0]$, $K_1(E,k_s) \equiv \left[\frac{2m_2^* E'}{\hbar^2} G(E-V_0,\alpha_2,\Delta_2)+k_s^2\right]^{1/2}$, $E' \equiv V_0 - E$, V_0 is the potential barrier encountered by the electron ($V_0 \equiv |E_{g_2} - E_{g_1}|$), $\alpha_i \equiv 1/E_{g0i}$, Δ_0 is the interface width, $k_s^2 = k_x^2 + k_y^2$,

$$\gamma(E,k_s) = K_2(E,k_s)[b_0 - \Delta_0],\ K_2(E,k_s) \equiv \left[\frac{2m_1^* E}{\hbar^2} G(E,\alpha_1,\Delta_1) - k_s^2\right]^{1/2},$$

and

$$\varepsilon(E,k_s) \equiv \left[\frac{K_1(E,k_s)}{K_2(E,k_s)} - \frac{K_2(E,k_s)}{K_1(E,k_s)}\right].$$

Therefore, the total electron concentration per unit volume in III–V quantum dot SL is given by

$$n_0 = \frac{2g_v}{d_x d_y d_z} \sum_{n_x=1}^{n_{x\max}} \sum_{n_y=1}^{n_{y\max}} \sum_{n_z=1}^{n_{z\max}} F_{-1}(\overline{\eta_9}), \tag{3.3}$$

where

$$\overline{\eta_9} \equiv \frac{E_{FQDSLGI} - \overline{E_{QD10}}}{k_B T},$$

in which $E_{FQDSLGI}$ is the Fermi energy in this case, $\overline{E_{QD10}}$ is the root of the equation

$$\left(\frac{\pi n_z}{d_z}\right)^2 = \left[\frac{1}{L_0^2}\left[\cos^{-1}\left[\frac{1}{2} f_{10}(E,n_x,n_y)\right]\Big|_{E=\overline{E_{QD10}}}\right]^2 - \left(\frac{\pi n_x}{d_x}\right)^2 - \left(\frac{\pi n_y}{d_y}\right)^2\right] \tag{3.4}$$

and

$$f_{10}(E,n_x,n_y) \equiv \Big[2\cosh\{\bar\beta(E,n_x,n_y)\}\ \cos\{\bar\gamma(E,n_x,n_y)\}$$

$$+\bar\varepsilon(E,n_x,n_y)\sinh\{\bar\beta(E,n_x,n_y)\}\sin\{\bar\gamma(E,n_x,n_y)\}$$

$$+\Delta_0\Bigg[\left(\frac{\{\bar{K}_1(E,n_x,n_y)\}^2}{\bar{K}_2(E,n_x,n_y)}-3\bar{K}_2(E,n_x,n_y)\right)$$
$$\times\cosh\{\bar{\beta}(E,n_x,n_y)\}\sin\{\bar{\gamma}(E,n_x,n_y)\}$$
$$+\left(3\bar{K}_1(E,n_x,n_y)-\frac{\{\bar{K}_2(E,n_x,n_y)\}^2}{\bar{K}_1(E,n_x,n_y)}\right)$$
$$\times\sinh\{\bar{\beta}(E,n_x,n_y)\}\cos\{\bar{\gamma}(E,n_x,n_y)\}\Bigg]$$
$$+\Delta_0\Bigg[2\left(\{\bar{K}_1(E,n_x,n_y)\}^2-\{\bar{K}_2(E,n_x,n_y)\}^2\right)$$
$$\times\cosh\{\bar{\beta}(E,n_x,n_y)\}\cos\{\bar{\gamma}(E,n_x,n_y)\}$$
$$+\frac{1}{12}\Bigg[\frac{5\{\bar{K}_2(E,n_x,n_y)\}^3}{\bar{K}_1(E,n_x,n_y)}+\frac{5\{\bar{K}_1(E,n_x,n_y)\}^3}{\bar{K}_2(E,n_x,n_y)}$$
$$-34\bar{K}_2(E,n_x,n_y)\bar{K}_1(E,n_x,n_y)\Bigg]$$
$$\times\sinh\{\bar{\beta}(E,n_x,n_y)\}\sin\{\bar{\gamma}(E,n_x,n_y)\}\Bigg]\Bigg],$$
$$\bar{\beta}(E,n_x,n_y)\equiv\bar{K}_1(E,n_x,n_y)[a_0-\Delta_0],$$

$$\bar{K}_1(E,n_x,n_y)\equiv\left[\frac{2m_2^*E'}{\hbar^2}G(E-V_0,\alpha_2,\Delta_2)+\left\{\left(\frac{\pi n_x}{d_x}\right)^2+\left(\frac{\pi n_y}{d_y}\right)^2\right\}\right]^{1/2},$$
$$\bar{\gamma}(E,n_x,n_y)=\bar{K}_2(E,n_x,n_y)[b_0-\Delta_0],$$
$$\bar{K}_2(E,n_x,n_y)\equiv\left[\frac{2m_1^*E}{\hbar^2}G(E,\alpha_1,\Delta_1)-\left\{\left(\frac{\pi n_x}{d_x}\right)^2+\left(\frac{\pi n_y}{d_y}\right)^2\right\}\right]^{1/2}$$

and
$$\bar{\varepsilon}(E,n_x,n_y)\equiv\left[\frac{\bar{K}_1(E,n_x,n_y)}{\bar{K}_2(E,n_x,n_y)}-\frac{\bar{K}_2(E,n_x,n_y)}{\bar{K}_1(E,n_x,n_y)}\right].$$

Thus, combining (1.13) and (3.3), the TPSM in this case can be expressed as

$$G_0=\frac{\pi^2k_B}{3e}\left[\sum_{n_x=1}^{n_{x\max}}\sum_{n_y=1}^{n_{y\max}}\sum_{n_z=1}^{n_{z\max}}F_{-1}(\overline{\eta_9})\right]^{-1}\left[\sum_{n_x=1}^{n_{x\max}}\sum_{n_y=1}^{n_{y\max}}\sum_{n_z=1}^{n_{z\max}}F_{-2}(\overline{\eta_9})\right]. \quad (3.5)$$

3.2.2 Magnetothermopower in II–VI Quantum Dot Superlattices with Graded Interfaces

The electron dispersion laws of the constituent materials of II–VI SLs are given by [26]

$$E = \frac{\hbar^2 k_s^2}{2m_{\perp,1}^*} + \frac{\hbar^2 k_z^2}{2m_{\parallel,1}^*} + C_0 k_s \tag{3.6}$$

and

$$\frac{\hbar^2 k^2}{2m_2^*} = EG\left(E, E_{g02}, \Delta_2\right), \tag{3.7}$$

where $m_{\perp,1}^*$ and $m_{\parallel,1}^*$ are the transverse and longitudinal effective electron masses, respectively, at the edge of the conduction band for the first material. The energy–wave-vector dispersion relation of the conduction electrons in II–VI SLs with graded interfaces can be expressed as

$$\cos\left(L_0 k\right) = \frac{1}{2}\Phi_1\left(E, k_s\right), \tag{3.8}$$

where

$$\begin{aligned}
\Phi_1(E,k_s) \equiv \Big[& 2\cosh\{\beta_1(E,k_s)\}\cos\{\gamma_1(E,k_s)\} \\
& + \varepsilon_1(E,k_s)\sinh\{\beta_1(E,k_s)\}\sin\{\gamma_1(E,k_s)\} \\
& + \Delta_0\Bigg[\left(\frac{\{K_3(E,k_s)\}^2}{K_4(E,k_s)} - 3K_4(E,k_s)\right)\cosh\{\beta_1(E,k_s)\}\sin\{\gamma_1(E,k_s)\} \\
& + \left(3K_3(E,k_s) - \frac{\{K_4(E,k_s)\}^2}{K_3(E,k_s)}\right)\sinh\{\beta_1(E,k_s)\}\cos\{\gamma_1(E,k_s)\}\Bigg] \\
& + \Delta_0\Big[2\left(\{K_3(E,k_s)\}^2 - \{K_4(E,k_s)\}^2\right)\cosh\{\beta_1(E,k_s)\}\cos\{\gamma_1(E,k_s)\} \\
& + \frac{1}{12}\left[\frac{5\{K_3(E,k_s)\}^3}{K_4(E,k_s)} + \frac{5\{K_4(E,k_s)\}^3}{K_3(E,k_s)} - 34K_4(E,k_s)K_3(E,k_s)\right] \\
& \times \sinh\{\beta_1(E,k_s)\}\sin\{\gamma_1(E,k_s)\}\Big]\Big],
\end{aligned}$$

$$\begin{aligned}
\beta_1(E,k_s) &\equiv K_3(E,k_s)\left[a_0 - \Delta_0\right], \\
K_3(E,k_s) &\equiv \left[\frac{2m_2^* E'}{\hbar^2}G(E - V_0, \alpha_2, \Delta_2) + k_s^2\right]^{1/2}, \\
\gamma_1(E,k_s) &= K_4(E,k_s)\left[b_0 - \Delta_0\right],
\end{aligned}$$

$$K_4(E,k_s) \equiv \left[\frac{2m^*_{\parallel,1}}{\hbar^2}\left[E - \frac{\hbar^2 k_s^2}{2m^*_{\perp,1}} \mp C_0 k_s\right]\right]^{1/2},$$

and

$$\varepsilon_1(E,k_s) \equiv \left[\frac{K_3(E,k_s)}{K_4(E,k_s)} - \frac{K_4(E,k_s)}{K_3(E,k_s)}\right].$$

The total electron concentration per unit volume is given by

$$n_0 = \frac{g_v}{d_x d_y d_z}\sum_{n_x=1}^{n_{x\max}}\sum_{n_y=1}^{n_{y\max}}\sum_{n_z=1}^{n_{z\max}} F_{-1}(\overline{\eta_{10}}), \tag{3.9}$$

where

$$\overline{\eta_{10}} \equiv \frac{E_{\text{FQDSLGI}} - \overline{E_{\text{QD11}}}}{k_B T}$$

and $\overline{E_{\text{QD11}}}$ is the root of the equation

$$\left(\frac{\pi n_z}{d_z}\right)^2 = \left[\frac{1}{L_0^2}\left[\cos^{-1}\left[\frac{1}{2}f_{11}(E,n_x,n_y)\right]\Big|_{E=\overline{E_{\text{QD11}}}}\right]^2 - \left(\frac{\pi n_x}{d_x}\right)^2 - \left(\frac{\pi n_y}{d_y}\right)^2\right] \tag{3.10}$$

where

$$\begin{aligned}
f_{11}(E,n_x,n_y) \equiv {} & \Big[2\cosh\{\bar{\beta}_1(E,n_x,n_y)\}\cos\{\bar{\gamma}_1(E,n_x,n_y)\} + \bar{\varepsilon}_1(E,n_x,n_y) \\
& \times \sinh\{\bar{\beta}_1(E,n_x,n_y)\}\sin\{\bar{\gamma}_1(E,n_x,n_y)\} \\
& + \Delta_0\Bigg[\left(\frac{\{\bar{K}_3(E,n_x,n_y)\}^2}{\bar{K}_4(E,n_x,n_y)} - 3\bar{K}_4(E,n_x,n_y)\right) \\
& \times \cosh\{\bar{\beta}_1(E,n_x,n_y)\}\sin\{\bar{\gamma}_1(E,n_x,n_y)\} \\
& + \left(3\bar{K}_3(E,n_x,n_y) - \frac{\{\bar{K}_4(E,n_x,n_y)\}^2}{\bar{K}_3(E,n_x,n_y)}\right) \\
& \times \sin\{\bar{\beta}_1(E,n_x,n_y)\}\cos\{\bar{\gamma}_1(E,n_x,n_y)\}\Bigg] \\
& + \Delta_0\Big[2\left(\{\bar{K}_3(E,n_x,n_y)\}^2 - \{\bar{K}_4(E,n_x,n_y)\}^2\right) \\
& \times \cosh\{\bar{\beta}_1(E,n_x,n_y)\}\cos\{\bar{\gamma}_1(E,n_x,n_y)\} \\
& + \frac{1}{12}\Bigg[\frac{5\{\bar{K}_3(E,n_x,n_y)\}^3}{\bar{K}_4(E,n_x,n_y)} + \frac{5\{\bar{K}_4(E,n_x,n_y)\}^3}{\bar{K}_3(E,n_x,n_y)} \\
& - 34\bar{K}_4(E,n_x,n_y)\bar{K}_3(E,n_x,n_y)\Bigg]\sinh\{\bar{\beta}_1(E,n_x,n_y)\} \\
& \times \sin\{\bar{\gamma}_1(E,k_s)n_x,n_y\}\Big]\Big],
\end{aligned}$$

$$\bar{\beta}_1(E,n_x,n_y) \equiv \bar{K}_3(E,n_x,n_y)[a_0-\Delta_0],$$

$$\bar{K}_3(E,n_x,n_y) \equiv \left[\frac{2m_2^* E'}{\hbar^2} G(E-V_0,\alpha_2,\Delta_2) + \left\{\left(\frac{\pi n_x}{d_x}\right)^2 + \left(\frac{\pi n_y}{d_y}\right)^2\right\}\right]^{1/2},$$

$$\bar{\gamma}_1(E,n_x,n_y) = \bar{K}_4(E,n_x,n_y)[b_0-\Delta_0],$$

$$\bar{K}_4(E,n_x,n_y) \equiv \left[\frac{2m_{\parallel,1}^*}{\hbar^2}\left[E-\hbar^2\left\{\left(\frac{\pi n_x}{d_x}\right)^2+\left(\frac{\pi n_y}{d_y}\right)^2\right\}(2m_{\perp,1}^*)^{-1} \mp C_0\left\{\left(\frac{\pi n_x}{d_x}\right)^2+\left(\frac{\pi n_y}{d_y}\right)^2\right\}^{1/2}\right]\right]^{1/2},$$

and

$$\bar{\varepsilon}_1(E,n_x,n_y) \equiv \left[\frac{\bar{K}_3(E,n_x,n_y)}{\bar{K}_4(E,n_x,n_y)} - \frac{\bar{K}_4(E,n_x,n_y)}{\bar{K}_3(E,n_x,n_y)}\right].$$

Thus, combining (1.13) and (3.9), the TPSM in this case can be expressed as

$$G_0 = \frac{\pi^2 k_B}{3e}\left[\sum_{n_x=1}^{n_{x\max}}\sum_{n_y=1}^{n_{y\max}}\sum_{n_z=1}^{n_{z\max}} F_{-1}(\overline{\eta_{10}})\right]^{-1}\left[\sum_{n_x=1}^{n_{x\max}}\sum_{n_y=1}^{n_{y\max}}\sum_{n_z=1}^{n_{z\max}} F_{-2}(\overline{\eta_{10}})\right]. \tag{3.11}$$

3.2.3 Magnetothermopower in IV–VI Quantum Dot Superlattices with Graded Interfaces

The **E–k** dispersion relation of the conduction electrons of the constituent materials of the IV–VI SLs can be expressed [40] as

$$E = a_i k_s^2 + b_i k_z^2 + \left[\left[c_i k_s^2 + d_i k_z^2\right] + \left(e_i k_s^2 + f_i k_y^2 + \frac{E_{g0i}}{2}\right)^2\right]^{1/2} - \frac{E_{g0i}}{2}, \tag{3.12}$$

where

$$a_i \equiv \left[\frac{\hbar^2}{2m_{\perp,i}^-}\right], \quad b_i \equiv \left(\frac{\hbar^2}{2m_{\parallel,i}^-}\right),$$

$$c_i \equiv P_{\perp,i}^2,\ d_i \equiv P_{\perp,i}^2, \quad e_i \equiv \left[\frac{\hbar^2}{2m_{\perp,i}^+}\right], \text{ and } f_i \equiv \left(\frac{\hbar^2}{2m_{\parallel,i}^+}\right).$$

The electron dispersion law in IV–VI SLs with graded interfaces can be expressed as

$$\cos\left(L_0 k\right) = \frac{1}{2}\Phi_2\left(E, k_s\right), \tag{3.13}$$

where

$$\begin{aligned}
&\Phi_2\left(E, k_s\right) \equiv \\
&\Big[2\cosh\left\{\beta_2\left(E, k_s\right)\right\}\cos\left\{\gamma_2\left(E, k_s\right)\right\} + \varepsilon_2\left(E, k_s\right)\sinh\left\{\beta_2\left(E, k_s\right)\right\}\sin\left\{\gamma_2\left(E, k_s\right)\right\} \\
&+ \Delta_0\Bigg[\left(\frac{\left\{K_5\left(E, k_s\right)\right\}^2}{K_6\left(E, k_s\right)} - 3K_6\left(E, k_s\right)\right)\cosh\left\{\beta_2\left(E, k_s\right)\right\}\sin\left\{\gamma_2\left(E, k_s\right)\right\} \\
&+ \left(3K_5\left(E, k_s\right) - \frac{\left\{K_6\left(E, k_s\right)\right\}^2}{K_5\left(E, k_s\right)}\right)\sinh\left\{\beta_2\left(E, k_s\right)\right\}\cos\left\{\gamma_2\left(E, k_s\right)\right\}\Bigg] \\
&+ \Delta_0\Bigg[2\left(\left\{K_5\left(E, k_s\right)\right\}^2 - \left\{K_6\left(E, k_s\right)\right\}^2\right)\cosh\left\{\beta_2\left(E, k_s\right)\right\}\cos\left\{\gamma_2\left(E, k_s\right)\right\} \\
&+ \frac{1}{12}\left[\frac{5\left\{K_5\left(E, k_s\right)\right\}^3}{K_6\left(E, k_s\right)} + \frac{5\left\{K_6\left(E, k_s\right)\right\}^3}{K_5\left(E, k_s\right)} - 34K_6\left(E, k_s\right)K_5\left(E, k_s\right)\right] \\
&\times \sinh\left\{\beta_2\left(E, k_s\right)\right\}\sin\left\{\gamma_2\left(E, k_s\right)\right\}\Bigg]\Big],
\end{aligned}$$

$$\beta_2\left(E, k_s\right) \equiv K_5\left(E, k_s\right)\left[a_0 - \Delta_0\right],$$

$$\begin{aligned}
K_5\left(E, k_x, k_y\right) \equiv \Big[&\left[\left(E - V_0\right)^2 H_{32} + \left(E - V_0\right)H_{42}\left(k_x, k_y\right) + H_{52}\left(k_x, k_y\right)\right]^{1/2} \\
&- \left[\left(E - V_0\right)H_{12} + H_{22}\left(k_x, k_y\right)\right]\Big]^{1/2},
\end{aligned}$$

$$\gamma_2\left(E, k_s\right) = K_6\left(E, k_s\right)\left[b_0 - \Delta_0\right],$$

$$\begin{aligned}
K_6\left(E, k_x, k_y\right) \equiv \Big[&\left[EH_{11} + H_{21}\left(k_x, k_y\right)\right] - \left[H_{31}E^2 + EH_{41}\left(k_x, k_y\right)\right. \\
&\left. + H_{51}\left(k_x, k_y\right)\right]^{1/2}\Big]^{1/2},
\end{aligned}$$

$$H_{1i} = b_i . \left(b_i^2 - f_i^2\right)^{-1}, i = 1 \, and \; 2,$$

$$\begin{aligned}
H_{2i}\left(k_x, k_y\right) = \left[2(b_i^2 - f_i^2)\right]^{-1}&\left[E_{g_i} b_i + d_i + f_i E_{g_i}\right. \\
&\left. + 2\left(c_i f_i - a_i b_i\right)\left(k_x^2 + k_y^2\right)\right],
\end{aligned}$$

$$H_{3i} = \frac{f_i^2}{\left(b_i^2 - f_i^2\right)^2},$$

$$H_{4i}\left(k_x, k_y\right) = \left[4(b_i^2 - f_i^2)^2\right]^{-1}\left[4b_i^2 E_{g_i} + 4b_i d_i + 4b_i f_i E_{g_i} + 4f_i^2 E_{g_i} + 8\left(k_x^2 + k_y^2\right)\left[b_i^2 a_i + C_i f_i b_i - a_i^2 b_i\right]\right]$$

$$H_{5i}\left(k_x, k_y\right) \equiv \left[4(b_i^2 - f_i^2)^2\right]^{-1}\Bigg[\left(k_x^2 + k_y^2\right)^2\left[-8a_i b_i C_i f_i + 4b_i^2 C_i^2 + 4f_i^2 a_i^2 - 4f_i^2 C_i^2\right] + \left(k_x^2 + k_y^2\right)\left[8d_i C_i f_i - 4a_i b_i d_i - 4a_i b_i f_i E_{g_i} + 4b_i^2 C_i + 4b_i^2 e_i E_{g_i} - 4a_i f_i^2 E_{g_i} - 4f_i^2 e_i E_{g_i}\right] + \left[E_{g_i}^2 b_i^2 + d_i^2 + f_i^2 E_{g_i}^2 + 2E_{g_i} f_i d_i\right]\Bigg]$$

and

$$\varepsilon_2\left(E, k_s\right) \equiv \left[\frac{K_5\left(E, k_s\right)}{K_6\left(E, k_s\right)} - \frac{K_6\left(E, k_s\right)}{K_5\left(E, k_s\right)}\right].$$

The total electron concentration per unit volume is given by

$$n_0 = \frac{2g_v}{d_x d_y d_z}\sum_{n_x=1}^{n_{x\max}}\sum_{n_y=1}^{n_{y\max}}\sum_{n_{z\min}}^{n_{z\max}} F_{-1}\left(\overline{\eta_{11}}\right), \tag{3.14}$$

where $\overline{\eta_{11}} \equiv \frac{E_{FQDSLGI} - \overline{E_{QD12}}}{k_B T}$ and $\overline{E_{QD12}}$ is the root of the equation

$$\left(\frac{\pi n_z}{d_z}\right)^2 = \left[\frac{1}{L_0^2}\left[\cos^{-1}\left[\frac{1}{2} f_{12}\left(E, n_x, n_y\right)\Big|_{E=\overline{E_{QD12}}}\right]\right]^2 - \left(\frac{\pi n_x}{d_x}\right)^2 - \left(\frac{\pi n_y}{d_y}\right)^2\right], \tag{3.15}$$

$$f_{12}\left(E, n_x, n_y\right) \equiv \Bigg[2\cosh\left\{\bar{\beta}_2\left(E, n_x, n_y\right)\right\}\cos\left\{\bar{\gamma}_2\left(E, n_x, n_y\right)\right\} + \bar{\varepsilon}_2\left(E, n_x, n_y\right) \times \sinh\left\{\bar{\beta}_2\left(E, n_x, n_y\right)\right\}\sin\left\{\bar{\gamma}_2\left(E, n_x, n_y\right)\right\} + \Delta_0\Bigg[\left(\frac{\left\{\bar{K}_5\left(E, n_x, n_y\right)\right\}^2}{\bar{K}_6\left(E, n_x, n_y\right)} - 3\bar{K}_6\left(E, n_x, n_y\right)\right) \times \cosh\left\{\bar{\beta}_2\left(E, n_x, n_y\right)\right\}\sin\left\{\bar{\gamma}_2\left(E, n_x, n_y\right)\right\} + \left(3\bar{K}_5\left(E, n_x, n_y\right) - \frac{\left\{\bar{K}_6\left(E, n_x, n_y\right)\right\}^2}{\bar{K}_5\left(E, n_x, n_y\right)}\right)$$

$$\times \sinh\left\{\bar{\beta}_2\left(E,n_x,n_y\right)\right\}\cos\left\{\bar{\gamma}_2\left(E,n_x,n_y\right)\right\}\Big]$$

$$+\Delta_0\Big[2\left(\left\{\bar{K}_5\left(E,n_x,n_y\right)\right\}^2-\left\{\bar{K}_6\left(E,n_x,n_y\right)\right\}^2\right)$$

$$\times \cosh\left\{\bar{\beta}_2\left(E,n_x,n_y\right)\right\}\cos\left\{\bar{\gamma}_2\left(E,n_x,n_y\right)\right\}$$

$$+\frac{1}{12}\Big[\frac{5\left\{\bar{K}_5\left(E,n_x,n_y\right)\right\}^3}{\bar{K}_6\left(E,n_x,n_y\right)}+\frac{5\left\{\bar{K}_6\left(E,n_x,n_y\right)\right\}^3}{\bar{K}_5\left(E,n_x,n_y\right)}$$

$$-34\bar{K}_6\left(E,n_x,n_y\right)\bar{K}_5\left(E,n_x,n_y\right)\Big]$$

$$\times \sin\left\{\bar{\beta}_2\left(E,n_x,n_y\right)\right\}\sin\left\{\bar{\gamma}_2\left(E,n_x,n_y\right)\right\}\Big]\Big],$$

$$\bar{\beta}_2\left(E,n_x,n_y\right)\equiv\bar{K}_5\left(E,n_x,n_y\right)\left[a_0-\Delta_0\right],$$

$$\bar{K}_5\left(E,n_x,n_y\right)\equiv\Big[\Big[(E-V_0)^2H_{32}+(E-V_0)\bar{H}_{42}\left(n_x,n_y\right)+\bar{H}_{52}\left(n_x,n_y\right)\Big]^{1/2}$$

$$-\Big[(E-V_0)H_{12}+\bar{H}_{22}\left(n_x,n_y\right)\Big]\Big]^{1/2},$$

$$\bar{H}_{2i}\left(n_x,n_y\right)=\left[2(b_i^2-f_i^2)\right]^{-1}\left[E_{g_i}b_i+d_i+f_iE_{g_i}\right.$$
$$\left.+2\left(C_if_i-a_ib_i\right)\varphi_1\left(n_x,n_y\right)\right],$$

$$\varphi_1\left(n_x,n_y\right)=\left[(\pi n_x/d_x)^2+(\pi n_y/d_y)^2\right]$$

$$\bar{H}_{4i}\left(n_x,n_y\right)=\left[4(b_i^2-f_i^2)^2\right]^{-1}\left[4b_i^2E_{g_i}+4b_id_i+4b_if_iE_{g_i}+4f_i^2E_{g_i}\right.$$
$$\left.+8\varphi_1\left(n_x,n_y\right)\left[b_i^2a_i+C_if_ib_i-a_i^2b_i\right]\right]$$

$$\bar{H}_{5i}\left(n_x,n_y\right)\equiv\left[4(b_i^2-f_i^2)^2\right]^{-1}\left[\varphi_1^2\left(n_x,n_y\right)\left[-8a_ib_iC_if_i+4b_i^2C_i^2+4f_i^2a_i^2\right.\right.$$
$$\left.-4f_i^2C_i^2\right]+\varphi_1\left(n_x,n_y\right)\left[8d_iC_if_i-4a_ib_id_i-4a_ib_if_iE_{g_i}\right.$$
$$\left.\left.+4b_i^2C_i+4b_i^2e_iE_{g_i}-4a_if_i^2E_{g_i}-4f_i^2e_iE_{g_i}\right]\right]$$
$$+\left[E_{g_i}^2b_i^2+d_i^2+f_i^2E_{g_i}^2+2E_{g_i}f_id_i\right],$$

$$\bar{\gamma}_2\left(E,n_x,n_y\right)=\bar{K}_6\left(E,n_x,n_y\right)\left[b_0-\Delta_0\right],$$

$$\bar{\varepsilon}_2\left(E,n_x,n_y\right)\equiv\left[\frac{\bar{K}_5\left(E,n_x,n_y\right)}{\bar{K}_6\left(E,n_x,n_y\right)}-\frac{\bar{K}_6\left(E,n_x,n_y\right)}{\bar{K}_5\left(E,n_x,n_y\right)}\right],$$

and

$$\bar{K}_6\left(E,n_x,n_y\right)\equiv\Big[\left[EH_{11}+\bar{H}_{21}\left(n_x,n_y\right)\right]-\left[H_{31}E^2+E\bar{H}_{41}\left(n_x,n_y\right)+\bar{H}_{51}\left(n_x,n_y\right)\right]^{1/2}\Big]^{1/2}.$$

Thus, combining (1.13) and (3.14), the TPSM in this case can be expressed as

$$G_0=\frac{\pi^2k_{\mathrm{B}}}{3e}\left[\sum_{n_x=1}^{n_{x_{\max}}}\sum_{n_y=1}^{n_{y_{\max}}}\sum_{n_{z_{\min}}}^{n_{z_{\max}}}F_{-1}\left(\overline{\eta_{11}}\right)\right]^{-1}\left[\sum_{n_x=1}^{n_{x_{\max}}}\sum_{n_y=1}^{n_{y_{\max}}}\sum_{n_{z_{\min}}}^{n_{z_{\max}}}F_{-2}\left(\overline{\eta_{11}}\right)\right]. \quad (3.16)$$

3.2.4 Magnetothermopower in HgTe/CdTe Quantum Dot Superlattices with Graded Interfaces

The dispersion relation of the conduction electrons of the constituent materials of HgTe/CdTe SLs can be expressed [35] as

$$E=\frac{\hbar^2k^2}{2m_1^*}+\frac{3\left|e\right|^2k}{128\varepsilon_{\mathrm{sc}}}, \quad (3.17)$$

$$\frac{\hbar^2k^2}{2m_2^*}=EG\left(E,E_{g02},\Delta_2\right). \quad (3.18)$$

The electron energy dispersion law in HgTe/CdTe SL is given by

$$\cos\left(L_0k\right)=\frac{1}{2}\Phi_3\left(E,k_s\right), \quad (3.19)$$

where

$$\begin{aligned}\Phi_3\left(E,k_s\right)\equiv\Big[&2\cosh\left\{\beta_3\left(E,k_s\right)\right\}\cos\left\{\gamma_3\left(E,k_s\right)\right\}+\varepsilon_3\left(E,k_s\right)\sinh\left\{\beta_3\left(E,k_s\right)\right\}\\&\times\sin\left\{\gamma_3\left(E,k_s\right)\right\}+\Delta_0\Big[\left(\frac{\left\{K_7\left(E,k_s\right)\right\}^7}{K_8\left(E,k_s\right)}-3K_8\left(E,k_s\right)\right)\\&\times\cosh\left\{\beta_3\left(E,k_s\right)\right\}\sin\left\{\gamma_3\left(E,k_s\right)\right\}\\&+\left(3K_7\left(E,k_s\right)-\frac{\left\{K_8\left(E,k_s\right)\right\}^2}{K_7\left(E,k_s\right)}\right)\sinh\left\{\beta_3\left(E,k_s\right)\right\}\cos\left\{\gamma_3\left(E,k_s\right)\right\}\Big]\end{aligned}$$

$$+ \Delta_0\Big[2\left(\{K_7(E,k_s)\}^2 - \{K_8(E,k_s)\}^2\right)\cosh\{\beta_3(E,k_s)\}$$
$$\times\cos\{\gamma_3(E,k_s)\} + \frac{1}{12}\Big[\frac{5\{K_8(E,k_s)\}^3}{K_7(E,k_s)} + \frac{5\{K_7(E,k_s)\}^3}{K_8(E,k_s)}$$
$$-34K_7(E,k_s)\,K_8(E,k_s)\Big]\sinh\{\beta_3(E,k_s)\}\sin\{\gamma_3(E,k_s)\}\Big]\Big],$$

$$\beta_3(E,k_s) \equiv K_7(E,k_s)\,[a_0 - \Delta_0],$$
$$K_7(E,k_s) \equiv \left[\frac{2m_2^* E'}{\hbar^2} G(E - V_0, E_{g02}, \Delta_2) + k_s^2\right]^{1/2},$$
$$\gamma_3(E,k_s) = K_8(E,k_s)\,[b_0 - \Delta_0],$$
$$K_8(E,k_x,k_y) \equiv \left[\frac{B_0^2 + 2AE - B_0\sqrt{B_0^2 + 4AE}}{2A^2} - k_s^2\right]^{1/2},$$

$$B_0 = \frac{3|e|^2}{128\varepsilon_{sc}},$$
$$A = \frac{\hbar^2}{2m_1^*},$$

and

$$\varepsilon_3(E,k_s) \equiv \left[\frac{K_7(E,k_s)}{K_8(E,k_s)} - \frac{K_8(E,k_s)}{K_7(E,k_s)}\right].$$

The total electron concentration per unit volume is given by

$$n_0 = \frac{2g_v}{d_x d_y d_z} \sum_{n_x=1}^{n_{x\max}} \sum_{n_y=1}^{n_{y\max}} \sum_{n_z=1}^{n_{z\max}} F_{-1}(\overline{\eta_{12}}), \tag{3.20}$$

where $\overline{\eta_{12}} \equiv \frac{E_{FQDSLGI} - \overline{E_{QD13}}}{k_B T}$ and $\overline{E_{QD13}}$ is the root of the equation

$$\left(\frac{\pi n_z}{d_z}\right)^2 = \left[\frac{1}{L_0^2}\left[\cos^{-1}\left[\frac{1}{2} f_{13}(E,n_x,n_y)\Big|_{E=\overline{E_{QD13}}}\right]\right]^2 - \left(\frac{\pi n_x}{d_x}\right)^2 - \left(\frac{\pi n_y}{d_y}\right)^2\right], \tag{3.21}$$

where

$$f_{13}(E,n_x,n_y) \equiv \Big[2\cosh\{\bar\beta_3(E,n_x,n_y)\}\cos\{\bar\gamma_3(E,n_x,n_y)\}$$
$$+ \bar\varepsilon_3(E,n_x,n_y)\sinh\{\bar\beta_3(E,n_x,n_y)\}\sin\{\bar\gamma_3(E,n_x,n_y)\}$$

$$+\Delta_0\Bigg[\Bigg(\frac{\{\bar{K}_7(E,n_x,n_y)\}^7}{\bar{K}_8(E,n_x,n_y)}-3\bar{K}_8(E,n_x,n_y)\Bigg)$$

$$\times\cosh\{\bar{\beta}_3(E,n_x,n_y)\}\sin\{\bar{\gamma}_3(E,n_x,n_y)\}$$

$$+\Bigg(3\bar{K}_7(E,n_x,n_y)-\frac{\{\bar{K}_8(E,n_x,n_y)\}^2}{\bar{K}_7(E,n_x,n_y)}\Bigg)$$

$$\times\sinh\{\bar{\beta}_3(E,n_x,n_y)\}\cos\{\bar{\gamma}_3(E,n_x,n_y)\}\Bigg]$$

$$+\Delta_0\Big[2\Big(\{\bar{K}_7(E,n_x,n_y)\}^2-\{\bar{K}_8(E,n_x,n_y)\}^2\Big)$$

$$\times\cosh\{\bar{\beta}_3(E,n_x,n_y)\}\cos\{\bar{\gamma}_3(E,n_x,n_y)\}$$

$$+\frac{1}{12}\Bigg[\frac{5\{\bar{K}_8(E,n_x,n_y)\}^3}{\bar{K}_7(E,n_x,n_y)}+\frac{5\{\bar{K}_7(E,n_x,n_y)\}^3}{\bar{K}_8(E,n_x,n_y)}$$

$$-34\bar{K}_7(E,n_x,n_y)\bar{K}_8(E,n_x,n_y)\Bigg]$$

$$\times\sin\{\bar{\beta}_3(E,n_x,n_y)\}\sin\{\bar{\gamma}_3(E,n_x,n_y)\}\Big]\Big],$$

$$\bar{\beta}_3(E,n_x,n_y)\equiv\bar{K}_7(E,n_x,n_y)[a_0-\Delta_0],$$

$$\bar{K}_7(E,n_x,n_y)\equiv\left[\frac{2m_2^*E'}{\hbar^2}G(E-V_0,E_{g02},\Delta_2)+\left\{\left(\frac{\pi n_x}{d_x}\right)^2+\left(\frac{\pi n_y}{d_y}\right)^2\right\}\right]^{1/2},$$

$$\bar{\gamma}_3(E,n_x,n_y)=\bar{K}_8(E,n_x,n_y)[b_0-\Delta_0],$$

$$\bar{K}_8(E,n_x,n_y)\equiv\left[\frac{B_0^2+2AE-B_0\sqrt{B_0^2+4AE}}{2A^2}-\left\{\left(\frac{\pi n_x}{d_x}\right)^2+\left(\frac{\pi n_y}{d_y}\right)^2\right\}\right]^{1/2},$$

and

$$\bar{\varepsilon}_3(E,n_x,n_y)\equiv\left[\frac{\bar{K}_7(E,n_x,n_y)}{\bar{K}_8(E,n_x,n_y)}-\frac{\bar{K}_8(E,n_x,n_y)}{\bar{K}_7(E,n_x,n_y)}\right].$$

Thus, combining (1.13) and (3.20), the TPSM in this case can be expressed as

$$G_0=\frac{\pi^2k_B}{3e}\left[\sum_{n_x=1}^{n_{x\max}}\sum_{n_y=1}^{n_{y\max}}\sum_{n_z=1}^{n_{z\max}}F_{-1}(\overline{\eta_{12}})\right]^{-1}\left[\sum_{n_x=1}^{n_{x\max}}\sum_{n_y=1}^{n_{y\max}}\sum_{n_z=1}^{n_{z\max}}F_{-2}(\overline{\eta_{12}})\right].\tag{3.22}$$

3.2.5 Magnetothermopower in III–V Quantum Dot Effective Mass Superlattices

Following Sasaki [37], the electron dispersion law in III–V effective mass superlattices (EMSLs) can be written as

$$k_x^2 = \left[\frac{1}{L_0^2}\left\{\cos^{-1}\left(f\left(E,k_y,k_z\right)\right)\right\}^2 - k_\perp^2\right], \tag{3.23}$$

in which $f\left(E,k_y,k_z\right) = a_1\cos\left[a_0C_1\left(E,k_\perp\right) + b_0D_1\left(E,k_\perp\right)\right] - a_2\cos\left[a_0C_1\left(E,k_\perp\right) - b_0D_1\left(E,k_\perp\right)\right]$, $k_\perp^2 = k_y^2 + k_z^2$,

$$a_1 = \left[\sqrt{\frac{m_2^*}{m_1^*}} + 1\right]^2\left[4\left(\frac{m_2^*}{m_1^*}\right)^{1/2}\right]^{-1},$$

$$a_2 = \left[-1 + \sqrt{\frac{m_2^*}{m_1^*}}\right]^2\left[4\left(\frac{m_2^*}{m_1^*}\right)^{1/2}\right]^{-1},$$

$$C_1\left(E,k_\perp\right) \equiv \left[\left(\frac{2m_1^*E}{\hbar^2}\right)G\left(E,E_{g_{01}},\Delta_1\right) - k_\perp^2\right]^{1/2},$$

and

$$D_1\left(E,k_\perp\right) \equiv \left[\left(\frac{2m_2^*E}{\hbar^2}\right)G\left(E,E_{g_{02}},\Delta_2\right) - k_\perp^2\right]^{1/2}.$$

The total electron concentration per unit volume is given by

$$n_0 = \frac{2g_v}{d_xd_yd_z}\sum_{n_x=1}^{n_{x_{\max}}}\sum_{n_y=1}^{n_{y_{\max}}}\sum_{n_z=1}^{n_{z_{\max}}}F_{-1}\left(\overline{\eta_{13}}\right), \tag{3.24}$$

where

$$\overline{\eta_{13}} \equiv \frac{E_{\mathrm{FQDSLEM}} - \overline{E_{\mathrm{QD14}}}}{k_BT},$$

E_{FQDSLEM} is the Fermi energy in this case, and $\overline{E_{\mathrm{QD14}}}$ is the root of the equation

$$\left(\frac{\pi n_x}{d_x}\right)^2 = \left[\frac{1}{L_0^2}\left[\cos^{-1}\left[f_{14}\left(E,n_y,n_z\right)\Big|_{E=\overline{E_{\mathrm{QD14}}}}\right]\right]^2 - \left(\frac{\pi n_y}{d_y}\right)^2 - \left(\frac{\pi n_z}{d_z}\right)^2\right], \tag{3.25}$$

in which $f_{14}\left(E,n_y,n_z\right) = a_1\cos\left[a_0\bar{C}_1\left(E,n_y,n_z\right) + b_0\bar{D}_1\left(E,n_y,n_z\right)\right] - a_2\cos\left[a_0\bar{C}_1\left(E,n_y,n_z\right) - b_0\bar{D}_1\left(E,n_y,n_z\right)\right]$,

$$\bar{C}_1\left(E,n_y,n_z\right)\equiv\left[\left(\frac{2m_1^*E}{\hbar^2}\right)G\left(E,E_{g01},\Delta_1\right)-\left\{\left(\frac{\pi n_y}{d_y}\right)^2+\left(\frac{\pi n_z}{d_z}\right)^2\right\}\right]^{1/2},$$

and

$$\bar{D}_1\left(E,n_y,n_z\right)\equiv\left[\left(\frac{2m_2^*E}{\hbar^2}\right)G\left(E,E_{g02},\Delta_2\right)-\left\{\left(\frac{\pi n_y}{d_y}\right)^2+\left(\frac{\pi n_z}{d_z}\right)^2\right\}\right]^{1/2}.$$

Thus, combining (1.13) and (3.24), the TPSM in this case can be expressed as

$$G_0=\frac{\pi^2k_B}{3e}\left[\sum_{n_x=1}^{n_{x\max}}\sum_{n_y=1}^{n_{y\max}}\sum_{n_z=1}^{n_{z\max}}F_{-1}\left(\overline{\eta_{13}}\right)\right]^{-1}\left[\sum_{n_x=1}^{n_{x\max}}\sum_{n_y=1}^{n_{y\max}}\sum_{n_z=1}^{n_{z\max}}F_{-2}\left(\overline{\eta_{13}}\right)\right]. \tag{3.26}$$

3.2.6 Magnetothermopower in II–VI Quantum Dot Effective Mass Superlattices

Following Sasaki [24], the electron dispersion law in II–VI EMSLs can be written as

$$k_z^2=\left[\frac{1}{L_0^2}\left\{\cos^{-1}\left(f_1\left(E,k_x,k_y\right)\right)\right\}^2-k_s^2\right], \tag{3.27}$$

in which $f_1\left(E,k_x,k_y\right)=a_3\cos\left[a_0C_2\left(E,k_s\right)+b_0D_2\left(E,k_s\right)\right]-a_4\cos\left[a_0C_2\left(E,k_s\right)-b_0D_2\left(E,k_s\right)\right]$, $k_s^2=k_x^2+k_y^2$,

$$a_3=\left[\sqrt{\frac{m_2^*}{m_{\parallel,1}^*}}+1\right]^2\left[4\left(\frac{m_2^*}{m_{\parallel,1}^*}\right)^{1/2}\right]^{-1},$$

$$a_4=\left[-1+\sqrt{\frac{m_2^*}{m_{\parallel,1}^*}}\right]^2\left[4\left(\frac{m_2^*}{m_{\parallel,1}^*}\right)^{1/2}\right]^{-1},$$

$$C_2\left(E,k_s\right)\equiv\left(\frac{2m_{\parallel,1}^*}{\hbar^2}\right)^{1/2}\left[E-\frac{\hbar^2k_s^2}{2m_{\perp,1}^*}\mp C_0k_s\right]^{1/2},$$

and

$$D_2\left(E,k_s\right)\equiv\left[\left(\frac{2m_2^*}{\hbar^2}\right)EG\left(E,E_{g02},\Delta_2\right)-k_s^2\right]^{1/2}.$$

The total electron concentration per unit volume is given by

$$n_0 = \frac{g_v}{d_x d_y d_z} \sum_{n_x=1}^{n_{x\max}} \sum_{n_y=1}^{n_{y\max}} \sum_{n_z=1}^{n_{z\max}} F_{-1}\left(\overline{\eta_{14}}\right), \tag{3.28}$$

where

$$\overline{\eta_{14}} \triangleq \frac{E_{\text{FQDSLEM}} - \overline{E_{\text{QD15}}}}{k_B T}$$

and $\overline{E_{\text{QD15}}}$ is the root of the equation

$$\left(\frac{\pi n_z}{d_z}\right)^2 = \left[\frac{1}{L_0^2}\left[\cos^{-1}\left[f_{15}\left(E, n_x, n_y\right)\Big|_{E=\overline{E_{\text{QD15}}}}\right]\right]^2 - \left(\frac{\pi n_x}{d_x}\right)^2 - \left(\frac{\pi n_y}{d_y}\right)^2\right], \tag{3.29}$$

in which $f_{15}\left(E, n_x, n_y\right) = a_3 \cos\left[a_0 \bar{C}_2\left(E, n_x, n_y\right) + b_0 \bar{D}_2\left(E, n_x, n_y\right)\right] - a_4 \cos\left[a_0 \bar{C}_2\left(E, n_x, n_y\right) - b_0 \bar{D}_2\left(E, n_x, n_y\right)\right]$,

$$\bar{C}_2\left(E, n_x, n_y\right) \equiv \left(\frac{2m_{\|,1}^*}{\hbar^2}\right)^{1/2}\left[E - \hbar^2\left\{\left(\frac{\pi n_x}{d_x}\right)^2 + \left(\frac{\pi n_y}{d_y}\right)^2\right\}\left(2m_{\perp,1}^*\right)^{-1} \mp C_0\left\{\left(\frac{\pi n_x}{d_x}\right)^2 + \left(\frac{\pi n_y}{d_y}\right)^2\right\}^{1/2}\right]^{1/2},$$

and

$$\bar{D}_2\left(E, n_x, n_y\right) \equiv \left[\left(\frac{2m_2^*}{\hbar^2}\right) E G\left(E, E_{g02}, \Delta_2\right) - \left\{\left(\frac{\pi n_x}{d_x}\right)^2 + \left(\frac{\pi n_y}{d_y}\right)^2\right\}\right]^{1/2}.$$

Thus, combining (1.13) and (3.28), the TPSM in this case can be expressed as

$$G_0 = \frac{\pi^2 k_B}{3e}\left[\sum_{n_x=1}^{n_{x\max}} \sum_{n_y=1}^{n_{y\max}} \sum_{n_z=1}^{n_{z\max}} F_{-1}\left(\overline{\eta_{14}}\right)\right]^{-1}\left[\sum_{n_x=1}^{n_{x\max}} \sum_{n_y=1}^{n_{y\max}} \sum_{n_z=1}^{n_{z\max}} F_{-2}\left(\overline{\eta_{14}}\right)\right]. \tag{3.30}$$

3.2.7 Magnetothermopower in IV–VI Quantum Dot Effective Mass Superlattices

Following Sasaki [37], the electron dispersion law in IV–VI, EMSLs can be written as

$$k_x^2 = \left[\frac{1}{L_0^2}\left\{\cos^{-1}\left(f_2\left(E, k_y, k_z\right)\right)\right\}^2 - k_\perp^2\right] \tag{3.31}$$

in which $f_2(E,k_y,k_z) = a_5\cos[a_0C_3(E,k_y,k_z) + b_0D_3(E,k_y,k_z)] - a_6\cos[a_0C_3(E,k_y,k_z) - b_0D_3(E,k_y,k_z)]$,

$$a_5 = \left[\sqrt{\frac{m_2^*}{m_1^*}} + 1\right]^2 \left[4\left(\frac{m_2^*}{m_1^*}\right)^{1/2}\right]^{-1},$$

$$m_i^* = \left[\frac{\hbar^2}{2\{a_i^2 - C_i^2\}}\right]\Big[a_i - [a_iC_i + a_ie_iE_{g_i} + C_i^2E_{g_i}] \times [E_{g_i}^2a_i^2 + C_i^2 + e_i^2E_{g_i}^2 + 2C_ia_iE_{g_i} + 2E_{g_i}e_iC_i + 2e_ia_iE_{g_i}^2]^{-1/2}\Big]$$

$$C_3(E,k_y,k_z) \equiv \left[\left[EH_{11} + H_{21}(k_y,k_z)\right] - \left[E^2H_{31} + EH_{41}(k_y,k_z) + H_{51}(k_y,k_z)\right]^{1/2}\right]^{1/2},$$

$$D_3(E,k_y,k_z) \equiv \left[\left[EH_{12} + H_{22}(k_y,k_z)\right] - \left[E^2H_{32} + EH_{42}(k_y,k_z) + H_{52}(k_y,k_z)\right]^{1/2}\right]^{1/2},$$

and

$$a_6 = \left[-1 + \sqrt{\frac{m_2^*}{m_1^*}}\right]^2 \left[4\left(\frac{m_2^*}{m_1^*}\right)^{1/2}\right]^{-1}.$$

The total electron concentration per unit volume is given by

$$n_0 = \frac{2g_v}{d_xd_yd_z}\sum_{n_x=1}^{n_{x\max}}\sum_{n_y=1}^{n_{y\max}}\sum_{n_{z\min}}^{n_{z\max}} F_{-1}(\overline{\eta_{15}}), \tag{3.32}$$

where $\overline{\eta_{15}} \equiv \frac{E_{\mathrm{FQDSLEM}} - \overline{E_{\mathrm{QD16}}}}{k_BT}$ and $\overline{E_{\mathrm{QD16}}}$ is the root of the equation

$$\left(\frac{\pi n_x}{d_x}\right)^2 = \left[\frac{1}{L_0^2}\left[\cos^{-1}\left[f_{16}(E,n_y,n_z)\Big|_{E=\overline{E_{\mathrm{QD16}}}}\right]\right]^2 - \left(\frac{\pi n_y}{d_y}\right)^2 - \left(\frac{\pi n_z}{d_z}\right)^2\right], \tag{3.33}$$

where

$$f_{16}(E,n_y,n_z) = a_5\cos[a_0\bar{C}_3(E,n_y,n_z) + b_0\bar{D}_3(E,n_y,n_z)] - a_6\cos[a_0\bar{C}_3(E,n_y,n_z) - b_0\bar{D}_3(E,n_y,n_z)],$$

$$\bar{C}_3(E,n_y,n_z) \equiv \left[\left[EH_{11}+\bar{H}_{21}(n_y,n_z)\right] -\left[E^2H_{31}+E\bar{H}_{41}(n_y,n_z)+\bar{H}_{51}(n_y,n_z)\right]^{1/2}\right]^{1/2},$$

$$\bar{D}_3(E,n_y,n_z) \equiv \left[\left[EH_{12}+\bar{H}_{22}(n_y,n_z)\right] -\left[E^2H_{32}+E\bar{H}_{42}(n_y,n_z)+\bar{H}_{52}(n_y,n_z)\right]^{1/2}\right]^{1/2}.$$

Thus, combining (1.13) and (3.32), the TPSM in this case can be expressed as

$$G_0 = \frac{\pi^2 k_B}{3e}\left[\sum_{n_x=1}^{n_{x\max}}\sum_{n_y=1}^{n_{y\max}}\sum_{n_{z\min}}^{n_{z\max}} F_{-1}(\overline{\eta_{15}})\right]^{-1}\left[\sum_{n_x=1}^{n_{x\max}}\sum_{n_y=1}^{n_{y\max}}\sum_{n_{z\min}}^{n_{z\max}} F_{-2}(\overline{\eta_{15}})\right] \quad (3.34)$$

3.2.8 Magnetothermopower in HgTe/CdTe Quantum Dot Effective Mass Superlattices

Following Sasaki [37], the electron dispersion law in HgTe/CdTe EMSLs can be written as

$$k_x^2 = \left[\frac{1}{L_0^2}\left\{\cos^{-1}\left(f_3(E,k_y,k_z)\right)\right\}^2 - k_\perp^2\right], \quad (3.35)$$

in which $f_3(E,k_\perp) = a_7\cos\left[a_0C_4(E,k_\perp)+b_0D_4(E,k_\perp)\right] - a_8\cos\left[a_0C_4(E,k_\perp)-b_0D_4(E,k_\perp)\right]$,

$$a_7 = \left[\sqrt{\frac{m_2^*}{m_1^*}}+1\right]^2\left[4\left(\frac{m_2^*}{m_1^*}\right)^{1/2}\right]^{-1},$$

$$a_8 = \left[-1+\sqrt{\frac{m_2^*}{m_1^*}}\right]^2\left[4\left(\frac{m_2^*}{m_1^*}\right)^{1/2}\right]^{-1},$$

$$C_4(E,k_\perp) \equiv \left[\frac{B_0^2+2AE-B_0\sqrt{B_0^2+4AE}}{2A^2} - k_\perp^2\right]^{1/2},$$

and

$$D_4(E,k_\perp) \equiv \left[\left(\frac{2m_2^*E}{\hbar^2}\right)G(E,E_{g02},\Delta_2) - k_\perp^2\right]^{1/2}.$$

The total electron concentration per unit volume is given by

$$n_0 = \frac{2g_v}{d_x d_y d_z} \sum_{n_x=1}^{n_{x\max}} \sum_{n_y=1}^{n_{y\max}} \sum_{n_z=1}^{n_{z\max}} F_{-1}\left(\overline{\eta_{16}}\right), \tag{3.36}$$

where $\overline{\eta_{16}} \equiv \frac{E_{FQDSLEM} - \overline{E_{QD17}}}{k_B T}$ and $\overline{E_{QD17}}$ is the root of

$$\left(\frac{\pi n_x}{d_x}\right)^2 = \left[\frac{1}{L_0^2}\left[\cos^{-1}\left[f_{17}\left(E, n_y, n_z\right)\Big|_{E=\overline{E_{QD17}}}\right]\right]^2 - \left(\frac{\pi n_y}{d_y}\right)^2 - \left(\frac{\pi n_z}{d_z}\right)^2\right], \tag{3.37}$$

where $f_{17}\left(E, n_y, n_z\right) = a_7 \cos\left[a_0 \bar{C}_4\left(E, n_y, n_z\right) + b_0 \bar{D}_4\left(E, n_y, n_z\right)\right] - a_8 \cos\left[a_0 \bar{C}_4\left(E, n_y, n_z\right) - b_0 \bar{D}_4\left(E, n_y, n_z\right)\right]$,

$$\bar{C}_4\left(E, n_y, n_z\right) \equiv \left[\frac{B_0^2 + 2AE - B_0\sqrt{B_0^2 + 4AE}}{2A^2} - \left\{\left(\frac{\pi n_y}{d_y}\right)^2 + \left(\frac{\pi n_z}{d_z}\right)^2\right\}\right]^{1/2}$$

and

$$\bar{D}_4\left(E, n_y, n_z\right) \equiv \left[\left(\frac{2m_2^* E}{\hbar^2}\right) G\left(E, E_{g02}, \Delta_2\right) - \left\{\left(\frac{\pi n_y}{d_y}\right)^2 + \left(\frac{\pi n_z}{d_z}\right)^2\right\}\right]^{1/2}.$$

Thus, combining (1.13) and (3.36), the TPSM in this case can be expressed as

$$G_0 = \frac{\pi^2 k_B}{3e}\left[\sum_{n_x=1}^{n_{x\max}} \sum_{n_y=1}^{n_{y\max}} \sum_{n_z=1}^{n_{z\max}} F_{-1}\left(\overline{\eta_{16}}\right)\right]^{-1}\left[\sum_{n_x=1}^{n_{x\max}} \sum_{n_y=1}^{n_{y\max}} \sum_{n_z=1}^{n_{z\max}} F_{-2}\left(\overline{\eta_{16}}\right)\right]. \tag{3.38}$$

3.3 Results and Discussion

Using the band constants from Table 1.1, the normalized TPSM in this case in $HgTe/Hg_{1-x}Cd_xTe$ (using (3.3) and (3.5)), CdS/ZnSe (using (3.9) and (3.11)), PbSe/PbTe (using (3.14) and (3.16)), and HgTe/CdTe (using (3.20) and (3.22)) quantum dot superlattices with graded interfaces have been plotted as a function of film thicykness as shown by curves (a), (b), (c), and (d), respectively, in Fig. 3.1. Figure 3.2 demonstrates the normalized TPSM for the said quantized structures as a function of impurity concentration. Figure 3.3 exhibits the normalized TPSM as a function of film thickness in $HgTe/Hg_{1-x}Cd_xTe$ (using (3.24) and (3.26)),

Table 3.1 The carrier statistics and the thermoelectric power under large magnetic field in III–V, II–VI, IV–VI, and HgTe/CdTe quantum dot superlattices with graded interfaces and also aforementioned quantum dot effective mass superlattices

Type of materials	Carrier statistics	TPSM
1. III–V quantum dot superlattices with graded interfaces	$n_0 = \frac{2g_v}{d_x d_y d_z} \sum_{n_x=1}^{n_{x\max}} \sum_{n_y=1}^{n_{y\max}} \sum_{n_z=1}^{n_{z\max}} F_{-1}(\overline{\eta_9}) \quad (3.3)$	On the basis of (3.3), $G_0 = \frac{\pi^2 k_B}{3e} \left[\sum_{n_x=1}^{n_{x\max}} \sum_{n_y=1}^{n_{y\max}} \sum_{n_z=1}^{n_{z\max}} F_{-1}(\overline{\eta_9}) \right]^{-1} \left[\sum_{n_x=1}^{n_{x\max}} \sum_{n_y=1}^{n_{y\max}} \sum_{n_z=1}^{n_{z\max}} F_{-2}(\overline{\eta_9}) \right] \quad (3.5)$
2. II–VI quantum dot superlattices with graded interfaces	$n_0 = \frac{g_v}{d_x d_y d_z} \sum_{n_x=1}^{n_{x\max}} \sum_{n_y=1}^{n_{y\max}} \sum_{n_z=1}^{n_{z\max}} F_{-1}(\overline{\eta_{10}}) \quad (3.9)$	On the basis of (3.9), $G_0 = \frac{\pi^2 k_B}{3e} \left[\sum_{n_x=1}^{n_{x\max}} \sum_{n_y=1}^{n_{y\max}} \sum_{n_z=1}^{n_{z\max}} F_{-1}(\overline{\eta_{10}}) \right]^{-1} \left[\sum_{n_x=1}^{n_{x\max}} \sum_{n_y=1}^{n_{y\max}} \sum_{n_z=1}^{n_{z\max}} F_{-2}(\overline{\eta_{10}}) \right] \quad (3.11)$
3. IV–VI quantum dot superlattices with graded interfaces	$n_0 = \frac{2g_v}{d_x d_y d_z} \sum_{n_x=1}^{n_{x\max}} \sum_{n_y=1}^{n_{y\max}} \sum_{n_{z\min}}^{n_{z\max}} F_{-1}(\overline{\eta_{11}}) \quad (3.14)$	On the basis of (3.14), $G_0 = \frac{\pi^2 k_B}{3e} \left[\sum_{n_x=1}^{n_{x\max}} \sum_{n_y=1}^{n_{y\max}} \sum_{n_{z\min}}^{n_{z\max}} F_{-1}(\overline{\eta_{11}}) \right]^{-1} \left[\sum_{n_x=1}^{n_{x\max}} \sum_{n_y=1}^{n_{y\max}} \sum_{n_{z\min}}^{n_{z\max}} F_{-2}(\overline{\eta_{11}}) \right] \quad (3.16)$
4. HgTe/CdTe quantum dot superlattices with graded interfaces	$n_0 = \frac{2g_v}{d_x d_y d_z} \sum_{n_x=1}^{n_{x\max}} \sum_{n_y=1}^{n_{y\max}} \sum_{n_z=1}^{n_{z\max}} F_{-1}(\overline{\eta_{12}}) \quad (3.20)$	On the basis of (3.20), $G_0 = \frac{\pi^2 k_B}{3e} \left[\sum_{n_x=1}^{n_{x\max}} \sum_{n_y=1}^{n_{y\max}} \sum_{n_z=1}^{n_{z\max}} F_{-1}(\overline{\eta_{12}}) \right]^{-1} \left[\sum_{n_x=1}^{n_{x\max}} \sum_{n_y=1}^{n_{y\max}} \sum_{n_z=1}^{n_{z\max}} F_{-2}(\overline{\eta_{12}}) \right] \quad (3.22)$

5. III–V quantum dot effective mass superlattice	$n_0 = \frac{2g_v}{d_x d_y d_z} \sum_{n_x=1}^{n_{x\max}} \sum_{n_y=1}^{n_{y\max}} \sum_{n_z=1}^{n_{z\max}} F_{-1}(\overline{\eta_{13}})$ (3.24)	On the basis of (3.24), $G_0 = \frac{\pi^2 k_B}{3e} \left[\sum_{n_x=1}^{n_{x\max}} \sum_{n_y=1}^{n_{y\max}} \sum_{n_z=1}^{n_{z\max}} F_{-1}(\overline{\eta_{13}}) \right]^{-1} \left[\sum_{n_x=1}^{n_{x\max}} \sum_{n_y=1}^{n_{y\max}} \sum_{n_z=1}^{n_{z\max}} F_{-2}(\overline{\eta_{13}}) \right]$ (3.26)
6. II–VI quantum dot effective mass superlattice	$n_0 = \frac{g_v}{d_x d_y d_z} \sum_{n_x=1}^{n_{x\max}} \sum_{n_y=1}^{n_{y\max}} \sum_{n_z=1}^{n_{z\max}} F_{-1}(\overline{\eta_{14}})$ (3.28)	On the basis of (3.28), $G_0 = \frac{\pi^2 k_B}{3e} \left[\sum_{n_x=1}^{n_{x\max}} \sum_{n_y=1}^{n_{y\max}} \sum_{n_z=1}^{n_{z\max}} F_{-1}(\overline{\eta_{14}}) \right]^{-1} \left[\sum_{n_x=1}^{n_{x\max}} \sum_{n_y=1}^{n_{y\max}} \sum_{n_z=1}^{n_{z\max}} F_{-2}(\overline{\eta_{14}}) \right]$ (3.30)
7. IV–VI quantum dot effective mass superlattice	$n_0 = \frac{2g_v}{d_x d_y d_z} \sum_{n_x=1}^{n_{x\max}} \sum_{n_y=1}^{n_{y\max}} \sum_{n_{z\min}}^{n_{z\max}} F_{-1}(\overline{\eta_{15}})$ (3.32)	On the basis of (3.32), $G_0 = \frac{\pi^2 k_B}{3e} \left[\sum_{n_x=1}^{n_{x\max}} \sum_{n_y=1}^{n_{y\max}} \sum_{n_{z\min}}^{n_{z\max}} F_{-1}(\overline{\eta_{15}}) \right]^{-1} \left[\sum_{n_x=1}^{n_{x\max}} \sum_{n_y=1}^{n_{y\max}} \sum_{n_{z\min}}^{n_{z\max}} F_{-2}(\overline{\eta_{15}}) \right]$ (3.34)
8. HgTe/CdTe quantum dot effective mass superlattices	$n_0 = \frac{2g_v}{d_x d_y d_z} \sum_{n_x=1}^{n_{x\max}} \sum_{n_y=1}^{n_{y\max}} \sum_{n_z=1}^{n_{z\max}} F_{-1}(\overline{\eta_{16}})$ (3.36)	On the basis of (3.36), $G_0 = \frac{\pi^2 k_B}{3e} \left[\sum_{n_x=1}^{n_{x\max}} \sum_{n_y=1}^{n_{y\max}} \sum_{n_z=1}^{n_{z\max}} F_{-1}(\overline{\eta_{16}}) \right]^{-1} \left[\sum_{n_x=1}^{n_{x\max}} \sum_{n_y=1}^{n_{y\max}} \sum_{n_z=1}^{n_{z\max}} F_{-2}(\overline{\eta_{16}}) \right]$ (3.38)

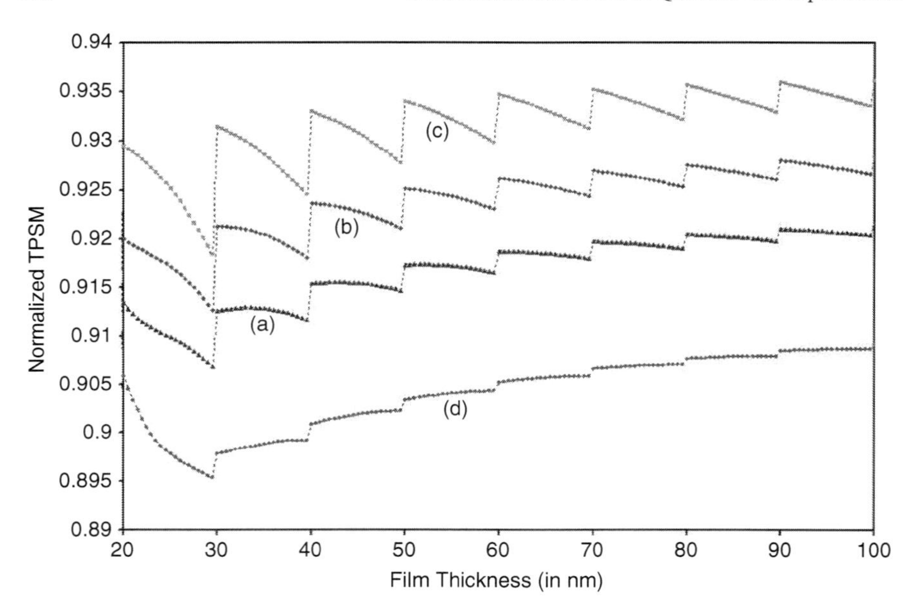

Fig. 3.1 Plot of the TPSM in (**a**) HgTe/$Hg_{1-x}Cd_xTe$, (**b**) CdS/ZnSe, (**c**) PbSe/PbTe, and (**d**) HgTe/CdTe quantum dot superlattices with graded interfaces as a function of film thickness

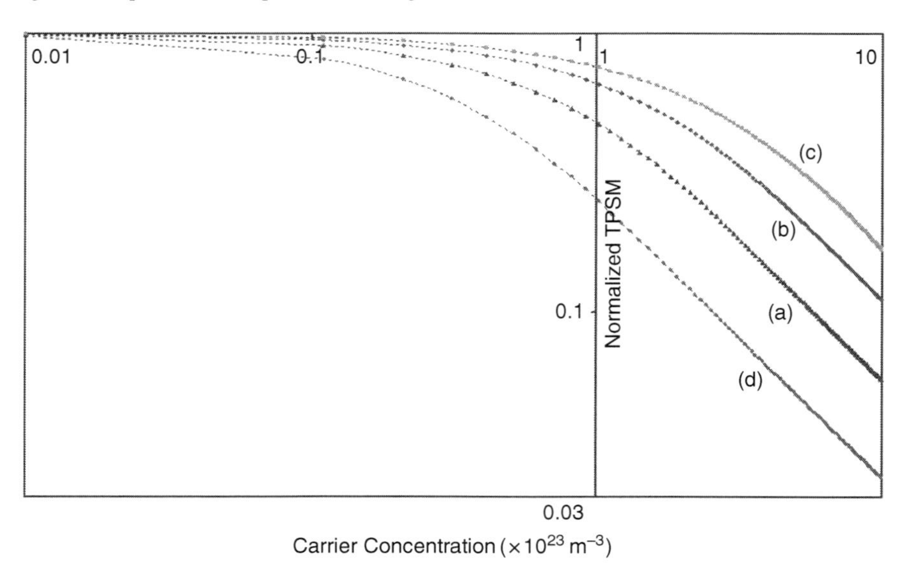

Fig. 3.2 Plot of the TPSM in (**a**) HgTe/$Hg_{1-x}Cd_xTe$, (**b**) CdS/ZnSe, (**c**) PbSe/PbTe, and (**d**) HgTe/CdTe quantum dot superlattices with graded interfaces as a function of carrier concentration

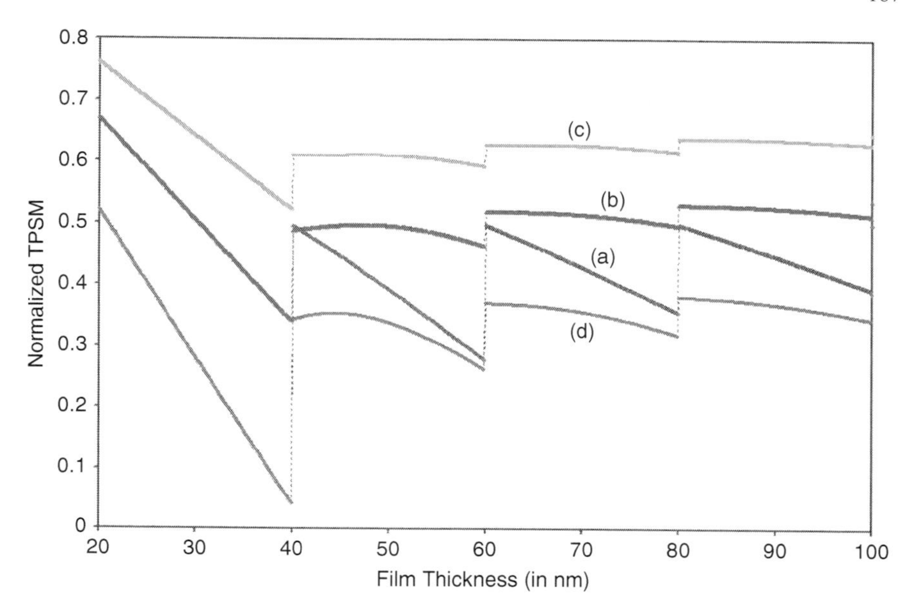

Fig. 3.3 Plot of the TPSM in (**a**) $HgTe/Hg_{1-x}Cd_xTe$, (**b**) CdS/ZnSe, (**c**) PbSe/PbTe, and (**d**) HgTe/CdTe quantum dot effective mass superlattices as a function of film thickness

CdS/ZnSe (using (3.28) and (3.30)), PbSe/PbTe (using (3.32) and (3.34)), and HgTe/CdTe (using (3.36) and (3.38)) quantum dot EMSLs as shown by curves (a), (b), (c), and (d), respectively. The normalized TPSM in $HgTe/Hg_{1-x}Cd_xTe$, CdS/ZnSe, PbSe/PbTe, and HgTe/CdTe quantum dot EMSLs has been plotted as a function of electron concentration in Fig. 3.4.

It appears from Fig. 3.1 that the TPSM in $HgTe/Hg_{1-x}Cd_xTe$, CdS/ZnSe, PbSe/PbTe, and HgTe/CdTe quantum dot superlattices with graded interfaces increases with increasing film thickness, exhibiting quantum jumps for fixed values of film thickness depending on the values of the energy band constants of the particular quantized structures. It is observed from Fig. 3.2 that the TPSM in quantum dots of aforementioned superlattices decreases with increasing carrier degeneracy and differ widely for large values of same, whereas for relatively small values of electron concentration, the TPSM exhibits a converging behavior. From Fig. 3.3, it is observed that the TPSM in $HgTe/Hg_{1-x}Cd_xTe$, CdS/ZnSe, PbSe/PbTe, and HgTe/CdTe quantum dot EMSLs oscillates with increasing film thickness. From Fig. 3.4, it appears that the TPSM of the aforementioned superlattices decreases with increasing concentration. It should be noted that all types of variations of TPSM with respect to thickness and concentration are basically band structure dependent.

It may further be noted that the TPSM of a two-dimensional electron gas in the presence of a periodic potential has already been formulated in the literature [41]. The SL is a three-dimensional system under periodic potential. There is a radical difference in the dispersion relations of the 3D quantized structures and the corresponding carrier energy spectra of the 2D systems. From the dispersion

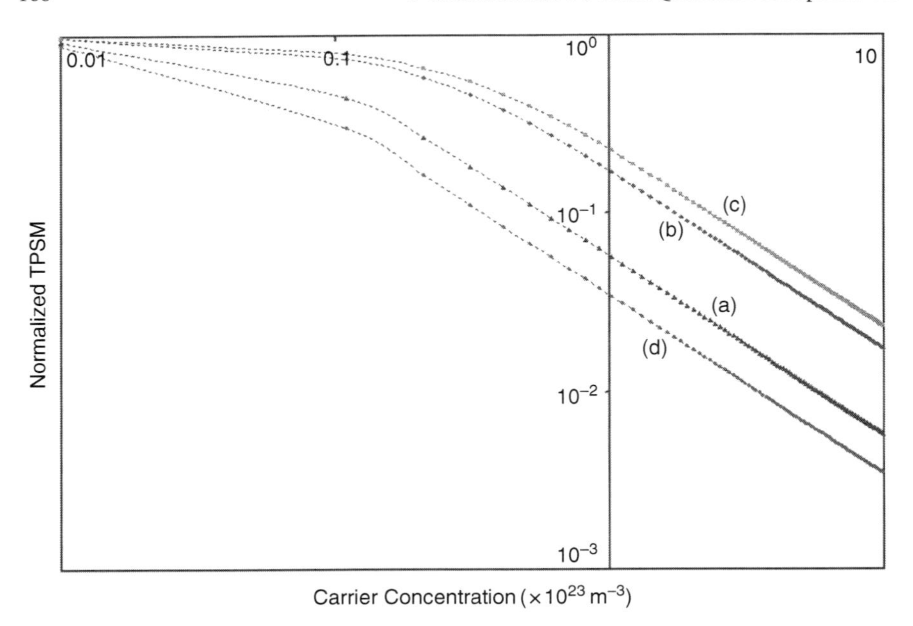

Fig. 3.4 Plot of the TPSM in (**a**) $HgTe/Hg_{1-x}Cd_xTe$, (**b**) CdS/ZnSe, (**c**) PbSe/PbTe, and (**d**) HgTe/CdTe quantum dot effective mass superlattices as a function of carrier concentration

relations of various superlattices as discussed in this chapter, the energy spectra of the various other types of low-dimensional systems can be formulated and the corresponding TPSMs can also be investigated. The results will be fundamentally different in all cases due to system asymmetry together with the change in the respective wave functions exhibiting new physical features in the respective cases. Therefore, it appears that the dispersion law and the corresponding wave function play a cardinal role in formulating any electronic property of any electronic material, since they change in a fundamental way in the presence of dimension reduction. Consequently, the derivations and the respective physical interpretations of the different transport quantities change radically [42, 43].

It is imperative to state that our investigations excludes the many-body, hot electron, spin, broadening, and the allied quantum dot and SL effects in this simplified theoretical formalism due to the absence of proper analytical techniques for including them for the generalized systems as considered here. Our simplified approach will be appropriate for the purpose of comparison when the methods of tackling the formidable problems after inclusion of the said effects for the generalized systems emerge. The inclusion of the said effects would certainly increase the accuracy of the results although the qualitative features of the TPSM would not change in the presence of the aforementioned effects. For the purpose of condensed presentation, the carrier statistics and the TPSM in III–V, II–VI, IV–VI, and HgTe/CdTe quantum dot superlattices with graded interfaces and also aforementioned quantum dot EMSLs have been presented in Table 3.1.

3.4 Open Research Problems

(R3.1) Investigate the DTP, PTP, and Z in the absence of magnetic field by considering all types of scattering mechanisms for III–V, II–VI, IV–VI, and HgTe/CdTe superlattices with graded interfaces and also the EMSLs of the aforementioned materials with the appropriate dispersion relations as formulated in this chapter.

(R3.2) Investigate the DTP, PTP, and Z in the absence of magnetic field by considering all types of scattering mechanisms for strained layer, random, short period, Fibonacci, polytype, and saw-toothed superlattices, respectively.

(R3.3) Investigate the DTP, PTP, and Z in the absence of magnetic field by considering all types of scattering mechanisms for (R3.1) and (R3.2) under an arbitrarily oriented (a) non-uniform electric field and (b) alternating electric field, respectively.

(R3.4) Investigate the DTP, PTP, and Z by considering all types of scattering mechanisms for (R3.1) and (R3.2) under an arbitrarily oriented alternating magnetic field by including broadening and the electron spin, respectively.

(R3.5) Investigate the DTP, PTP, and Z by considering all types of scattering mechanisms for (R3.1) and (R3.2) under an arbitrarily oriented alternating magnetic field and crossed alternating electric field by including broadening and the electron spin, respectively.

(R3.6) Investigate the DTP, PTP, and Z by considering all types of scattering mechanisms for (R3.1) and (R3.2) under an arbitrarily oriented alternating magnetic field and crossed alternating non-uniform electric field by including broadening and the electron spin, respectively.

(R3.7) Investigate the DTP, PTP, and Z in the absence of magnetic field for all types of superlattices as considered in this chapter under exponential, Kane, Halperin, Lax, and Bonch-Bruevich band tails [42], respectively.

(R3.8) Investigate the DTP, PTP and Z in the absence of magnetic field for the problem as defined in (R3.7) under an arbitrarily oriented (a) non-uniform electric field and (b) alternating electric field, respectively.

(R3.9) Investigate the DTP, PTP, and Z for the problem as defined in (R3.7) under an arbitrarily oriented alternating magnetic field by including broadening and the electron spin, respectively.

(R3.10) Investigate the DTP, PTP, and Z for the problem as defined in (R3.7) under an arbitrarily oriented alternating magnetic field and crossed alternating electric field by including broadening and the electron spin, respectively.

(R3.11) Investigate the problems as defined in (R3.1)–(R3.10) for all types of quantum dot superlattices as discussed in this chapter.

(R3.12) Investigate the problems as defined in (R3.1)–(R3.10) for all types of quantum dot superlattices as discussed in this chapter in the presence of strain.

(R3.13) Introducing new theoretical formalisms, investigate all the problems of this chapter in the presence of hot electron effects.

(R3.14) Investigate the influence of deep traps and surface states separately for all the appropriate problems of this chapter after proper modifications.

References

1. L.V. Keldysh, Sov. Phys. Solid State **4**, 1658 (1962)
2. L. Esaki, R. Tsu, IBM J. Res. Dev. **14**, 61 (1970)
3. G. Bastard, *Wave Mechanics Applied to Heterostructures*, (Editions de Physique, Les Ulis, France, 1990)
4. E.L. Ivchenko, G. Pikus, *Superlattices and other Heterostructures*, (Springer-Berlin, 1995)
5. R. Tsu, *Superlattices to Nanoelectronics*, (Elsevier, The Netherlands, 2005)
6. P. Fürjes, Cs. Dücs, M. Ádám, J. Zettner, I. Bársony, Superlatt. Microstruct. **35**, 455 (2004)
7. T. Borca-Tasciuc, D. Achimov, W.L. Liu, G. Chen, H.-W. Ren, C.-H. Lin, S.S. Pei, Microscale Thermophysical Eng. **5**, 225 (2001)
8. B.S. Williams, Nat. Photonics **1**, 517 (2007)
9. A. Kosterev, G. Wysocki, Y. Bakhirkin, S. So, R. Lewicki, F. Tittel, R.F. Curl, Appl. Phys. B **90**, 165 (2008)
10. M.A. Belkin, F. Capasso, F. Xie, A. Belyanin, M. Fischer, A. Wittmann, J. Faist, Appl. Phys. Lett. **92**, 201101 (2008)
11. G.J. Brown, F. Szmulowicz, R. Linville, A. Saxler, K. Mahalingam, C.-H. Lin, C.H. Kuo, W.Y. Hwang, IEEE Photonics Technol. Lett. **12**, 684 (2000)
12. H.J. Haugan, G.J. Brown, L. Grazulis, K. Mahalingam, D.H. Tomich, Physics E: Low-dimensional Systems and Nanostructures **20**, 527 (2004)
13. S.A. Nikishin, V.V. Kuryatkov, A. Chandolu, B.A. Borisov, G.D. Kipshidze, I. Ahmad, M. Holtz, H. Temkin, Jpn. J. Appl. Phys. **42**, L1362 (2003)
14. Y.-K. Su, H.-C. Wang, C.-L. Lin, W.-B. Chen, S.-M. Chen, Jpn. J. Appl. Phys. **42**, L751 (2003)
15. C.H. Liu, Y.K. Su, L.W. Wu, S.J. Chang, R.W. Chuang, Semicond. Sci. Technol. **18**, 545 (2003)
16. S.-B. Che, I. Nomura, A. Kikuchi, K. Shimomura, K. Kishino, Phys. Stat. Sol. (b) **229**, 1001 (2002)
17. C.P. Endres, F. Lewen, T.F. Giesen, S. Schlemmer, D.G. Paveliev, Y.I. Koschurinov, V.M. Ustinov, A.E. Zhucov, Rev. Sci. Instrum. **78**, 043106 (2007).
18. F. Klappenberger, K.F. Renk, P. Renk, B. Rieder, Y.I. Koshurinov, D.G. Pavelev, V. Ustinov, A. Zhukov, N. Maleev, A. Vasilyev, Appl. Phys. Letts. **84**, 3924 (2004).
19. X. Jin, Y. Maeda, T. Saka, M. Tanioku, S. Fuchi, T. Ujihara, Y. Takeda, N. Yamamoto, Y. Nakagawa, A. Mano, S. Okumi, M. Yamamoto, T. Nakanishi, H. Horinaka, T. Kato, T. Yasue, T. Koshikawa, J. Crys. Growth **310**, 5039 (2008)
20. X. Jin, N. Yamamoto, Y. Nakagawa, A. Mano, T. Kato, M. Tanioku, T. Ujihara, Y. Takeda, S. Okumi, M. Yamamoto, T. Nakanishi, T. Saka, H. Horinaka, T. Kato, T. Yasue, T. Koshikawa, Appl. Phys. Express **1**, 045002 (2008)
21. B.H. Lee, K.H. Lee, S. Im, M.M. Sung, Org. Electron. **9**, 1146 (2008)
22. P.-H. Wu, Y.-K. Su, I.-L. Chen, C.-H. Chiou, J.-T. Hsu, W.-R. Chen, Jpn. J. Appl. Phys. **45**, L647 (2006)
23. A.C. Varonides, Renewable Energy **33**, 273 (2008)
24. M. Walther, G. Weimann, Phys. Stat. Sol. (b) **203**, 3545 (2006)
25. R. Rehm, M. Walther, J. Schmitz, J. Fleiβner, F. Fuchs, J. Ziegler, W. Cabanski, Opto-Electron. Rev. **14**, 19 (2006)
26. R. Rehm, M. Walther, J. Scmitz, J. Fleissner, J. Ziegler, W. Cabanski, R. Breiter, Electron. Lett. **42**, 577 (2006)
27. G.J. Brown, F. Szmulowicz, H. Haugan, K. Mahalingam, S. Houston, Microelectronics J. **36**, 256 (2005)
28. K.V. Vaidyanathan, R.A. Jullens, C.L. Anderson, H.L. Dunlap, Solid State Electron.**26**, 717 (1983)
29. B. A. Wilson, IEEE, J. Quantum Electron. **24**, 1763 (1988)
30. M. Krichbaum, P. Kocevar, H. Pascher, G. Bauer, IEEE J. Quantum Electron. **24**, 717 (1988)
31. J.N. Schulman, T. C. McGill, Appl. Phys. Lett. **34**, 663 (1979)
32. H. Kinoshita, T. Sakashita, H. Fajiyasu, J. Appl. Phys. **52**, 2869 (1981)
33. L. Ghenin, R.G. Mani, J.R. Anderson, J.T. Cheung, Phys. Rev. B **39**, 1419 (1989)

34. C.A. Hoffman, J.R. Mayer, F.J. Bartoli, J.W. Han, J.W. Cook, J.F. Schetzina, J.M. Schubman, Phys. Rev. B **39**, 5208 (1989)
35. V.A. Yakovlev, Sov. Phys. Semicond. **13**, 692 (1979)
36. E.O. Kane., J. Phys. Chem. Solids **1**, 249 (1957)
37. H. Sasaki, Phys. Rev. B **30**, 7016 (1984)
38. H.X. Jiang, J.Y. Lin, J. Appl. Phys. **61**, 624 (1987)
39. J.J. Hopfield, J. Phys. Chem. Solids **15**, 97 (1960)
40. G.M.T. Foley, P.N. Langenberg, Phys. Rev. B **15B**, 4850 (1977)
41. F.M. Peters, P. Vasilopoulos, Phys. Rev. B **46**, 4667 (1992)
42. B.R. Nag, *Electron Transport in Compound Semiconductors,* Springer Series in Solid-state Sciences, vol 11 (Springer-Verlag, Germany, 1980)
43. K.P. Ghatak, S.N. Biswas, Sol. State Electron. **37**, 1437 (1994)

Chapter 4
Thermoelectric Power in Quantum Wire Superlattices Under Large Magnetic Field

4.1 Introduction

In Chap. 3, the TPSM has been investigated under large magnetic field in quantum dot superlattices having various band structures. In this chapter, the TPSM has been studied in III–V, II–VI, IV–VI, and HgTe/CdTe quantum wire superlattices (QWSL) with graded interfaces from Sects. 4.2.1 to 4.2.4. Sections 4.2.5–4.2.8 contain the investigation of the magnetothermopower in III–V, II–VI, IV–VI, and HgTe/CdTe quantum wire effective mass superlattices. The results and discussion is present in Sect. 4.3 and Sect. 4.4 includes open research problems.

4.2 Theoretical Background

4.2.1 Magnetothermopower in III–V Quantum Wire Superlattices with Graded Interfaces

The electron dispersion law in III–V QWSL can be written following (3.4) as [1]

$$(k_z)^2 = \left[\frac{1}{L_0^2}\left[\cos^{-1}\left[\frac{1}{2}f_{10}\left(E, n_x, n_y\right)\right]\right]^2 - \left(\frac{\pi n_x}{d_x}\right)^2 - \left(\frac{\pi n_y}{d_y}\right)^2\right], \quad (4.1)$$

where $f_{10}(E, n_x, n_y)$ has been defined in connection with (3.4) of Chap. 3.

Considering only the lowest miniband, since in an actual SL only the lowest miniband is significantly populated at low temperatures, where the quantum effects become prominent, the relation between the 1D electron concentration per unit length and the Fermi energy in the present case can be written as

$$n_0 = \left(\frac{2g_v}{\pi}\right)\sum_{n_x=1}^{n_{x\max}}\sum_{n_y=1}^{n_{y\max}}\left[T_{41}\left(n_x, n_y, E_{FQWSLGI}\right) + T_{42}\left(n_x, n_y, E_{FQWSLGI}\right)\right], \quad (4.2)$$

where

$$T_{41}(n_x, n_y, E_{\text{FQWSLGI}}) \equiv \left[\frac{1}{L_0^2} \left[\cos^{-1} \left[\frac{1}{2} f_{10} \left(E_{\text{FQWSLGI}}, n_x, n_y \right) \right] \right]^2 - \left(\frac{\pi n_x}{d_x} \right)^2 - \left(\frac{\pi n_y}{d_y} \right)^2 \right]^{1/2},$$

E_{FQWSLGI} is the Fermi energy in this case,

$$T_{42}(n_x, n_y, E_{\text{FQWSLGI}}) \equiv \sum_{r=1}^{s} L(r) \left[T_{41} \left(n_x, n_y, E_{\text{FQWSLGI}} \right) \right],$$

and

$$L(r) \equiv \left[2 \left(k_{\text{B}} T \right)^{2r} \left(1 - 2^{1-2r} \right) \zeta(2r) \frac{\partial^{2r}}{\partial E_{\text{FQWSL}}^{2r}} \right].$$

The use of (1.13) and (4.2) leads to the expression of the TPSM in this case as

$$G_0 = \left(\frac{\pi^2 k_{\text{B}}^2 T}{3e} \right) \times \frac{\sum_{n_x=1}^{n_{x_{\max}}} \sum_{n_y=1}^{n_{y_{\max}}} \left[\left\{ T_{41} \left(n_x, n_y, E_{\text{FQWSLGI}} \right) \right\}' + \left\{ T_{42} \left(n_x, n_y, E_{\text{FQWSLGI}} \right) \right\}' \right]}{\sum_{n_x=1}^{n_{x_{\max}}} \sum_{n_y=1}^{n_{y_{\max}}} \left[T_{41} \left(n_x, n_y, E_{\text{FQWSLGI}} \right) + T_{42}(n_x, n_y, E_{\text{FQWSLGI}}) \right]}. \tag{4.3}$$

4.2.2 Magnetothermopower in II–VI Quantum Wire Superlattices with Graded Interfaces

The electron dispersion law in II–VI QWSL, can be written as

$$(k_z)^2 = \left[\frac{1}{L_0^2} \left[\cos^{-1} \left[\frac{1}{2} f_{11} \left(E, n_x, n_y \right) \right] \right]^2 - \left(\frac{\pi n_x}{d_x} \right)^2 - \left(\frac{\pi n_y}{d_y} \right)^2 \right], \tag{4.4}$$

where $f_{11}(E, n_x, n_y)$ has been defined in connection with (3.10) of Chap. 3.

The electron concentration per unit length in this case can be expressed as

$$n_0 = \left(\frac{2 g_{\text{v}}}{\pi} \right) \sum_{n_x=1}^{n_{x_{\max}}} \sum_{n_y=1}^{n_{y_{\max}}} \left[T_{43} \left(n_x, n_y, E_{\text{FQWSLGI}} \right) + T_{44} \left(n_x, n_y, E_{\text{FQWSLGI}} \right) \right], \tag{4.5}$$

where

$$T_{43}\left(n_x,n_y,E_{\mathrm{FQWSLGI}}\right)\equiv\left[\frac{1}{L_0^2}\left[\cos^{-1}\left[\frac{1}{2}f_{11}\left(E_{\mathrm{FQWSLGI}},n_x,n_y\right)\right]\right]^2-\left(\frac{\pi n_x}{d_x}\right)^2-\left(\frac{\pi n_y}{d_y}\right)^2\right]^{1/2}$$

and

$$T_{44}\left(n_x,n_y,E_{\mathrm{FQWSLGI}}\right)\equiv\sum_{r=1}^{s}L(r)\left[T_{43}\left(n_x,n_y,E_{\mathrm{FQWSLGI}}\right)\right].$$

The use of (1.13) and (4.5) leads to the expression of the TPSM as

$$G_0=\left(\frac{\pi^2k_B^2T}{3e}\right)\times\frac{\sum\limits_{n_x=1}^{n_{x\max}}\sum\limits_{n_y=1}^{n_{y\max}}\left[\left\{T_{43}\left(n_x,n_y,E_{\mathrm{FQWSLGI}}\right)\right\}'+\left\{T_{44}\left(n_x,n_y,E_{\mathrm{FQWSLGI}}\right)\right\}'\right]}{\sum\limits_{n_x=1}^{n_{x\max}}\sum\limits_{n_y=1}^{n_{y\max}}\left[T_{43}\left(n_x,n_y,E_{\mathrm{FQWSLGI}}\right)+T_{44}\left(n_x,n_y,E_{\mathrm{FQWSLGI}}\right)\right]}. \tag{4.6}$$

4.2.3 *Magnetothermopower in IV–VI Quantum Wire Superlattices with Graded Interfaces*

The electron dispersion law in IV–VI QWSL can be written as

$$(k_z)^2=\left[\frac{1}{L_0^2}\left[\cos^{-1}\left[\frac{1}{2}f_{12}\left(E,n_x,n_y\right)\right]\right]^2-\left(\frac{\pi n_x}{d_x}\right)^2-\left(\frac{\pi n_y}{d_y}\right)^2\right], \tag{4.7}$$

where $f_{12}\left(E,n_x,n_y\right)$ has been defined in connection with (3.15) of Chap. 3.

The electron concentration per unit length in this case can be expressed as

$$n_0=\left(\frac{2g_v}{\pi}\right)\sum_{n_x=1}^{n_{x\max}}\sum_{n_y=1}^{n_{y\max}}\left[T_{45}\left(n_x,n_y,E_{\mathrm{FQWSLGI}}\right)+T_{46}\left(n_x,n_y,E_{\mathrm{FQWSLGI}}\right)\right], \tag{4.8}$$

where

$$T_{45}\left(n_x,n_y,E_{\mathrm{FQWSLGI}}\right)\equiv\left[\frac{1}{L_0^2}\left[\cos^{-1}\left[\frac{1}{2}f_{12}\left(E_{\mathrm{FQWSLGI}},n_x,n_y\right)\right]\right]^2-\left(\frac{\pi n_x}{d_x}\right)^2-\left(\frac{\pi n_y}{d_y}\right)^2\right]^{1/2}$$

and

$$T_{46}\left(n_x, n_y, E_{\mathrm{FQWSL}}\right) \equiv \sum_{r=1}^{s} L(r)\left[T_{45}\left(n_x, n_y, E_{\mathrm{FQWSL}}\right)\right].$$

The use of (1.13) and (4.8) leads to the expression of the TPSM as

$$G_0 = \left(\frac{\pi^2 k_{\mathrm{B}}^2 T}{3e}\right) \times \frac{\sum\limits_{n_x=1}^{n_{x_{\max}}} \sum\limits_{n_y=1}^{n_{y_{\max}}} \left[\left\{T_{45}\left(n_x, n_y, E_{\mathrm{FQWSLGI}}\right)\right\}' + \left\{T_{46}\left(n_x, n_y, E_{\mathrm{FQWSLGI}}\right)\right\}'\right]}{\sum\limits_{n_x=1}^{n_{x_{\max}}} \sum\limits_{n_y=1}^{n_{y_{\max}}} \left[T_{45}\left(n_x, n_y, E_{\mathrm{FQWSLGI}}\right) + T_{46}\left(n_x, n_y, E_{\mathrm{FQWSLGI}}\right)\right]}. \tag{4.9}$$

4.2.4 *Magnetothermopower in HgTe/CdTe Quantum Wire Superlattices with Graded Interfaces*

The electron dispersion law in HgTe/CdTe QWSL can be written as

$$(k_z)^2 = \left[\frac{1}{L_0^2}\left[\cos^{-1}\left[\frac{1}{2} f_{13}\left(E, n_x, n_y\right)\right]\right]^2 - \left(\frac{\pi n_x}{d_x}\right)^2 - \left(\frac{\pi n_y}{d_y}\right)^2\right], \tag{4.10}$$

where $f_{13}(E, n_x, n_y)$ has been defined in connection with (3.21) of Chap. 3.

The electron concentration per unit length in this case can be expressed as

$$n_0 = \left(\frac{2g_{\mathrm{v}}}{\pi}\right) \sum_{n_x=1}^{n_{\max}} \sum_{n_y=1}^{n_{y_{\max}}} \left[T_{47}\left(n_x, n_y, E_{\mathrm{FQWSLGI}}\right) + T_{48}\left(n_x, n_y, E_{\mathrm{FQWSLGI}}\right)\right], \tag{4.11}$$

where

$$T_{47}\left(n_x, n_y, E_{\mathrm{FQWSLGI}}\right) \equiv \left[\frac{1}{L_0^2}\left[\cos^{-1}\left[\frac{1}{2} f_{13}\left(E_{\mathrm{FQWSLGI}}, n_x, n_y\right)\right]\right]^2 - \left(\frac{\pi n_x}{d_x}\right)^2 - \left(\frac{\pi n_y}{d_y}\right)^2\right]^{1/2}$$

and

$$T_{48}\left(n_x, n_y, E_{\mathrm{FQWSLGI}}\right) \equiv \sum_{r=1}^{s} L(r)\left[T_{47}\left(n_x, n_y, E_{\mathrm{FQWSLGI}}\right)\right].$$

The use of (1.13) and (4.11) leads to the expression of the TPSM as

$$G_0 = \left(\frac{\pi^2 k_B^2 T}{3e}\right) \times \frac{\sum_{n_x=1}^{n_{\max}} \sum_{n_y=1}^{n_{y\max}} \left[\left\{T_{47}\left(n_x, n_y, E_{FQWSLGI}\right)\right\}' + \left\{T_{48}\left(n_x, n_y, E_{FQWSLGI}\right)\right\}'\right]}{\sum_{n_x=1}^{n_{\max}} \sum_{n_y=1}^{n_{y\max}} \left[T_{47}\left(n_x, n_y, E_{FQWSLGI}\right) + T_{48}\left(n_x, n_y, E_{FQWSLGI}\right)\right]}. \tag{4.12}$$

4.2.5 *Magnetothermopower in III–V Quantum Wire Effective Mass Superlattices*

The electron dispersion law in III–V QW effective mass SL, can be written as

$$(k_x)^2 = \left[\frac{1}{L_0^2}\left[\cos^{-1}\left[f_{14}\left(E, n_y, n_z\right)\right]\right]^2 - \left(\frac{\pi n_y}{d_y}\right)^2 - \left(\frac{\pi n_z}{d_z}\right)^2\right], \tag{4.13}$$

where $f_{14}(E, n_y, n_z)$ has been defined in connection with (3.25) of Chap. 3.

The electron concentration per unit length in this case can be expressed as

$$n_0 = \left(\frac{2g_v}{\pi}\right) \sum_{n_y=1}^{n_{y\max}} \sum_{n_z=1}^{n_{z\max}} \left[T_{49}\left(n_y, n_z, E_{FQWSLEM}\right) + T_{50}\left(n_y, n_z, E_{FQWSLEM}\right)\right], \tag{4.14}$$

where

$$T_{49}\left(n_y, n_z, E_{FQWSLEM}\right) \equiv \left[\frac{1}{L_0^2}\left[\cos^{-1}\left[f_{14}(E_{FQWSLEM}, n_y, n_z)\right]\right]^2 - \left(\frac{\pi n_y}{d_y}\right)^2 - \left(\frac{\pi n_z}{d_z}\right)^2\right]^{1/2},$$

$E_{FQWSLEM}$ is the Fermi energy in the present case, and

$$T_{50}\left(n_y, n_z, E_{FQWSLEM}\right) \equiv \sum_{r=1}^{s} L(r)\left[T_{49}(n_y, n_z, E_{FQWSLEM})\right].$$

The use of (1.13) and (4.14) leads to the expression of the TPSM as

$$G_0 = \left(\frac{\pi^2 k_B^2 T}{3e}\right) \times \frac{\sum_{n_y=1}^{n_{y\max}} \sum_{n_z=1}^{n_{z\max}} \left[\left\{T_{49}\left(n_y, n_z, E_{FQWSLEM}\right)\right\}' + \left\{T_{50}\left(n_y, n_z, E_{FQWSLEM}\right)\right\}'\right]}{\sum_{n_y=1}^{n_{y\max}} \sum_{n_z=1}^{n_{z\max}} \left[T_{49}\left(n_y, n_z, E_{FQWSLEM}\right) + T_{50}\left(n_y, n_z, E_{FQWSLEM}\right)\right]}. \tag{4.15}$$

4.2.6 *Magnetothermopower in II–VI Quantum Wire Effective Mass Superlattices*

The electron dispersion law in II–VI QW effective mass SL can be written as

$$(k_z)^2 = \left[\frac{1}{L_0^2}\left[\cos^{-1}\left[f_{15}\left(E, n_x, n_y\right)\right]\right]^2 - \left(\frac{\pi n_x}{d_x}\right)^2 - \left(\frac{\pi n_y}{d_y}\right)^2\right], \tag{4.16}$$

where $f_{15}(E, n_x, n_y)$ has been defined in connection with (3.29) of Chap. 3.

The electron concentration per unit length in this case can be expressed as

$$n_0 = \left(\frac{2g_v}{\pi}\right) \sum_{n_x=1}^{n_{x\max}} \sum_{n_y=1}^{n_{y\max}} \left[T_{51}\left(n_x, n_y, E_{FQWSLEM}\right) + T_{52}\left(n_x, n_y, E_{FQWSLEM}\right)\right], \tag{4.17}$$

where

$$T_{51}\left(n_x, n_y, E_{FQWSLEM}\right) \equiv \left[\frac{1}{L_0^2}\left[\cos^{-1}\left[f_{15}(E_{FQWSLEM}, n_x, n_y)\right]\right]^2 - \left(\frac{\pi n_x}{d_x}\right)^2 - \left(\frac{\pi n_y}{d_y}\right)^2\right]^{1/2}$$

and

$$T_{52}(n_x, n_y, E_{FQWSLEM}) \equiv \sum_{r=1}^{s} L(r)\left[T_{51}\left(n_x, n_y, E_{FQWSLEM}\right)\right].$$

The use of (1.13) and (4.17) leads to the expression of the TPSM as

$$G_0 = \left(\frac{\pi^2 k_B^2 T}{3e}\right) \times \frac{\sum_{n_x=1}^{n_{x\max}} \sum_{n_y=1}^{n_{y\max}} \left[\left\{T_{51}\left(n_x, n_y, E_{FQWSLEM}\right)\right\}' + \left\{T_{52}\left(n_x, n_y, E_{FQWSLEM}\right)\right\}'\right]}{\sum_{n_x=1}^{n_{x\max}} \sum_{n_y=1}^{n_{y\max}} \left[T_{51}\left(n_x, n_y, E_{FQWSLEM}\right) + T_{52}\left(n_x, n_y, E_{FQWSLEM}\right)\right]}. \tag{4.18}$$

4.2.7 *Magnetothermopower in IV–VI Quantum Wire Effective Mass Superlattices*

The electron dispersion law in IV–VI QW effective mass SL can be written as

$$k_x^2 = \left[\frac{1}{L_0^2}\left[\cos^{-1}\left[f_{16}\left(E, n_y, n_z\right)\right]\right]^2 - \left(\frac{\pi n_y}{d_y}\right)^2 - \left(\frac{\pi n_z}{d_z}\right)^2\right], \tag{4.19}$$

where $f_{16}(E, n_y, n_z)$ has been defined in connection with (3.33) of Chap. 3.

The electron concentration per unit length in this case can be expressed as

$$n_0 = \left(\frac{2g_v}{\pi}\right) \sum_{n_y=1}^{n_{y\max}} \sum_{n_z=1}^{n_{z\max}} \left[T_{53}\left(n_y, n_z, E_{FQWSLEM}\right) + T_{54}\left(n_y, n_z, E_{FQWSLEM}\right)\right], \tag{4.20}$$

where

$$T_{53}\left(n_y, n_z, E_{FQWSLEM}\right) \equiv \left[\frac{1}{L_0^2}\left[\cos^{-1}\left[f_{16}\left(E_{FQWSLEM}, n_y, n_z\right)\right]\right]^2 - \left(\frac{\pi n_y}{d_y}\right)^2 - \left(\frac{\pi n_z}{d_z}\right)^2\right]^{1/2}$$

and

$$T_{54}\left(n_y, n_z, E_{FQWSLEM}\right) \equiv \sum_{r=1}^{s} L(r)\left[T_{53}\left(n_y, n_z, E_{FQWSLEM}\right)\right].$$

Thus, (1.13) and (4.20) leads to the expression of the TPSM as

$$G_0 = \left(\frac{\pi^2 k_B^2 T}{3e}\right) \times \frac{\sum_{n_y=1}^{n_{y\max}} \sum_{n_z=1}^{n_{z\max}} \left[\left\{T_{53}\left(n_y, n_z, E_{FQWSLEM}\right)\right\}' + \left\{T_{54}\left(n_y, n_z, E_{FQWSLEM}\right)\right\}'\right]}{\sum_{n_y=1}^{n_{y\max}} \sum_{n_z=1}^{n_{z\max}} \left[T_{53}\left(n_y, n_z, E_{FQWSLEM}\right) + T_{54}\left(n_y, n_z, E_{FQWSLEM}\right)\right]}. \tag{4.21}$$

4.2.8 *Magnetothermopower in HgTe/CdTe Quantum Wire Effective Mass Superlattices*

The electron dispersion law in HgTe/CdTe QW effective mass SL can be written as

$$k_x^2 = \left[\frac{1}{L_0^2}[\cos^{-1}[f_{17}\left(E, n_y, n_z\right)]]^2 - \left(\frac{\pi n_y}{d_y}\right)^2 - \left(\frac{\pi n_z}{d_z}\right)^2\right], \tag{4.22}$$

where $f_{17}(E, n_y, n_z)$ has been defined in connection with (3.37) of Chap. 3.

The electron concentration per unit length in this case can be expressed as

$$n_0 = \left(\frac{2g_v}{\pi}\right) \sum_{n_y=1}^{n_{y\max}} \sum_{n_z=1}^{n_{z\max}} \left[T_{55}\left(n_y, n_z, E_{FQWSLEM}\right) + T_{56}\left(n_y, n_z, E_{FQWSLEM}\right)\right], \tag{4.23}$$

where

$$T_{55}\left(n_y, n_z, E_{FQWSLEM}\right) \equiv \left[\frac{1}{L_0^2}\left[\cos^{-1}\left[f_{17}\left(E_{FQWSLEM}, n_y, n_z\right)\right]\right]^2 - \left(\frac{\pi n_y}{d_y}\right)^2 - \left(\frac{\pi n_z}{d_z}\right)^2\right]^{1/2}$$

and

$$T_{56}\left(n_y, n_z, E_{FQWSLEM}\right) \equiv \sum_{r=1}^{s} L(r)[T_{55}(n_y, n_z, E_{FQWSLEM})].$$

The use of (1.13) and (4.23) leads to the expression of the TPSM as

$$G_0 = \left(\frac{\pi^2 k_B^2 T}{3e}\right) \times \frac{\sum_{n_y=1}^{n_{y\max}} \sum_{n_z=1}^{n_{z\max}} \left[\{T_{55}(n_y, n_z, E_{FQWSLEM})\}' + \{T_{56}(n_y, n_z, E_{FQWSLEM})\}' \right]}{\sum_{n_y=1}^{n_{y\max}} \sum_{n_z=1}^{n_{z\max}} \left[T_{55}(n_y, n_z, E_{FQWSLEM}) + T_{56}(n_y, n_z, E_{FQWSLEM}) \right]}. \tag{4.24}$$

4.3 Result and Discussions

Using (4.2) and (4.3) and the band constants from Table 1.1, the TPSM in QW III–V SLs (taking $GaAs/Ga_{1-x}Al_xAs$ and $In_xGa_{1-x}As/InP$ QW SLs) with graded interfaces has been plotted as functions of the film thickness and impurity concentration at 10 K, respectively, as shown in Figs. 4.1 and 4.2, respectively.

The thermoelectric power has been normalized to the value $\pi^2 k_B/3e$ with respect to all the figures. The TPSM has been plotted for (a) CdS/ZnSe with $C_0 = 0$, (b) CdS/ZnSe with $C_0 \neq 0$ (using (4.5) and (4.6)) (c) HgTe/CdTe (using

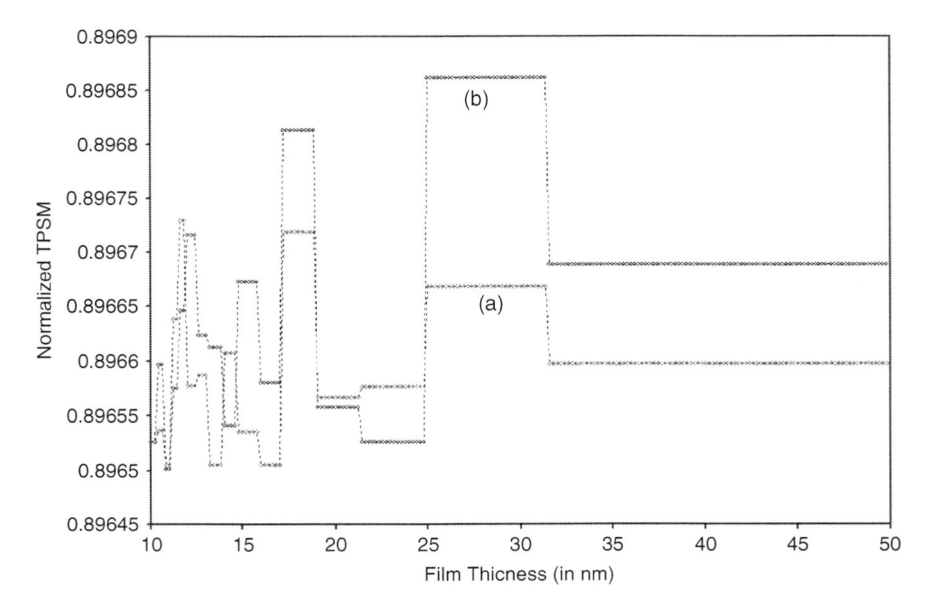

Fig. 4.1 Plot of the normalized TPSM in (**a**) $GaAs/Ga_{1-x}Al_xAs$ and (**b**) $In_xGa_{1-x}As/InP$ quantum wire superlattices with graded interfaces as a function of film thickness

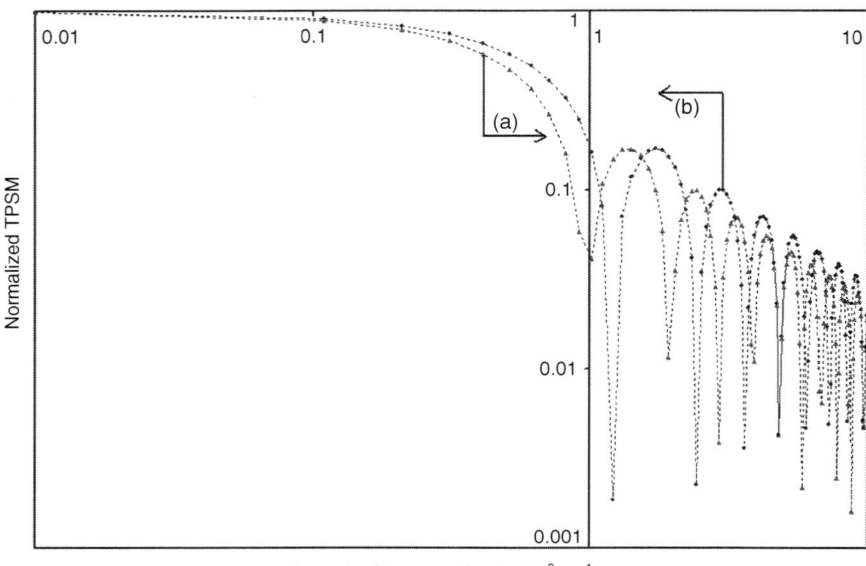

Fig. 4.2 Plot of the normalized TPSM in (**a**) $GaAs/Ga_{1-x}Al_xAs$ and (**b**) $In_xGa_{1-x}As/InP$ quantum wire superlattices with graded interfaces as a function of impurity concentration

(4.11) and (4.12)), and (d) PbSe/PbTe (using (4.8) and (4.9)) QWSL with graded interfaces as functions of film thickness and impurity concentration in Figs. 4.3 and 4.4, respectively. The TPSM in $GaAs/Ga_{1-x}Al_xAs$, HgTe/CdTe, CdS/ZnSe, $HgTe/Hg_{1-x}Cd_xTe$, and PbSe/PbTe quantum wire effective mass superlattices has been plotted as functions of film thickness and impurity concentration in Figs. 4.5 and 4.6, respectively.

The effect of size quantization is clearly exhibited by Figs. 4.1 and 4.3, in which the composite fluctuations are due to the combined influence of the Landau quantization effect (due to magnetic field) with the size quantization effect. It also appears from the same figures that the TPSM bears step functional dependency function of film thickness due to the Van Hove Singularity [2]. Since the Fermi level decreases with the increase in the film thickness, the thermoelectric power increases [3]. This physical fact also governs the nature of oscillatory variation of all the curves, where the change in film thickness with respect to TPSM for all types of superlattices appears. The thermoelectric power changes with film thickness in oscillatory manner, where the nature of oscillations is totally different. It should also be noted that the TPSM decreases with the increasing carrier degeneracy exhibiting different types of oscillations as is observed from Fig. 4.2. It may be also noted that due to the confinement of carriers along two orthogonal directions, the TPSM exhibits the composite oscillations in Figs. 4.1 and 4.3, while in Figs. 4.2 and 4.4, the absence of composite oscillation are due to the suppression of the size quantization number along one direction by another.

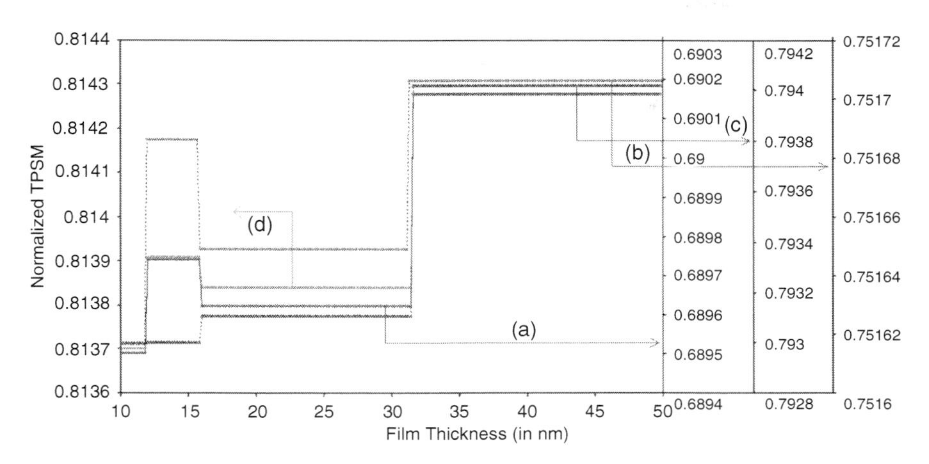

Fig. 4.3 Plot of the TPSM in (**a**) CdS/ZnSe with $C_0 = 0$, (**b**) CdS/ZnSe with $C_0 \neq 0$, (**c**) HgTe/CdTe, and (**d**) PbSe/PbTe quantum wire superlattices with graded interfaces as a function of film thickness

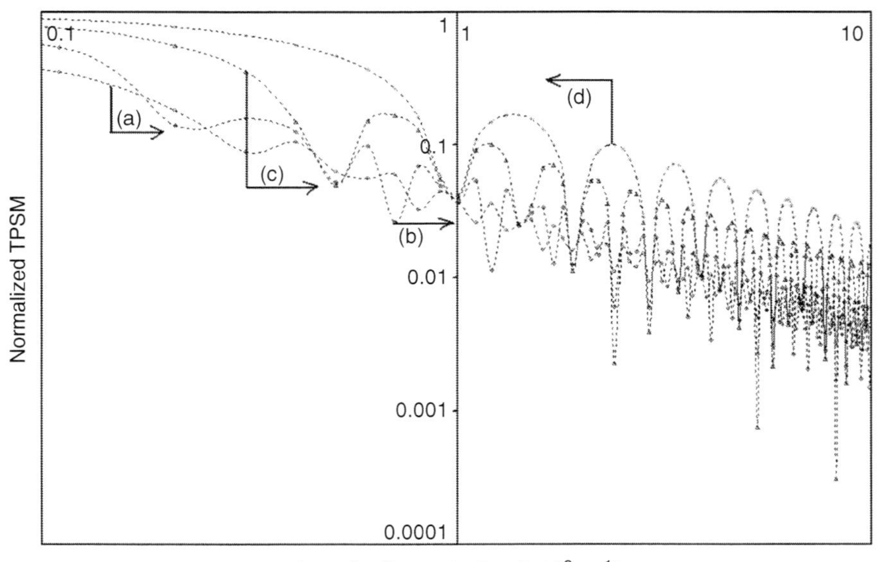

Fig. 4.4 Plot of the TPSM in (**a**) CdS/ZnSe with $C_0 = 0$, (**b**) CdS/ZnSe with $C_0 \neq 0$, (**c**) HgTe/CdTe, and (**d**) PbSe/PbTe quantum wire superlattices with graded interfaces as a function of impurity concentration

It appears from Fig. 4.5 that the TPSM in $GaAs/Ga_{1-x}Al_xAs$, CdS/ZnSe, HgTe/CdTe, and PbSe/PbTe quantum wire effective mass superlattices also exhibits such composite oscillations with increasing film thickness. The nature of oscillation in effective mass SLs are radically different than that of the corresponding graded

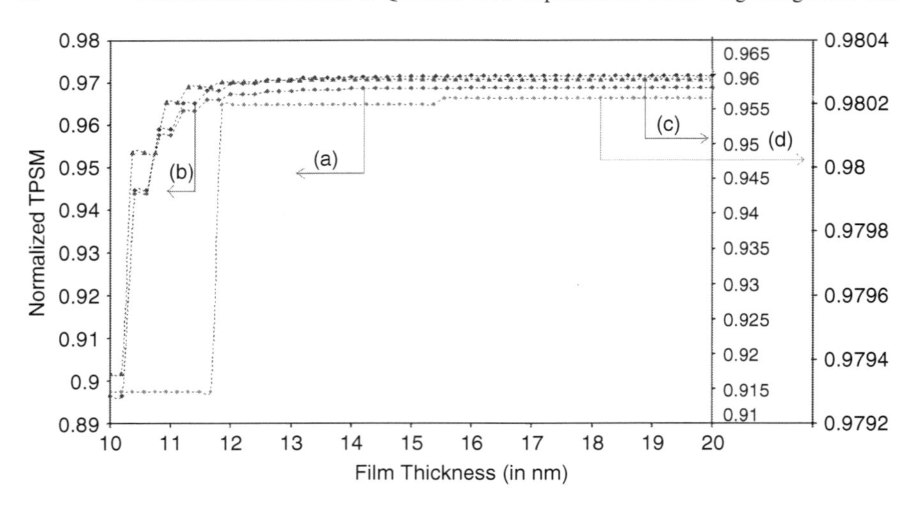

Fig. 4.5 Plot of the TPSM in (**a**) $GaAs/Ga_{1-x}Al_xAs$, (**b**) CdS/ZnSe, (**c**) HgTe/CdTe, and (**d**) PbSe/PbTe quantum wire effective mass superlattices as a function of film thickness

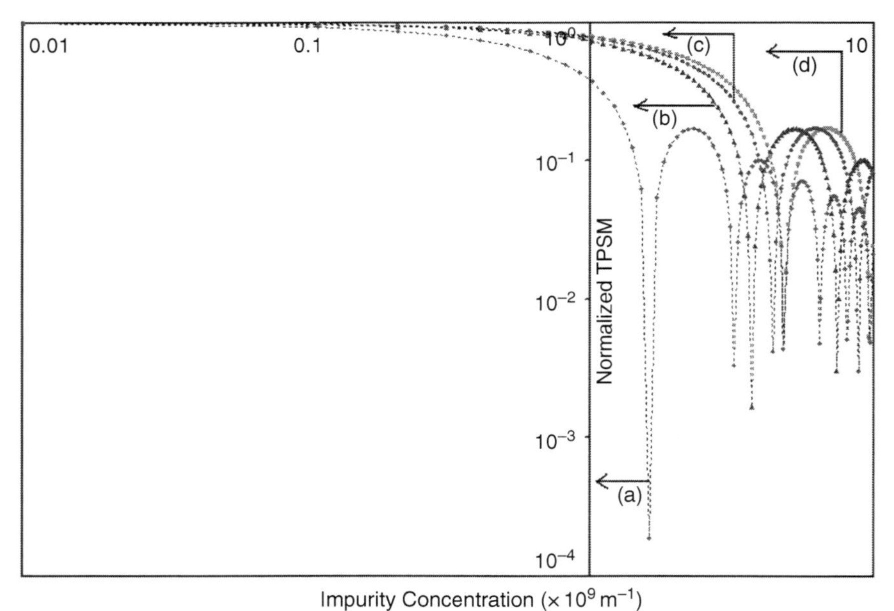

Fig. 4.6 Plot of the TPSM in (**a**) $GaAs/Ga_{1-x}Al_xAs$, (**b**) CdS/ZnSe, (**c**) HgTe/CdTe, and (**d**) PbSe/PbTe quantum wire effective mass superlattices as a function of impurity concentration

Table 4.1 The carrier statistics and the thermoelectric power in quantum wire superlattices under large magnetic field

Type of materials	Carrier statistics	TPSM
1. III–V quantum wire superlattices with graded interfaces	$n_0 = \left(\frac{2g_v}{\pi}\right) \sum_{n_x=1}^{n_{x\max}} \sum_{n_y=1}^{n_{y\max}} \left[T_{41}(n_x, n_y, E_{FQWSLGI}) + T_{42}(n_x, n_y, E_{FQWSLGI})\right]$ (4.2)	On the basis of (4.2), $G_0 = \left(\frac{\pi^2 k_B^2 T}{3e}\right) \frac{\sum_{n_x=1}^{n_{x\max}} \sum_{n_y=1}^{n_{y\max}} \left[\{T_{41}(n_x, n_y, E_{FQWSLGI})\}' + \{T_{42}(n_x, n_y, E_{FQWSLGI})\}'\right]}{\sum_{n_x=1}^{n_{x\max}} \sum_{n_y=1}^{n_{y\max}} \left[T_{41}(n_x, n_y, E_{FQWSLGI}) + T_{42}(n_x, n_y, E_{FQWSLGI})\right]}$ (4.3)
2. II–VI quantum wire superlattices with graded interfaces	$n_0 = \left(\frac{2g_v}{\pi}\right) \sum_{n_x=1}^{n_{x\max}} \sum_{n_y=1}^{n_{y\max}} \left[T_{43}(n_x, n_y, E_{FQWSLGI}) + T_{44}(n_x, n_y, E_{FQWSLGI})\right]$ (4.5)	On the basis of (4.5), $G_0 = \left(\frac{\pi^2 k_B^2 T}{3e}\right) \frac{\sum_{n_x=1}^{n_{x\max}} \sum_{n_y=1}^{n_{y\max}} \left[\{T_{43}(n_x, n_y, E_{FQWSLGI})\}' + \{T_{44}(n_x, n_y, E_{FQWSLGI})\}'\right]}{\sum_{n_x=1}^{n_{x\max}} \sum_{n_y=1}^{n_{y\max}} \left[T_{43}(n_x, n_y, E_{FQWSLGI}) + T_{44}(n_x, n_y, E_{FQWSLGI})\right]}$ (4.6)
3. IV–VI quantum wire superlattices with graded interfaces	$n_0 = \left(\frac{2g_v}{\pi}\right) \sum_{n_x=1}^{n_{x\max}} \sum_{n_y=1}^{n_{y\max}} \left[T_{45}(n_x, n_y, E_{FQWSLGI}) + T_{46}(n_x, n_y, E_{FQWSLGI})\right]$ (4.8)	On the basis of (4.8), $G_0 = \left(\frac{\pi^2 k_B^2 T}{3e}\right) \frac{\sum_{n_x=1}^{n_{x\max}} \sum_{n_y=1}^{n_{y\max}} \left[\{T_{45}(n_x, n_y, E_{FQWSLGI})\}' + \{T_{46}(n_x, n_y, E_{FQWSLGI})\}'\right]}{\sum_{n_x=1}^{n_{x\max}} \sum_{n_y=1}^{n_{y\max}} \left[T_{45}(n_x, n_y, E_{FQWSLGI}) + T_{46}(n_x, n_y, E_{FQWSLGI})\right]}$ (4.9)
4. HgTe/CdTe quantum wire superlattices with graded interfaces	$n_0 = \left(\frac{2g_v}{\pi}\right) \sum_{n_x=1}^{n_{\max}} \sum_{n_y=1}^{n_{y\max}} \left[T_{47}(n_x, n_y, E_{FQWSLGI}) + T_{48}(n_x, n_y, E_{FQWSLGI})\right]$ (4.11)	On the basis of (4.11), $G_0 = \left(\frac{\pi^2 k_B^2 T}{3e}\right) \frac{\sum_{n_x=1}^{n_{\max}} \sum_{n_y=1}^{n_{y\max}} \left[\{T_{47}(n_x, n_y, E_{FQWSLGI})\}' + \{T_{48}(n_x, n_y, E_{FQWSLGI})\}'\right]}{\sum_{n_x=1}^{n_{\max}} \sum_{n_y=1}^{n_{y\max}} \left[T_{47}(n_x, n_y, E_{FQWSLGI}) + T_{48}(n_x, n_y, E_{FQWSLGI})\right]}$ (4.12)

(continued)

Table 4.1 (Continued)

Type of materials	Carrier statistics	TPSM
5. III–V quantum wire effective mass superlattice	$n_0 = \left(\frac{2g_v}{\pi}\right)\sum_{n_y=1}^{n_{y\max}}\sum_{n_z=1}^{n_{z\max}} \left[T_{49}(n_y, n_z, E_{FQWSLEM}) + T_{50}(n_y, n_z, E_{FQWSLEM})\right]$ (4.14)	On the basis of (4.14), $G_0 = \left(\frac{\pi^2 k_B^2 T}{3e}\right) \frac{\sum_{n_y=1}^{n_{y\max}}\sum_{n_z=1}^{n_{z\max}}\left[\{T_{49}(n_y, n_z, E_{FQWSLEM})\}' + \{T_{50}(n_y, n_z, E_{FQWSLEM})\}'\right]}{\sum_{n_y=1}^{n_{y\max}}\sum_{n_z=1}^{n_{z\max}}\left[T_{49}(n_y, n_z, E_{FQWSLEM}) + T_{50}(n_y, n_z, E_{FQWSLEM})\right]}$ (4.15)
6. II–VI quantum wire effective mass superlattice	$n_0 = \left(\frac{2g_v}{\pi}\right)\sum_{n_x=1}^{n_{x\max}}\sum_{n_y=1}^{n_{y\max}} \left[T_{51}(n_x, n_y, E_{FQWSLEM}) + T_{52}(n_x, n_y, E_{FQWSLEM})\right]$ (4.17)	On the basis of (4.17), $G_0 = \left(\frac{\pi^2 k_B^2 T}{3e}\right) \frac{\sum_{n_x=1}^{n_{x\max}}\sum_{n_y=1}^{n_{y\max}}\left[\{T_{51}(n_x, n_y, E_{FQWSLEM})\}' + \{T_{52}(n_x, n_y, E_{FQWSLEM})\}'\right]}{\sum_{n_x=1}^{n_{x\max}}\sum_{n_y=1}^{n_{y\max}}\left[T_{51}(n_x, n_y, E_{FQWSLEM}) + T_{52}(n_x, n_y, E_{FQWSLEM})\right]}$ (4.18)
7. IV–VI quantum wire effective mass superlattice	$n_0 = \left(\frac{2g_v}{\pi}\right)\sum_{n_y=1}^{n_{y\max}}\sum_{n_z=1}^{n_{z\max}} \left[T_{53}(n_y, n_z, E_{FQWSLEM}) + T_{54}(n_y, n_z, E_{FQWSLEM})\right]$ (4.20)	On the basis of (4.20), $G_0 = \left(\frac{\pi^2 k_B^2 T}{3e}\right) \frac{\sum_{n_y=1}^{n_{y\max}}\sum_{n_z=1}^{n_{z\max}}\left[\{T_{53}(n_y, n_z, E_{FQWSLEM})\}' + \{T_{54}(n_y, n_z, E_{FQWSLEM})\}'\right]}{\sum_{n_y=1}^{n_{y\max}}\sum_{n_z=1}^{n_{z\max}}\left[T_{53}(n_y, n_z, E_{FQWSLEM}) + T_{54}(n_y, n_z, E_{FQWSLEM})\right]}$ (4.21)
8. HgTe/CdTe quantum wire effective mass superlattices	$n_0 = \left(\frac{2g_v}{\pi}\right)\sum_{n_y=1}^{n_{y\max}}\sum_{n_z=1}^{n_{z\max}} \left[T_{55}(n_y, n_z, E_{FQWSLEM}) + T_{56}(n_y, n_z, E_{FQWSLEM})\right]$ (4.23)	On the basis of (4.23), $G_0 = \left(\frac{\pi^2 k_B^2 T}{3e}\right) \frac{\sum_{n_y=1}^{n_{y\max}}\sum_{n_z=1}^{n_{z\max}}\left[\{T_{55}(n_y, n_z, E_{FQWSLEM})\}' + \{T_{56}(n_y, n_z, E_{FQWSLEM})\}'\right]}{\sum_{n_y=1}^{n_{y\max}}\sum_{n_z=1}^{n_{z\max}}\left[T_{55}(n_y, n_z, E_{FQWSLEM}) + T_{56}(n_y, n_z, E_{FQWSLEM})\right]}$ (4.24)

interfaces which is the direct signature of the difference in band structure in the respective cases as found from all the respective corresponding figures.

From Fig. 4.6, we observe that the TPSM in the aforementioned case decreases with increasing impurity concentration and differ widely for large values of impurity concentration, whereas for relatively small values of the carrier degeneracy, the TPSM converges to a single value in the whole range of the impurity concentration considered.

For the purpose of condensed presentation, the carrier statistics and the corresponding TPSM has been given in the Table 4.1

4.4 Open Research Problem

(R4.1) Investigate all the appropriate problems of Chap. 3 for all types of QWSL in the presence of strain.

References

1. K.P. Ghatak, S. Bhattacharya, D. De, *Einstein Relation in Compound Semiconductor and Their Nanostructures*, Springer Series in Materials Science, vol 116 (Springer, Germany, 2008)
2. K. Seeger, *Semiconductor Physics, an Introduction*, 9th edn. (Springer, Germany, 2004)
3. B.R. Nag, *Electron Transport in Compound Semiconductors*, Springer Series in Solid State Sciences, vol 11 (Springer, Germany, 1980)

Part II
Thermoelectric Power Under Magnetic Quantization in Macro and Microelectronic Materials

Chapter 5
Thermoelectric Power in Macroelectronic Materials Under Magnetic Quantization

5.1 Introduction

It is well known that under magnetic quantization the motion of the electrons in semiconductors parallel to the direction of the quantizing magnetic field is not affected, although the area of the wave-vector space perpendicular to the same gets quantized leading to the formation of Landau subbands. The electronic properties of semiconductors in the presence of magnetic quantization have been investigated in the literature for the last few decades [1–68]. It is interesting to note that TPSM in semiconductors under strong magnetic quantization has relatively been less investigated in the literature. In this chapter, we shall investigate the same in nonlinear optical, III–V, II–VI, bismuth, IV–VI, and stressed materials in Sects. 5.2.1–5.2.6, respectively. Sections 5.3 and 5.4 contain results and discussion and open research problems pertinent to this chapter.

5.2 Theoretical Background

5.2.1 *Magnetothermopower in Nonlinear Optical Materials*

In the presence of an arbitrarily oriented quantizing magnetic field B along k_z' direction which makes an angle θ with k_z axis and lies in the $k_x - k_z$ plane, the dispersion relation of the conduction electrons in nonlinear optical materials in the present case can be expressed extending the method as given by Wallace [69] as

$$\gamma(E) = \overline{M}_{\pm}(n, E, \theta) + \overline{a}_0(E, \theta)(k_z')^2, \tag{5.1}$$

where

$$\overline{M}_{\pm}(n,E,\theta) \equiv \frac{2eB}{\hbar}\left(n+\frac{1}{2}\right)\left[f_1(E)\{f_1(E)\cos^2\theta + f_2(E)\sin^2\theta\}\right]^{1/2}$$
$$\pm\left[\frac{eB\hbar E_g}{6}\left\{\frac{(E_g+\Delta_\perp)}{m_\perp^*\left(E_g+\frac{2}{3}\Delta_\perp\right)}\right\}^{1/2}\right]$$
$$\times\left[\left(E+E_g+\delta+\left[\frac{\Delta_{||}^2-\Delta_\perp^2}{3\Delta_{||}}\right]\right)^2\left\{\frac{\Delta_{||}^2(E_g+\Delta_\perp)\cos^2\theta}{m_\perp^*\left(E_g+\frac{2}{3}\Delta_\perp\right)}\right\}\right.$$
$$\left.+\left\{\frac{(E+E_g)^2(E_g+\Delta_{||})\Delta_\perp^2\sin^2\theta}{m_{||}^*\left(E_g+\frac{2}{3}\Delta_{||}\right)}\right\}\right]^{1/2},$$

n (= 0, 1, 2, 3, ...) is the Landau quantum number,

$$\overline{a}_0(E,\theta) \equiv \frac{(f_1(E)f_2(E))}{(f_1(E)\cos^2\theta + f_2(E)\sin^2\theta)},$$

and $k_z' = k_z\cos\theta + k_x\sin\theta$.

The use of (5.1) leads to the expression of electron concentration per unit volume in this case as [70]

$$n_0 = \frac{g_v eB}{2\pi^2\hbar}\sum_{n=0}^{n_{\max}}[T_{51}(n,E_{FB}) + T_{52}(n,E_{FB})], \tag{5.2}$$

where

$$T_{51}(n,E_{FB}) \equiv \left[\frac{\gamma(E_{FB}) - \overline{M}_{\pm}(n,E_{FB},\theta)}{\overline{a}_0(E_{FB},\theta)}\right]^{1/2},$$

E_{FB} is the Fermi energy in the presence of magnetic quantization as measured from the edge of the conduction band in the vertically upward direction in the absence of any quantization, $T_{52}(n,E_{FB}) \equiv \sum_{r=1}^{s} Q(r)\,[T_{51}(n,E_{FB})]$, and

$$Q(r) \equiv \left[2(k_BT)^{2r}(1-2^{1-2r})\zeta(2r)\frac{\partial^{2r}}{\partial E_{FB}^{2r}}\right].$$

Thus, using (1.13) and (5.2), the TPSM in this case assumes the form

$$G_0 = \left(\frac{\pi^2k_B^2T}{3e}\right)\left[\frac{\sum_{n=0}^{n_{\max}}\left[\{T_{51}(n,E_{FB})\}' + \{T_{52}(n,E_{FB})\}'\right]}{\sum_{n=0}^{n_{\max}}[T_{51}(n,E_{FB}) + T_{52}(n,E_{FB})]}\right]. \tag{5.3}$$

5.2.2 Magnetothermopower in Kane Type III–V Materials

1. Under the conditions $\delta = 0$, $\Delta_{\|} = \Delta_{\perp} = \Delta$, and $m_{\|}^{*} = m_{\perp}^{*} = m^{*}$, (5.1) assumes the form

$$I(E) = \left(n + \frac{1}{2}\right)\hbar\omega_0 + \frac{\hbar^2 k_z^2}{2m^*} \pm eB\hbar\Delta\left[6m^*\left(E + E_g + \frac{2}{3}\Delta\right)\right]^{-1}, \quad (5.4)$$

where $I(E)$ has already been defined in connection with (1.16) of Chap. 1 and $\omega_0 = eB/m^*$. Equation (5.4) is the magnetodispersion relation of the conduction electrons of III–V materials [71].

Thus, the electron concentration per unit volume can be written as [70]

$$n_0 = \frac{g_v eB\sqrt{2m^*}}{2\pi^2 h^2}\sum_{n=0}^{n_{\max}}\left[T_{53}(n, E_{FB}) + T_{54}(n, E_{FB})\right], \quad (5.5)$$

where

$$T_{53}(n, E_{FB}) \equiv \left[I(E_{FB}) - \left(n + \frac{1}{2}\right)\hbar\omega_0 \mp \frac{eB\hbar\Delta}{6m^*\left(E_{FB} + E_g + \frac{2}{3}\Delta\right)}\right]^{1/2}$$

and

$$T_{54}(n, E_{FB}) \equiv \sum_{r=1}^{s} Q(r) T_{35}(n, E_{FB}).$$

Using (1.13) and (5.5), the TPSM in this case can be expressed as

$$G_0 = \left(\frac{\pi^2 k_B^2 T}{3e}\right)\left[\frac{\sum_{n=0}^{n_{\max}}\left[\{T_{53}(n, E_{FB})\}' + \{T_{54}(n, E_{FB})\}'\right]}{\sum_{n=0}^{n_{\max}}\left[T_{53}(n, E_{FB}) + T_{54}(n, E_{FB})\right]}\right]. \quad (5.6)$$

2. Under the condition $\Delta \gg E_{g_0}$, (5.4) can be expressed as

$$E(1 + \alpha E) = \left(n + \frac{1}{2}\right)\hbar\omega_0 + \left(\hbar^2 k_z^2/2m^*\right) \pm \frac{1}{2}\mu_0 g^* B \quad (5.7)$$

where $\mu_0 = (e\hbar/2m_0)$ is known as the Bohr magnetron, g^* is the magnitude of the band edge g-factor and is equal to (m_0/m^*) in accordance with the two band model of Kane [71].

Thus, the electron concentration per unit volume can be written as [70]

$$n_0 = \frac{g_v eB\sqrt{2m^*}}{2\pi^2\hbar^2}\sum_{n=0}^{n_{\max}}\left[T_{55}(n, E_{FB}) + T_{56}(n, E_{FB})\right] \quad (5.8)$$

where

$$T_{55}(n, E_{FB}) \equiv \left[E_{FB}(1+\alpha E_{FB}) - \left(n+\frac{1}{2}\right)\hbar\omega_0 \pm \frac{1}{2}g^*\mu_0 B\right]^{1/2}$$

and $T_{56}(n, E_{FB}) \equiv \sum_{r=1}^{s} Q(r) T_{39}(n, E_{FB})$.

Using (1.13) and (5.8), the TPSM in this case assumes the form

$$G_0 = \left(\frac{\pi^2 k_B^2 T}{3e}\right)\left[\frac{\sum_{n=0}^{n_{max}}\left[\{T_{55}(n, E_{FB})\}' + \{T_{56}(n, E_{FB})\}'\right]}{\sum_{n=0}^{n_{max}}\left[T_{55}(n, E_{FB}) + T_{56}(n, E_{FB})\right]}\right]. \tag{5.9}$$

Under the condition $\alpha E_{FB} \ll 1$, (5.8) gets simplified as [70]

$$n_0 = \frac{g_v N_C \theta_{B1}}{2}\left[\sum_{n=0}^{n_{max}}\frac{1}{\sqrt{a_{01}}}\left[\left(1+\frac{3}{2}\alpha b_{01}\right)F_{-1/2}(\eta_B) + \frac{3}{4}\alpha k_B T F_{1/2}(\eta_B)\right]\right], \tag{5.10}$$

where $N_C = 2\left(2\pi m^* k_B T/h^2\right)^{3/2}$, $\theta_{B1} = \frac{\hbar\omega_0}{k_B T}$,

$$a_{01} \equiv \left[1+\alpha\left(n+\frac{1}{2}\right)\hbar\omega_0 \pm \frac{1}{2}g^*\mu_0 B\right],$$

$$b_{01} \equiv (a_{01})^{-1}\left[\left(n+\frac{1}{2}\right)\hbar\omega_0 \pm \frac{1}{2}g^*\mu_0 B\right],$$

and

$$\eta_B = \frac{E_{FB} - b_{01}}{k_B T}.$$

Using (5.10) and (1.13), the TPSM in this case can be written as

$$G_0 = \frac{\pi^2 k_B}{3e}\left[\frac{\sum_{n=0}^{n_{max}}\frac{1}{\sqrt{a_{01}}}\left[\left(1+\frac{3}{2}\alpha b_{01}\right)F_{-3/2}(\eta_B) + \frac{3}{4}\alpha k_B T F_{-1/2}(\eta_B)\right]}{\sum_{n=0}^{n_{max}}\frac{1}{\sqrt{a_{01}}}\left[\left(1+\frac{3}{2}\alpha b_{01}\right)F_{-1/2}(\eta_B) + \frac{3}{4}\alpha k_B T F_{1/2}(\eta_B)\right]}\right]. \tag{5.11}$$

The expressions for the electron concentration and the TPSM under the condition $\alpha \to 0$ can, respectively, be written as

$$n_0 = \frac{g_v N_C \theta_{B1}}{2}\sum_{n=0}^{n_{max}} F_{-1/2}(\overline{\eta}_B) \tag{5.12}$$

where

$$\overline{\eta}_{\mathrm{B}} \equiv (k_{\mathrm{B}}T)^{-1}\left[E_{\mathrm{FB}} - \left(n + \frac{1}{2}\right)\hbar\omega_0 \mp \frac{1}{2}g^*\mu_0 B\right]$$

and

$$G_0 = \frac{\pi^2 k_{\mathrm{B}}}{3e}\left[\sum_{n=0}^{n_{\max}} F_{-1/2}\left(\overline{\eta}_{\mathrm{B}}\right)\right]^{-1}\left[\sum_{n=0}^{n_{\max}} F_{-3/2}\left(\overline{\eta}_{\mathrm{B}}\right)\right]. \tag{5.13}$$

Under the condition of non-degeneracy, (5.13) gets simplified to the well-known form as given by (2.37).

5.2.3 Magnetothermopower in II–VI Materials

The Hamiltonian of the conduction electron of II–VI semiconductors in the presence of a quantizing magnetic field B along z-direction assumes the form

$$\hat{H}_{\mathrm{B}} = \frac{(\hat{p}_x)^2}{2m_\perp^*} + \frac{\left(\hat{p}_y - |e|\,B\hat{x}\right)^2}{2m_\perp^*} \pm \frac{C_0}{\hbar}\left[(\hat{p}_x)^2 + \left(\hat{p}_y - |e|\,B\hat{x}\right)^2\right]^{1/2} + \frac{(\hat{p}_z)^2}{2m_{\parallel}^*} \tag{5.14}$$

where the "hats" denote the respective operators.

The application of the operator method leads to the magnetodispersion relation of the carriers of II–VI semiconductors, including spin, as [70]

$$E = \frac{\hbar^2 k_z^2}{2m_{\parallel}^*} + \phi_{\pm}(n), \tag{5.15}$$

where

$$\phi_{\pm}(n) = \frac{\hbar e B}{m_\perp^*}\left(n + \frac{1}{2}\right) \pm C_0\left[\frac{2eB}{\hbar}\left(n + \frac{1}{2}\right)\right]^{1/2} \pm \frac{1}{2}g^*\mu_0 B.$$

Combining (5.15) with the occupation probability factor, the electron concentration per unit volume can be written as [70]

$$n_0 = \frac{g_{\mathrm{v}} e B\sqrt{2\pi m_{\parallel}^* k_{\mathrm{B}}T}}{h^2}\sum_{n=0}^{n_{\max}} F_{-1/2}(\theta_3), \quad \theta_3 \equiv \frac{E_{\mathrm{FB}} - \phi_{\pm}(n)}{k_{\mathrm{B}}T}. \tag{5.16}$$

Thus, using (1.13) and (5.16), the TPSM for the II–VI materials in the presence of a quantizing magnetic field along z-direction assumes the form

$$G_0 = \frac{\pi^2 k_{\mathrm{B}}}{3e}\left[\sum_{n=0}^{n_{\max}} F_{-1/2}(\theta_3)\right]^{-1}\left[\sum_{n=0}^{n_{\max}} F_{-3/2}(\theta_3)\right]. \tag{5.17}$$

5.2.4 Magnetothermopower in Bismuth

5.2.4.1 The McClure and Choi model

The carrier energy spectrum in accordance with the McClure and Choi model in the presence of a quantizing magnetic field B along z-direction up to the first order by including spin effects can be expressed as [72, 73]

$$E\,(1+\alpha E)=\left(n+\frac{1}{2}\right)\hbar\omega(E)+\left(n^2+1+n\right)\frac{\alpha\hbar^2\omega^2(E)}{4}+\frac{\hbar^2k_z^2}{2m_3}\left[1-\frac{\alpha\left(n+\frac{1}{2}\right)\hbar\omega(E)}{2}\right]\pm\frac{1}{2}g^*\mu_0B, \quad (5.18)$$

where

$$\omega(E)\equiv\frac{eB}{\sqrt{m_1m_2}}\left[1+\alpha E\left(1-\frac{m_2}{m_2'}\right)\right]^{1/2}.$$

The use of (5.18) leads to the expression of electron concentration per unit volume as [70]

$$n_0=\frac{g_veB\sqrt{2m_3}}{2\pi^2\hbar^2}\sum_{n=0}^{n_{\max}}\Big[T_{57}\,(n,E_{FB})+T_{58}\,(n,E_{FB})\Big], \quad (5.19)$$

where

$$T_{57}\,(n,E_{FB})\equiv\left[1-\frac{\alpha\left(n+\frac{1}{2}\right)\hbar\omega\,(E_{FB})}{2}\right]^{-1/2}\left[E_{FB}\,(1+\alpha E_{FB})-\left(n+\frac{1}{2}\right)\hbar\omega\,(E_{FB})-\left(n^2+n+1\right)\frac{\alpha\hbar^2\omega^2\,(E_{FB})}{4}\mp\frac{1}{2}g^*\mu_0B\right]^{1/2}$$

and $T_{58}\,(n,E_{FB})\equiv\sum_{r=1}^{s}Q(r)\,[T_{57}\,(n,E_{FB})]$.

Thus, using (5.19) and (1.13), the TPSM in accordance with the McClure and Choi model in this case can be written as

$$G_0=\left(\frac{\pi^2k_B^2T}{3e}\right)\left[\frac{\sum_{n=0}^{n_{\max}}\left[\{T_{57}\,(n,E_{FB})\}'+\{T_{58}\,(n,E_{FB})\}'\right]}{\sum_{n=0}^{n_{\max}}\left[T_{57}\,(n,E_{FB})+T_{58}\,(n,E_{FB})\right]}\right]. \quad (5.20)$$

5.2.4.2 The Cohen Model

The application of the above method in Cohen model leads to the electron energy spectrum in Bi in the presence of quantizing magnetic field B along z-direction as [74]

$$E\left(1+\alpha E\right)=\left(n+\frac{1}{2}\right)\hbar\omega(E)\pm\frac{1}{2}g^{*}\mu_{0}B+\frac{3}{8}\alpha\left(n^{2}+n+\frac{1}{2}\right)\hbar^{2}\omega^{2}(E)+\frac{\hbar^{2}k_{z}^{2}}{2m_{3}} \tag{5.21}$$

Thus, the electron concentration per unit volume assumes the form [70]

$$n_{0}=\frac{\left(eBg_{\mathrm{v}}\right)\left(\sqrt{2m_{3}}\right)}{2\pi^{2}\hbar^{2}}\sum_{n=0}^{n_{\max}}\left[T_{59}\left(n,E_{\mathrm{FB}}\right)+T_{60}\left(n,E_{\mathrm{FB}}\right)\right] \tag{5.22}$$

where

$$T_{59}\left(n,E_{\mathrm{FB}}\right)\equiv\left[E_{\mathrm{FB}}\left(1+\alpha E_{\mathrm{FB}}\right)-\left(n+\frac{1}{2}\right)\hbar\omega\left(E_{\mathrm{FB}}\right)\mp\frac{1}{2}g^{*}\mu_{0}B-\frac{3}{8}\alpha\left(n^{2}+n+\frac{1}{2}\right)\hbar^{2}\omega^{2}\left(E_{\mathrm{FB}}\right)\right]^{1/2},$$

$$\omega\left(E_{\mathrm{FB}}\right)\equiv\frac{eB}{\sqrt{m_{1}m_{2}}}\left[1+\alpha E_{\mathrm{FB}}\left(1-\frac{m_{2}}{m_{2}'}\right)\right]^{1/2},$$

and $T_{60}\left(n,E_{\mathrm{FB}}\right)\equiv\sum_{r=1}^{s}Q\left(r\right)\left[T_{59}\left(n,E_{\mathrm{FB}}\right)\right]$.

Hence, combining (5.22) with (1.13), the TPSM can be expressed as

$$G_{0}=\left(\frac{\pi^{2}k_{\mathrm{B}}^{2}T}{3e}\right)\left[\frac{\sum_{n=0}^{n_{\max}}\left[\left\{T_{59}\left(n,E_{\mathrm{FB}}\right)\right\}'+\left\{T_{60}\left(n,E_{\mathrm{FB}}\right)\right\}'\right]}{\sum_{n=0}^{n_{\max}}\left[T_{59}\left(n,E_{\mathrm{FB}}\right)+T_{60}\left(n,E_{\mathrm{FB}}\right)\right]}\right] \tag{5.23}$$

5.2.4.3 The Lax Model

For this model, the magnetodispersion relation can be written as

$$E\left(1+\alpha E\right)=\left(n+\frac{1}{2}\right)\hbar\omega_{03}+\left(\hbar^{2}k_{z}^{2}/2m_{3}^{*}\right)\pm\frac{1}{2}g^{*}\mu_{0}B, \tag{5.24}$$

where

$$\omega_{03}=\frac{eB}{\sqrt{m_{1}m_{2}}}.$$

The electron concentration per unit volume can be expressed as [70]

$$n_0 = \frac{(eBg_v)\left(\sqrt{2m_3}\right)}{2\pi^2\hbar^2} \sum_{n=0}^{n_{\max}} \left[T_{61}(n, E_{FB}) + T_{62}(n, E_{FB})\right] \tag{5.25}$$

where

$$T_{61}(n, E_{FB}) \equiv \left[E_{FB}(1+\alpha E_{FB}) - \left(n+\frac{1}{2}\right)\hbar\omega_{03} \pm \frac{1}{2}g^*\mu_0 B\right]^{1/2}$$

and $T_{62}(n, E_{FB}) \equiv \sum_{r=1}^{s} Q(r) T_{61}(n, E_{FB})$.

Thus, combining (1.13) and (5.25), the TPSM in this case assumes the form

$$G_0 = \left(\frac{\pi^2 k_B^2 T}{3e}\right)\left[\frac{\sum_{n=0}^{n_{\max}} \left[\{T_{61}(n, E_{FB})\}' + \{T_{62}(n, E_{FB})\}'\right]}{\sum_{n=0}^{n_{\max}} [T_{61}(n, E_{FB}) + T_{62}(n, E_{FB})]}\right]. \tag{5.26}$$

5.2.5 Magnetothermopower in IV–VI Materials

It is well known that the Cohen model (which finds use in bismuth) also describes the carriers of the IV–VI compounds where the energy band constants correspond to the said compounds. Equations (5.22) and (5.23) are applicable in this regard.

5.2.6 Magnetothermopower in Stressed Materials

The simplified expression of the electron energy spectrum in stressed Kane type semiconductors in the presence of an arbitrarily oriented quantizing magnetic field B, which makes angles α_1, β_1, and γ_1 with k_x, k_y, and k_z axes, respectively, can be written as [70] as

$$1 - \left[k_z'\right]^2 [I_2(E)]^{-1} = I_3(n, E), \tag{5.27}$$

where

$$I_2(E) \equiv \left[a^*(E)\right]^2 \cos^2\alpha_1 + \left[b^*(E)\right]^2 \cos^2\beta_1 + \left[c^*(E)\right]^2 \cos^2\gamma_1$$

and

$$I_3(n, E) \equiv \left(\frac{2eB}{\hbar}\right)\left(n+\frac{1}{2}\right)\left[\left[a^*(E)\right]\left[b^*(E)\right]\left[c^*(E)\right]\right]^{-1}\left[I_2(E)\right]^{1/2}.$$

The electron concentration per unit volume can be expressed as [70]

$$n_0 = \frac{g_v eB}{\pi^2 \hbar} \sum_{n=0}^{n_{\max}} \left[T_{63}(n, E_{FB}) + T_{64}(n, E_{FB}) \right], \tag{5.28}$$

where

$$T_{63}(n, E_{FB}) \equiv \sqrt{I_2(E_{FB})} \left[\sqrt{1 - [I_3(n, E_{FB})]} \right]$$

and

$$T_{64}(n, E_{FB}) \equiv \sum_{r=1}^{s} L(r)\, T_{63}(n, E_{FB}).$$

Combining (1.13) and (5.28), the TPSM in this case assumes the form

$$G_0 = \left(\frac{\pi^2 k_B^2 T}{3e} \right) \left[\frac{\sum_{n=0}^{n_{\max}} \left[\{T_{63}(n, E_{FB})\}' + \{T_{64}(n, E_{FB})\}' \right]}{\sum_{n=0}^{n_{\max}} \left[T_{63}(n, E_{FB}) + T_{64}(n, E_{FB}) \right]} \right]. \tag{5.29}$$

Finally, we infer that under stress-free condition together with the substitution $B_2^2 \equiv 3\hbar^2 E_{g_0}/4m^*$, (5.28) and (5.29) get simplified to (5.8) and (5.9), respectively.

5.3 Results and Discussion

Using (5.2) and (5.3) and the energy band constants as given in Chap. 14, in Fig. 5.1 the normalized magnetothermopower has been plotted as a function of inverse magnetic field for Cd_3As_2 as shown in plot (a) of Fig. 5.1 where the plot (b) represents the case for $\delta = 0$ and has been drawn to assess the influence of crystal field splitting on the magnetothermopower in Cd_3As_2. The plots (c) and (d) in the same figure refer to the three and two band models of Kane (using (5.5); (5.6) and (5.8); (5.9), respectively), whereas the plot (e) exhibits the parabolic energy bands (using (5.12) and (5.13)). Figure 5.2 exhibits the plot of the normalized magnetothermopower as a function of impurity concentration for Cd_3As_2 for all cases of Fig. 5.1. The plot (a) of Fig. 5.3 shows the variation of the normalized magnetothermopower in Cd_3As_2as a function of orientation of the quantizing magnetic field for $\delta \neq 0$ and the plot (b) refers for $\delta = 0$. Figures 5.4–5.6 represent the variation of magnetothermopower as functions of inverse quantizing magnetic field, impurity concentration and angular orientation of the quantizing magnetic field for $CdGeAs_2$ for the respective cases of Figs. 5.1–5.3. It should be noted that under varying magnetic field, the concentration has been set to the value of 10^{24} m^{-3}, while, under varying electron concentration, the magnetic field is fixed to 2 T. In all the figures, the numerical value of the TPSM has been normalized to $(\pi^2 k_B^2 T/3e)$. It appears from Figs. 5.1 and 5.4 that the magnetothermopower oscillates with $1/B$. It is well known that

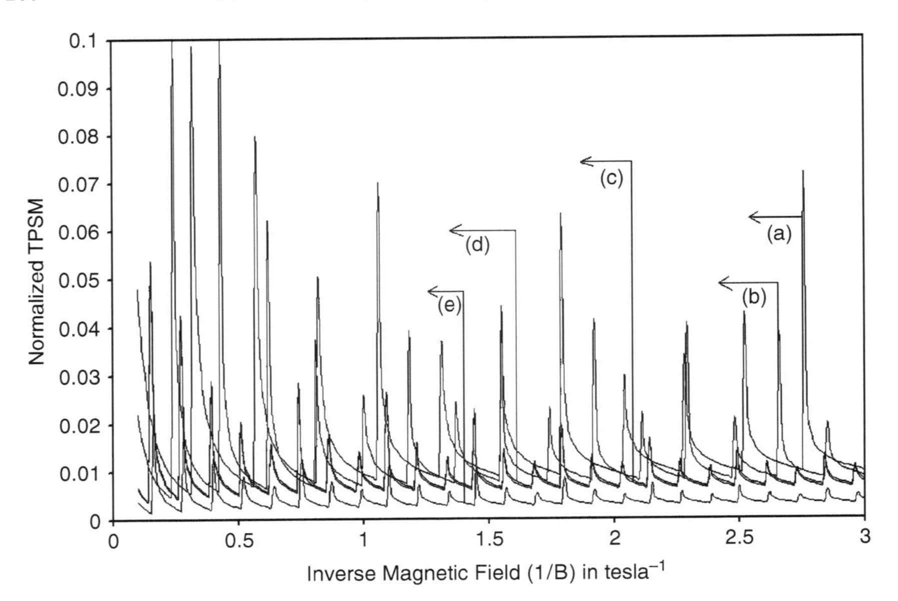

Fig. 5.1 Plot of the normalized TPSM as a function of inverse magnetic field for Cd_3As_2 in accordance with the (**a**) generalized band model ($\delta \neq 0$), (**b**) $\delta = 0$, (**c**) three and (**d**) two band models of Kane together with parabolic energy bands (**e**)

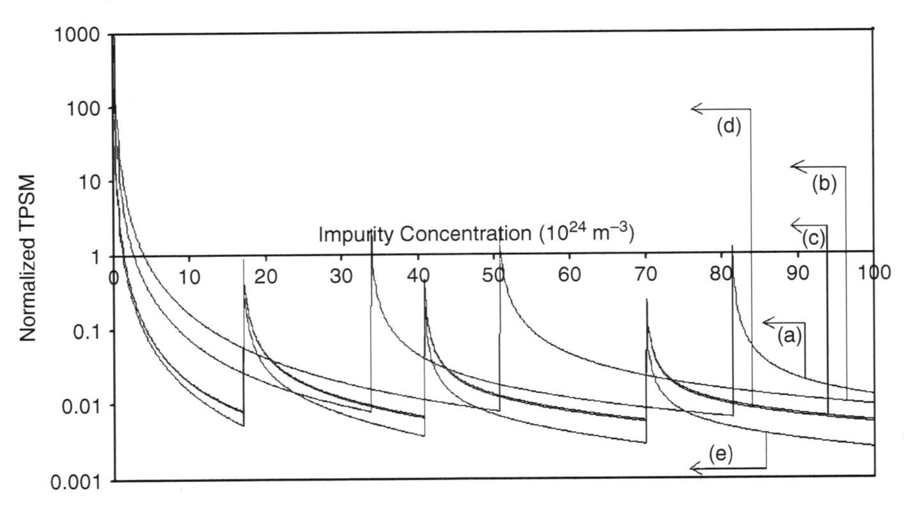

Fig. 5.2 Plot of the normalized TPSM as a function of impurity concentration for Cd_3As_2 for all cases of Fig. 5.1

density-of-states in semiconductors under magnetic quantization exhibits oscillatory dependence with inverse quantizing magnetic field, which is being reflected in this case. In fact, all electronic properties of electronic materials in the presence of quantizing magnetic field exhibit periodic variation with inverse quantizing

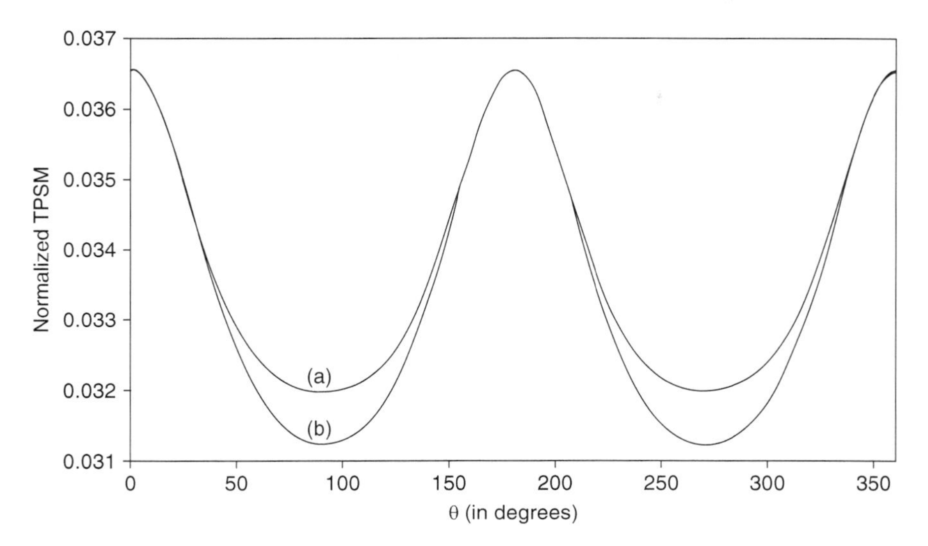

Fig. 5.3 Plot of the normalized TPSM as a function of angular orientation of the quantizing magnetic field for Cd_3As_2 for (**a**) $\delta \neq 0$ and (**b**) $\delta = 0$

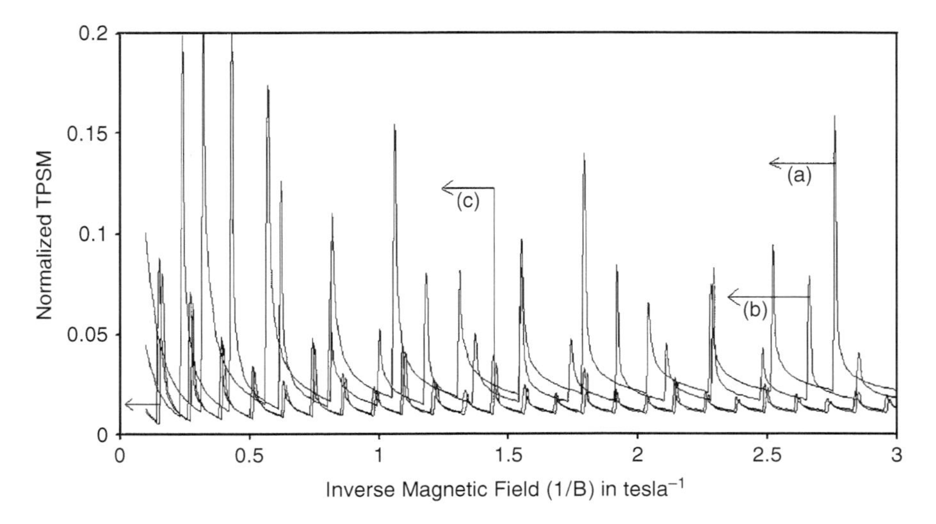

Fig. 5.4 Plot of the normalized TPSM as a function of inverse magnetic field for $CdGeAs_2$ in accordance with the generalized band model (**a**) $\delta \neq 0$, (**b**) $\delta = 0$, (**c**) three and (**d**) two band models of Kane

magnetic field. The origin of oscillations of the magnetothermopower is the same as that of the Shubnikov de Haas oscillations. The influence of crystal field splitting on the magnetothermopower can easily be conjectured by comparing the appropriate plots of Figs. 5.1 and 5.4. Besides, the differences among three and two band models of Kane together with parabolic energy bands for magnetothermopower of

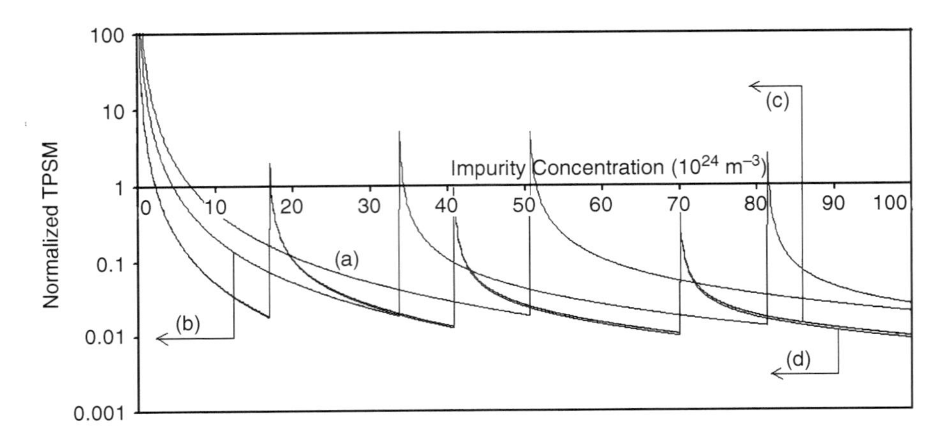

Fig. 5.5 Plot of the normalized TPSM as a function of impurity concentration for $CdGeAs_2$ for all cases of Fig. 5.4

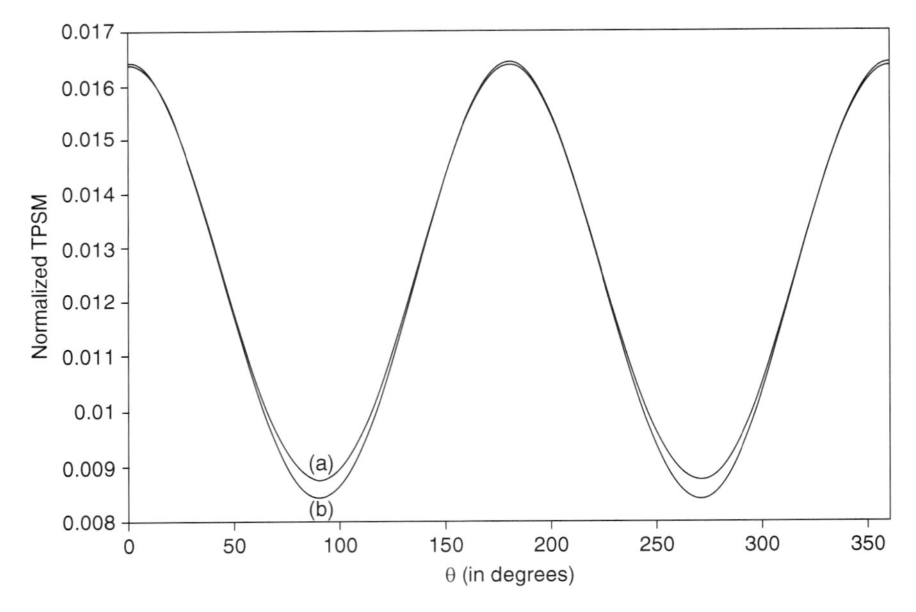

Fig. 5.6 Plot of the normalized TPSM as a function of angular orientation of the quantizing magnetic field for $CdGeAs_2$ for (**a**) $\delta \neq 0$ and (**b**) $\delta = 0$

Cd_3As_2 and $CdGeAs_2$ can easily be assessed by comparing the appropriate plots of Figs. 5.1 and 5.3. From Figs. 5.2 and 5.5, it appears that magnetothermopower oscillates with impurity concentration in Cd_3As_2 and $CdGeAs_2$ with different numerical values exhibiting the signature of the SdH effect. Although the rates of variations are different, the influence of spectral constants on all types of band models follows the same trend as observed in Figs. 5.2 and 5.5, respectively. From Figs. 5.3 and 5.6, it

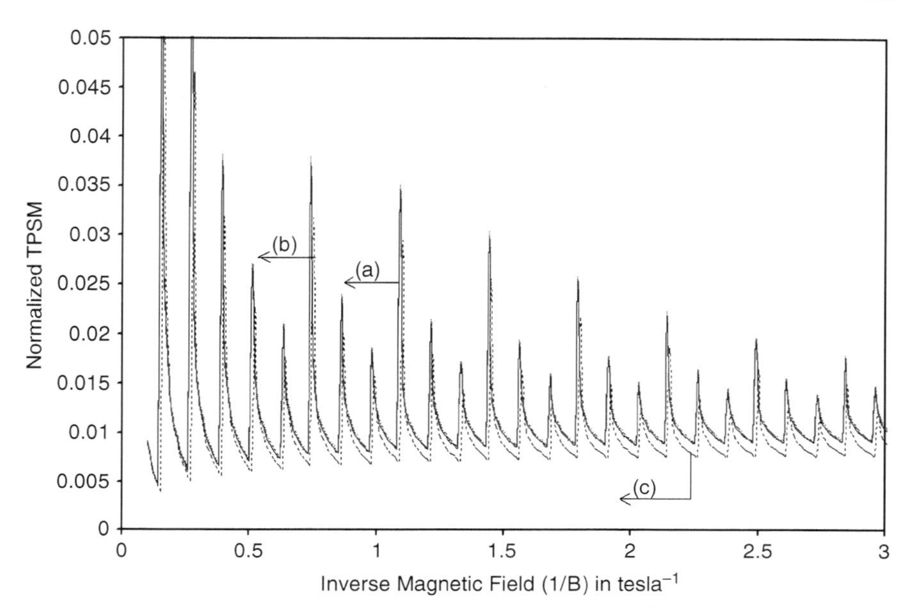

Fig. 5.7 Plot of the normalized TPSM as a function of inverse magnetic field for InAs in accordance with the (**a**) three and (**b**) two band models of Kane together with (**c**) the parabolic energy bands

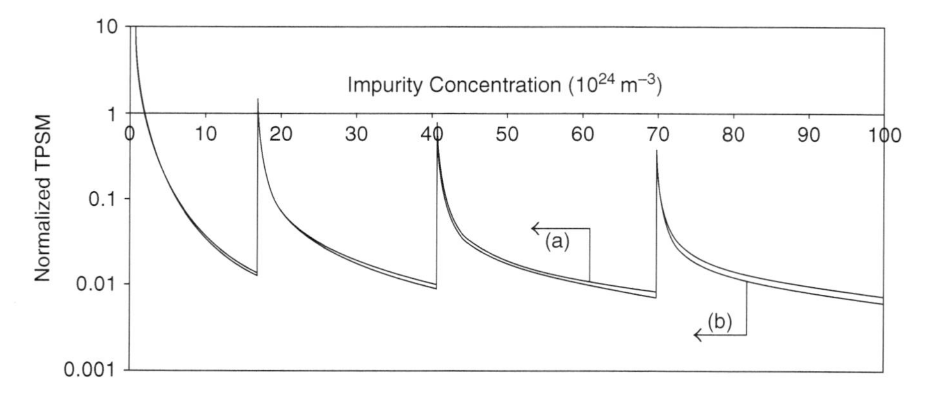

Fig. 5.8 Plot of the normalized TPSM as a function of inverse magnetic field for InSb in accordance with the (**a**) three and (**b**) two band models of Kane together with (**c**) the parabolic energy bands

appears that the magnetothermopower shows sinusoidal dependence with increasing θ and the variation is periodically repeated which appears from the said figures. For three and two band models of Kane together with parabolic energy bands, the magnetothermopower becomes θ invariant and for this reason these plots are not shown in Figs. 5.3 and 5.6, respectively. Using (5.5); (5.6) and (5.8); (5.9) and (5.12); (5.13) for three and two band models of Kane together with parabolic energy bands, the

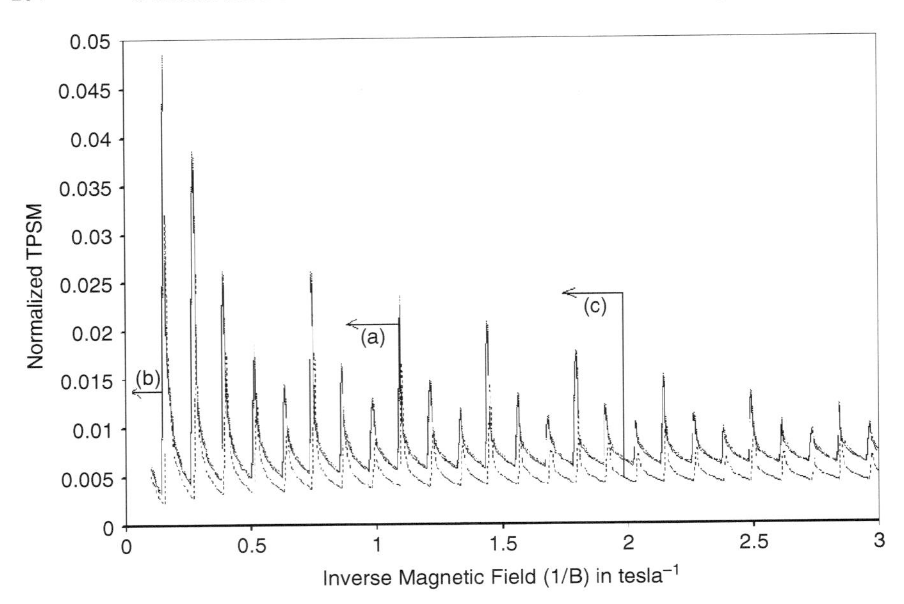

Fig. 5.9 Plot of the normalized TPSM as a function of impurity concentration for InAs in accordance with the (**a**) three and (**b**) two band energy models of Kane

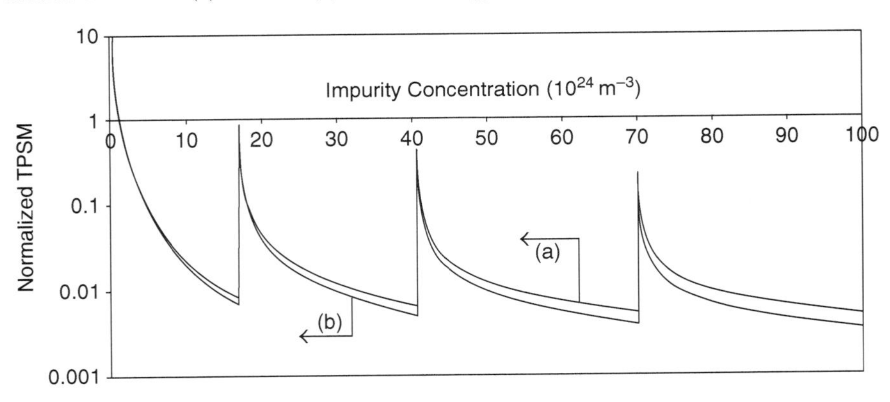

Fig. 5.10 Plot of the normalized TPSM as a function of impurity concentration for InSb in accordance with the (**a**) three and (**b**) two band models of Kane

normalized magnetothermopower for InAs and InSb as a function of $1/B$ has been plotted in Figs. 5.7 and 5.8, respectively. The normalized TPSM as a function of impurity concentration for three and two band models of Kane for InAs and InSb has been plotted in Figs. 5.9 and 5.10, respectively. It appears from the numerical values that the influence of the three band model of Kane in the energy spectrum of III–V, ternary, and quaternary compounds are difficult to distinguish from that of the two band model of Kane. Using (5.16) and (5.17), the normalized magnetothermopower has been plotted as a function of $1/B$ for p-CdS in Fig. 5.11 where the

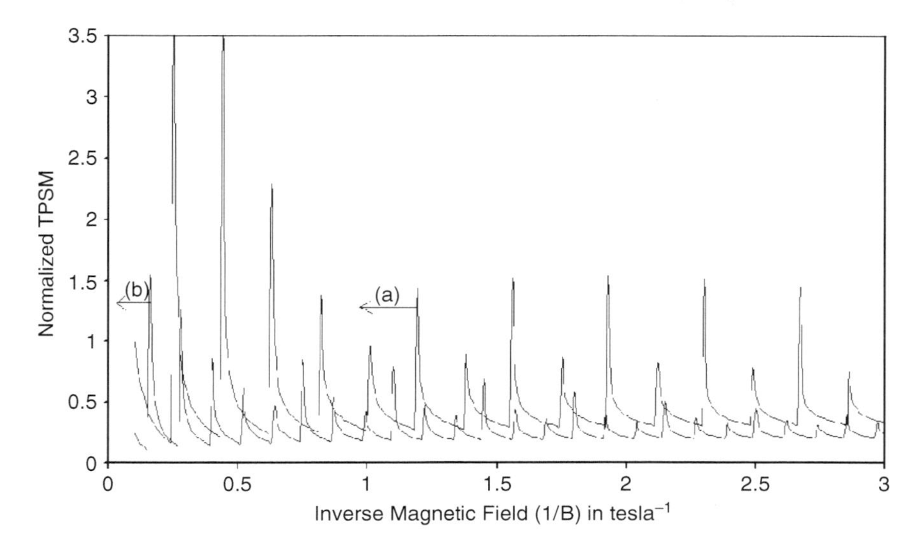

Fig. 5.11 Plot of the normalized TPSM as a function of inverse magnetic field in p-CdS for (**a**) $C_0 = 0$ and (**b**) $C_0 \neq 0$

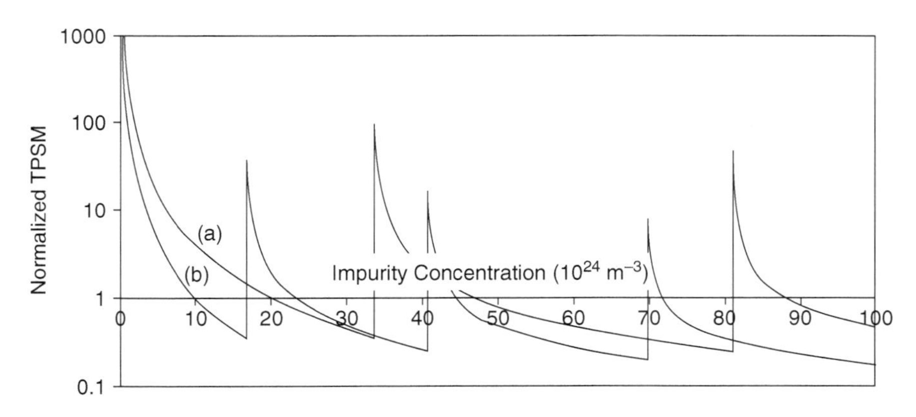

Fig. 5.12 Plot of the normalized TPSM as a function of impurity concentration field in p-CdS for (**a**) $C_0 = 0$ and (**b**) $C_0 \neq 0$

plots (a) and (b) are valid for $C_0 = 0$ and $C_0 \neq 0$, respectively. Figure 5.12 exhibits the plot of the same as a function of impurity concentration for all cases of Fig. 5.11. The influence of the term C_0 which represents the splitting of the two-spin states by the spin orbit coupling and the crystalline field is apparent from Figs. 5.11 and 5.12.

The normalized magnetothermopower in bismuth in accordance with the models of McClure and Choi (using (5.19) and (5.20)), Hybrid, Cohen (using (5.22) and (5.23)), Lax (using (5.25) and (5.26)), and parabolic ellipsoidal has been plotted in Figs. 5.13 and 5.14, respectively, as functions of inverse quantizing magnetic field and impurity concentration, respectively. Figures 5.15 and 5.16 exhibit the plots of

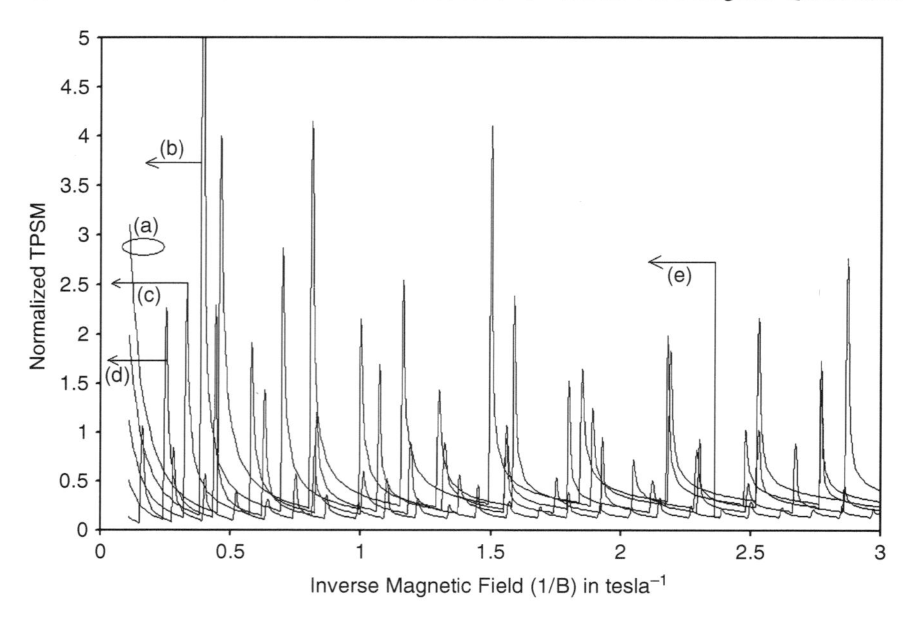

Fig. 5.13 Plot of the normalized TPSM as a function of inverse magnetic field for bismuth in accordance with the (**a**) McClure and Choi, (**b**) Hybrid, (**c**) Cohen, (**d**) Lax, and (**e**) parabolic energy bands

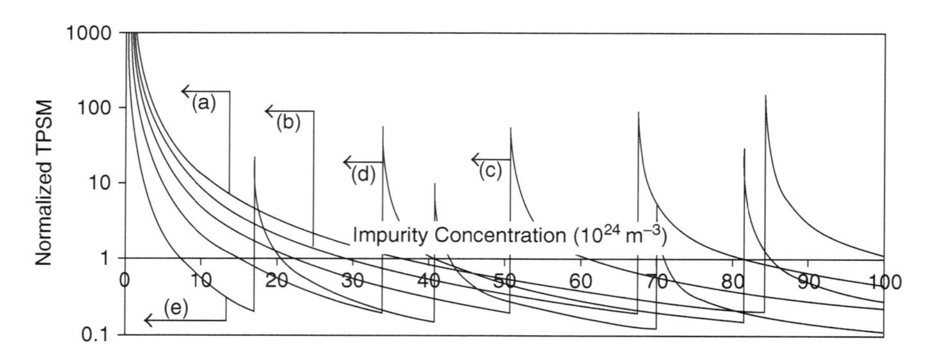

Fig. 5.14 Plot of the normalized TPSM as a function of impurity concentration for bismuth for all cases of Fig. 5.13

normalized TPSM in this case as functions of inverse quantizing magnetic field and impurity concentration for PbSnTe (using (5.22) and (5.23)). Figures 5.17 and 5.18 demonstrate the same for stressed InSb (using (5.28) and (5.29)) and it appears that the influence of stress leads to the enhancement of the TPSM in this case. The influence of spin splitting has not been considered in obtaining the oscillatory plots since the peaks in all the figures would increase in number with decrease in amplitude if spin splitting term is included in the respective numerical computations without

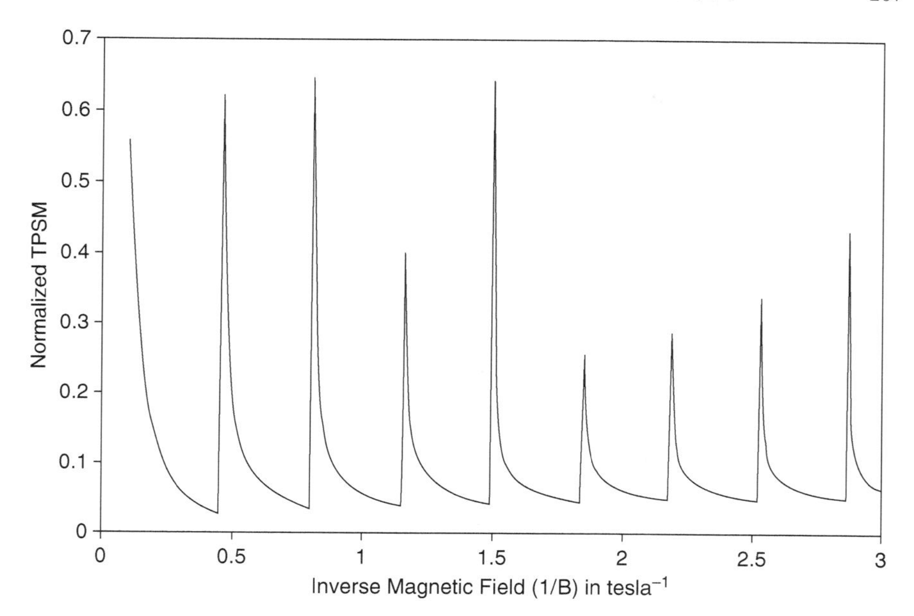

Fig. 5.15 Plot of the normalized TPSM as a function of inverse magnetic field for PbSnTe

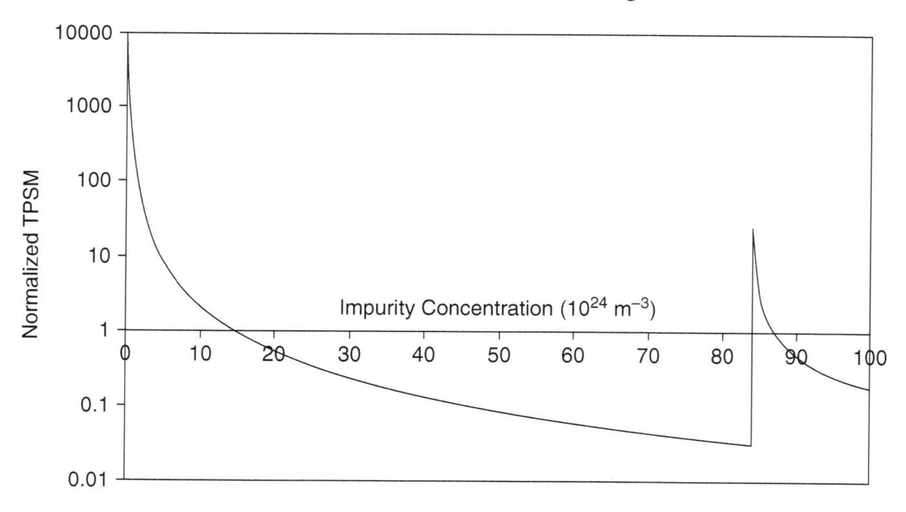

Fig. 5.16 Plot of the normalized TPSM as a function of impurity concentration for PbSnTe

introducing new physics. The effect of collision broadening has not been taken into account in this simplified analysis, although the effects of collisions are usually small at low temperatures, the sharpness of the amplitude of the oscillatory plots would somewhat be reduced by collision broadening. Nevertheless, the present analysis would remain valid qualitatively since the effects of collision broadening can usually be taken into account by an effective increase in temperature. Although in a

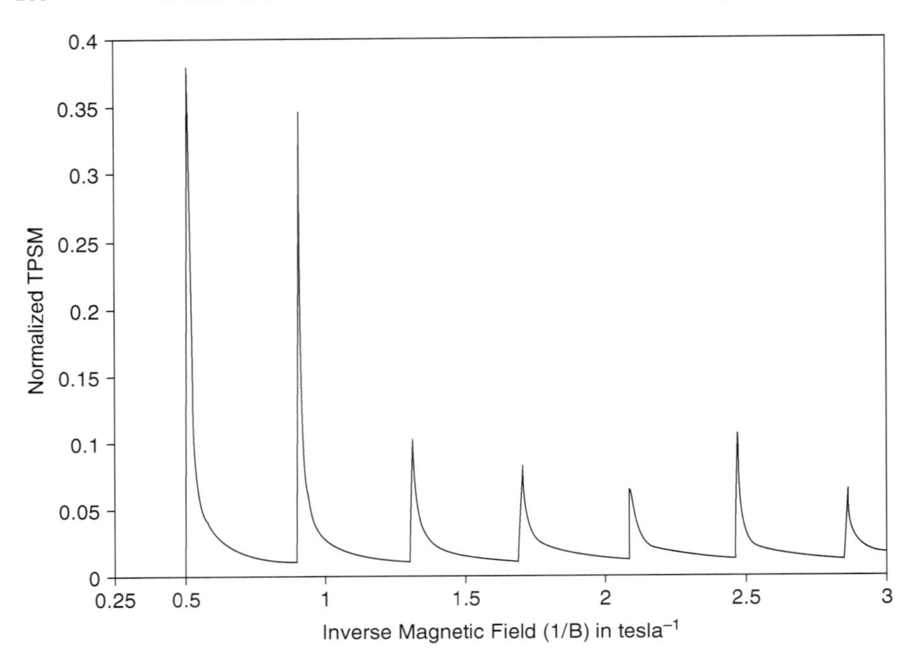

Fig. 5.17 Plot of the normalized TPSM as a function of inverse magnetic field for stressed InSb

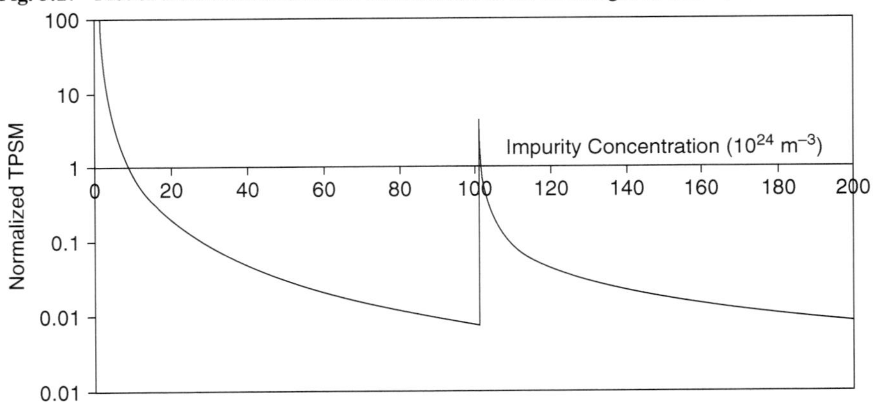

Fig. 5.18 Plot of the normalized TPSM as a function of impurity concentration for stressed InSb

more rigorous statement the effect of electron–electron interaction should be considered along with the self-consistent procedure, the simplified analysis as presented in this chapter exhibits the basic qualitative features of the TPSM in the present case for degenerate materials having various band structures under the magnetic quantization with reasonable accuracy. For the purpose of condensed presentation, the carrier statistics and the TPSM pertinent to this chapter have been presented in Table 5.1.

Table 5.1 The carrier statistics and the thermoelectric power under magnetic quantization in macroelectronic materials of nonlinear optical, III–V, II–VI, bismuth, IV–VI, and stressed materials

Type of materials	Carrier statistics	TPSM
1. Nonlinear optical materials	$n_0 = \frac{g_v eB}{2\pi^2\hbar}\sum_{n=0}^{n_{max}}[T_{51}(n,E_{FB}) + T_{52}(n,E_{FB})]$ (5.2)	On the basis of (5.2), $G_0 = \left(\frac{\pi^2 k_B^2 T}{3e}\right)\left[\frac{\sum_{n=0}^{n_{max}}[\{T_{51}(n,E_{FB})\}' + \{T_{52}(n,E_{FB})\}']}{\sum_{n=0}^{n_{max}}[T_{51}(n,E_{FB}) + T_{52}(n,E_{FB})]}\right]$ (5.3)
2. Kane type III–V materials	Under the conditions $\delta = 0$, $\Delta_{\|\|} = \Delta_{\perp} = \Delta$ and $m_{\|\|}^* = m_{\perp}^* = m^*$, $n_0 = \frac{g_v eB\sqrt{2m^*}}{2\pi^2\hbar^2}\sum_{n=0}^{n_{max}}\left[T_{53}(n,E_{FB}) + T_{54}(n,E_{FB})\right]$ (5.5)	On the basis of (5.5), $G_0 = \left(\frac{\pi^2 k_B^2 T}{3e}\right)\left[\frac{\sum_{n=0}^{n_{max}}[\{T_{53}(n,E_{FB})\}' + \{T_{54}(n,E_{FB})\}']}{\sum_{n=0}^{n_{max}}[T_{53}(n,E_{FB}) + T_{54}(n,E_{FB})]}\right]$ (5.6)
	Under the condition $\Delta \gg E_{g_0}$, $n_0 = \frac{g_v eB\sqrt{2m^*}}{2\pi^2 h^2}\sum_{n=0}^{n_{max}}\left[T_{55}(n,E_{FB}) + T_{56}(n,E_{FB})\right]$ (5.8)	On the basis of (5.8), $G_0 = \left(\frac{\pi^2 k_B^2 T}{3e}\right)\left[\frac{\sum_{n=0}^{n_{max}}[\{T_{55}(n,E_{FB})\}' + \{T_{56}(n,E_{FB})\}']}{\sum_{n=0}^{n_{max}}[T_{55}(n,E_{FB}) + T_{56}(n,E_{FB})]}\right]$ (5.9)
	Under the condition $\alpha E_{FB} \ll 1$, $n_0 = \frac{g_v N_C \theta_{B1}}{2}\left[\sum_{n=0}^{n_{max}}\frac{1}{\sqrt{a_{01}}}\left[(1+\frac{3}{2}\alpha b_{01})\right]\right]F_{-1/2}(\eta_B) + \frac{3}{4}\alpha k_B T F_{1/2}(\eta_B)\Big]\Big]$ (5.10)	On the basis of (5.10), $G_0 = \frac{\pi^2 k_B}{3e}\left[\frac{\sum_{n=0}^{n_{max}}\frac{1}{\sqrt{a_{01}}}\left[\left(1+\frac{3}{2}\alpha b_{01}\right)F_{-3/2}(\eta_B) + \frac{3}{4}\alpha k_B T F_{-1/2}(\eta_B)\right]}{\sum_{n=0}^{n_{max}}\frac{1}{\sqrt{a_{01}}}\left[\left(1+\frac{3}{2}\alpha b_{01}\right)F_{-1/2}(\eta_B) + \frac{3}{4}\alpha k_B T F_{1/2}(\eta_B)\right]}\right]$ (5.11)
	Under the condition $\alpha \to 0$, $n_0 = \frac{g_v N_C \theta_{B1}}{2}\sum_{n=0}^{n_{max}} F_{-1/2}(\bar{\eta}_B)$ (5.12)	On the basis of (5.12), $G_0 = \frac{\pi^2 k_B}{3e}\left[\sum_{n=0}^{n_{max}} F_{-1/2}(\bar{\eta}_B)\right]^{-1}\left[\sum_{n=0}^{n_{max}} F_{-3/2}(\bar{\eta}_B)\right]$ (5.13)

(continued)

Table 5.1 (Continued)

Type of materials	Carrier statistics	TPSM
3. II–VI materials	$n_0 = \frac{g_v eB\sqrt{2\pi m_{\|}^* k_B T}}{h^2} \sum_{n=0}^{n_{max}} F_{-1/2}(\theta_3)$ (5.16)	On the basis of (5.16), $G_0 = \frac{\pi^2 k_B}{3e}\left[\sum_{n=0}^{n_{max}} F_{-1/2}(\theta_3)\right]^{-1}\left[\sum_{n=0}^{n_{max}} F_{-3/2}(\theta_3)\right]$ (5.17)
4. Bismuth	The McClure and Choi model $n_0 = \frac{g_v eB\sqrt{2m_3}}{2\pi^2\hbar^2}\sum_{n=0}^{n_{max}}\left[T_{57}(n, E_{FB}) + T_{58}(n, E_{FB})\right]$ (5.19)	On the basis of (5.19), $G_0 = \left(\frac{\pi^2 k_B^2 T}{3e}\right)\left[\frac{\sum_{n=0}^{n_{max}}[\{T_{57}(n, E_{FB})\}' + \{T_{58}(n, E_{FB})\}']}{\sum_{n=0}^{n_{max}}[T_{57}(n, E_{FB}) + T_{58}(n, E_{FB})]}\right]$ (5.20)
	The Cohen model $n_0 = \frac{(eBg_v)(\sqrt{2m_3})}{2\pi^2\hbar^2}\sum_{n=0}^{n_{max}}\left[T_{59}(n, E_{FB}) + T_{60}(n, E_{FB})\right]$ (5.22)	On the basis of (5.22), $G_0 = \frac{\pi^2 k_B^2 T}{3e}\left[\frac{\sum_{n=0}^{n_{max}}[\{T_{59}(n, E_{FB})\}' + \{T_{60}(n, E_{FB})\}']}{\sum_{n=0}^{n_{max}}[T_{59}(n, E_{FB}) + T_{60}(n, E_{FB})]}\right]$ (5.23)
	The Lax model $n_0 = \frac{(eBg_v)(\sqrt{2m_3})}{2\pi^2\hbar^2}\sum_{n=0}^{n_{max}}\left[T_{61}(n, E_{FB}) + T_{62}(n, E_{FB})\right]$ (5.25)	On the basis of (5.25), $G_0 = \left(\frac{\pi^2 k_B^2 T}{3e}\right)\left[\frac{\sum_{n=0}^{n_{max}}[\{T_{61}(n, E_{FB})\}' + \{T_{62}(n, E_{FB})\}']}{\sum_{n=0}^{n_{max}}[T_{61}(n, E_{FB}) + T_{62}(n, E_{FB})]}\right]$ (5.26)
5. IV–VI materials	The expression of n_0 in this case is given by (5.22) in which the constants of the energy band spectrum correspond to the carriers of the IV–VI semiconductors	The expression of TPSM in this case is given by (5.23) in which the constants of the energy band spectrum correspond to the carriers of the IV–VI semiconductors
6. Stressed materials	$n_0 = \frac{g_v eB}{\pi^2 h}\sum_{n=0}^{n_{max}}[T_{63}(n, E_{FB}) + T_{64}(n, E_{FB})]$ (5.28)	On the basis of (5.28), $G_0 = \left(\frac{\pi^2 k_B^2 T}{3e}\right)\left[\frac{\sum_{n=0}^{n_{max}}[\{T_{63}(n, E_{FB})\}' + \{T_{64}(n, E_{FB})\}']}{\sum_{n=0}^{n_{max}}[T_{63}(n, E_{FB}) + T_{64}(n, E_{FB})]}\right]$ (5.29)

5.4 Open Research Problems

(R5.1) Investigate the DTP, PTP, and Z both in the presence and in the absence of an arbitrarily oriented quantizing magnetic field by considering all types of scattering mechanisms including broadening and the electron spin (applicable under magnetic quantization) for all the bulk materials whose unperturbed carrier energy spectra are defined in Chap. 1.

(R5.2) Investigate the DTP, PTP, and Z by considering all types of scattering mechanisms in the presence of quantizing magnetic field under an arbitrarily oriented (a) nonuniform electric field and (b) alternating electric field, respectively, for all the materials whose unperturbed carrier energy spectra are defined in Chap. 1 by including spin and broadening, respectively.

(R5.3) Investigate the DTP, PTP, and Z by considering all types of scattering mechanisms under an arbitrarily oriented alternating quantizing magnetic field by including broadening and the electron spin for all the materials whose unperturbed carrier energy spectra are defined in Chap. 1.

(R5.4) Investigate the DTP, PTP, and Z by considering all types of scattering mechanisms under an arbitrarily oriented alternating quantizing magnetic field and crossed alternating electric field by including broadening and the electron spin for all the materials whose unperturbed carrier energy spectra are defined in Chap. 1.

(R5.5) Investigate the DTP, PTP, and Z by considering all types of scattering mechanisms under an arbitrarily oriented alternating quantizing magnetic field and crossed alternating nonuniform electric field by including broadening and the electron spin for all the materials whose unperturbed carrier energy spectra are defined in Chap. 1.

(R5.6) Investigate the DTP, PTP, and Z in the presence and absence of an arbitrarily oriented quantizing magnetic field by considering all types of scattering mechanisms under exponential, Kane, Halperin, Lax, and Bonch–Bruevich band tails [69] for all the materials whose unperturbed carrier energy spectra are defined in Chap. 1 by including spin and broadening (applicable under magnetic quantization).

(R5.7) Investigate the DTP, PTP, and Z in the presence of an arbitrarily oriented quantizing magnetic field by considering all types of scattering mechanisms for all the materials as defined in (R5.6) under an arbitrarily oriented (a) nonuniform electric field and (b) alternating electric field, respectively, whose unperturbed carrier energy spectra are defined in Chap. 1.

(R5.8) Investigate the DTP, PTP, and Z by considering all types of scattering mechanisms for all the materials as described in (R5.6) under an arbitrarily oriented alternating quantizing magnetic field by including broadening and the electron spin whose unperturbed carrier energy spectra are defined in Chap. 1.

(R5.9) Investigate the DTP, PTP, and Z by considering all types of scattering mechanisms as discussed in (R5.6) under an arbitrarily oriented alternating quantizing magnetic field and crossed alternating electric field by including broadening and the electron spin for all the materials whose unperturbed carrier energy spectra are defined in Chap. 1.

(R5.10) Investigate all the appropriate problems of this chapter after proper modifications introducing new theoretical formalisms for functional, negative refractive index, macromolecular, organic, and magnetic materials.

(R5.11) Investigate all the appropriate problems of this chapter for p-InSb, p-CuCl and stressed semiconductors having diamond structure valence bands whose dispersion relations of the carriers in bulk materials are given by Cunningham [74], Yekimov et al. [75], and Roman et al. [76], respectively.

References

1. N. Miura, *Physics of Semiconductors in High Magnetic Fields*, Series on Semiconductor Science and Technology (Oxford University Press, USA, 2007)
2. K.H.J Buschow, F.R. de Boer, *Physics of Magnetism and Magnetic Materials* (Springer, New York, 2003)
3. D. Sellmyer, R. Skomski (eds.), *Advanced Magnetic Nanostructures* (Springer, New York, 2005)
4. J.A.C. Bland, B. Heinrich (ed.), *Ultrathin Magnetic Structures III: Fundamentals of Nanomagnetism (Pt. 3)* (Springer-Verlag, Germany, 2005)
5. B.K. Ridley, *Quantum Processes in semiconductors*, 4th edn. (Oxford publications, Oxford, 1999)
6. J.H. Davies, *Physics of low dimensional semiconductors* (Cambridge University Press, UK, 1998)
7. S. Blundell, *Magnetism in Condensed Matter*, Oxford Master Series in Condensed Matter Physics (Oxford University Press, USA, 2001)
8. C. Weisbuch, B. Vinter, *Quantum Semiconductor Structures: Fundamentals and Applications* (Academic Publishers, USA, 1991)
9. D. Ferry, *Semiconductor Transport* (CRC, USA, 2000)
10. M. Reed (ed.), *Semiconductors and Semimetals: Nanostructured Systems* (Academic Press, USA, 1992)
11. T. Dittrich, *Quantum Transport and Dissipation* (Wiley-VCH Verlag GmbH, Germany, 1998)
12. A.Y. Shik, *Quantum Wells: Physics & Electronics of Two-Dimensional Systems* (World Scientific, USA, 1997)
13. K.P. Ghatak, M. Mondal, Zietschrift fur Naturforschung A **41a**, 881 (1986)
14. K.P. Ghatak, M. Mondal, J. Appl. Phys. **62**, 922 (1987)
15. K.P. Ghatak, S.N. Biswas, Phys. Stat. Sol. (b) **140**, K107 (1987)
16. K.P. Ghatak, M. Mondal, J. Magn. Magn. Mater. **74**, 203 (1988)
17. K.P. Ghatak, M. Mondal, Phys. Stat. Sol. (b) **139**, 195 (1987)
18. K.P. Ghatak, M. Mondal, Phys. Stat. Sol. (b) **148**, 645 (1988)
19. K.P. Ghatak, B. Mitra, A. Ghoshal, Phys. Stat. Sol. (b) **154**, K121 (1989)
20. K.P. Ghatak, S.N. Biswas, J. Low Temp. Phys. **78**, 219 (1990)
21. K.P. Ghatak, M. Mondal, Phys. Stat. Sol. (b) **160**, 673 (1990)
22. K.P. Ghatak, B. Mitra, Phys. Lett. A **156**, 233 (1991)
23. K.P. Ghatak, A. Ghoshal, B. Mitra, Nuovo Cimento D **13D**, 867 (1991)
24. K.P. Ghatak, M. Mondal, Phys. Stat. Sol. (b) **148**, 645 (1989)
25. K.P. Ghatak, B. Mitra, Int. J. Electron. **70**, 345 (1991)
26. K.P. Ghatak, S.N. Biswas, J. Appl. Phys. **70**, 299 (1991)
27. K.P. Ghatak, A. Ghoshal, Phys. Stat. Sol. (b) **170**, K27 (1992)
28. K.P. Ghatak, Nuovo Cimento D **13D**, 1321 (1992)

29. K.P. Ghatak, B. Mitra, Int. J. Electron. **72**, 541 (1992)
30. K.P. Ghatak, S.N. Biswas, Nonlinear Opt. **4**, 347 (1993)
31. K.P. Ghatak, M. Mondal, Phys. Stat. Sol. (b) **175**, 113 (1993)
32. K.P. Ghatak, S.N. Biswas, Nonlinear Opt. **4**, 39 (1993)
33. K.P. Ghatak, B. Mitra, Nuovo Cimento **15D**, 97 (1993)
34. K.P. Ghatak, S.N. Biswas, Nanostruct. Mater. **2**, 91 (1993)
35. K.P. Ghatak, M. Mondal, Phys. Stat. Sol. (b) **185**, K5 (1994)
36. K.P. Ghatak, B. Goswami, M. Mitra, B. Nag, Nonlinear Opt. **16**, 9 (1996)
37. K.P. Ghatak, M. Mitra, B. Goswami, B. Nag, Nonlinear Opt. **16**, 167 (1996)
38. K.P. Ghatak, D.K. Basu, B. Nag, J. Phys. Chem. Solid **58**, 133 (1997)
39. K. P. Ghatak, B. Nag, Nanostruct. Mater. **10**, 923 (1998)
40. D. Roy Choudhury, A.K. Choudhury, K.P. Ghatak, A.N. Chakravarti, Phys. Stat. Sol. (b) **98**, K141 (1980)
41. A.N. Chakravarti, K.P. Ghatak, A. Dhar and S. Ghosh, Phys. Stat. Sol. (b) **105**, K55 (1981)
42. A.N. Chakravarti, A.K. Choudhury, K.P. Ghatak, Phys. Stat. Sol. (a) **63**, K97 (1981)
43. A.N. Chakravarti, A.K. Choudhury, K.P. Ghatak, S. Ghosh, A. Dhar, J. Appl. Phys. **25**, 105 (1981)
44. A.N. Chakravarti, K.P. Ghatak, G.B. Rao, K.K. Ghosh, Phys. Stat. Sol. (b) **112**, 75 (1982)
45. A.N. Chakravarti, K.P. Ghatak, K.K. Ghosh, H.M. Mukherjee, Phys. Stat. Sol. (b) **116**, 17 (1983)
46. M. Mondal, K.P. Ghatak, Phys. Stat. Sol. (b) **133**, K143 (1984)
47. M. Mondal, K.P. Ghatak, Phys. Stat. Sol. (b) **126**, K47 (1984)
48. M. Mondal, K.P. Ghatak, Phys. Stat. Sol. (b) **126**, K41 (1984)
49. M. Mondal, K.P. Ghatak, Phys. Stat. Sol. (b) **129**, K745 (1985)
50. M. Mondal, K.P. Ghatak, Phys. Scr. **31**, 615 (1985)
51. M. Mondal, K.P. Ghatak, Phys. Stat. Sol. (b) **135**, 239 (1986)
52. M. Mondal, K.P. Ghatak, Phys. Stat. Sol. (b) **93**, 377 (1986)
53. M. Mondal, K.P. Ghatak, Phys. Stat. Sol. (b) **135**, K21 (1986)
54. M. Mondal, S. Bhattacharyya, K.P. Ghatak, Appl. Phys. A **42A**, 331 (1987)
55. S.N. Biswas, N. Chattopadhyay, K.P. Ghatak, Phys. Stat. Sol. (b) **141**, K47 (1987)
56. B. Mitra, K.P. Ghatak, Phys. Stat. Sol. (b) **149**, K117 (1988)
57. B. Mitra, A. Ghoshal, K.P. Ghatak, Phys. Stat. Sol. (b) **150**, K67 (1988)
58. M. Mondal, K.P. Ghatak, Phys. Stat. Sol. (b) **147**, K179 (1988)
59. M. Mondal, K.P. Ghatak, Phys. Stat. Sol. (b) **146**, K97 (1988)
60. B. Mitra, A. Ghoshal, K.P. Ghatak, Phys. Stat. Sol. (b) **153**, K209 (1989)
61. B. Mitra, K. P. Ghatak, Phys. Lett. **142A**, 401 (1989)
62. B. Mitra, A. Ghoshal, K.P. Ghatak, Phys. Stat. Sol. (b) **154**, K147 (1989)
63. B. Mitra, K.P. Ghatak, Sol. State Electron. **32**, 515 (1989)
64. B. Mitra, A. Ghoshal, K.P. Ghatak, Phys. Stat. Sol. (b) **155**, K23 (1989)
65. B. Mitra, K.P. Ghatak, Phys. Lett. **135A**, 397 (1989)
66. B. Mitra, K.P. Ghatak, Phys. Lett. A **146A**, 357 (1990)
67. B. Mitra, K.P. Ghatak, Phys. Stat. Sol. (b) **164**, K13 (1991)
68. S.N. Biswas, K.P. Ghatak, Int. J. Electron. **70**, 125 (1991)
69. P.R. Wallace, Phys. Stat. Sol. (b), **92**, 49 (1979)
70. K.P. Ghatak, S. Bhattacharya, D. De, *Einstein Relation in Compound Semiconductors and Their Nanostructures*, Springer Series in Materials Science, vol 116 (Springer-Verlag, Germany, 2008)
71. B.R. Nag, *Electron Transport in Compound Semiconductors*, Springer Series in Solid-State Sciences, vol 11 (Springer-Verlag, Germany, 1980)
72. C.C. Wu, C.J. Lin, J. Low Temp. Phys. **57**, 469 (1984)
73. M.H. Chen, C.C. Wu, C.J. Lin, J. Low Temp. Phys. **55**, 127 (1984)
74. R.W. Cunningham, Phys. Rev. **167**, 761 (1968)
75. A.I. Yekimov, A.A. Onushchenko, A.G. Plyukhin, L. Efros, J. Exp. Theor. Phys. **88**, 1490 (1985)
76. B.J. Roman, A.W. Ewald, Phys. Rev. B **5**, 3914 (1972)

Chapter 6
Thermoelectric Power in Superlattices Under Magnetic Quantization

6.1 Introduction

In this chapter, we shall study the thermoelectric power under magnetic quantization in III–V, II–VI, IV–VI, and HgTe/CdTe superlattices with graded interfaces in Sects. 6.2.1–6.2.4 of the theoretical background. In Sects. 6.2.5–6.2.8, we have investigated the same for III–V, II–VI, IV–VI, and HgTe/CdTe effective mass superlattices, respectively. In Sects. 6.2.9–6.2.16, we have studied the thermoelectric power in the presence of a quantizing magnetic field for quantum wells of the aforementioned superlattices. Sections 6.3 and 6.4 contain results and discussion and open research problems, respectively.

6.2 Theoretical Background

6.2.1 Magnetothermopower in III–V Superlattices with Graded Interfaces

In the presence of a quantizing magnetic field B along z-direction, the simplified magnetodispersion relation can be, written following (3.2) as [1]

$$k_z = \frac{1}{L_0}\left[\rho(n,E) - \left\{\frac{2eB}{\hbar}L_0^2\left(n+\frac{1}{2}\right)\right\}\right]^{1/2}, \tag{6.1}$$

where

$$\rho(n,E) = \left[\cos^{-1}\left\{\frac{1}{2}\psi(n,E)\right\}\right]^2$$

n is the Landau quantum number,

$$\psi(n,E) = \Bigg[2\cosh\{\beta(n,E)\}\cos\{\gamma(n,E)\} + \varepsilon(n,E)\sinh\{\beta(n,E)\}\sin\{\gamma(n,E)\}$$
$$+ \Delta_0\Bigg[\left(\frac{\{K_1(n,E)\}^2}{K_2(n,E)} - 3K_2(n,E)\right)\cosh\{\beta(n,E)\}\sin\{\gamma(n,E)\}$$
$$+ \left(3K_1(n,E) - \frac{\{K_2(n,E)\}^2}{K_1(n,E)}\right)\sinh\{\beta(n,E)\}\cos\{\gamma(n,E)\}\Bigg]$$
$$+ \Delta_0\Bigg[2(\{K_1(n,E)\}^2 - \{K_2(n,E)\}^2)\cosh\{\beta(n,E)\}\cos\{\gamma(n,E)\}$$
$$+ \frac{1}{12}\Big(\frac{5\{K_1(n,E)\}^3}{K_2(n,E)} + \frac{5\{K_2(n,E)\}^3}{K_1(n,E)}$$
$$- \{34K_2(n,E)K_1(n,E)\}\Big)\sinh\{\beta(n,E)\}\sin\{\gamma(n,E)\}\Bigg]\Bigg]$$

$$\beta(n,E) \equiv K_1(n,E)\,[a_0 - \Delta_0]$$
$$K_1(n,E) \equiv \left[\frac{2m_2^* E'}{h^2} G(E - V_0, \alpha_2, \Delta_2) + \frac{2|e|B}{h}\left(n + \frac{1}{2}\right)\right]^{1/2},$$
$$\gamma(n,E) = K_2(n,E)\,[b_0 - \Delta_0]$$
$$K_2(n,E) \equiv \left[\frac{2m_1^* E}{h^2} G(E, \alpha_1, \Delta_1) - \left\{\frac{2eB}{h}\left(n + \frac{1}{2}\right)\right\}\right]^{1/2},$$

and

$$\varepsilon(n,E) \equiv \left[\frac{K_1(n,E)}{K_2(n,E)} - \frac{K_2(n,E)}{K_1(n,E)}\right].$$

Considering only the lowest miniband, since in an actual SL only the lowest miniband is significantly populated at low temperatures, where the quantum effects become prominent, the electron concentration per unit volume in this case can be written as

$$n_0 = \left(\frac{eBg_v}{\pi^2 h L_0}\right)\sum_{n=0}^{n_{\max}}[T_{61}(n, E_{\mathrm{FSLBGI}}) + T_{62}(n, E_{\mathrm{FSLBGI}})], \tag{6.2}$$

where

$$T_{61}(n, E_{\mathrm{FSLBGI}}) \equiv \left[\rho(E_{\mathrm{FSLBGI}}, n) - \left\{\frac{2eB}{h}\left(n + \frac{1}{2}\right)L_0^2\right\}\right]^{1/2},$$

E_{FSLBGI} is the Fermi energy in this case,

$$T_{62}(n, E_{\mathrm{FSLBGI}}) \equiv \sum_{r=1}^{s} W(r)[T_{61}(n, E_{\mathrm{FSLBGI}})],$$

and

$$W(r) \equiv 2(k_B T)^{2r}(1 - 2^{1-2r})\zeta(2r)\frac{\partial^{2r}}{\partial E_{FSLBGI}^{2r}}.$$

The use of (1.13) and (6.2) leads to the expression of the thermoelectric power under magnetic quantization as

$$G_0 = \left(\frac{\pi^2 k_B^2 T}{3e}\right) \frac{\sum_{n=0}^{n_{max}} \left[\{T_{61}(n, E_{FSLBGI})\}' + \{T_{62}(n, E_{FSLBGI})\}'\right]}{\sum_{n=0}^{n_{max}} \left[T_{61}(n, E_{FSLBGI}) + T_{62}(n, E_{FSLBGI})\right]}. \tag{6.3}$$

6.2.2 *Magnetothermopower in II–VI Superlattices with Graded Interfaces*

In the presence of a quantizing magnetic field B along z-direction, the simplified magnetodispersion relation in this case can be written following (3.8) as

$$k_z^2 = \frac{1}{L_0^2}\left[\rho_1(n, E) - \left\{\frac{2eB}{h} L_0^2 \left(n + \frac{1}{2}\right)\right\}\right], \tag{6.4}$$

where

$$\rho_1(n, E) = \left[\cos^{-1}\left\{\frac{1}{2}\psi_1(n, E)\right\}\right]^2,$$

$$\begin{aligned}
\psi_1(n, E) = \Bigg[& 2\cosh\{\beta_1(n, E)\}\cos\{\gamma_1(n, E)\} \\
& + \varepsilon_1(n, E)\sinh\{\beta_1(n, E)\}\sin\{\gamma_1(n, E)\} \\
& + \Delta_0\Bigg[\left(\frac{\{K_3(n, E)\}^2}{K_4(n, E)} - 3K_4(n, E)\right)\cosh\{\beta_1(n, E)\}\sin\{\gamma_1(n, E)\} \\
& + \left(3K_3(n, E) - \frac{\{K_4(n, E)\}^2}{K_3(n, E)}\right)\sinh\{\beta_1(n, E)\}\cos\{\gamma_1(n, E)\}\Bigg] \\
& + \Delta_0\Bigg[2\left(\{K_3(n, E)\}^2 - \{K_4(n, E)\}^2\right)\cosh\{\beta_1(n, E)\}\cos\{\gamma_1(n, E)\} \\
& + \frac{1}{12}\left(\frac{5\{K_3(n, E)\}^3}{K_4(n, E)} + \frac{5\{K_4(n, E)\}^3}{K_3(n, E)} - \{34K_4(n, E)K_3(n, E)\}\right) \\
& \times \sinh\{\beta_1(n, E)\}\sin\{\gamma_1(n, E)\}\Bigg]\Bigg]
\end{aligned}$$

$$\beta_1(n, E) \equiv K_3(n, E)[a_0 - \Delta_0],$$

$$K_3(n, E) \equiv \left[\frac{2m_2^*}{\hbar^2}E'G(E - V_0, \alpha_2, \Delta_2) + \frac{2eB}{\hbar}\left(n + \frac{1}{2}\right)\right]^{1/2},$$

$$\gamma_1(n,E) = K_4(n,E)[b_0 - \Delta_0],$$

$$K_4(n,E) \equiv \left[\frac{2m^*_{\|,1}}{\hbar^2}\left[E - \frac{\hbar eB}{m^*_{\perp,1}}\left(n+\frac{1}{2}\right) \mp C_0\left\{\frac{2eB}{\hbar}\left(n+\frac{1}{2}\right)\right\}^{1/2}\right]\right]^{1/2}.$$

and

$$\varepsilon_1(n,E) \equiv \left[\frac{K_3(n,E)}{K_4(n,E)} - \frac{K_4(n,E)}{K_3(n,E)}\right].$$

The electron concentration per unit volume in this case can be expressed as

$$n_0 = \left(\frac{eBg_v}{2\pi^2 hL_0}\right)\sum_{n=0}^{n_{\max}}[T_{63}(n,E_{\mathrm{FSLBGI}}) + T_{64}(n,E_{\mathrm{FSLBGI}})], \tag{6.5}$$

where

$$T_{63}(n,E_{\mathrm{FSLBGI}}) \equiv \left[\rho_1(n,E_{\mathrm{FSLBGI}}) - \left\{\frac{2eB}{h}\left(n+\frac{1}{2}\right)L_0^2\right\}\right]^{1/2}$$

and

$$T_{64}(n,E_{\mathrm{FSLBGI}}) \equiv \sum_{r=1}^{s} W(r)[T_{63}(n,E_{\mathrm{FSLBGI}})].$$

The use of (1.13) and (6.5) leads to the expression of the thermoelectric power under strong magnetic quantization as

$$G_0 = \left(\frac{\pi^2 k_B^2 T}{3e}\right)\frac{\sum\limits_{n=0}^{n_{\max}}[\{T_{63}(n,E_{FSLBGI})\}' + \{T_{64}(n,E_{\mathrm{FSLBGI}})\}']}{\sum\limits_{n=0}^{n_{\max}}[T_{63}(n,E_{FSLBGI}) + T_{64}(n,E_{\mathrm{FSLBGI}})]}. \tag{6.6}$$

6.2.3 Magnetothermopower in IV–VI Superlattices with Graded Interfaces

The simplified magnetodispersion relation in this case can be written following (3.13) as

$$k_z = \frac{1}{L_0}\left[\rho_2(n,E) - \left\{\frac{2eB}{h}L_0^2\left(n+\frac{1}{2}\right)\right\}\right]^{1/2}, \tag{6.7}$$

where

$$\rho_2(n,E) = \left[\cos^{-1}\left\{\frac{1}{2}\psi_2(n,E)\right\}\right]^2,$$

$$\psi_2(n,E) = \Big[2\cosh\{\beta_2(n,E)\}\cos\{\gamma_2(n,E)\}$$
$$+\varepsilon_2(n,E)\sinh\{\beta_2(n,E)\}\sin\{\gamma_2(n,E)\}$$
$$+\Delta_0\Big[\left(\frac{\{K_5(n,E)\}^2}{K_6(n,E)} - 3K_6(n,E)\right)\cosh\{\beta_2(n,E)\}\sin\{\gamma_2(n,E)\}$$
$$+\left(3K_5(n,E) - \frac{\{K_6(n,E)\}^2}{K_5(n,E)}\right)\sinh\{\beta_2(n,E)\}\cos\{\gamma_2(n,E)\}\Big]$$
$$+\Delta_0\Big[2\left(\{K_5(n,E)\}^2 - \{K_6(n,E)\}^2\right)\cosh\{\beta_2(n,E)\}\cos\{\gamma_2(n,E)\}$$
$$+\frac{1}{12}\left(\frac{5\{K_5(n,E)\}^3}{K_6(n,E)} + \frac{5\{K_6(n,E)\}^3}{K_5(n,E)} - \{34K_6(n,E)K_5(n,E)\}\right)$$
$$\times\sinh\{\beta_2(n,E)\}\sin\{\gamma_2(n,E)\}\Big]\Big],$$

$$\beta_2(n,E) \equiv K_5(n,E)[a_0 - \Delta_0],$$
$$K_5(n,E) \equiv \Big[\left[(E-V_0)^2 H_{32} + (E-V_0)H_{42}(n) + H_{52}(n)\right]^{1/2}$$
$$-\left[(E-V_0)H_{12} + H_{22}(n)\right]\Big]^{1/2},$$
$$\gamma_2(n,E) = K_6(n,E)[b_0 - \Delta_0],$$
$$H_{4i}(n) = \left[4(b_i^2 - f_i^2)^2\right]^{-1}\left[4b_i^2E_{g_i} + 4b_id_i + 4b_if_iE_{g_i} + 4f_i^2E_{g_i}\right.$$
$$\left.+8\varphi_6(n)\left[b_i^2a_i + C_if_ib_i - a_i^2b_i\right]\right]\varphi_6(n) = \left\{\frac{2eB}{\hbar}\left(n+\frac{1}{2}\right)\right\}$$
$$H_{5i}(n) \equiv \left[4(b_i^2 - f_i^2)^2\right]^{-1}\Big[(\varphi_6(n))^2\left[-8a_ib_iC_if_i + 4b_i^2C_i^2 + 4f_i^2a_i^2\right.$$
$$\left.-4f_i^2C_i^2\right] + \varphi_6(n)\left[8d_iC_if_i - 4a_ib_id_i - 4a_ib_if_iE_{g_i} + 4b_i^2C_i\right.$$
$$\left.+4b_i^2e_iE_{g_i} - 4a_if_i^2E_{g_i} - 4f_i^2e_iE_{g_i}\right]\Big] + \Big[E_{g_i}^2b_i^2 + d_i^2$$
$$+f_i^2E_{g_i}^2 + 2E_{g_i}f_id_i\Big]\Big]$$
$$H_{2i}(n) = \left[2(b_i^2 - f_i^2)\right]^{-1}\left[E_{g_i}b_i + d_i + f_iE_{g_i} + 2(C_if_i - a_ib_i)\varphi_6(n)\right].$$

and

$$\varepsilon_2(n,E) \equiv \left[\frac{K_5(n,E)}{K_6(n,E)} - \frac{K_6(n,E)}{K_5(n,E)}\right].$$

The electron concentration per unit volume in this case can be expressed as

$$n_0 = \left(\frac{eBg_v}{\pi^2 hL_0}\right)\sum_{n=0}^{n_{\max}}\left[T_{65}(n,E_{\mathrm{FSLBGI}}) + T_{66}(n,E_{\mathrm{FSLBGI}})\right], \qquad (6.8)$$

where

$$T_{65}(n,E_{\mathrm{FSLBGI}}) \equiv \left[\rho_2(n,E_{\mathrm{FSLBGI}}) - \left\{\frac{2eB}{h}\left(n+\frac{1}{2}\right)L_0^2\right\}\right]^{1/2}$$

and

$$T_{66}(n, E_{\mathrm{FSLBGI}}) \equiv \sum_{r=1}^{s} W(r)[T_{65}(n, E_{\mathrm{FSLBGI}})].$$

The use of (1.13) and (6.8) leads to the expression of the thermoelectric power under strong magnetic quantization as

$$G_0 = \left(\frac{\pi^2 k_B^2 T}{3e}\right) \frac{\sum_{n=0}^{n_{\max}} [\{T_{65}(n, E_{\mathrm{FSLBGI}})\}' + \{T_{66}(n, E_{\mathrm{FSLBGI}})\}']}{\sum_{n=0}^{n_{\max}} [T_{65}(n, E_{\mathrm{FSLBGI}}) + T_{66}(n, E_{\mathrm{FSLBGI}})]}. \quad (6.9)$$

6.2.4 Magnetothermopower in HgTe/CdTe Superlattices with Graded Interfaces

The magnetodispersion law in this case can be written following (3.19) as

$$k_z = \frac{1}{L_0}\left[\rho_3(n, E) - \left\{\frac{2eB}{h} L_0^2 \left(n + \frac{1}{2}\right)\right\}\right]^{1/2}, \quad (6.10)$$

where

$$\rho_3(n, E) = \left[\cos^{-1}\left\{\frac{1}{2}\psi_3(n, E)\right\}\right]^2,$$

$$\begin{aligned}
\psi_3(n, E) = \Bigg[& 2\cosh\{\beta_3(n, E)\}\cos\{\gamma_3(n, E)\} \\
& + \varepsilon_3(n, E)\sinh\{\beta_3(n, E)\}\sin\{\gamma_3(n, E)\} \\
& + \Delta_0\Bigg[\left(\frac{\{K_7(n, E)\}^2}{K_8(n, E)} - 3K_8(n, E)\right)\cosh\{\beta_3(n, E)\}\sin\{\gamma_3(n, E)\} \\
& + \left(3K_7(n, E) - \frac{\{K_8(n, E)\}^2}{K_7(n, E)}\right)\sinh\{\beta_3(n, E)\}\cos\{\gamma_3(n, E)\}\Bigg] \\
& + \Delta_0\Bigg[2\left(\{K_7(n, E)\}^2 - \{K_8(n, E)\}^2\right)\cosh\{\beta_3(n, E)\}\cos\{\gamma_3(n, E)\} \\
& + \frac{1}{12}\Bigg(\frac{5\{K_7(n, E)\}^3}{K_8(n, E)} + \frac{5\{K_8(n, E)\}^3}{K_7(n, E)} \\
& - \{34K_8(n, E)K_7(n, E)\}\Bigg)\sinh\{\beta_3(n, E)\}\sin\{\gamma_3(n, E)\}\Bigg]\Bigg],
\end{aligned}$$

$$\beta_3(n, E) \equiv K_7(n, E)[a_0 - \Delta_0],$$

$$K_7(n, E) \equiv \left[\left(\frac{2m_2^* E'}{h^2}\right) G(E - V_0, \alpha_2, \Delta_2) + \left\{\frac{2|e|B}{h}\left(n + \frac{1}{2}\right)\right\}\right]^{1/2},$$

$$\gamma_3(n, E) = K_8(n, E)[b_0 - \Delta_0],$$

$$K_8(n,E) \equiv \left[\frac{B_0^2 + 2AE - B_0\sqrt{B_0^2 + 4AE}}{2A^2} - \left\{ \frac{2|e|B}{h}\left(n + \frac{1}{2}\right) \right\} \right]^{1/2},$$

and

$$\varepsilon_3(n,E) \equiv \left[\frac{K_7(n,E)}{K_8(n,E)} - \frac{K_8(n,E)}{K_7(n,E)} \right].$$

The electron concentration per unit volume in this case can be expressed as

$$n_0 = \left(\frac{eBg_v}{\pi^2 h L_0} \right) \sum_{n=0}^{n_{\max}} [T_{67}(n, E_{\mathrm{FSLBGI}}) + T_{68}(n, E_{\mathrm{FSLBGI}})], \tag{6.11}$$

where

$$T_{67}(n, E_{\mathrm{FSLBSI}}) \equiv \left[\rho_3(n, E_{\mathrm{FSLBSI}}) - \left\{ \frac{2eB}{h}\left(n + \frac{1}{2}\right) L_0^2 \right\} \right]^{1/2}$$

and

$$T_{68}(n, E_{\mathrm{FSLBSI}}) \equiv \sum_{r=1}^{s} W(r)\,[T_{67}(n, E_{\mathrm{FSLBSI}})].$$

The use of (1.13) and (6.11) leads to the expression of the thermoelectric power under strong magnetic quantization as

$$G_0 = \left(\frac{\pi^2 k_B^2 T}{3e} \right) \frac{\sum_{n=0}^{n_{\max}} [\{T_{67}(n, E_{\mathrm{FSLBGI}})\}' + \{T_{68}(n, E_{\mathrm{FSLBGI}})\}']}{\sum_{n=0}^{n_{\max}} [T_{67}(n, E_{\mathrm{FSLBGI}}) + T_{68}(n, E_{\mathrm{FSLBGI}})]}. \tag{6.12}$$

6.2.5 Magnetothermopower in III–V Effective Mass Superlattices

In the presence of an external quantizing magnetic field along x-direction, the simplified magnetodispersion law in this case can be written following (3.23) as

$$k_x^2 = [\rho_4(n,E)] \tag{6.13}$$

in which

$$\rho_4(n,E) = \frac{1}{L_0^2}\left[\cos^{-1}(\bar{f}(n,E))\right]^2 - \left\{ \frac{2eB}{h}\left(n + \frac{1}{2}\right) \right\},$$

$$\bar{f}(n,E) = a_1 \cos[a_0 C_1(n,E) + b_0 D_1(n,E)] - a_2 \cos[a_0 C_1(n,E) - b_0 D_1(n,E)],$$

$$C_1(n, E) \equiv \left[\left(\frac{2m_1^* E}{h^2}\right) G(E, E_{g_1}, \Delta_1) - \left\{\frac{2eB}{h}\left(n + \frac{1}{2}\right)\right\}\right]^{1/2},$$

and

$$D_1(n, E) \equiv \left[\left(\frac{2m_2^* E}{h^2}\right) G(E, E_{g_2}, \Delta_2) - \left\{\frac{2eB}{h}\left(n + \frac{1}{2}\right)\right\}\right]^{1/2}.$$

The electron concentration per unit volume in this case can be expressed as

$$n_0 = \left(\frac{eBg_v}{\pi^2 h}\right) \sum_{n=0}^{n_{\max}} [T_{69}(n, E_{\mathrm{FSLBEM}}) + T_{610}(n, E_{\mathrm{FSLBEM}})], \tag{6.14}$$

where $T_{69}(n, E_{\mathrm{FSLBEM}}) \equiv [\rho_4(n, E_{\mathrm{FSLBEM}})]^{1/2}$, E_{FSLBEM} is the Fermi energy in the present case and $T_{610}(n, E_{\mathrm{FSLBEM}}) \equiv \sum_{r=1}^{s} W(r)\,[T_{69}(n, E_{\mathrm{FSLBEM}})]$.

The use of (1.13) and (6.14) leads to the expression of the thermoelectric power under strong magnetic quantization as

$$G_0 = \left(\frac{\pi^2 k_B^2 T}{3e}\right) \frac{\sum\limits_{n=0}^{n_{\max}} [\{T_{69}(n, E_{\mathrm{FSLBEM}})\}' + \{T_{610}(n, E_{\mathrm{FSLBEM}})\}']}{\sum\limits_{n=0}^{n_{\max}} [T_{69}(n, E_{\mathrm{FSLBEM}}) + T_{610}(n, E_{\mathrm{FSLBEM}})]}. \tag{6.15}$$

6.2.6 Magnetothermopower in II–VI Effective Mass Superlattices

Under magnetic quantization along z-direction, the simplified magnetodispersion law can be expressed following (3.27) as

$$k_z^2 = [\rho_5(n, E)] \tag{6.16}$$

in which

$$\rho_5(n, E) = \frac{1}{L_0^2}[\cos^{-1}(\bar{f}_1(n, E))]^2 - \left\{\frac{2eB}{h}\left(n + \frac{1}{2}\right)\right\},$$

$$\begin{aligned}\bar{f}_1(n, E) &= a_3 \cos[a_0 C_2(n, E) + b_0 D_2(n, E)] \\ &\quad - a_4 \cos[a_0 C_2(n, E) - b_0 D_2(n, E)],\end{aligned}$$

$$C_0(E, k_\perp) \equiv \left(\frac{2m_{\parallel,1}^*}{h^2}\right)^{1/2}\left[E - \frac{heB}{m_{\perp,1}^*}\left(n + \frac{1}{2}\right) \mp C_0\left\{\frac{2eB}{h}\left(n + \frac{1}{2}\right)\right\}^{1/2}\right]^{1/2},$$

and

$$D_2(n,E) \equiv \left[\left(\frac{2m_2^*}{h^2}\right) EG(E, E_{g_2}, \Delta_2) - \frac{2eB}{h}\left(n+\frac{1}{2}\right)\right]^{1/2}.$$

The electron concentration per unit volume in this case can be expressed as

$$n_0 = \left(\frac{eBg_v}{2\pi^2 h}\right)\sum_{n=0}^{n_{max}}[T_{611}(n, E_{FSLBEM}) + T_{612}(n, E_{FSLBEM})] \quad (6.17)$$

where

$$T_{611}(n, E_{FSLBEM}) \equiv [\rho_5(n, E_{FSLBEM})]^{1/2}$$

and

$$T_{612}(n, E_{FSLBEM}) \equiv \sum_{r=1}^{s} W(r)[T_{611}(n, E_{FSLBEM})].$$

Thus, using (1.13) and (6.17) leads to the expression of the thermoelectric power under strong magnetic quantization as

$$G_0 = \left(\frac{\pi^2 k_B^2 T}{3e}\right)\frac{\sum_{n=0}^{n_{max}}[\{T_{611}(n, E_{FSLBEM})\}' + \{T_{612}(n, E_{FSLBEM})\}']}{\sum_{n=0}^{n_{max}}[T_{611}(n, E_{FSLBEM}) + T_{612}(n, E_{FSLBEM})]}. \quad (6.18)$$

6.2.7 Magnetothermopower in IV–VI Effective Mass Superlattices

Thus, in the presence of a quantizing magnetic field along x-direction, the simplified magnetodispersion law in this case can be written following (3.31) as

$$k_x^2 = [\rho_6(n,E)] \quad (6.19)$$

in which

$$\rho_6(n,E) = \frac{1}{L_0^2}[\cos^{-1}(\bar{f}_2(n,E))]^2 - \left\{\frac{2eB}{h}\left(n+\frac{1}{2}\right)\right\},$$

$$\bar{f}_2(n,E) = a_5\cos[a_0C_3(n,E) + b_0D_3(n,E)] - a_6\cos[a_0C_3(n,E) - b_0D_3(n,E)],$$

$$C_3(n,E) \equiv \left[[EH_{11} + H_{21}(n)] - [E^2H_{31} + EH_{41}(n) + H_{51}(n)]^{1/2}\right]^{1/2},$$

and

$$D_3(n,E) \equiv \left[[EH_{12} + H_{22}(n)] - [E^2H_{32} + EH_{42}(n) + H_{52}(n)]^{1/2}\right]^{1/2}.$$

The electron concentration per unit volume in this case can be expressed as

$$n_0 = \left(\frac{eBg_v}{\pi^2 h}\right)\sum_{n=0}^{n_{\max}}[T_{613}(n, E_{FSLBEM}) + T_{614}(n, E_{FSLBEM})], \tag{6.20}$$

where

$$T_{613}(n, E_{FSLBEM}) \equiv [\rho_6(n, E_{FSLBEM})]^{1/2}$$

and

$$T_{614}(n, E_{FSLBEM}) \equiv \sum_{r=1}^{s} W(r)[T_{613}(n, E_{FSLBEM})].$$

Thus, using (1.13) and (6.20) leads to the expression of the thermoelectric power under strong magnetic quantization as

$$G_0 = \left(\frac{\pi^2 k_B^2 T}{3e}\right)\frac{\sum_{n=0}^{n_{\max}}[\{T_{613}(n, E_{FSLBEM})\}' + \{T_{614}(n, E_{FSLBEM})\}']}{\sum_{n=0}^{n_{\max}}[T_{613}(n, E_{FSLBEM}) + T_{614}(n, E_{FSLBEM})]}. \tag{6.21}$$

6.2.8 Magnetothermopower in HgTe/CdTe Effective Mass Superlattices

In the presence of an external magnetic field along x-direction, the simplified magnetodispersion law in this case can be written following (3.35) as

$$k_x^2 = [\rho_7(n,E)] \tag{6.22}$$

in which

$$\rho_7(n,E) = \left\{\frac{1}{L_0^2}[\cos^{-1}(\bar{f}_3(n,E))]^2\right\} - \left\{\frac{2eB}{h}\left(n+\frac{1}{2}\right)\right\},$$

$$\bar{f}_3(n,E) = a_7\cos[a_0C_4(n,E) + b_0D_4(n,E)] - a_8\cos[a_0C_4(n,E) - b_0D_4(n,E)],$$

$$C_4(n,E) \equiv \left[\frac{B_0^2 + 2AE - B_0\sqrt{B_0^2 + 4AE}}{2A^2} - \left\{\frac{2eB}{h}\left(n+\frac{1}{2}\right)\right\}\right]^{1/2},$$

and

$$D_4(n, E) \equiv \left[\left(\frac{2m_2^* E}{h^2} \right) G(E, E_{g_2}, \Delta_2) - \left\{ \frac{2eB}{h} \left(n + \frac{1}{2} \right) \right\} \right]^{1/2}.$$

The electron concentration per unit volume in this case can be expressed as

$$n_0 = \left(\frac{eBg_v}{\pi^2 h} \right) \sum_{n=0}^{n_{\max}} [T_{615}(n, E_{\text{FSLBEM}}) + T_{616}(n, E_{\text{FSLBEM}})], \tag{6.23}$$

where

$$T_{615}(n, E_{\text{FSLBEM}}) \equiv [\rho_6(n, E_{\text{FSLBEM}})]^{1/2}$$

and

$$T_{616}(n, E_{\text{FSLBEM}}) \equiv \sum_{r=1}^{s} W(r) \, [T_{615}(n, E_{\text{FSLBEM}})].$$

The use of (1.13) and (6.23) leads to the expression of the thermoelectric power under strong magnetic quantization as

$$G_0 = \left(\frac{\pi^2 k_B^2 T}{3e} \right) \frac{\sum_{n=0}^{n_{\max}} [\{T_{615}(n, E_{\text{FSLBEM}})\}' + \{T_{616}(n, E_{\text{FSLBEM}})\}']}{\sum_{n=0}^{n_{\max}} [T_{615}(n, E_{\text{FSLBEM}}) + T_{616}(n, E_{\text{FSLBEM}})]}. \tag{6.24}$$

6.2.9 *Magnetothermopower in III–V Quantum Well Superlattices with Graded Interfaces*

The electron dispersion law in this case can be written following (6.1) as

$$\left(\frac{n_z \pi}{d_z} \right) = \frac{1}{L_0} \left[\rho(n, E_{61}) - \left\{ \frac{2eB}{h} L_0^2 \left(n + \frac{1}{2} \right) \right\} \right]^{1/2}, \tag{6.25}$$

where E_{61} is the totally quantized energy in this case and $\rho(n, E_{61})$ in this case should be obtained by replacing E by E_{61} in the definition of $\rho(n, E)$ as given below (6.1).

The electron concentration per unit volume in this case assumes the form

$$n_0 = \frac{eBg_v}{\pi h} \sum_{n_z=1}^{n_{z\max}} \sum_{n=0}^{n_{\max}} F_{-1}(\eta_{61}), \tag{6.26}$$

where $\eta_{61} \equiv (k_B T)^{-1}(E_{\text{FSLBQWGI}} - E_{61})$ and E_{FSLBQWGI} is the Fermi energy in the present case.

Therefore, using (1.13) and (6.26), the TPSM in this case is given by

$$G_0 = \frac{\pi^2 k_B}{3e}\left[\sum_{n_z=1}^{n_{z\max}}\sum_{n=0}^{n_{\max}} F_{-1}(\eta_{61})\right]^{-1}\left[\sum_{n_z=1}^{n_{z\max}}\sum_{n=0}^{n_{\max}} F_{-2}(\eta_{61})\right]. \tag{6.27}$$

6.2.10 Magnetothermopower in II–VI Quantum Well Superlattices with Graded Interfaces

The electron energy spectrum under magnetic quantization in II–VI quantum well superlattices with graded interfaces can be expressed following (6.4) as

$$\left(\frac{n_z\pi}{d_z}\right)^2 = \frac{1}{L_0^2}\left[\rho_1(n, E_{62}) - \left\{\frac{2eB}{h}L_0^2\left(n+\frac{1}{2}\right)\right\}\right], \tag{6.28}$$

where E_{62} is the totally quantized energy in this case and $\rho_1\,(n, E_{62})$ in this case should be obtained by replacing E by E_{62} in the definition of $\rho_1(n, E)$ as given below (6.4).

The electron concentration per unit volume in this case assumes the form

$$n_0 = \frac{eBg_v}{\pi h}\sum_{n_z=1}^{n_{z\max}}\sum_{n=0}^{n_{\max}} F_{-1}(\eta_{62}), \tag{6.29}$$

where $\eta_{62} \equiv (k_B T)^{-1}(E_{FSLBQWGI} - E_{62})$.

Therefore, using (1.13) and (6.29), the TPSM in this case is given by

$$G_0 = \frac{\pi^2 k_B}{3e}\left[\sum_{n_z=1}^{n_{z\max}}\sum_{n=0}^{n_{\max}} F_{-1}(\eta_{62})\right]^{-1}\left[\sum_{n_z=1}^{n_{z\max}}\sum_{n=0}^{n_{\max}} F_{-2}(\eta_{62})\right]. \tag{6.30}$$

6.2.11 Magnetothermopower in IV–VI Quantum Well Superlattices with Graded Interfaces

The electron energy spectrum can be expressed following (6.7) as

$$\left(\frac{\pi n_z}{d_z}\right) = \frac{1}{L_0}\left[\rho_2(n, E_{63}) - \left\{\frac{2eB}{h}L_0^2\left(n+\frac{1}{2}\right)\right\}\right]^{1/2}, \tag{6.31}$$

where E_{63} is the totally quantized energy in this case and $\rho_2(n, E_{63})$ in this case should be obtained by replacing E by E_{63} in the definition of $\rho_2(n, E_{63})$ as given below (6.7).

The electron concentration per unit volume in this case assumes the form

$$n_0 = \frac{eBg_v}{\pi h} \sum_{n_z=1}^{n_{z\max}} \sum_{n=0}^{n_{\max}} F_{-1}(\eta_{63}), \tag{6.32}$$

where $\eta_{63} \equiv (k_B T)^{-1}(E_{FSLBQWGI} - E_{63})$.

Therefore, using (1.13) and (6.32), the TPSM in this case is given by

$$G_0 = \frac{\pi^2 k_B}{3e} \left[\sum_{n_z=1}^{n_{z\max}} \sum_{n=0}^{n_{\max}} F_{-1}(\eta_{63}) \right]^{-1} \left[\sum_{n_z=1}^{n_{z\max}} \sum_{n=0}^{n_{\max}} F_{-2}(\eta_{63}) \right]. \tag{6.33}$$

6.2.12 *Magnetothermopower in HgTe/CdTe Quantum Well Superlattices with Graded Interfaces*

Following (6.10), the electron dispersion law in this case assumes the form

$$\left(\frac{n_z \pi}{d_z}\right) = \frac{1}{L_0} \left[\rho_3(n, E_{64}) - \left\{ \frac{2eB}{h} L_0^2 \left(n + \frac{1}{2}\right) \right\} \right]^{1/2}, \tag{6.34}$$

where E_{64} is the totally quantized energy in this case and $\rho_3(n, E_{64})$ in this case should be obtained by replacing E by E_{64} in the definition of $\rho_3(n, E_{64})$ as given below (6.10).

The electron concentration per unit volume in this case can be expressed as

$$n_0 = \frac{eBg_v}{\pi h} \sum_{n_z=1}^{n_{z\max}} \sum_{n=0}^{n_{\max}} F_{-1}(\eta_{64}), \tag{6.35}$$

where $\eta_{64} \equiv (k_B T)^{-1}(E_{FSLBQWGI} - E_{64})$.

Therefore, using (1.13) and (6.35), the TPSM in this case is given by

$$G_0 = \frac{\pi^2 k_B}{3e} \left[\sum_{n_z=1}^{n_{z\max}} \sum_{n=0}^{n_{\max}} F_{-1}(\eta_{64}) \right]^{-1} \left[\sum_{n_z=1}^{n_{z\max}} \sum_{n=0}^{n_{\max}} F_{-2}(\eta_{64}) \right]. \tag{6.36}$$

6.2.13 *Magnetothermopower in III–V Quantum Well-Effective Mass Superlattices*

The electron energy spectrum under magnetic quantization in III–V quantum well-effective mass superlattices can be written following (6.13) as

$$\left(\frac{\pi n_x}{d_x}\right)^2 = [\rho_4(n, E_{65})], \tag{6.37}$$

where E_{65} is the totally quantized energy in this case and $\rho_4(n, E_{65})$ in this case should be obtained by replacing E by E_{65} in the definition of $\rho_4(n, E_{65})$ as given below (6.13).

The electron concentration per unit volume in this case assumes the form

$$n_0 = \frac{eBg_v}{\pi h} \sum_{n=0}^{n_{\max}} \sum_{n_x=1}^{n_{x\max}} F_{-1}(\eta_{65}), \tag{6.38}$$

where $\eta_{65} \equiv (k_B T)^{-1}(E_{FSLBQWEM} - E_{65})$ and $E_{FSLBQWEM}$ is the Fermi energy in the present case.

Therefore, using (1.13) and (6.38), the TPSM in this case is given by

$$G_0 = \frac{\pi^2 k_B}{3e} \left[\sum_{n=0}^{n_{\max}} \sum_{n_x=1}^{n_{x\max}} F_{-1}(\eta_{65})\right]^{-1} \left[\sum_{n=0}^{n_{\max}} \sum_{n_x=1}^{n_{x\max}} F_{-2}(\eta_{65})\right]. \tag{6.39}$$

6.2.14 *Magnetothermopower in II–VI Quantum Well-Effective Mass Superlattices*

The electron energy spectrum in II–VI quantum well-effective mass superlattices in the presence of a quantizing magnetic field can be expressed following (6.16) as

$$\left(\frac{n_z \pi}{d_z}\right)^2 = [\rho_5(n, E_{66})], \tag{6.40}$$

where E_{66} is the totally quantized energy in this case and $\rho_5(n, E_{66})$ in this case should be obtained by replacing E by E_{66} in the definition of $\rho_5(n, E_{66})$ as given below (6.16).

The electron concentration per unit volume in this case assumes the form

$$n_0 = \frac{eBg_v}{\pi h} \sum_{n=0}^{n_{\max}} \sum_{n_z=1}^{n_{z\max}} F_{-1}(\eta_{66}), \tag{6.41}$$

where $\eta_{66} \equiv (k_B T)^{-1}(E_{FSLBQWEM} - E_{66})$.

Therefore, using (1.13) and (6.41), the TPSM in this case is given by

$$G_0 = \frac{\pi^2 k_B}{3e}\left[\sum_{n=0}^{n_{\max}}\sum_{n_z=1}^{n_{z\max}} F_{-1}(\eta_{66})\right]^{-1}\left[\sum_{n=0}^{n_{\max}}\sum_{n_z=1}^{n_{z\max}} F_{-2}(\eta_{66})\right]. \quad (6.42)$$

6.2.15 *Magnetothermopower in IV–VI Quantum Well-Effective Mass Superlattices*

The electron energy spectrum in IV–VI quantum well-effective mass superlattices in the presence of a quantizing magnetic field can be expressed following (6.16) as

$$\left(\frac{\pi n_x}{d_x}\right)^2 = [\rho_6(n, E_{67})], \quad (6.43)$$

where E_{67} is the totally quantized energy in this case, and $\rho_6(n, E_{67})$ in this case should be obtained by replacing E by E_{67} in the definition of $\rho_6(n, E_{67})$ as given below (6.19).

The electron concentration per unit volume in this case assumes the form

$$n_0 = \frac{eBg_v}{\pi h}\sum_{n=0}^{n_{\max}}\sum_{n_z=1}^{n_{z\max}} F_{-1}(\eta_{67}), \quad (6.44)$$

where $\eta_{67} \equiv (k_B T)^{-1}(E_{FSLBQWEM} - E_{67})$.

Therefore, using (1.13) and (6.44), the TPSM in this case is given by

$$G_0 = \frac{\pi^2 k_B}{3e}\left[\sum_{n=0}^{n_{\max}}\sum_{n_z=1}^{n_{z\max}} F_{-1}(\eta_{67})\right]^{-1}\left[\sum_{n=0}^{n_{\max}}\sum_{n_z=1}^{n_{z\max}} F_{-2}(\eta_{67})\right]. \quad (6.45)$$

6.2.16 *Magnetothermopower in HgTe/CdTe Quantum Well-Effective Mass Superlattices*

The electron energy spectrum in IV–VI quantum well-effective mass superlattices in the presence of a quantizing magnetic field can be expressed following (6.16) as

$$\left(\frac{\pi n_x}{d_x}\right)^2 = [\rho_7(n, E_{68})], \quad (6.46)$$

where E_{68} is the totally quantized energy in this case and $\rho_7(n, E_{68})$ in this case should be obtained by replacing E by E_{68} in the definition of $\rho_7(n, E_{68})$ as given below (6.22).

The electron concentration per unit volume in this case (n_0) assumes the form

$$n_0 = \frac{eBg_v}{\pi h} \sum_{n=0}^{n_{\max}} \sum_{n_z=1}^{n_{z\max}} F_{-1}(\eta_{68}), \tag{6.47}$$

where $\eta_{68} \equiv (k_B T)^{-1}(E_{\text{FSLBQWEM}} - E_{68})$.

Therefore, using (1.13) and (6.47), the TPSM in this case is given by

$$G_0 = \frac{\pi^2 k_B}{3e} \left[\sum_{n=0}^{n_{\max}} \sum_{n_z=1}^{n_{z\max}} F_{-1}(\eta_{68}) \right]^{-1} \left[\sum_{n=0}^{n_{\max}} \sum_{n_z=1}^{n_{z\max}} F_{-2}(\eta_{68}) \right]. \tag{6.48}$$

6.3 Results and Discussion

Using Table 6.1, we have plotted in Figs. 6.1 and 6.2 the magnetothermoelectric power as functions of inverse quantizing magnetic field and impurity concentration, respectively, for (a) HgTe/CdTe (using (6.11) and (6.12)), (b) PbTe/PbSnTe (using (6.8) and (6.9)), (c) CdS/CdTe (using (6.5) and (6.6)), and (d) GaAs/$Ga_{1-x}Al_xAs$ (using (6.2) and (6.3)) superlattices with graded interfaces. With decreasing magnetic field intensity, the thermoelectric power increases periodically as a result of SdH periodicity.

However, with increasing impurity concentration, the thermoelectric power increases to some extent exhibiting spikes for higher values, a result which already been discussed in Chap. 5. It appears that the TPSM is lower in magnitude for GaAs/$Ga_{1-x}Al_xAs$ and higher in magnitude for HgTe/CdTe for all the cases. In Figs. 6.3 and 6.4, the magnetothermoelectric power as functions of inverse quantizing magnetic field and impurity concentration for (a) HgTe/CdTe (using (6.23) and (6.24)), (b) PbTe/PbSnTe (using (6.20) and (6.21)), (c) CdS/CdTe (using (6.17) and (6.18)), and (d) GaAs/$Ga_{1-x}Al_xAs$ (using (6.14) and (6.15)) effective mass superlattices structures. The concentration has been fixed at a value 10^{22}m^{-3} for varying magnetic field intensity while 10 T was fixed for varying impurity concentration. With decreasing magnetic field intensity, the thermoelectric power increases periodically as a result of SdH periodicity. However, with increasing impurity concentration, the thermoelectric power decreases.

In Figs. 6.5 and 6.6, the magnetothermoelectric power as functions of film thickness and 2D carrier concentration for HgTe/CdTe (using (6.35) and (6.36)), PbTe/PbSnTe (using (6.32) and (6.33)), CdS/CdTe (using (6.29) and (6.30)), and (d) GaAs/$Ga_{1-tx}Al_xAs$ (using (6.26) and (6.27)) for QWSLs with graded interfaces. It appears that the TPSM in this case signatures an increasing step like variation

Table 6.1 The carrier statistics and the thermoelectric power under magnetic quantization in III–V, II–VI, IV–VI, and HgTe/CdTe superlattices with graded interfaces; III–V, II–VI, IV–VI, and HgTe/CdTe effective mass superlattices; III–V, II–VI, IV–VI, and HgTe/CdTe quantum well superlattices with graded interfaces; and III–V, II–VI, IV–VI, and HgTe/CdTe quantum well-effective mass superlattices

Type of materials	Carrier statistics	TPSM
1. III–V superlattices with graded interfaces	$n_0 = \left(\frac{eBg_v}{\pi^2 hL_0}\right)\sum_{n=0}^{n_{max}}[T_{61}(n, E_{FSLBGI}) + T_{62}(n, E_{FSLBGI})]$ (6.2)	On the basis of (6.2), $G_0 = \left(\frac{\pi^2 k_B^2 T}{3e}\right)\frac{\sum_{n=0}^{n_{max}}[\{T_{61}(n, E_{FSLBGI})\}' + \{T_{62}(n, E_{FSLBGI})\}']}{\sum_{n=0}^{n_{max}}[T_{61}(n, E_{FSLBGI}) + T_{62}(n, E_{FSLBGI})]}$ (6.3)
2. II–VI superlattices with graded interfaces	$n_0 = \left(\frac{eBg_v}{2\pi^2 hL_0}\right)\sum_{n=0}^{n_{max}}[T_{63}(n, E_{FSLBGI}) + T_{64}(n, E_{FSLBGI})]$ (6.5)	On the basis of (6.5), $G_0 = \left(\frac{\pi^2 k_B^2 T}{3e}\right)\frac{\sum_{n=0}^{n_{max}}[\{T_{63}(n, E_{FSLBGI})\}' + \{T_{64}(n, E_{FSLBGI})\}']}{\sum_{n=0}^{n_{max}}[T_{63}(n, E_{FSLBGI}) + T_{64}(n, E_{FSLBGI})]}$ (6.6)
3. IV–VI superlattices with graded interfaces	$n_0 = \left(\frac{eBg_v}{\pi^2 hL_0}\right)\sum_{n=0}^{n_{max}}[T_{65}(n, E_{FSLBGI}) + T_{66}(n, E_{FSLBGI})]$ (6.8)	On the basis of (6.8), $G_0 = \left(\frac{\pi^2 k_B^2 T}{3e}\right)\frac{\sum_{n=0}^{n_{max}}[\{T_{65}(n, E_{FSLBGI})\}' + \{T_{66}(n, E_{FSLBGI})\}']}{\sum_{n=0}^{n_{max}}[T_{65}(n, E_{FSLBGI}) + T_{66}(n, E_{FSLBGI})]}$ (6.9)
4. HgTe/CdTe superlattices with graded interfaces	$n_0 = \left(\frac{eBg_v}{\pi^2 hL_0}\right)\sum_{n=0}^{n_{max}}[T_{67}(n, E_{FSLBGI}) + T_{68}(n, E_{FSLBGI})]$ (6.11)	On the basis of (6.11), $G_0 = \left(\frac{\pi^2 k_B^2 T}{3e}\right)\frac{\sum_{n=0}^{n_{max}}[\{T_{67}(n, E_{FSLBGI})\}' + \{T_{68}(n, E_{FSLBGI})\}']}{\sum_{n=0}^{n_{max}}[T_{67}(n, E_{FSLBGI}) + T_{68}(n, E_{FSLBGI})]}$ (6.12)

(continued)

Table 6.1 (Continued)

Type of materials	Carrier statistics	TPSM
5. III–V effective mass superlattice	$n_0 = \left(\frac{eBg_v}{\pi^2 h}\right) \sum_{n=0}^{n_{max}} [T_{69}(n, E_{FSLBEM}) + T_{610}(n, E_{FSLBEM})]$ (6.14)	On the basis of (6.14), $G_0 = \left(\frac{\pi^2 k_B^2 T}{3e}\right) \frac{\sum_{n=0}^{n_{max}} [\{T_{69}(n, E_{FSLBEM})\}' + \{T_{610}(n, E_{FSLBEM})\}']}{\sum_{n=0}^{n_{max}} [T_{69}(n, E_{FSLBEM}) + T_{610}(n, E_{FSLBEM})]}$ (6.15)
6. II–VI effective mass superlattice	$n_0 = \left(\frac{eBg_v}{2\pi^2 h}\right) \sum_{n=0}^{n_{max}} [T_{611}(n, E_{FSLBEM}) + T_{612}(n, E_{FSLBEM})]$ (6.17)	On the basis of (6.17), $G_0 = \left(\frac{\pi^2 k_B^2 T}{3e}\right) \frac{\sum_{n=0}^{n_{max}} [\{T_{611}(n, E_{FSLBEM})\}' + \{T_{612}(n, E_{FSLBEM})\}']}{\sum_{n=0}^{n_{max}} [T_{611}(n, E_{FSLBEM}) + T_{612}(n, E_{FSLBEM})]}$ (6.18)
7. IV–VI effective mass superlattice	$n_0 = \left(\frac{eBg_v}{\pi^2 h}\right) \sum_{n=0}^{n_{max}} [T_{613}(n, E_{FSLBEM}) + T_{614}(n, E_{FSLBEM})]$ (6.20)	On the basis of (6.20), $G_0 = \left(\frac{\pi^2 k_B^2 T}{3e}\right) \frac{\sum_{n=0}^{n_{max}} [\{T_{613}(n, E_{FSLBEM})\}' + \{T_{614}(n, E_{FSLBEM})\}']}{\sum_{n=0}^{n_{max}} [T_{613}(n, E_{FSLBEM}) + T_{614}(n, E_{FSLBEM})]}$ (6.21)
8. HgTe/CdTe effective mass superlattices	$n_0 = \left(\frac{eBg_v}{\pi^2 h}\right) \sum_{n=0}^{n_{max}} [T_{615}(n, E_{FSLBEM}) + T_{616}(n, E_{FSLBEM})]$ (6.23)	On the basis of (6.23), $G_0 = \left(\frac{\pi^2 k_B^2 T}{3e}\right) \frac{\sum_{n=0}^{n_{max}} [\{T_{615}(n, E_{FSLBEM})\}' + \{T_{616}(n, E_{FSLBEM})\}']}{\sum_{n=0}^{n_{max}} [T_{615}(n, E_{FSLBEM}) + T_{616}(n, E_{FSLBEM})]}$ (6.24)
9. III–V quantum well superlattices with graded interfaces	$n_0 = \frac{eBg_v}{\pi h} \sum_{n_z=1}^{n_{z\max}} \sum_{n=0}^{n_{max}} F_{-1}(\eta_{61})$ (6.26)	On the basis of (6.26), $G_0 = \frac{\pi^2 k_B}{3e} \left[\sum_{n_z=1}^{n_{z\max}} \sum_{n=0}^{n_{max}} F_{-1}(\eta_{61})\right]^{-1} \left[\sum_{n_z=1}^{n_{z\max}} \sum_{n=0}^{n_{max}} F_{-2}(\eta_{61})\right]$ (6.27)

10. II–VI quantum well superlattices with graded interfaces	$n_0 = \frac{eBg_v}{\pi h} \sum_{n_z=1}^{n_{z\max}} \sum_{n=0}^{n_{\max}} F_{-1}(\eta_{62})$ (6.29)	On the basis of (6.29), $G_0 = \frac{\pi^2 k_B}{3e} \left[\sum_{n_z=1}^{n_{z\max}} \sum_{n=0}^{n_{\max}} F_{-1}(\eta_{62}) \right]^{-1} \left[\sum_{n_z=1}^{n_{z\max}} \sum_{n=0}^{n_{\max}} F_{-2}(\eta_{62}) \right]$ (6.30)
11. IV–VI quantum well superlattices with graded interfaces	$n_0 = \frac{eBg_v}{\pi h} \sum_{n_z=1}^{n_{z\max}} \sum_{n=0}^{n_{\max}} F_{-1}(\eta_{63})$ (6.32)	On the basis of (6.32), $G_0 = \frac{\pi^2 k_B}{3e} \left[\sum_{n_z=1}^{n_{z\max}} \sum_{n=0}^{n_{\max}} F_{-1}(\eta_{63}) \right]^{-1} \left[\sum_{n_z=1}^{n_{z\max}} \sum_{n=0}^{n_{\max}} F_{-2}(\eta_{63}) \right]$ (6.33)
12. HgTe/CdTe quantum well superlattices with graded interfaces	$n_0 = \frac{eBg_v}{\pi h} \sum_{n_z=1}^{n_{z\max}} \sum_{n=0}^{n_{\max}} F_{-1}(\eta_{64})$ (6.35)	On the basis of (6.35), $G_0 = \frac{\pi^2 k_B}{3e} \left[\sum_{n_z=1}^{n_{z\max}} \sum_{n=0}^{n_{\max}} F_{-1}(\eta_{64}) \right]^{-1} \left[\sum_{n_z=1}^{n_{z\max}} \sum_{n=0}^{n_{\max}} F_{-2}(\eta_{64}) \right]$ (6.36)
13. III–V quantum well-effective mass superlattice	$n_0 = \frac{eBg_v}{\pi h} \sum_{n=0}^{n_{\max}} \sum_{n_x=1}^{n_{x\max}} F_{-1}(\eta_{65})$ (6.38)	On the basis of (6.38), $G_0 = \frac{\pi^2 k_B}{3e} \left[\sum_{n=0}^{n_{\max}} \sum_{n_x=1}^{n_{x\max}} F_{-1}(\eta_{65}) \right]^{-1} \left[\sum_{n=0}^{n_{\max}} \sum_{n_x=1}^{n_{x\max}} F_{-2}(\eta_{65}) \right]$ (6.39)

(continued)

Table 6.1 (Continued)

Type of materials	Carrier statistics	TPSM
14. II–VI quantum well-effective mass superlattice	$n_0 = \frac{eBg_v}{\pi h} \sum_{n=0}^{n_{max}} \sum_{n_z=1}^{n_{zmax}} F_{-1}(\eta_{66})$ (6.41)	On the basis of (6.41), $G_0 = \frac{\pi^2 k_B}{3e} \left[\sum_{n=0}^{n_{max}} \sum_{n_z=1}^{n_{zmax}} F_{-1}(\eta_{66}) \right]^{-1} \left[\sum_{n=0}^{n_{max}} \sum_{n_z=1}^{n_{zmax}} F_{-2}(\eta_{66}) \right]$ (6.42)
15. IV–VI quantum well-effective mass superlattice	$n_0 = \frac{eBg_v}{\pi h} \sum_{n=0}^{n_{max}} \sum_{n_z=1}^{n_{zmax}} F_{-1}(\eta_{67})$ (6.44)	On the basis of (6.44), $G_0 = \frac{\pi^2 k_B}{3e} \left[\sum_{n=0}^{n_{max}} \sum_{n_z=1}^{n_{zmax}} F_{-1}(\eta_{67}) \right]^{-1} \left[\sum_{n=0}^{n_{max}} \sum_{n_z=1}^{n_{zmax}} F_{-2}(\eta_{67}) \right]$ (6.45)
16. HgTe/CdTe quantum well-effective mass superlattices	$n_0 = \frac{eBg_v}{\pi h} \sum_{n=0}^{n_{max}} \sum_{n_z=1}^{n_{zmax}} F_{-1}(\eta_{68})$ (6.47)	On the basis of (6.47), $G_0 = \frac{\pi^2 k_B}{3e} \left[\sum_{n=0}^{n_{max}} \sum_{n_z=1}^{n_{zmax}} F_{-1}(\eta_{68}) \right]^{-1} \left[\sum_{n=0}^{n_{max}} \sum_{n_z=1}^{n_{zmax}} F_{-2}(\eta_{68}) \right]$ (6.48)

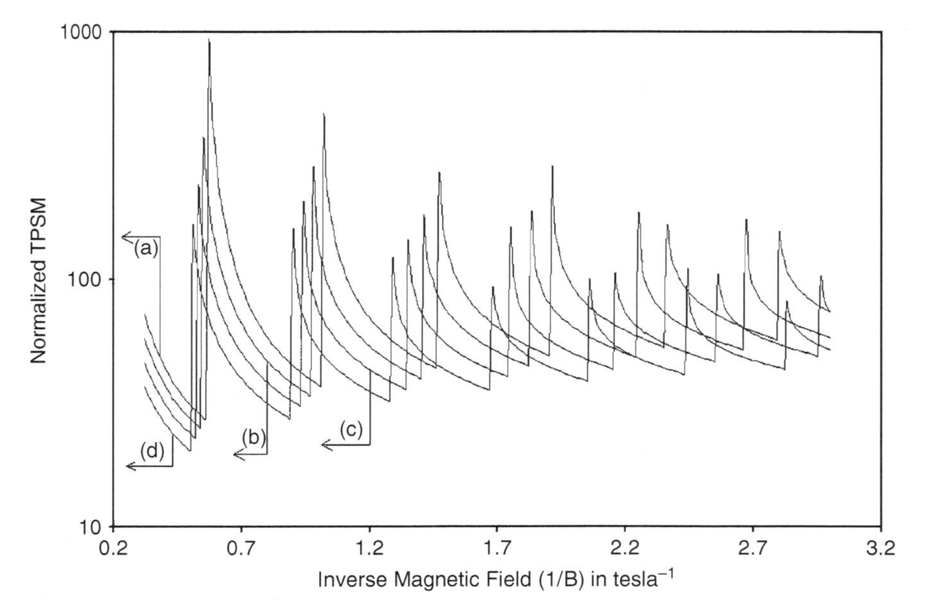

Fig. 6.1 The plot of the TPSM as a function of inverse quantizing magnetic field for (**a**) HgTe/CdTe, (**b**) PbTe/PbSnTe, (**c**) CdS/CdTe, and (**d**) GaAs/$Ga_{1-tx}Al_xAs$ and superlattices with graded interfaces

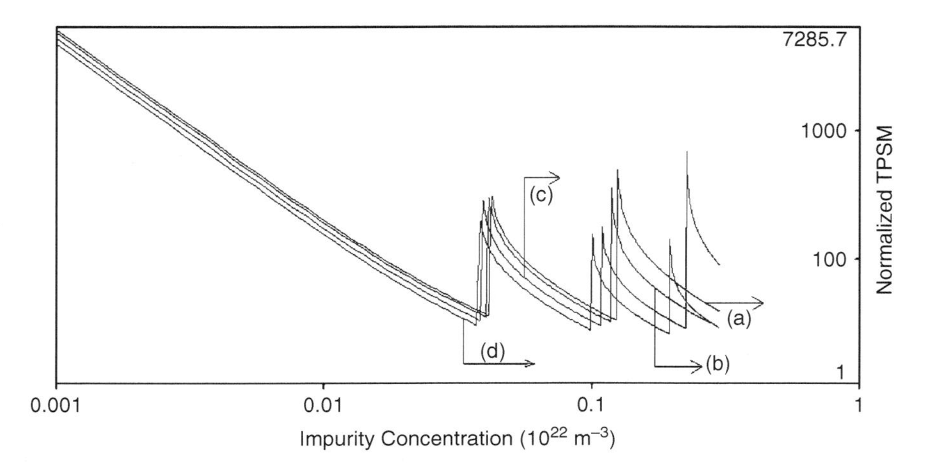

Fig. 6.2 Plot of the TPSM as a function of impurity concentration for all the cases of Fig. 6.1

with increasing film thickness and decreases with increasing 2D carrier concentration. In Figs. 6.7 and 6.8, the magnetothermoelectric power as function of film thickness and 2D carrier concentration for HgTe/CdTe (using (6.47) and (6.48)), PbTe/PbSnTe (using (6.44) and (6.45)), CdS/CdTe (using (6.41) and (6.42)), and (d) GaAs/$Ga_{1-tx}Al_xAs$ (using (6.38) and (6.39)) for QW effective mass SLs. It

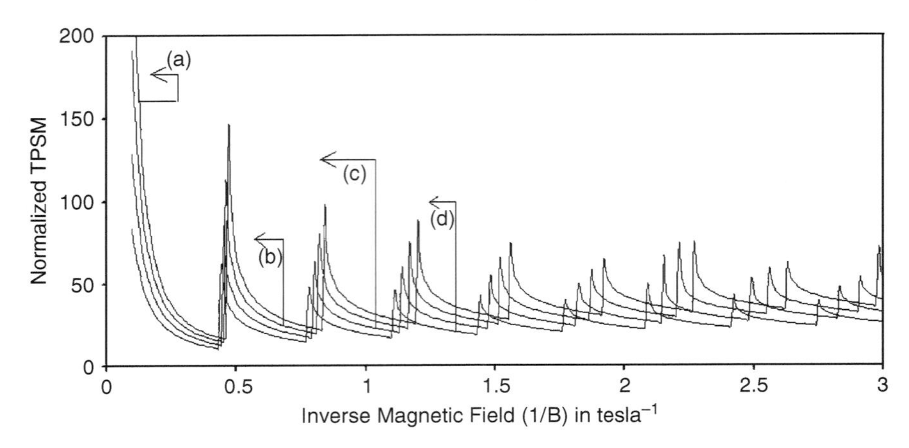

Fig. 6.3 Plot of the TPSM as a function of inverse quantizing magnetic field for (**a**) HgTe/CdTe, (**b**) PbTe/PbSnTe, (**c**) CdS/CdTe, and (**d**) GaAs/$Ga_{1-tx}Al_xAs$ effective mass superlattices

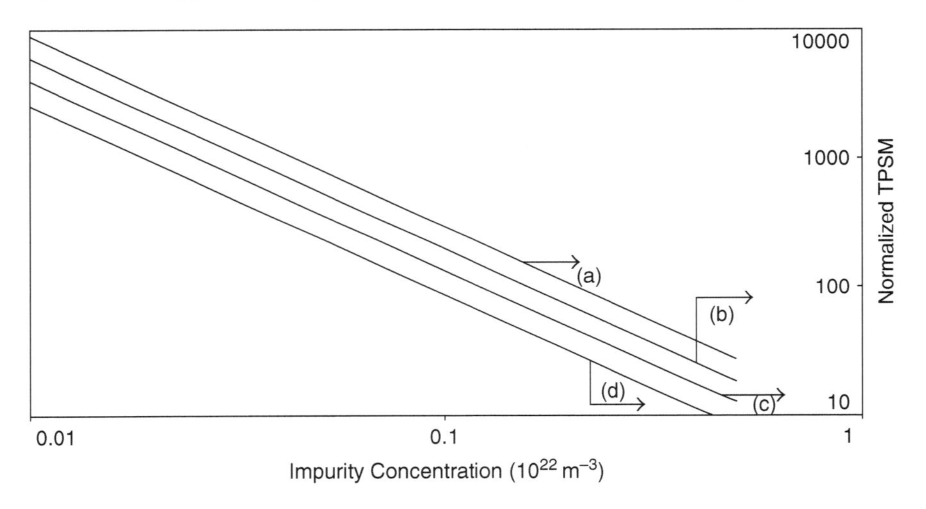

Fig. 6.4 Plot of the TPSM as a function of impurity concentration for all the cases of Fig. 6.3

appears from Figs. 6.7 and 6.8 and Figs. 6.5 and 6.6 that the nature of variations of the TPSM for all types of QW effective mass SLs does not differ widely as compared with the corresponding QWSLs with graded interfaces. For the purpose of condensed presentation the carrier concentration and the corresponding TPSM for this chapter have been presented in Table 6.1.

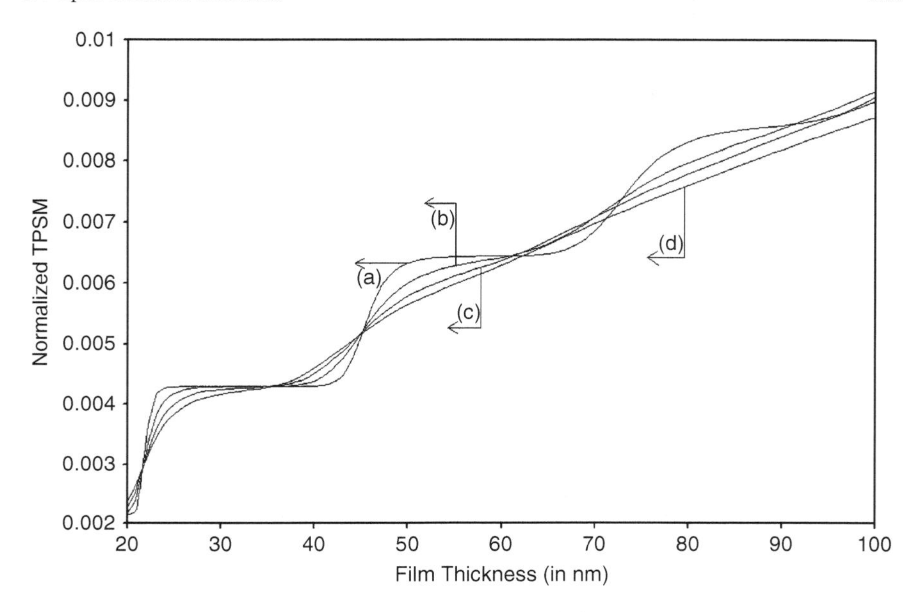

Fig. 6.5 Plot of the normalized TPSM as a function of film thickness for (**a**) HgTe/CdTe, (**b**) PbTe/PbSnTe, (**c**) CdS/CdTe, and (**d**) $GaAs/Ga_{1-x}Al_xAs$ quantum well superlattices with graded interfaces

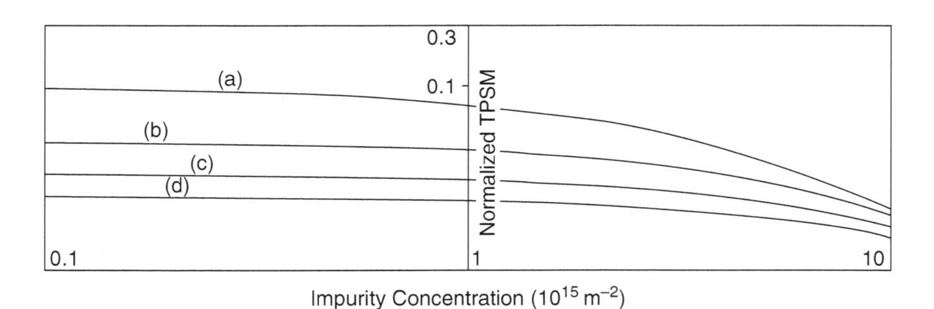

Fig. 6.6 Plot of the normalized TPSM as a function of impurity concentration for all the cases of Fig. 6.6

6.4 Open Research Problems

(R6.1) Investigate the DTP, PTP, and Z in the absence of magnetic field by considering all types of scattering mechanisms for III–V, II–VI, IV–VI, and HgTe/CdTe quantum well and quantum wire superlattices with graded interfaces and also the effective mass superlattices of the aforementioned materials.

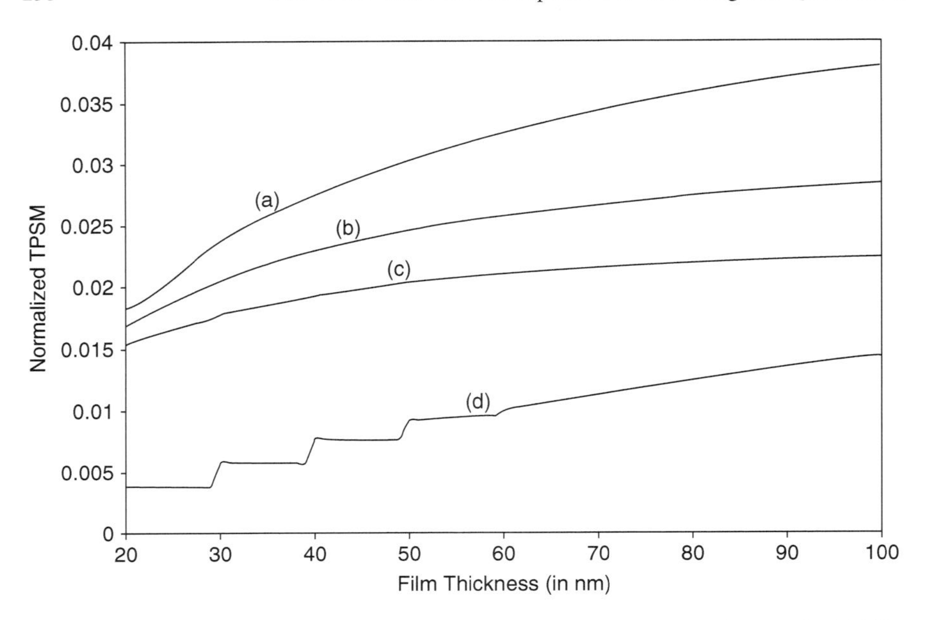

Fig. 6.7 Plot of the normalized TPSM as a function of film thickness for (**a**) HgTe/CdTe, (**b**) PbTe/PbSnTe, (**c**) CdS/CdTe, and (**d**) GaAs/$Ga_{1-tx}Al_xAs$ quantum well-effective mass superlattices

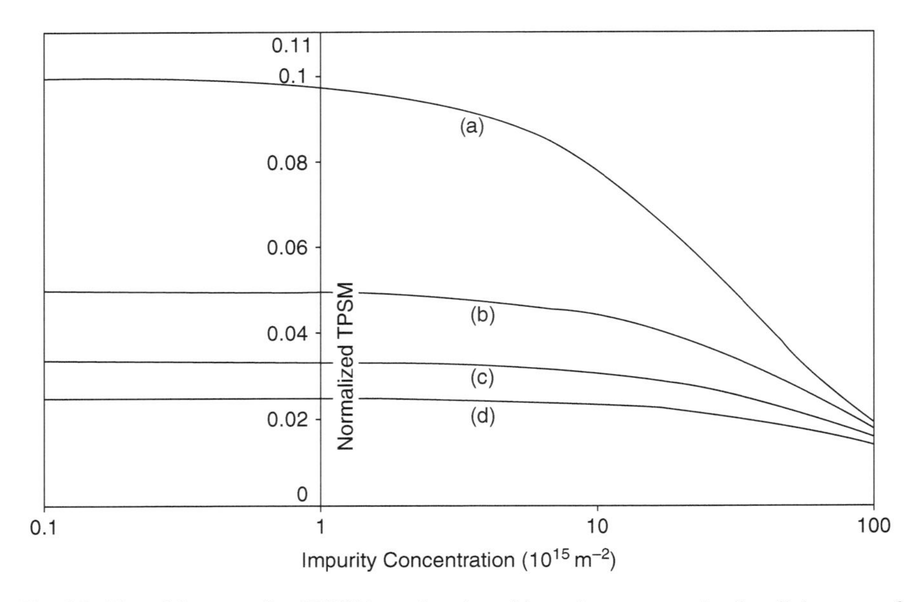

Fig. 6.8 Plot of the normalized TPSM as a function of impurity concentration for all the cases of Fig. 6.7

(R6.2) Investigate the DTP, PTP, and Z in the absence of magnetic field by considering all types of scattering mechanisms for strained layer, random, short period and Fibonacci, polytype and saw-tooth quantum well and quantum wire superlattices.

(R6.3) Investigate the DTP, PTP, and Z in the presence of an arbitrarily oriented quantizing magnetic field in the presence of spin and broadening by considering all types of scattering mechanisms for (R6.1) and (R6.2) under an arbitrarily oriented (a) nonuniform electric field and (b) alternating electric field, respectively.

(R6.4) Investigate the DTP, PTP, and Z by considering all types of scattering mechanisms for (R6.1) and (R6.2) under an arbitrarily oriented alternating magnetic field by including broadening and the electron spin, respectively.

(R6.5) Investigate the DTP, PTP, and Z by considering all types of scattering mechanisms for (R6.1) and (R6.2) under an arbitrarily oriented quantizing alternating magnetic field and crossed alternating electric field by including broadening and the electron spin, respectively.

(R6.6) Investigate the DTP, PTP, and Z by considering all types of scattering mechanisms for (R6.1) and (R6.2) under an arbitrarily oriented alternating quantizing magnetic field and crossed alternating nonuniform electric field by including broadening and the electron spin, respectively.

(R6.7) Investigate the DTP, PTP, and Z in the absence of magnetic field for all types of quantum well and quantum wire superlattices as considered in this chapter under exponential, Kane, Halperin, Lax, and Bonch-Bruevich band tails [2], respectively.

(R6.8) Investigate the DTP, PTP, and Z in the presence of quantizing magnetic field including spin and broadening for the problem as defined in (R6.7) under an arbitrarily oriented (a) non-uniform electric field and (b) alternating electric field, respectively.

(R6.9) Investigate the DTP, PTP, and Z for the problem as defined in (R6.7) under an arbitrarily oriented alternating quantizing magnetic field by including broadening and the electron spin, respectively.

(R6.10) Investigate the DTP, PTP, and Z for the problem as defined in (R6.7) under an arbitrarily oriented alternating quantizing magnetic field and crossed alternating electric field by including broadening and the electron spin, respectively.

(R6.11) Investigate all the appropriate problems as defined in (R6.1) to (R6.10) for all types of quantum dot superlattices.

(R6.12) Investigate all the appropriate problems as defined in (R6.1) to (R6.10) for all types of quantum dot superlattices in the presence of strain.

(R6.13) Introducing new theoretical formalisms, investigate all the problems of this chapter in the presence of hot electron effects.

(R6.14) Investigate the influence of deep traps and surface states separately for all the appropriate problems of this chapter after proper modifications.

References

1. K.P. Ghatak, S. Bhattacharya, D. De, in *Einstein Relation in Compound Semiconductors and Their Nanostructures*. Springer Series in Materials Science, vol 116 (Springer, Germany, 2008)
2. B.R. Nag, in *Electron Transport in Compound Semiconductors*. Springer Series in Solid State Sciences, vol 11 (Springer, Germany, 1980)

Chapter 7
Thermoelectric Power in Ultrathin Films Under Magnetic Quantization

7.1 Introduction

In this chapter, we shall study the thermoelectric power in UFs of nonlinear optical materials under magnetic quantization on the basis of the generalized dispersion relation as given in Chap. 1 in Sect. 7.2.1 of theoretical background of this chapter. The three and two band models of Kane are special cases of the generalized dispersion relations of nonlinear optical compounds. In Sect. 7.2.2, we shall investigate the thermoelectric power under strong magnetic quantization and III–V materials on the basis of three and two band models of Kane together with parabolic energy bands. Section 7.2.3 explores the thermoelectric power in the presence of a quantizing magnetic field in UFs of II–VI materials. Section 7.2.4 presents the study of the thermoelectric power in UFs of bismuth under magnetic quantization in accordance with the models of McClure and Choi, Cohen, Lax, and ellipsoidal parabolic, respectively. The thermoelectric power in the presence of a quantizing magnetic field in UFs of stressed compounds has been studied in Sect. 7.2.5. Sections 7.3 and 7.4 contain results and discussion and open research problems, respectively.

7.2 Theoretical Background

7.2.1 Magnetothermopower in Ultrathin Films of Nonlinear Optical Materials

The energy spectrum of the conduction electrons in nonlinear optical materials in the presence of a quantizing magnetic field B along z-direction can be written from (5.1) as [1]

$$\gamma(E) = \frac{2eB}{\hbar}\left(n+\frac{1}{2}\right) f_1(E) + f_2(E)k_z^2 \pm \frac{eB\hbar\Delta_{||}E_{g0}}{6}\left[\frac{E_{g0}+\Delta_\perp}{m_\perp^*\left(E_{g0}+\frac{2}{3}\Delta_\perp\right)}\right] \times\left[E+E_{g0}+\delta+\frac{\Delta_{||}^2-\Delta_\perp^2}{3\Delta_{||}}\right]. \quad (7.1)$$

Therefore, the electron energy spectrum in UFs of nonlinear optical materials under magnetic quantization assumes the form

$$\gamma(E_{71}) = \frac{2eB}{\hbar}\left(n+\frac{1}{2}\right) f_1(E_{71}) + f_2(E_{71})\left(\frac{\pi n_z}{d_z}\right)^2 \pm \frac{eB\hbar\Delta_{||}E_{g0}}{6} \times\left[\frac{E_{g0}+\Delta_\perp}{m_\perp^*\left(E_{g0}+\frac{2}{3}\Delta_\perp\right)}\right]\left[E_{71}+E_{g0}+\delta+\frac{\Delta_{||}^2-\Delta_\perp^2}{3\Delta_{||}}\right], \quad (7.2)$$

where E_{71} is the totally quantized energy in this case.

The electron concentration per unit area can be written as

$$n_0 = \frac{g_v eB}{h}\sum_{n=0}^{n_{\max}}\sum_{n_z=1}^{n_{z\max}} F_{-1}(\eta_{71}), \quad (7.3)$$

where $\eta_{71} = (k_B T)^{-1}\left[E_{FBUF} - E_{71}\right]$ and E_{FBUF} is the Fermi energy in UFs in the presence of magnetic quantization as measured from the edge of the conduction band in the vertically upward direction in the absence of any quantization.

Combining (1.13) and (7.3), the thermoelectric power under magnetic quantization in UFs of nonlinear optical materials can be expressed as

$$G_0 = \frac{\pi^2 k_B}{3e}\left[\sum_{n=0}^{n_{\max}}\sum_{n_z=1}^{n_{z\max}} F_{-1}(\eta_{71})\right]^{-1}\left[\sum_{n=0}^{n_{\max}}\sum_{n_z=1}^{n_{z\max}} F_{-2}(\eta_{71})\right]. \quad (7.4)$$

7.2.2 Magnetothermopower in Ultrathin Films of Kane Type III–V Materials

(a) The energy spectrum of the conduction electrons in UFs of Kane type III–V materials under magnetic quantization can be written, in accordance with the three band model of Kane, following (5.4) as

$$I(E_{72}) = \left(n+\frac{1}{2}\right)\hbar\omega_0 + \frac{\hbar^2}{2m^*}\left(\frac{\pi n_z}{d_z}\right)^2 \pm \frac{eB\hbar\Delta}{6m^*\left(E_{72}+E_{g0}+\frac{2}{3}\Delta\right)}, \quad (7.5)$$

where E_{72} is the totally quantized energy in this case.

The electron concentration per unit area can be expressed as

$$n_0 = \frac{g_v e B}{h} \sum_{n=0}^{n_{\max}} \sum_{n_z=1}^{n_{z\max}} F_{-1}(\eta_{72}), \tag{7.6}$$

where $\eta_{72} = (k_B T)^{-1} [E_{FBUF} - E_{72}]$.
Combining (1.13) and (7.6), the thermoelectric power under magnetic quantization in UFs of III–V materials can be written as

$$G_0 = \frac{\pi^2 k_B}{3e} \left[\sum_{n=0}^{n_{\max}} \sum_{n_z=1}^{n_{z\max}} F_{-1}(\eta_{72}) \right]^{-1} \left[\sum_{n=0}^{n_{\max}} \sum_{n_z=1}^{n_{z\max}} F_{-2}(\eta_{72}) \right]. \tag{7.7}$$

(b) In accordance with the two band model, the electron energy spectrum in UFs of III–V materials under magnetic quantization can be expressed following (5.7) as

$$E_{73}(1 + \alpha E_{73}) = \left(n + \frac{1}{2}\right)\hbar\omega_0 + \frac{\hbar^2}{2m^*}\left(\frac{\pi n_z}{d_z}\right)^2 \pm \frac{1}{2}\mu_0 g^* B, \tag{7.8}$$

where E_{73} is the totally quantized energy in this case.
The electron concentration per unit area can be written as

$$n_0 = \frac{g_v e B}{h} \sum_{n=0}^{n_{\max}} \sum_{n_z=1}^{n_{z\max}} F_{-1}(\eta_{73}), \tag{7.9}$$

where $\eta_{73} = (k_B T)^{-1} [E_{FBUF} - E_{73}]$.
Combining (1.13) and (7.9), the thermoelectric power in the presence of a quantizing magnetic field in UFs of III–V materials assumes the form

$$G_0 = \frac{\pi^2 k_B}{3e} \left[\sum_{n=0}^{n_{\max}} \sum_{n_z=1}^{n_{z\max}} F_{-1}(\eta_{73}) \right]^{-1} \left[\sum_{n=0}^{n_{\max}} \sum_{n_z=1}^{n_{z\max}} F_{-2}(\eta_{73}) \right]. \tag{7.10}$$

(c) The magnetodispersion law in UFs of semiconductors having parabolic energy bands assumes the form

$$E_{74} = \left(n + \frac{1}{2}\right)\hbar\omega_0 + \frac{\hbar^2}{2m^*}\left(\frac{\pi n_z}{d_z}\right)^2 \pm \frac{1}{2}\mu_0 g^* B, \tag{7.11}$$

where E_{74} is the totally quantized energy in this case.

The electron concentration per unit area can be expressed as

$$n_0 = \frac{g_v eB}{h} \sum_{n=0}^{n_{\max}} \sum_{n_z=1}^{n_{z\max}} F_{-1}(\eta_{74}), \tag{7.12}$$

where $\eta_{74} = (k_B T)^{-1} [E_{FBUF} - E_{74}]$.

Combining (1.13) and (7.12), the thermoelectric power in this case assumes the form

$$G_0 = \frac{\pi^2 k_B}{3e} \left[\sum_{n=0}^{n_{\max}} \sum_{n_z=1}^{n_{z\max}} F_{-1}(\eta_{74}) \right]^{-1} \left[\sum_{n=0}^{n_{\max}} \sum_{n_z=1}^{n_{z\max}} F_{-2}(\eta_{74}) \right]. \tag{7.13}$$

7.2.3 Magnetothermopower in Ultrathin Films of II–VI Materials

The magnetodispersion relation of the carriers in UFs of II–VI materials can be written following (5.15) as

$$E_{75,\pm} = \phi_\pm(n) + \frac{\hbar^2}{2m_\parallel^*} \left(\frac{\pi n_z}{d_z} \right)^2, \tag{7.14}$$

where $E_{75,+}$ is the totally quantized energy in this case.

The electron concentration per unit area can be expressed as

$$n_0 = \frac{g_v eB}{h} \sum_{n=0}^{n_{\max}} \sum_{n_z=1}^{n_{z\max}} \left[F_{-1}\left(\eta_{75,+}\right) + F_{-1}\left(\eta_{75,-}\right) \right], \tag{7.15}$$

where $\eta_{75,\pm} = (k_B T)^{-1} \left[E_{FBUF} - E_{75,\pm} \right]$.

Combining (1.13) and (7.15), the thermoelectric power under magnetic quantization in UFs of III–V materials can be written as

$$G_0 = \frac{\pi^2 k_B}{3e} \left[\sum_{n=0}^{n_{\max}} \sum_{n_z=1}^{n_{z\max}} \left[F_{-1}\left(\eta_{75,+}\right) + F_{-1}\left(\eta_{75,-}\right) \right] \right]^{-1} \left[\sum_{n=0}^{n_{\max}} \sum_{n_z=1}^{n_{z\max}} \left[F_{-2}\left(\eta_{75,+}\right) + F_{-2}\left(\eta_{75,-}\right) \right] \right]. \tag{7.16}$$

7.2.4 *Magnetothermopower in Ultrathin Films of Bismuth*

7.2.4.1 The McClure and Choi Model

The magnetodispersion relation of the carriers in UFs of bismuth in accordance with the McClure and Choi model can be expressed following (5.18) as

$$E_{76}\left(1+\alpha E_{76}\right)=\left(n+\frac{1}{2}\right)\hbar\omega(E_{76})+\left(n^2+1+n\right)+\left(\frac{\hbar^2}{2m_3}\right)\left(\frac{\pi n_z}{d_z}\right)^2 \left[1-\frac{\alpha\left(n+\frac{1}{2}\right)\hbar\omega\left(E_{76}\right)}{2}\right]\pm\frac{1}{2}\left|g^*\right|\mu_0 B, \tag{7.17}$$

where E_{76} is the totally quantized energy in this case and

$$\omega(E_{76})\equiv\frac{eB}{\sqrt{m_1m_2}}\left[1+\alpha E_{76}\left(1-\frac{m_2}{m_2'}\right)\right].$$

The electron concentration per unit area in this case assumes the form

$$n_0=\frac{g_{\mathrm{v}}eB}{h}\sum_{n=0}^{n_{\max}}\sum_{n_z=1}^{n_{z\max}}F_{-1}(\eta_{76}), \tag{7.18}$$

where $\eta_{76}=(k_{\mathrm{B}}T)^{-1}\left[E_{\mathrm{FBUF}}-E_{76}\right]$.

Combining (1.13) and (7.18), the thermoelectric power under magnetic quantization in UFs of bismuth can be written as

$$G_0=\frac{\pi^2k_{\mathrm{B}}}{3e}\left[\sum_{n=0}^{n_{\max}}\sum_{n_z=1}^{n_{z\max}}F_{-1}\left(\eta_{76}\right)\right]^{-1}\left[\sum_{n=0}^{n_{\max}}\sum_{n_z=1}^{n_{z\max}}F_{-2}\left(\eta_{76}\right)\right]. \tag{7.19}$$

7.2.4.2 The Cohen Model

The magnetodispersion relation of the carriers in UFs of bismuth in accordance with the Cohen model can be expressed following (5.21) as

$$E_{77}\left(1+\alpha E_{77}\right)=\left(n+\frac{1}{2}\right)\hbar\omega(E_{77})\pm\frac{1}{2}g^*\mu_0 B +\frac{3}{8}\alpha\left(n^2+n+\frac{1}{2}\right)\hbar^2\omega^2(E_{77})+\left(\frac{\hbar^2}{2m_3}\right)\left(\frac{\pi n_z}{d_z}\right)^2, \tag{7.20}$$

where E_{77} is the totally quantized energy in this case and

$$\omega(E_{77}) \equiv \frac{eB}{\sqrt{m_1 m_2}}\left[1+\alpha E_{77}\left(1-\frac{m_2}{m_2'}\right)\right].$$

The electron concentration per unit area in this case assumes the form

$$n_0 = \frac{g_v eB}{h}\sum_{n=0}^{n_{\max}}\sum_{n_z=1}^{n_{z\max}} F_{-1}(\eta_{77}), \tag{7.21}$$

where $\eta_{77} = (k_B T)^{-1}[E_{FBUF} - E_{77}]$.

Combining (1.13) and (7.21), the thermoelectric power under magnetic quantization in UFs of bismuth can be written as

$$G_0 = \frac{\pi^2 k_B}{3e}\left[\sum_{n=0}^{n_{\max}}\sum_{n_z=1}^{n_{z\max}} F_{-1}(\eta_{77})\right]^{-1}\left[\sum_{n=0}^{n_{\max}}\sum_{n_z=1}^{n_{z\max}} F_{-2}(\eta_{77})\right]. \tag{7.22}$$

7.2.4.3 The Lax Model

The magnetodispersion relation of the carriers in UFs of bismuth in accordance with the Lax model can be expressed following (5.24) as

$$E_{78}(1+\alpha E_{78}) = \left(n+\frac{1}{2}\right)\hbar\omega_{03} + \left(\hbar^2\left(\frac{n_z\pi}{d_z}\right)^2(2m_3^*)^{-1}\right) \pm \frac{1}{2}g^*\mu_0 B, \tag{7.23}$$

where E_{78} is the totally quantized energy in this case.

The electron concentration per unit area in this case assumes the form

$$n_0 = \frac{g_v eB}{h}\sum_{n=0}^{n_{\max}}\sum_{n_z=1}^{n_{z\max}} F_{-1}(\eta_{78}), \tag{7.24}$$

where $\eta_{78} = (k_B T)^{-1}[E_{FBUF} - E_{78}]$.

Combining (1.13) and (7.24), the thermoelectric power under magnetic quantization in UFs of bismuth can be written as

$$G_0 = \frac{\pi^2 k_B}{3e}\left[\sum_{n=0}^{n_{\max}}\sum_{n_z=1}^{n_{z\max}} F_{-1}(\eta_{78})\right]^{-1}\left[\sum_{n=0}^{n_{\max}}\sum_{n_z=1}^{n_{z\max}} F_{-2}(\eta_{78})\right]. \tag{7.25}$$

7.2.5 *Magnetothermopower in Ultrathin Films of IV–VI Materials*

In the same tuning with Chap. 5, we can write that the carriers of the IV–VI materials can be described by Cohen model, where the energy band constants should correspond to the said compounds. Equations (7.21) and (7.22) are applicable in this context.

7.2.6 *Magnetothermopower in Ultrathin Films of Stressed Materials*

The dispersion relation of the conduction electrons in UFs of stressed materials in the presence of a quantizing magnetic field B along z-direction can be written following (1.105) as

$$\left(n+\frac{1}{2}\right)\hbar\omega_6(E_{79})+\left(\frac{\pi n_z}{d_z}\right)^2\left[c^*(E_{79})\right]^{-2}=1, \tag{7.26}$$

where $\omega_6(E_{79}) = eB\left[a^*(E_{79})b^*(E_{79})\right]^{-1}$ and E_{79} is the totally quantized energy in this case.

The electron concentration per unit area in this case assumes the form

$$n_0=\frac{2g_{\mathrm{v}}eB}{h}\sum_{n=0}^{n_{\max}}\sum_{n_z=1}^{n_{z\max}}F_{-1}(\eta_{79}), \tag{7.27}$$

where $\eta_{79} = (k_{\mathrm{B}}T)^{-1}\left[E_{\mathrm{FBUF}} - E_{79}\right]$.

Combining (1.13) and (7.27), the thermoelectric power under magnetic quantization in UFs of bismuth can be written as

$$G_0=\frac{\pi^2 k_{\mathrm{B}}}{3e}\left[\sum_{n=0}^{n_{\max}}\sum_{n_z=1}^{n_{z\max}}F_{-1}(\eta_{79})\right]^{-1}\left[\sum_{n=0}^{n_{\max}}\sum_{n_z=1}^{n_{z\max}}F_{-2}(\eta_{79})\right]. \tag{7.28}$$

It is interesting to note that under the condition of nondegeneracy, all the results for all the models converge into the result of classical TPSM equation as given in the Preface.

7.3 Results and Discussion

Using (7.3) and (7.4) in curve (b) of Fig. 7.1, the normalized thermoelectric power of ultrathin films of tetragonal materials (taking Cd_3As_2 as an example) have been plotted in curve (a) of Fig. 7.1 as a function of inverse quantizing magnetic field by

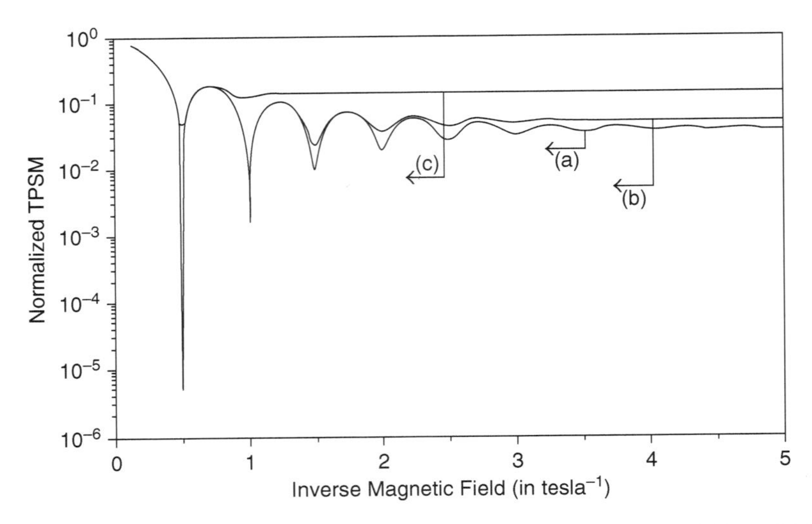

Fig. 7.1 Plot of the normalized TPSM as a function of inverse magnetic field for UFs of (**a**) Cd_3As_2 and (**b**) $CdGeAs_2$ in accordance with the generalized band model ($\delta \neq 0$). The plot (**c**) refers to n-InSb in accordance with the three band model of Kane ($n_0 = 10^{15}\,\mathrm{m}^{-2}$ and $d_z = 10\,\mathrm{nm}$)

taking $\delta \neq 0$ on the basis of the generalized energy band model of (7.2). The curve (b) shows the same dependence for nonlinear optical materials (taking $CdGeAs_2$ as an example) and has been plotted with $\delta \neq 0$. The curve (c) is valid for III–V materials (taking InSb as an example) and has been plotted by using (7.7) and (7.6), respectively. The three band energy model of Kane for InSb is valid for such highly nonparabolic material. The influence of energy band constants for the three aforementioned compounds can be estimated from the said curves. For all the figures of this chapter, lattice temperature has been taken as $T = 10\,\mathrm{K}$ and consequently for the purpose of simplified numerical computation we have considered only the first subband occupancy in connection with the quantization due to the Born–Von Karman boundary condition for various Landau levels due to the quantizing magnetic field. It appears that the thermoelectric power exhibits a periodic oscillation with increase in the magnetic field, which has also been discussed in Chap. 5.

In Fig. 7.2, we have plotted the normalized TPSM as a function of film thickness under constant magnetic field for all the cases of Fig. 7.1. The TPSM appears to exhibit composite oscillations because of the ad-mixture of size quantized levels with the Landau subbands. The nature of the variation of the TPSM from a staircase to the highly zigzag can be explained as the combined influence of the magnetic quantization with the size quantization. As the thickness starts lowering, the influence of the field decreases due to which the staircase variation is retrieved.

The TPSM as function of carrier concentration for said materials for both magnetic ($n = 0$) and size ($n_z = 1$) quantum limits has been plotted in Fig. 7.3 from which we can conclude that the TPSM decreases with carrier concentration for

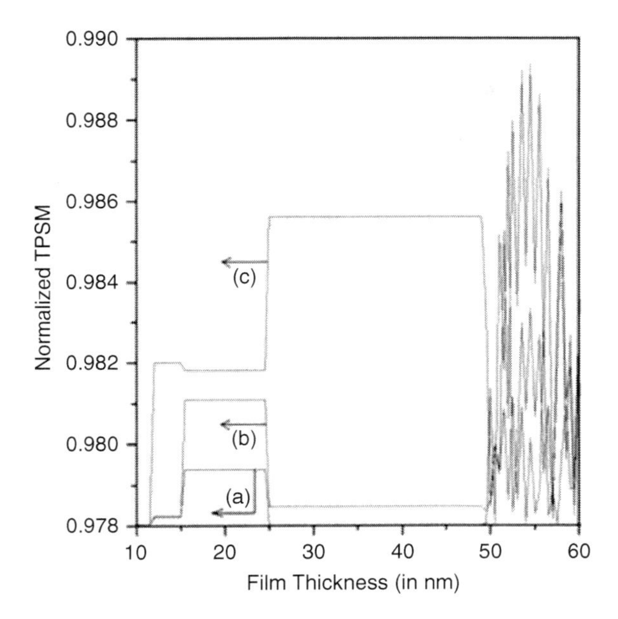

Fig. 7.2 Plot of the normalized TPSM as a function of film thickness for ultrathin films of (**a**) Cd_3As_2 and (**b**) $CdGeAs_2$ in accordance with the generalized band model ($\delta \neq 0$). The plot (**c**) refers to n − InSb in accordance with the three band model of Kane

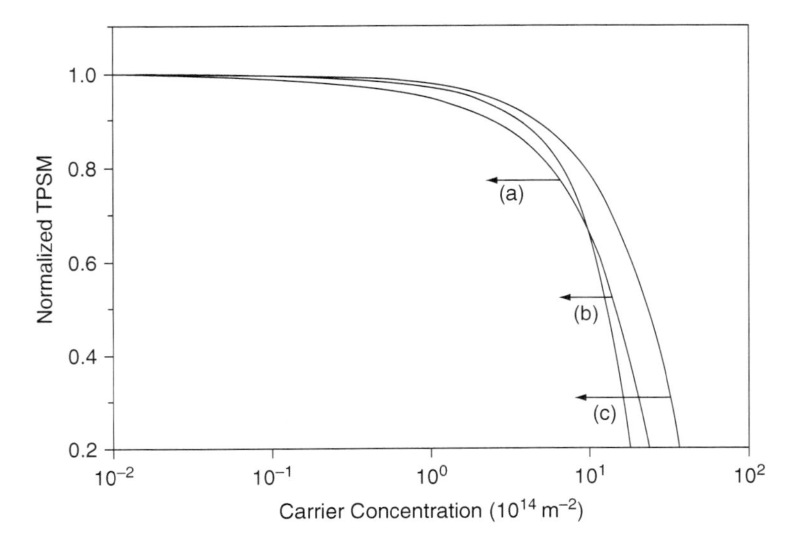

Fig. 7.3 Plot of the normalized TPSM as a function of carrier concentration for ultrathin films of (**a**) Cd_3As_2 and (**b**) $CdGeAs_2$ in accordance with the generalized band model ($\delta \neq 0$). The plot (**c**) refers to n − InSb in accordance with the three band model of Kane

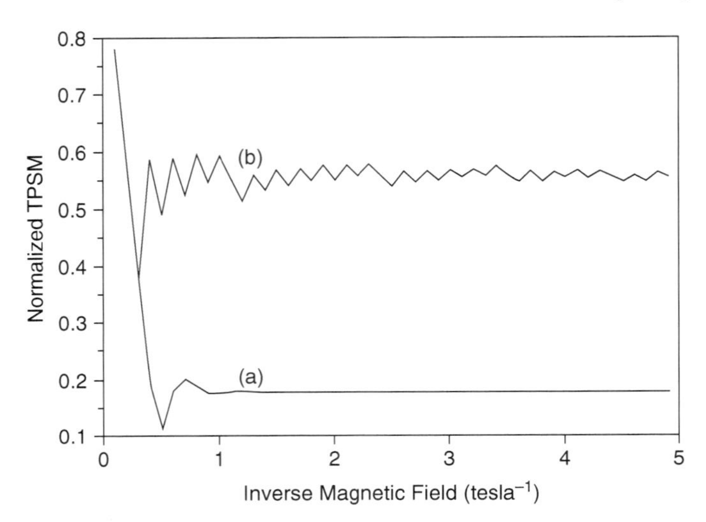

Fig. 7.4 Plot of the normalized TPSM as a function of inverse magnetic field for ultrathin films of (**a**) CdS ($C_0 \neq 0$) and (**b**) stressed InSb

relatively large values, whereas for the relatively low values of the carrier degeneracy, the magnetothermopower shows the converging tendency. It appears from Figs. 7.1 to 7.3 that InSb exhibits largest numerical TPSM as compared to Cd_3As_2 and $CdGeAs_2$ for UFs under magnetic quantization.

In Figs. 7.4–7.6, we have plotted the TPSM for ultrathin films of II–VI (using (7.15) and (7.16)) and stressed III–V materials (using (7.27) and (7.28)) as functions of inverse magnetic field, thickness, and carrier concentration, respectively. The film thickness for Figs. 7.4–7.6 are kept to 10 nm, while $B = 2$ T for Figs. 7.5 and 7.6, respectively. It appears from Figs. 7.4–7.6 that the normalized TPSM for UFs of stressed InSb exhibits higher numerical values as compared to the corresponding UFs of CdS. Figure 7.7 exhibits the plots of the normalized TPSM as function of inverse magnetic field for UFs of bismuth in accordance with the models of (a) McClure and Choi (using (7.18) and (7.19)) and (b) Cohen (using (7.21) and (7.22)), respectively.

Besides, the plot (c) in the same figure is valid for IV–VI materials (using PbTe as an example), whose carrier dispersion laws follow the Cohen model. Figures 7.8 and 7.9 demonstrate the said variations as a function of film thickness and carrier concentration, respectively. It appears that the bismuth exhibits higher TPSM than that of PbTe. For the purpose of simplicity the spin effects has been neglected in the computations. The inclusion of spin increases the number of oscillatory spikes by two with the decrement in amplitudes. All the plots have been normalized to the value $\left(\pi^2 k_B / 3e\right)$. The use of the data in the figures as presented in this chapter can also be used to compare the TPSM for other types of materials.

For the purpose of condensed presentation, the carrier concentration and the corresponding TPSM for this chapter have been presented in Table 7.1.

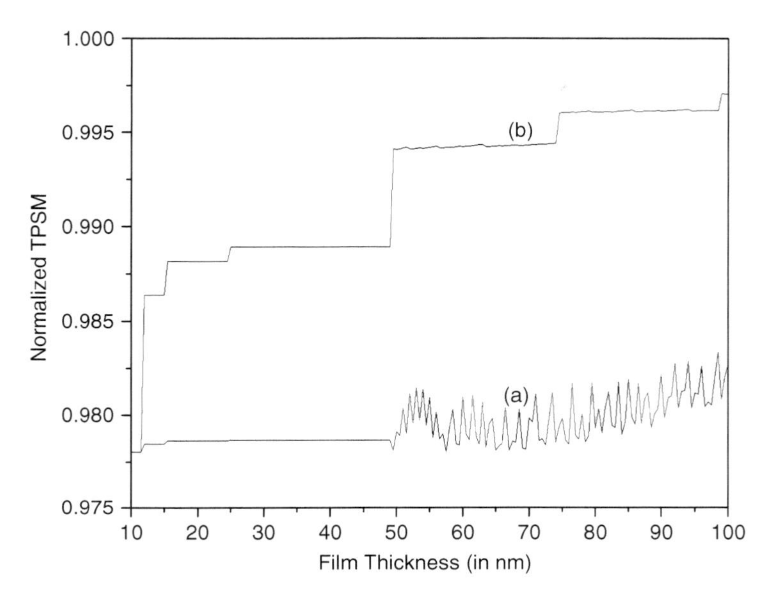

Fig. 7.5 Plot of the normalized TPSM as a function film thickness for ultrathin films of (**a**) CdS ($C_0 \neq 0$) and (**b**) stressed InSb

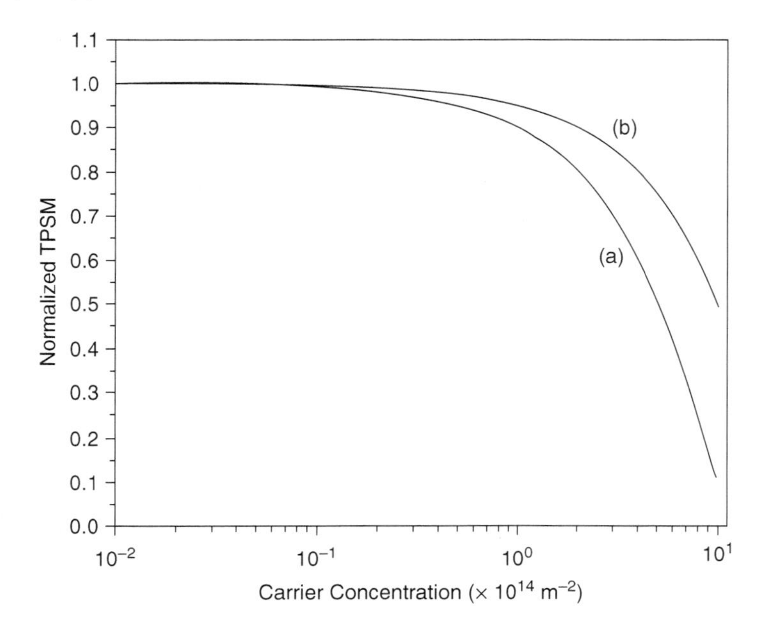

Fig. 7.6 Plot of the normalized TPSM as a function of carrier concentration for ultrathin films of (**a**) CdS ($C_0 \neq 0$) and (**b**) stressed InSb

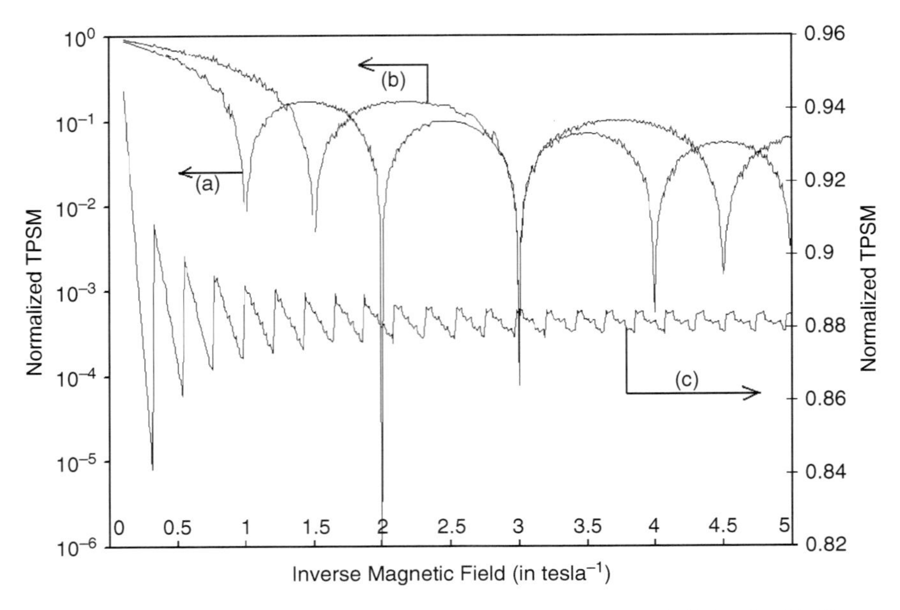

Fig. 7.7 Plot of the normalized TPSM as a function of inverse magnetic field for ultrathin films of bismuth in accordance with the (**a**) McClure and Choi and (**b**) Cohen models. The plot (**c**) refers to PbTe following Cohen model

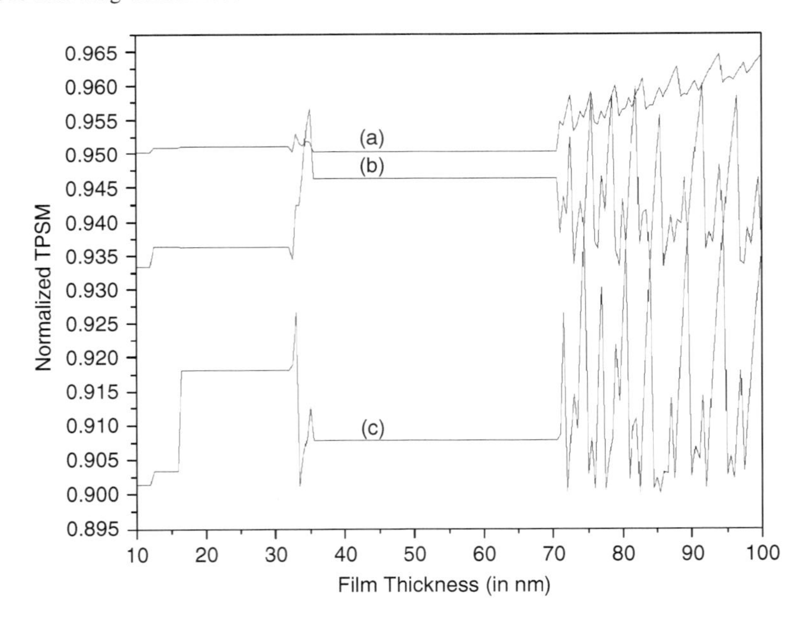

Fig. 7.8 Plot of the normalized TPSM as a function of film thickness for ultrathin films of bismuth in accordance with the (**a**) McClure and Choi and (**b**) Cohen models. The plot (**c**) refers to PbTe following Cohen model

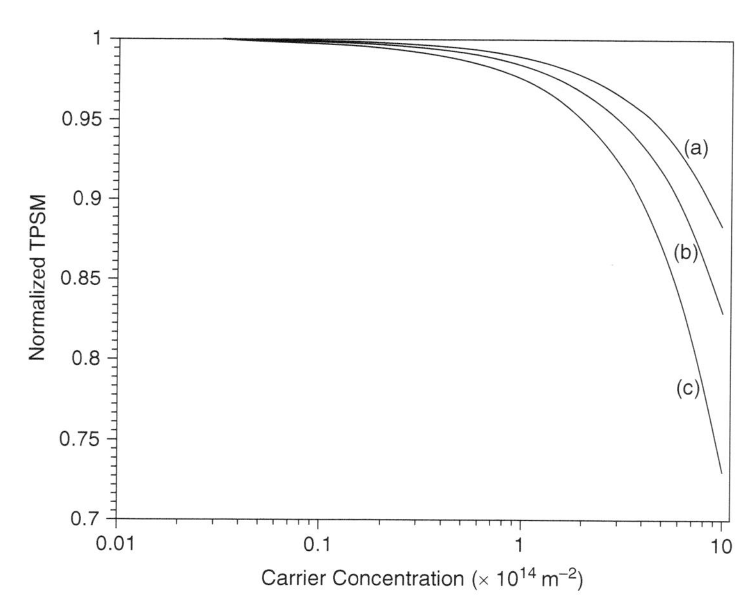

Fig. 7.9 Plot of the normalized TPSM as a function of carrier concentration for ultrathin films of bismuth in accordance with the (**a**) McClure and Choi and (**b**) Cohen models. The plot (**c**) refers to PbTe following Cohen model

7.4 Open Research Problems

(R7.1) Investigate the DTP, PTP, and Z in the presence of arbitrarily oriented quantizing magnetic field and in the presence of electron spin and broadening by considering all types of scattering mechanisms for UFs and by considering the presence of finite, parabolic, and circular potential wells applied separately for all the materials whose unperturbed carrier energy spectra are defined in Chap. 1.

(R7.2) Investigate (R7.1) in the presence of an additional arbitrarily oriented (a) nonuniform electric field and (b) alternating electric field, respectively, for all the materials whose unperturbed carrier energy spectra are defined in this Chap. 1 by considering all types of scattering mechanisms.

(R7.3) Investigate the DTP, PTP, and Z in the presence of arbitrarily oriented alternating quantizing magnetic field in the presence of electron spin and broadening by considering all types of scattering mechanisms for UFs by incorporating the presence of finite, parabolic, and circular potential wells applied separately for all the materials whose unperturbed carrier energy spectra are defined in Chap. 1.

(R7.4) Investigate the DTP, PTP, and Z under an arbitrarily oriented alternating quantizing magnetic field and crossed alternating electric field by including broadening and the electron spin for UFs of all the materials whose unperturbed carrier energy spectra are defined in Chap. 1 by considering all types of scattering mechanisms.

Table 7.1 The carrier statistics and the thermoelectric power under magnetic quantization in ultrathin films of nonlinear optical, Kane type III–V, II–VI, bismuth, IV–VI, and stressed materials

Type of materials	Carrier statistics	TPSM
1. Nonlinear optical materials	$n_0 = \frac{g_v eB}{h} \sum_{n=0}^{n_{max}} \sum_{n_z=1}^{n_{z\max}} F_{-1}(\eta_{71})$ (7.3)	On the basis of (7.3), $G_0 = \frac{\pi^2 k_B}{3e} \left[\sum_{n=0}^{n_{max}} \sum_{n_z=1}^{n_{z\max}} F_{-1}(\eta_{71}) \right]^{-1} \left[\sum_{n=0}^{n_{max}} \sum_{n_z=1}^{n_{z\max}} F_{-2}(\eta_{71}) \right]$ (7.4)
2. Kane type III–V materials	(a) Three band model of Kane: $n_0 = \frac{g_v eB}{h} \sum_{n=0}^{n_{max}} \sum_{n_z=1}^{n_{z\max}} F_{-1}(\eta_{72})$ (7.6)	On the basis of (7.6), $G_0 = \frac{\pi^2 k_B}{3e} \left[\sum_{n=0}^{n_{max}} \sum_{n_z=1}^{n_{z\max}} F_{-1}(\eta_{72}) \right]^{-1} \left[\sum_{n=0}^{n_{max}} \sum_{n_z=1}^{n_{z\max}} F_{-2}(\eta_{72}) \right]$ (7.7)
	(b) Two band model of Kane: $n_0 = \frac{g_v eB}{h} \sum_{n=0}^{n_{max}} \sum_{n_z=1}^{n_{z\max}} F_{-1}(\eta_{73})$ (7.9)	On the basis of (7.9), $G_0 = \frac{\pi^2 k_B}{3e} \left[\sum_{n=0}^{n_{max}} \sum_{n_z=1}^{n_{z\max}} F_{-1}(\eta_{73}) \right]^{-1} \left[\sum_{n=0}^{n_{max}} \sum_{n_z=1}^{n_{z\max}} F_{-2}(\eta_{73}) \right]$ (7.10)
	(c) Parabolic energy band: $n_0 = \frac{g_v eB}{h} \sum_{n=0}^{n_{max}} \sum_{n_z=1}^{n_{z\max}} F_{-1}(\eta_{74})$ (7.12)	On the basis of (7.12), $G_0 = \frac{\pi^2 k_B}{3e} \left[\sum_{n=0}^{n_{max}} \sum_{n_z=1}^{n_{z\max}} F_{-1}(\eta_{74}) \right]^{-1} \left[\sum_{n=0}^{n_{max}} \sum_{n_z=1}^{n_{z\max}} F_{-2}(\eta_{74}) \right]$ (7.13)
3. II–VI materials	$n_0 = \frac{g_v eB}{h} \sum_{n=0}^{n_{max}} \sum_{n_z=1}^{n_{z\max}} [F_{-1}(\eta_{75,+}) + F_{-1}(\eta_{75,-})$ (7.15)	On the basis of (7.15), $G_0 = \frac{\pi^2 k_B}{3e} \left[\sum_{n=0}^{n_{max}} \sum_{n_z=1}^{n_{z\max}} [F_{-1}(\eta_{75,+}) + F_{-1}(\eta_{75,-})] \right]^{-1} \left[\sum_{n=0}^{n_{max}} \sum_{n_z=1}^{n_{z\max}} [F_{-2}(\eta_{75,+}) + F_{-2}(\eta_{75,-})] \right]$ (7.16)

4. Bismuth	(a) The McClure and Choi model: $n_0 = \frac{g_v eB}{h} \sum_{n=0}^{n_{max}} \sum_{n_z=1}^{n_{zmax}} F_{-1}(\eta_{76})$ (7.18)	On the basis of (7.18), $G_0 = \frac{\pi^2 k_B}{3e} \left[\sum_{n=0}^{n_{max}} \sum_{n_z=1}^{n_{zmax}} F_{-1}(\eta_{76}) \right]^{-1} \left[\sum_{n=0}^{n_{max}} \sum_{n_z=1}^{n_{zmax}} F_{-2}(\eta_{76}) \right]$ (7.19)
	(b) The Cohen Model: $n_0 = \frac{g_v eB}{h} \sum_{n=0}^{n_{max}} \sum_{n_z=1}^{n_{zmax}} F_{-1}(\eta_{77})$ (7.21)	On the basis of (7.21), $G_0 = \frac{\pi^2 k_B}{3e} \left[\sum_{n=0}^{n_{max}} \sum_{n_z=1}^{n_{zmax}} F_{-1}(\eta_{77}) \right]^{-1} \left[\sum_{n=0}^{n_{max}} \sum_{n_z=1}^{n_{zmax}} F_{-2}(\eta_{77}) \right]$ (7.22)
	(c) The Lax Model: $n_0 = \frac{g_v eB}{h} \sum_{n=0}^{n_{max}} \sum_{n_z=1}^{n_{zmax}} F_{-1}(\eta_{78})$ (7.24)	On the basis of (7.24), $G_0 = \frac{\pi^2 k_B}{3e} \left[\sum_{n=0}^{n_{max}} \sum_{n_z=1}^{n_{zmax}} F_{-1}(\eta_{78}) \right]^{-1} \left[\sum_{n=0}^{n_{max}} \sum_{n_z=1}^{n_{zmax}} F_{-2}(\eta_{78}) \right]$ (7.25)
5. IV–VI materials	The expression of n_0 in this case is given by (7.21) in which the constants of the energy band spectrum correspond to carriers of the IV–VI semiconductors	The expression of TPSM in this case is given by (7.22) in which the contents of the energy band spectrum correspond to the the carriers of the IV–VI semiconductors
6. Stressed materials	$n_0 = \frac{2g_v eB}{h} \sum_{n=0}^{n_{max}} \sum_{n_z=1}^{n_{zmax}} F_{-1}(\eta_{79})$ (7.27)	On the basis of (7.27) $G_0 = \frac{\pi^2 k_B}{3e} \left[\sum_{n=0}^{n_{max}} \sum_{n_z=1}^{n_{zmax}} F_{-1}(\eta_{79}) \right]^{-1} \left[\sum_{n=0}^{n_{max}} \sum_{n_z=1}^{n_{zmax}} F_{-2}(\eta_{79}) \right]$ (7.28)

(R7.5) Investigate the DTP, PTP, and Z under an arbitrarily oriented alternating quantizing magnetic field and crossed alternating nonuniform electric field by including broadening and the electron spin whose for UFs of all the materials unperturbed carrier energy spectra are defined in Chap. 1 by considering all types of scattering mechanisms.

(R7.6) Investigate the DTP, PTP, and Z in the presence of a quantizing magnetic field under exponential, Kane, Halperin, Lax, and Bonch-Bruevich band tails [1] for UFs of all the materials whose unperturbed carrier energy spectra are defined in Chap. 1 by considering all types of scattering mechanisms.

(R7.7) Investigate the DTP, PTP, and Z in the presence of quantizing magnetic field for UFs of all the materials as defined in (R7.6) under an arbitrarily oriented (a) nonuniform electric field and (b) alternating electric field, respectively, by considering all types of scattering mechanisms.

(R7.8) Investigate the DTP, PTP, and Z for the UFs of all the materials as described in (R7.6) under an arbitrarily oriented alternating quantizing magnetic field by including broadening and the electron spin by considering all types of scattering mechanisms.

(R7.9) Investigate the DTP, PTP, and Z for UFs of all the materials as discussed in (R7.6) under an arbitrarily oriented alternating quantizing magnetic field and crossed alternating electric field by including broadening and the electron spin by considering all types of scattering mechanisms.

(R7.10) Investigate all the appropriate problems after proper modifications introducing new theoretical formalisms for all types of UFs of all the materials as discussed in (R7.6) for functional, negative refractive index, macromolecular, organic, and magnetic materials by considering all types of scattering mechanisms in the presence of strain.

(R7.11) Investigate all the appropriate problems of this chapter for all types of UFs for p – InSb, p-CuCl, and semiconductors having diamond structure valence bands whose dispersion relations of the carriers in bulk materials are given by Cunningham [2], Yekimov et al. [3], and Roman et al. [4], respectively, by considering all types of scattering mechanisms in the presence of strain.

(R7.12) Investigate the influence of deep traps and surface states separately for all the appropriate problems of all the chapters after proper modifications by considering all types of scattering mechanisms.

References

1. K.P. Ghatak, S. Bhattacharya, D. De, in *Einstein Relation in Compound Semiconductors and Their Nanostructures*. Springer Series in Materials Science, vol 116 (Springer, Germany, 2008)
2. R.W. Cunningham, Phys. Rev. **167**, 761 (1968)
3. A.I. Yekimov, A.A. Onushchenko, A.G. Plyukhin, Al.L. Efros, J. Exp. Theor. Phys. **88**, 1490 (1985)
4. B.J. Roman, A.W. Ewald, Phys. Rev. B **5**, 3914 (1972)

Part III
Thermoelectric Power Under Large Magnetic Field in Quantum Confined Optoelectronic Materials in the Presence of Light Waves

Chapter 8
Optothermoelectric Power in Ultrathin Films and Quantum Wires of Optoelectronic Materials Under Large Magnetic Field

8.1 Introduction

With the advent of semiconductor optoelectronics, there has been a considerable interest in studying the optical processes in semiconductors and their nanostructures [1]. It appears from the literature that the investigations have been carried out on the assumption that the carrier energy spectra are invariant quantities in the presence of intense light waves, which is not fundamentally true. The physical properties of semiconductors in the presence of light waves which change the basic dispersion relation have been relatively less investigated in the literature [2–4]. In this chapter, we shall study the thermoelectric power under large magnetic field in UFs and QWs of III–V, ternary, and quaternary materials in the presence of external photoexcitation on the basis of electron dispersion laws in the presence of external light waves.

Ternary and quaternary compounds enjoy the singular position in the entire spectrum of optoelectronic materials. It is well known that the ternary alloy $Hg_{1-x}Cd_xTe$ is a classic narrow gap compound. By adjusting the alloy composition, the band gap of this ternary alloy can be varied to cover the spectral range from 0.8 to over 30 μm [5]. $Hg_{1-x}Cd_xTe$ finds extensive applications in infrared detector materials and photovoltaic detector arrays in the 8–12 μm wave bands [6]. The above uses have generated the $Hg_{1-x}Cd_xTe$ technology for the experimental realization of high mobility single crystal with specially prepared surfaces. The same compound is the optimum choice for illuminating the narrow subband physics because the relevant material constants are within easy experimental reach [7]. It may be mentioned that the quaternary alloy $In_{1-x}Ga_xAs_yP_{1-y}$ lattice matched to InP also finds wide use in the fabrication of avalanche photodetectors [8], heterojunction lasers [9], light emitting diodes [10], and avalanche photodiodes [11]. Besides, the field effect transistors, detectors, switches, modulators, solar cells, filters, and new types of integrated optical devices are made from the quaternary systems [12].

In Sect. 8.2.1, the thermoelectric power under large magnetic field in the UFs of the said materials has been investigated in the presence of external photoexcitation whose unperturbed electron energy spectra are defined by the three and two band models of Kane together with parabolic energy bands by formulating respective dispersion relation in the presence of light waves. In Sect. 8.2.2, the same has been investigated for QWs. Sections 8.3 and 8.4 contain results and discussion and open research problems pertinent to this chapter, respectively.

8.2 Theoretical Background

8.2.1 Optothermoelectric Power in Ultrathin Films of Optoelectronic Materials Under Large Magnetic Field

The doubly degenerate wave functions $u_1(\vec{k},\vec{r})$ and $u_2(\vec{k},\vec{r})$ can be expressed as [13–15]

$$u_1(\vec{k},\vec{r}) = a_{k+}[(is)\downarrow'] + b_{k+}\left[\frac{X' - iY'}{\sqrt{2}}\uparrow'\right] + c_{k+}[Z'\downarrow'] \tag{8.1a}$$

and

$$u_2(\vec{k},\vec{r}) = a_{k-}[(is)\uparrow'] - b_{k-}\left[\frac{X' + iY'}{\sqrt{2}}\downarrow'\right] + c_{k-}[Z'\uparrow']. \tag{8.1b}$$

s is the s-type atomic orbital in both unprimed and primed coordinates, $\downarrow'$ indicates the spin down function in the primed coordinates,

$$a_{k\pm} \equiv \beta\left[E_{g0} - (\gamma_{0k\pm})^2(E_{g0} - \delta')\right]^{1/2}(E_{g0} + \delta')^{-1/2}\Big],$$

$$\beta \equiv \left[(6(E_{g0} + 2\Delta/3)(E_{g0} + \Delta))/\chi\right]^{1/2},$$

$$\chi \equiv \left(6E_{g0}^2 + 9E_{g0}\Delta + 4\Delta^2\right),$$

$$\gamma_{0k\pm} \equiv \left[\frac{(\xi_{1k} \mp E_{g0})}{2(\xi_{1k} + \delta')}\right]^{1/2},$$

$$\xi_{1k} \equiv E_c(\vec{k}) - E_v(\vec{k}) = E_{g0}\left[1 + 2\left(1 + \frac{m_c}{m_v}\right)\frac{\gamma(E)}{E_{g0}}\right]^{1/2},$$

$$\delta' \equiv \left(E_{g0}^2\Delta\right)(\chi)^{-1},$$

X', Y', and Z' are the p-type atomic orbitals in the primed coordinates, $\uparrow'$ indicates the spin-up function in the primed coordinates, $b_{k\pm} \equiv \rho\gamma_{0k\pm}$, $\rho \equiv (4\Delta^2/3\chi)^{1/2}$, $c_{k\pm} \equiv t\gamma_{0k\pm}$, and $t \equiv \left[6(E_{g0} + 2\Delta/3)^2/\chi\right]^{1/2}$.

We can therefore write the expression for the optical matrix element (OME) as

$$\text{OME} = \hat{p}_{cv}(\vec{k}) = < u_1(\vec{k},\vec{r})\left|\hat{p}\right|u_2(\vec{k},\vec{r}) > . \tag{8.1c}$$

Since the photon vector has no interaction in the same band for the study of interband optical transition, we can therefore write

$$\langle S\,|\hat{p}\,|S\,\rangle = \langle X\,|\hat{p}\,|X\,\rangle = \langle Y\,|\hat{p}\,|Y\,\rangle = \langle Z\,|\hat{p}\,|Z\,\rangle = 0 \quad \text{and}$$
$$\langle X\,|\hat{p}\,|Y\,\rangle = \langle Y\,|\hat{p}\,|Z\,\rangle = \langle Z\,|\hat{p}\,|X\,\rangle = 0.$$

There are finite interactions between the conduction band (CB) and the valance band (VB) and we can obtain

$$\left\langle S|\hat{P}|X\right\rangle = \hat{i}\cdot\hat{P}_x$$
$$\left\langle S|\hat{P}|Y\right\rangle = \hat{j}\cdot\hat{P}_y$$
$$\left\langle S|\hat{P}|Z\right\rangle = \hat{k}\cdot\hat{P}_z.$$

where $\hat{i}$, $\hat{j}$, and $\hat{k}$ are the unit vectors along x, y, and z axes, respectively.

It is well known that

$$\begin{bmatrix}\uparrow'\\ \downarrow'\end{bmatrix} = \begin{bmatrix} e^{-i\phi/2}\cos(\theta\,/\,2) & e^{i\phi/2}\sin(\theta\,/\,2)\\ -e^{-i\phi/2}\sin(\theta\,/\,2) & e^{i\phi/2}\cos(\theta\,/\,2)\end{bmatrix}\begin{bmatrix}\uparrow\\ \downarrow\end{bmatrix} \quad \text{and}$$
$$\begin{bmatrix}X'\\ Y'\\ Z'\end{bmatrix} = \begin{bmatrix}\cos\theta\cos\phi & \cos\theta\sin\phi & -\sin\theta\\ -\sin\phi & \cos\phi & 0\\ \sin\theta\cos\phi & \sin\theta\sin\phi & \cos\theta\end{bmatrix}\begin{bmatrix}X\\ Y\\ Z\end{bmatrix}.$$

Besides, the spin vector can be written as

$$\vec{S} = \frac{\hbar}{2}\vec{\sigma},$$

where

$$\sigma_x = \begin{bmatrix}0 & 1\\ 1 & 0\end{bmatrix},\ \sigma_y = \begin{bmatrix}0 & -i\\ i & 0\end{bmatrix},\ \text{and}\ \sigma_z = \begin{bmatrix}1 & 0\\ 0 & -1\end{bmatrix}.$$

From above, we can write

$$\hat{p}_{\mathrm{CV}}\left(\vec{k}\right) = \left\langle u_1(\vec{k},\vec{r})|\hat{P}|u_2(\vec{k},\vec{r})\right\rangle = \left\langle\left\{a_{k+}[(iS)\downarrow'] + b_{k+}\left[\left(\frac{X'-iY'}{\sqrt{2}}\right)\uparrow'\right]\right.\right.$$
$$\left.\left. + c_{k+}[Z'\downarrow']\right\}|\hat{P}|\left\{a_{k-}[(iS)\uparrow'] - b_{k-}\left[\left(\frac{X'+iY'}{\sqrt{2}}\right)\downarrow'\right] + c_{k-}[Z'\uparrow']\right\}\right\rangle.$$

Using above relations, we get

$$\begin{aligned}\hat{p}_{\mathrm{CV}}\left(\vec{k}\right) &= \left\langle u_1(\vec{k},\vec{r})|\hat{P}|u_2(\vec{k},\vec{r})\right\rangle\\ &= \frac{b_{k+}a_{k-}}{\sqrt{2}}\left\{\left\langle(X'-iY')|\hat{P}|iS\rangle\langle\uparrow'|\uparrow'\right\rangle\right\}\end{aligned}$$

$$+ \, c_{k_+} a_{k_-} \{ \langle Z' | \hat{P} | iS \rangle \langle \downarrow' | \uparrow' \rangle \} - \frac{a_{k_+} b_{k_-}}{\sqrt{2}} \{ \langle iS | \hat{P} | (X' + iY') \rangle \langle \downarrow' | \downarrow' \rangle \}$$
$$+ \, a_{k_+} c_{k_-} \{ \langle iS | \hat{P} | Z' \rangle \langle \downarrow' | \uparrow' \rangle \}. \tag{8.1d}$$

Now

$$\langle (X' - iY') | \hat{P} | iS \rangle = \langle (X') | \hat{P} | iS \rangle - \langle (iY') | \hat{P} | iS \rangle$$
$$= i \int u^*_{X'} \hat{P} S - \int -iu^*_{Y'} \hat{P} iu_X = i \langle X' | \hat{P} | S \rangle - \langle Y' | \hat{P} | S \rangle$$

From the above relations, for X', Y', and Z', we get

$$| \, X' \rangle = \cos\theta \cos\phi \, | \, X \rangle + \cos\theta \sin\phi \, | \, Y \rangle - \sin\theta \, | \, Z \rangle$$

Thus, $\left\langle X' | \hat{P} | S \right\rangle = \cos\theta \cos\phi \left\langle X | \hat{P} | S \right\rangle + \cos\theta \sin\phi \left\langle Y | \hat{P} | S \right\rangle - \sin\theta \left\langle Z | \hat{P} | S \right\rangle = \hat{P} \hat{r}_1$, where $\hat{r}_1 = \hat{i} \cos\theta \cos\phi + \hat{j} \cos\theta \sin\phi - \hat{k} \sin\theta$

$$\left| Y' \right\rangle = -\sin\phi \, |X\rangle + \cos\phi \, |Y\rangle + 0 \, |Z\rangle$$

Thus, $\left\langle Y' | \hat{P} | S \right\rangle = -\sin\phi \left\langle X | \hat{P} | S \right\rangle + \cos\phi \left\langle Y | \hat{P} | S \right\rangle + 0 \left\langle Z | \hat{P} | S \right\rangle = \hat{P} \hat{r}_2$, where $\hat{r}_2 = -\hat{i} \sin\phi + \hat{j} \cos\phi$ so that $\left\langle (X' - iY') | \hat{P} | S \right\rangle = \hat{P}(i\hat{r}_1 - \hat{r}_2)$.

Thus,

$$\frac{a_{k_-} b_{k_+}}{\sqrt{2}} \left\langle (X' - iY') | \hat{P} | S \right\rangle \left\langle \uparrow' | \uparrow' \right\rangle = \frac{a_{k_-} b_{k_+}}{\sqrt{2}} \hat{P}(i\hat{r}_1 - \hat{r}_2) \left\langle \uparrow' | \uparrow' \right\rangle. \tag{8.1e}$$

Now since

$$\left\langle iS | \hat{P} | \left(X' + iY' \right) \right\rangle = i \left\langle S | \hat{P} | X' \right\rangle - \left\langle S | \hat{P} | Y' \right\rangle = \hat{P}(i\hat{r}_1 - \hat{r}_2).$$

We can write

$$-\left[\frac{a_{k_+} b_{k_-}}{\sqrt{2}} \left\{ \left\langle iS | \hat{P} | (X' + iY') \right\rangle \left\langle \downarrow' | \downarrow' \right\rangle \right\} \right] = -\left[\frac{a_{k_+} b_{k_-}}{\sqrt{2}} \hat{P}(i\hat{r}_1 - \hat{r}_2) \left\langle \downarrow' | \downarrow' \right\rangle \right]. \tag{8.1f}$$

Similarly, we get

$$\left| Z' \right\rangle = \sin\theta \cos\phi \, |X\rangle + \sin\theta \sin\phi \, |Y\rangle + \cos\theta \, |Z\rangle \, .$$

So that $\left\langle Z' | \hat{P} | iS \right\rangle = i \left\langle Z' | \hat{P} | S \right\rangle = i \hat{P} \{ \sin\theta \cos\phi \hat{i} + \sin\theta \sin\phi \hat{j} + \cos\theta \hat{k} \} = i \hat{P} \hat{r}_3$, where $\hat{r}_3 = \hat{i} \sin\theta \cos\phi + \hat{j} \sin\theta \sin\phi + \hat{k} \cos\theta$.

Thus,

$$c_{k_+}a_{k_-}\left\langle Z'|\hat{P}|iS\right\rangle\langle\downarrow'|\uparrow'\rangle = c_{k_+}a_{k_-}i\,\hat{P}\hat{r}_3\langle\downarrow'|\uparrow'\rangle. \tag{8.1g}$$

Similarly, we can write

$$c_{k_-}a_{k_+}\left\langle iS|\hat{P}|Z'\right\rangle\langle\downarrow'|\uparrow'\rangle = c_{k_-}a_{k_+}i\,\hat{P}\hat{r}_3\langle\downarrow'|\uparrow'\rangle. \tag{8.1h}$$

Therefore, we obtain

$$\frac{a_{k_-}b_{k_+}}{\sqrt{2}}\left\{\left\langle(X'-iY')|\hat{P}|iS\right\rangle\langle\uparrow'|\uparrow'\rangle\right\} - \frac{a_{k_+}b_{k_-}}{\sqrt{2}}\left\{\left\langle iS|\hat{P}|(X'+iY')\right\rangle\langle\downarrow'|\downarrow'\rangle\right\}$$
$$= \frac{\hat{P}}{\sqrt{2}}(-a_{k_+}b_{k_-}\langle\downarrow'|\downarrow'\rangle + a_{k_-}b_{k_+}\langle\uparrow|\uparrow'\rangle)(i\hat{r}_1-\hat{r}_2) \tag{8.1i}$$

Also, we can write

$$c_{k_+}a_{k_-}\left\langle Z'|\hat{P}|iS\right\rangle\langle\downarrow'|\uparrow'\rangle + c_{k_-}a_{k_+}\left\langle iS|\hat{P}|Z'\right\rangle\langle\downarrow'|\uparrow'\rangle$$
$$= i\,\hat{P}(c_{k_+}a_{k_-} + c_{k_-}a_{k_+})\hat{r}_3\left[\langle\downarrow'|\downarrow'\rangle\right]. \tag{8.1j}$$

Combining the appropriate equations, we can further write

$$\hat{p}_{\mathrm{CV}}(\vec{k}) = \frac{\hat{P}}{\sqrt{2}}(i\hat{r}_1-\hat{r}_2)\left\{(b_{k_+}a_{k_-})\langle\uparrow'|\uparrow'\rangle - (b_{k_-}a_{k_+})\langle\downarrow'|\downarrow'\rangle\right\}$$
$$+ i\,\hat{P}\hat{r}_3(c_{k_+}a_{k_-} - c_{k_-}a_{k_+})\langle\downarrow'|\uparrow'\rangle. \tag{8.1k}$$

From the above relations, we obtain

$$\left.\begin{aligned}\uparrow' &= \mathrm{e}^{-\mathrm{i}\phi/2}\cos(\theta/2)\uparrow + \mathrm{e}^{\mathrm{i}\phi/2}\sin(\theta/2)\downarrow\\ \downarrow' &= -\mathrm{e}^{-\mathrm{i}\phi/2}\sin(\theta/2)\uparrow + \mathrm{e}^{\mathrm{i}\phi/2}\cos(\theta/2)\downarrow\end{aligned}\right\}. \tag{8.1l}$$

Therefore,

$$\langle\downarrow'|\uparrow'\rangle_x = -\sin(\theta/2)\cos(\theta/2)\langle\uparrow|\uparrow\rangle_x + \mathrm{e}^{-i\phi}\cos^2(\theta/2)\langle\downarrow|\uparrow\rangle_x$$
$$-e^{-i\varphi}\sin^2(\theta/2)\langle\uparrow|\downarrow\rangle_x + \sin(\theta/2)\cos(\theta/2)\langle\downarrow|\downarrow\rangle_x \tag{8.1m}$$

But we know from above that

$$\langle\uparrow|\uparrow\rangle_x = 0,\ \langle\uparrow|\downarrow\rangle = \frac{1}{2},\ \langle\downarrow|\uparrow\rangle_x = \frac{1}{2}\quad\text{and}\quad\langle\downarrow|\downarrow\rangle_x = 0.$$

Thus, from (8.1m), we get

$$\begin{aligned}\langle\downarrow'|\uparrow'\rangle_x &= \frac{1}{2}[\mathrm{e}^{-\mathrm{i}\phi}\cos^2(\theta/2) - \mathrm{e}^{\mathrm{i}\phi}\sin^2(\theta/2)] \\ &= \frac{1}{2}[(\cos\phi - i\sin\phi)\cos^2(\theta/2) - (\cos\phi + i\sin\phi)\sin^2(\theta/2)] \\ &= \frac{1}{2}[\cos\phi\cos\theta - i\sin\phi]. \end{aligned} \tag{8.1n}$$

Similarly, we obtain

$$\langle\downarrow'|\uparrow'\rangle_y = \frac{1}{2}[i\cos\phi + \sin\phi\cos\theta] \quad \text{and} \quad \langle\downarrow'|\uparrow'\rangle_z = \frac{1}{2}[-\sin\theta].$$

Therefore,

$$\begin{aligned}\langle\downarrow'|\uparrow'\rangle &= \hat{i}\langle\downarrow'|\uparrow'\rangle_x + \hat{j}\langle\downarrow'|\uparrow'\rangle_y + \hat{k}\langle\downarrow'|\uparrow'\rangle_z \\ &= \frac{1}{2}\left\{(\cos\theta\cos\phi - i\sin\phi)\hat{i} + (i\cos\phi + \sin\phi\cos\theta)\hat{j} - \sin\theta\hat{k}\right\} \\ &= \frac{1}{2}\left[\left\{(\cos\theta\cos\phi)\hat{i} + (\sin\phi\cos\theta)\hat{j} - \sin\theta\hat{k}\right\}\right. \\ &\quad \left. + i\left\{-\hat{i}\sin\phi + \hat{j}\cos\phi\right\}\right] \\ &= \frac{1}{2}[\hat{r}_1 + i\hat{r}_2] = -\frac{1}{2}i[i\hat{r}_1 - \hat{r}_2].\end{aligned}$$

Similarly, we can write

$$\langle\uparrow'|\uparrow'\rangle = \frac{1}{2}\left[\hat{i}\sin\theta\cos\phi + \hat{j}\sin\theta\sin\phi + \hat{k}\cos\theta\right] = \frac{1}{2}\hat{r}_3 \quad \text{and} \quad \langle\downarrow'|\downarrow'\rangle = -\frac{1}{2}\hat{r}_3.$$

Thus, combining the above results, we can write

$$\begin{aligned}\hat{p}_{\mathrm{CV}}(\vec{k}) &= \frac{\hat{P}}{\sqrt{2}}(i\hat{r}_1 - \hat{r}_2)\left\{(a_{k_-}b_{k_+})\langle\uparrow'|\uparrow'\rangle - (b_{k_-}a_{k_+})\langle\downarrow'|\downarrow'\rangle\right\} \\ &\quad + i\hat{P}\hat{r}_3\left\{(c_{k_+}a_{k_-} + c_{k_-}a_{k_+})\langle\downarrow'|\uparrow'\rangle\right\} = \frac{\hat{P}}{2}\hat{r}_3(i\hat{r}_1 - \hat{r}_2) \\ &\quad \left\{\left(\frac{a_{k_-}b_{k_+}}{\sqrt{2}} + \frac{b_{k_-}a_{k_+}}{\sqrt{2}}\right)\right\} + \frac{\hat{P}}{2}\hat{r}_3(i\hat{r}_1 - \hat{r}_2)\left\{(c_{k_+}a_{k_-} + c_{k_-}a_{k_+})\right\}.\end{aligned}$$

Thus,

$$\hat{p}_{\mathrm{CV}}\left(\vec{k}\right) = \frac{\hat{P}}{2}\hat{r}_3(i\hat{r}_1 - \hat{r}_2)\left\{a_{k_+}\left(\frac{b_{k_-}}{\sqrt{2}} + c_{k_-}\right) + a_{k_-}\left(\frac{b_{k_+}}{\sqrt{2}} + c_{k_+}\right)\right\}. \tag{8.1o}$$

We can write that $|\hat{r}_1| = |\hat{r}_2| = |\hat{r}_3| = 1$, also, $\hat{P}\hat{r}_3 = \hat{P}_x \sin\theta\cos\phi\hat{i} + \hat{P}_y \sin\theta \sin\phi\hat{j} + \hat{P}_z\cos\theta\hat{k}$, where $\hat{P} = \left\langle S|\hat{P}|X\right\rangle = \left\langle S|\hat{P}|Y\right\rangle = \left\langle S|\hat{P}|Z\right\rangle$, $\left\langle S|\hat{P}|X\right\rangle = \int u_C^*(0,\vec{r})\hat{P}u_{VX}(0,\vec{r})d^3r = \hat{P}_{CVX}(0)$, $\left\langle S|\hat{P}|Y\right\rangle = \hat{P}_{CVY}(0)$, and $\left\langle S|\hat{P}|Z\right\rangle = \hat{P}_{CVZ}(0)$.

Thus, $\hat{P} = \hat{P}_{CVX}(0) = \hat{P}_{CVY}(0) = \hat{P}_{CVZ}(0) = \hat{P}_{CV}(0)$, where $\hat{P}_{CV}(0) \equiv \int u_c^*(0,\vec{r})\hat{P}u_V(0,\vec{r})d^3r \equiv \hat{P}$.

For a plane polarized light wave, we have the polarization vector $\vec{\varepsilon}_s = \hat{k}$, when the light wave vector is traveling along the z-axis. Therefore, for a plane polarized light wave, we have considered $\vec{\varepsilon}_s = \hat{k}$.

Then, from (8.1o), we get

$$\left(\vec{\varepsilon}\cdot\hat{p}_{CV}(\vec{k})\right) = \vec{k}\cdot\frac{\hat{P}}{2}\hat{r}_3\left(i\hat{r}_1 - \hat{r}_2\right)\left[A\left(\vec{k}\right) + B\left(\vec{k}\right)\right]\cos\omega t \tag{8.1p}$$

and

$$\left.\begin{aligned} A(\vec{k}) &= a_{k-}\left(\frac{b_{k+}}{\sqrt{2}} + c_{k+}\right) \\ B(\vec{k}) &= a_{k+}\left(\frac{b_{k-}}{\sqrt{2}} + c_{k-}\right) \end{aligned}\right\} \tag{8.1q}$$

Thus,

$$\begin{aligned} \left|\vec{\varepsilon}\cdot\hat{p}_{cv}\left(\vec{k}\right)\right|^2 &= \left|\hat{k}\cdot\frac{\hat{P}}{2}\hat{r}_3\right|^2 |i\hat{r}_1 - \hat{r}_2|^2\left[A(\vec{k}) + B(\vec{k})\right]^2\cos^2\omega t \\ &= \frac{1}{4}\left|\hat{P}_z\cos\theta\right|^2\left[A\left(\vec{k}\right) + B\left(\vec{k}\right)\right]^2\cos^2\omega t. \end{aligned} \tag{8.1r}$$

So, the average value of $\left|\vec{\varepsilon}\cdot\hat{p}_{cv}(\vec{k})\right|^2$ for a plane polarized light wave is given by

$$\begin{aligned} \left\langle\left|\vec{\varepsilon}\cdot\hat{p}_{cv}\left(\vec{k}\right)\right|^2\right\rangle_{av} &= \frac{2}{4}\left|\hat{P}_z\right|^2\left[A(\vec{k}) + B(\vec{k})\right]^2\left(\int_0^{2\pi}d\phi\int_0^{\pi}\cos^2\theta\sin\theta d\theta\right) \\ \left(\frac{1}{2}\right) &= \frac{2\pi}{3}\left|\hat{P}_z\right|^2\left[A\left(\vec{k}\right) + B\left(\vec{k}\right)\right]^2 \end{aligned} \tag{8.1s}$$

where $\left|\hat{P}_z\right|^2 = \left(\frac{1}{2}\right)\left|\vec{k}\cdot\hat{p}_{cv}(0)\right|^2$ and

$$\left|\vec{k}\cdot\hat{p}_{cv}(0)\right|^2 = \frac{m^2}{4m_r}\frac{E_{g0}\left(E_{g0} + \Delta\right)}{\left(E_{g0} + \frac{2}{3}\Delta\right)}. \tag{8.1t}$$

We can express $A(\vec{k})$ and $B(\vec{k})$ in terms of constants of the energy spectra in the following way.

Substituting $a_{k_\pm}$, $b_{k_\pm}$, $c_{k_\pm}$, and $\gamma_{0k_\pm}$ in $A(\vec{k})$ and $B(\vec{k})$ in (8.1q) we get

$$A(\vec{k}) = \beta\left(t + \frac{\rho}{\sqrt{2}}\right)\left\{\left(\frac{E_{g0}}{E_{g0}+\delta'}\right)\gamma_{0k_+}^2 - \gamma_{0k_+}^2\gamma_{0k_-}^2\left(\frac{E_{g0}-\delta'}{E_{g0}+\delta'}\right)\right\}^{1/2} \quad (8.1u)$$

$$B(\vec{k}) = \beta\left(t + \frac{\rho}{\sqrt{2}}\right)\left\{\left(\frac{E_{g0}}{E_{g0}+\delta'}\right)\gamma_{0k_-}^2 - \gamma_{0k_+}^2\gamma_{0k_-}^2\left(\frac{E_{g0}-\delta'}{E_{g0}+\delta'}\right)\right\}^{1/2} \quad (8.1v)$$

in which

$$\gamma_{0k_+}^2 \equiv \frac{\xi_{1k}-E_{g0}}{2(\xi_{1k}+\delta')} \equiv \frac{1}{2}\left[1-\left(\frac{E_{g0}+\delta'}{\xi_{1k}+\delta'}\right)\right] \quad \text{and}$$

$$\gamma_{0k_-}^2 \equiv \frac{\xi_{1k}+E_{g0}}{2(\xi_{1k}+\delta')} \equiv \frac{1}{2}\left[1+\left(\frac{E_{g0}-\delta'}{\xi_{1k}+\delta'}\right)\right].$$

Substituting $x \equiv \xi_{1k}+\delta'$ in $\gamma_{0k_\pm}^2$, we can write

$$A(\vec{k}) = \beta\left(t + \frac{\rho}{\sqrt{2}}\right)\left\{\left(\frac{E_{g0}}{E_{g0}+\delta'}\right)\frac{1}{2}\left(1-\frac{E_{g0}+\delta'}{x}\right)\right.$$
$$\left. -\frac{1}{4}\left(\frac{E_{g0}-\delta'}{E_{g0}+\delta'}\right)\left(1-\frac{E_{g0}+\delta'}{x}\right)\left(1+\frac{E_{g0}-\delta'}{x}\right)\right\}^{1/2}.$$

Thus,

$$A(\vec{k}) = \frac{\beta}{2}\left(t+\frac{\rho}{\sqrt{2}}\right)\left\{1-\frac{2a_0}{x}+\frac{a_1}{x^2}\right\}^{1/2},$$

where $a_0 \equiv \left(E_{g0}^2+\delta'^2\right)\left(E_{g0}+\delta'\right)^{-1}$ and $a_1 \equiv \left(E_{g0}-\delta'\right)^2$.

After tedious algebra, one can show that

$$A(\vec{k}) = \frac{\beta}{2}\left(t+\frac{\rho}{\sqrt{2}}\right)\left(E_{g0}-\delta'\right)\left[\frac{1}{\xi_{1k}+\delta'}-\frac{1}{E_{g0}+\delta'}\right]^{1/2}$$
$$\left[\frac{1}{\xi_{1k}+\delta'}-\frac{\left(E_{g0}+\delta'\right)}{\left(E_{g0}-\delta'\right)^2}\right]^{1/2} \quad (8.1w)$$

Similarly, from (8.1v) can write

$$B(\vec{k}) = \beta\left(t + \frac{\rho}{\sqrt{2}}\right)\left\{\left(\frac{E_{g0}}{E_{g0}+\delta'}\right)\frac{1}{2}\left(1+\frac{E_{g0}-\delta'}{x}\right)\right.$$
$$\left. -\frac{1}{4}\left(\frac{E_{g0}-\delta'}{E_{g0}+\delta'}\right)\left(1-\frac{E_{g0}+\delta'}{x}\right)\left(1+\frac{E_{g0}-\delta'}{x}\right)\right\}^{1/2}.$$

So that finally we get

$$B\left(\vec{k}\right) = \frac{\beta}{2}\left(t + \frac{\rho}{\sqrt{2}}\right)\left(1 + \frac{E_{g0} - \delta'}{\xi_{1k} + \delta'}\right). \tag{8.1x}$$

In the presence of light wave, the Hamiltonian ($\hat{H}$) of an electron characterized by the vector potential $\vec{A}$ can be written following [15] as

$$\hat{H} = \left[\left|\left(\hat{p} + |e|\,\vec{A}\right)\right|^2 / \, 2m\right] + V(\vec{r}) \tag{8.1y}$$

in which $\hat{p}$ is the momentum operator, $V(\vec{r})$ is the crystal potential, and m is the free electron mass.

Equation (8.1y) can be expressed as

$$\hat{H} = \hat{H}_0 + \hat{H}' \tag{8.1A}$$

where

$$\hat{H}_0 = \frac{\hat{p}^2}{2m} + V(\vec{r}) \text{ and } \hat{H}' = \frac{|e|}{2m}\vec{A} \cdot \hat{p}. \tag{8.1B}$$

The perturbed Hamiltonian $\hat{H}'$ can be written as

$$\hat{H}' = \left(\frac{-i\hbar\,|e|}{2m}\right)\left(\vec{A} \cdot \nabla\right) \tag{8.1C}$$

where $i = \sqrt{-1}$ and $\hat{p} = -i\hbar\nabla$

The vector potential ($\vec{A}$) of the monochromatic light of plane wave can be expressed as

$$\vec{A} = A_0 \vec{\varepsilon}_s \cos(\vec{s}_0 \cdot \vec{r} - \omega t) \tag{8.1D}$$

where A_0 is the amplitude of the light wave, $\vec{\varepsilon}_s$ is the polarization vector, $\vec{s}_0$ is the momentum vector of the incident photon, $\vec{r}$ is the position vector, ω is the angular frequency of light wave, and t is the time scale. The matrix element of $\hat{H}'_{nl}$ between initial state, $\psi_l(\vec{q}, \vec{r})$ and final state $\psi_n(\vec{k}, \vec{r})$ in different bands can be written as

$$\hat{H}'_{nl} = \frac{|e|}{2m}\left\langle n\vec{k}\left|\vec{A} \cdot \hat{p}\right| l\vec{q}\right\rangle \tag{8.1E}$$

Using (8.1C) and (8.1D), we can rewrite (8.1E) as

$$\hat{H}'_{nl} = \left(\frac{-i\hbar\,|e|\,A_0}{4m}\right)\vec{\varepsilon}_s \cdot \left[\left\{\left\langle n\vec{k}\left|e^{(i\vec{s}_0\cdot\vec{r})}\nabla\right| l\vec{q}\right\rangle e^{-i\omega t}\right\} + \left\{\left\langle n\vec{k}\left|e^{(-i\vec{s}_0\cdot\vec{r})}\nabla\right| l\vec{q}\right\rangle e^{i\omega t}\right\}\right]. \tag{8.1F}$$

The first matrix element of (8.1F) can be written as

$$\left\langle n\vec{k}\left|\mathrm{e}^{(\mathrm{i}\vec{s}_0\cdot\vec{r})}\nabla\right|l\vec{q}\right\rangle = \int \mathrm{e}^{\left(\mathrm{i}\left[\vec{q}+\vec{s}_0-\vec{k}\right]\cdot\vec{r}\right)} i\vec{q}u_n^*\left(\vec{k},\vec{r}\right)u_l\left(\vec{q},\vec{r}\right)\mathrm{d}^3r + \int \mathrm{e}^{\left(\mathrm{i}\left[\vec{q}+\vec{s}_0-\vec{k}\right]\cdot\vec{r}\right)} u_n^*\left(\vec{k},\vec{r}\right)\nabla u_l\left(\vec{q},\vec{r}\right)\mathrm{d}^3r. \quad (8.1G)$$

The functions $u_n^*u_l$ and $u_n^*\nabla u_l$ are periodic. The integral over all space can be separated into a sum over unit cells times an integral over a single unit cell. It is assumed that the wave length of the electromagnetic wave is sufficiently large so that if $\vec{k}$ and $\vec{q}$ are within the Brillouin zone, $(\vec{q}+\vec{s}_0-\vec{k})$ is not a reciprocal lattice vector.

Therefore, we can write equation (8.1G) as

$$\begin{aligned} &< n\vec{k}\left|\mathrm{e}^{(\mathrm{i}\vec{s}_0\cdot\vec{r})}\nabla\right|l\vec{q}> \\ &= \left[\frac{(2\pi)^3}{\Omega}\right]\left\{i\vec{q}\delta\left(\vec{q}+\vec{s}_0-\vec{k}\right)\delta_{nl} + \delta\left(\vec{q}+\vec{s}_0-\vec{k}\right)\int_{\text{cell}} u_n^*\left(\vec{k},\vec{r}\right)\nabla u_l\left(\vec{q},\vec{r}\right)\mathrm{d}^3r\right\} \\ &= \left[\frac{(2\pi)^3}{\Omega}\right]\left\{\delta\left(\vec{q}+\vec{s}_0-\vec{k}\right)\int_{\text{cell}} u_n^*\left(\vec{k},\vec{r}\right)\nabla u_l\left(\vec{q},\vec{r}\right)\mathrm{d}^3r\right\}, \end{aligned} \quad (8.1H)$$

where Ω is the volume of the unit cell and $\int u_n^*(\vec{k},\vec{r})u_l(\vec{q},\vec{r})\mathrm{d}^3r = \delta(\vec{q}-\vec{k})\delta_{nl} = 0$, since $n \neq l$.

The delta function expresses the conservation of wave vector in the absorption of light wave and $\vec{s}_0$ is small compared to the dimension of a typical Brillouin zone and we set $\vec{q} = \vec{k}$.

From (8.1G) and (8.1H), we can write

$$\hat{H}'_{nl} = \frac{|e|\,A_0}{2m}\vec{\varepsilon}_{\mathrm{s}}\cdot\hat{p}_{nl}(\vec{k})\delta(\vec{q}-\vec{k})\cos(\omega t) \quad (8.1I)$$

where $\hat{p}_{nl}(\vec{k}) = -\mathrm{i}\hbar\int u_n^*\nabla u_l\mathrm{d}^3r = \int u_n^*(\vec{k},\vec{r})\hat{p}u_l(\vec{k},\vec{r})\mathrm{d}^3r$

Therefore, we can write

$$\hat{H}'_{nl} = \frac{|e|\,A_0}{2m}\vec{\varepsilon}\cdot\hat{p}_{nl}(\vec{k}) \quad (8.1J)$$

where $\vec{\varepsilon} = \vec{\varepsilon}_{\mathrm{s}}\cos\omega t$.

When a photon interacts with a semiconductor, the carriers (i.e., electrons) are generated in the bands which are followed by the inter-band transitions. For example, when the carriers are generated in the valence band, the carriers then make inter-band transition to the conduction band. The transition of the electrons within the same band, i.e., $\hat{H}'_{nn} = \left\langle n\vec{k}|\hat{H}'|n\vec{k}\right\rangle$, is neglected. Because, in such a case, i.e., when the carriers are generated within the same bands by photons, are lost by recombination within the aforementioned band resulting zero carriers.

Therefore,

$$< n\vec{k} \left| \hat{H}' \right| n\vec{k} > = 0 \tag{8.1K}$$

with $n = \text{c}$ stands for conduction band and $l = \text{v}$ stands for valance band, the energy equation for the conduction electron can approximately be written as

$$\gamma(E) = \left(\frac{\hbar^2 k^2}{2m^*}\right) + \frac{\left(\frac{|e|A_0}{2m}\right)^2 \left\langle \left| \vec{\varepsilon} \cdot \hat{p}_{\text{cv}}(\vec{k}) \right|^2 \right\rangle_{\text{av}}}{E_{\text{c}}(\vec{k}) - E_{\text{v}}(\vec{k})}, \tag{8.1L}$$

where $\gamma(E) \equiv E(aE + 1)(bE + 1)/(cE + 1)$, $a \equiv 1/E_{g_0}$, E_{g_0} is the unperturbed band gap, $b \equiv 1/\left(E_{g_0} + \Delta\right)$, $c \equiv 1/\left(E_{g_0} + 2\Delta/3\right)$, and $\left\langle \left| \vec{\varepsilon} \cdot \hat{p}_{\text{cv}}(\vec{k}) \right|^2 \right\rangle_{\text{av}}$ represents the average of the square of the optical matrix element (OME).

For the three band model of Kane, we can write,

$$\xi_{1k} = E_{\text{c}}(\vec{k}) - E_{\text{v}}(\vec{k}) = \left(E_{g_0}^2 + E_{g_0}\hbar^2 k^2/m_{\text{r}}\right)^{1/2}, \tag{8.1M}$$

where m_{r} is the reduced mass and is given by $m_{\text{r}}^{-1} = (m^*)^{-1} + m_{\text{v}}^{-1}$ and m_{v} is the effective mass of the heavy hole at the top of the valance band in the absence of any field.

Thus, combining the appropriate equations, we can write

$$\begin{aligned}
\left(\frac{|e|\,A_0}{2m}\right)^2 \frac{\left\langle \left| \vec{\varepsilon} \cdot \hat{p}_{\text{cv}}(\vec{k}) \right|^2 \right\rangle_{\text{av}}}{E_{\text{c}}(\vec{k}) - E_{\text{v}}(\vec{k})} &= \left(\frac{|e|\,A_0}{2m}\right)^2 \frac{2\pi}{3} \left| \vec{k} \cdot \hat{p}_{\text{cv}}(0) \right|^2 \frac{\beta^2}{4} \left(t + \frac{\rho}{\sqrt{2}}\right)^2 \\
&\times \frac{1}{\xi_{1k}} \left\{ \left(1 + \frac{E_{g_0} - \delta'}{\xi_{1k} + \delta'}\right) + (E_{g_0} - \delta') \left[\frac{1}{\xi_{1k} + \delta'} - \frac{1}{E_{g_0} + \delta'}\right]^{1/2} \right. \\
&\left. \times \left[\frac{1}{\xi_{1k} + \delta'} - \frac{E_{g_0} + \delta'}{\left(E_{g_0} - \delta'\right)^2}\right]^{1/2} \right\}^2.
\end{aligned} \tag{8.1N}$$

It is well known that [15]

$$A_0^2 = \frac{I\lambda^2}{2\pi^2 c^3 \sqrt{\varepsilon_{\text{sc}}\varepsilon_0}} \tag{8.1O}$$

where I is the light intensity of wavelength λ, ε_0 is the permittivity of free space, and c is the velocity of light. Thus, the simplified electron energy spectrum in III–V, ternary, and quaternary materials up to the second order in the presence of light waves can approximately be written as

$$\frac{\hbar^2 k^2}{2m^*} = \beta_0(E, \lambda) \tag{8.1P}$$

where $\beta_0(E,\lambda) \equiv [\gamma(E) - \theta_0(E,\lambda)]$,

$$\theta_0(E,\lambda) \equiv \frac{|e|^2}{96 m_r \pi c^3} \frac{I\lambda^2}{\sqrt{\varepsilon_{sc}\varepsilon_0}} \frac{E_{g0}(E_{g0}+\Delta)}{\left(E_{g0}+\frac{2}{3}\Delta\right)} \frac{\beta^2}{4} \left(t + \frac{\rho}{\sqrt{2}}\right)^2 \frac{1}{\phi_0(E)}$$
$$\left\{\left(1 + \frac{E_{g0}-\delta'}{\phi_0(E)+\delta'}\right) + (E_{g0}-\delta')\left[\frac{1}{\phi_0(E)+\delta'} - \frac{1}{E_{g0}+\delta'}\right]^{1/2}\right.$$
$$\left.\left[\frac{1}{\phi_0(E)+\delta'} - \frac{E_{g0}+\delta'}{(E_{g0}-\delta')^2}\right]^{1/2}\right\}^2,$$

and

$$\phi_0(E) \equiv E_{g0}\left(1 + 2\left(1 + \frac{m^*}{m_v}\right)\frac{\gamma(E)}{E_{g0}}\right)^{1/2}.$$

1. For the two band model of Kane, we have $\Delta \to 0$. Under this condition, $\gamma(E) \to E(1+aE) = \frac{\hbar^2 k^2}{2m^*}$. Since $\beta \to 1$, $t \to 1$, $\rho \to 0$, $\delta' \to 0$ for $\Delta \to 0$, from (8.1P), we can write the energy spectrum of III–V, ternary and quaternary materials in the presence of external photoexcitation whose unperturbed conduction electrons obey the two band model of Kane as

$$\frac{\hbar^2 k^2}{2m^*} = \tau_0(E,\lambda) \tag{8.2}$$

where $\tau_0(E,\lambda) \equiv E(1+aE) - B_0(E,\lambda)$,

$$B_0(E,\lambda) \equiv \frac{|e|^2 I\lambda^2 E_{g0}}{384\pi c^3 m_r \sqrt{\varepsilon_{sc}\varepsilon_0}} \frac{1}{\phi_1(E)} \left\{\left(1 + \frac{E_{g0}}{\phi_1(E)}\right) + E_{g0}\left[\frac{1}{\phi_1(E)} - \frac{1}{E_{g0}}\right]\right\}^2,$$
$$\phi_1(E) \equiv E_{g0}\left\{1 + \frac{2m^*}{m_r} aE(1+aE)\right\}^{1/2}.$$

2. For relatively wide band gap semiconductors, one can write $a \to 0$, $b \to 0$, $c \to 0$, and $\gamma(E) \to E$.
Thus, from (8.2), we get,

$$\frac{\hbar^2 k^2}{2m^*} = \rho_0(E,\lambda)$$
$$\rho_0(E,\lambda) \equiv E - \frac{|e|^2 I\lambda^2}{96\pi c^3 m_r \sqrt{\varepsilon_{sc}\varepsilon_0}}\left[1 + \left(\frac{2m^*}{m_r}\right) aE\right]^{-3/2}. \tag{8.3}$$

The dispersion relation of the 2D electrons in UFs of optoelectronic materials, the conduction electrons of whose bulk samples are defined by the dispersion relations as given by (8.1P), (8.2), and (8.3) can, respectively, be expressed as

$$k_x^2 + k_y^2 = \frac{2m^* \beta_0(E,\lambda)}{\hbar^2} - \left(\frac{\pi n_{z81}}{d_z}\right)^2, \tag{8.4}$$

$$k_x^2 + k_y^2 = \frac{2m^* \tau_0(E,\lambda)}{\hbar^2} - \left(\frac{\pi n_{z82}}{d_z}\right)^2, \tag{8.5}$$

$$k_x^2 + k_y^2 = \frac{2m^* \rho_0(E,\lambda)}{\hbar^2} - \left(\frac{\pi n_{z83}}{d_z}\right)^2, \tag{8.6}$$

where $n_{z8J}(J = 1, 2, 3)$ is the size quantum number.

The electron concentration per unit area in the presence of light waves assumes the form

$$n_0 = \left(\frac{m^* g_v}{\pi \hbar^2}\right) \sum_{n_{z81}=1}^{n_{z81\max}} [\phi_{81}(E_{F2DL}, n_{z81}) + \phi_{82}(E_{F2DL}, n_{z81})], \tag{8.7}$$

$$n_0 = \left(\frac{m^* g_v}{\pi \hbar^2}\right) \sum_{n_{z82}=1}^{n_{z82\max}} [\phi_{83}(E_{F2DL}, n_{z82}) + \phi_{84}(E_{F2DL}, n_{z82})], \tag{8.8}$$

$$n_0 = \left(\frac{m^* g_v}{\pi \hbar^2}\right) \sum_{n_{z83}=1}^{n_{z83\max}} [\phi_{85}(E_{F2DL}, n_{z83}) + \phi_{86}(E_{F2DL}, n_{z83})], \tag{8.9}$$

where E_{F2DL} is the Fermi energy in UFs in the presence of light waves as measured from the edge of the conduction band in the vertically upward direction in the absence of any quantization,

$$\phi_{81}(E_{F2DL}, n_{z81}) = \left[\frac{2m^*}{\hbar^2}\beta_0(E_{F2DL}, \lambda) - \left(\frac{\pi n_{z81}}{d_z}\right)^2\right],$$

$$\phi_{82}(E_{F2DL}, n_{z81}) = \sum_{r=1}^{s_0} Z_{r,Y}[\phi_{81}(E_{F2DL}, n_{z81})], \; Y = 2\text{DL},$$

$$\phi_{83}(E_{F2DL}, n_{z82}) = \left[\frac{2m^*}{\hbar^2}\tau(E_{F2DL}, \lambda) - \left(\frac{\pi n_{z82}}{d_z}\right)^2\right],$$

$$\phi_{84}(E_{F2DL}, n_{z82}) = \sum_{r=1}^{s_0} Z_{r,Y}[\phi_{83}(E_{F2DL}, n_{z82})],$$

$$\phi_{85}(E_{F2DL}, n_{z83}) = \left[\frac{2m^*}{\hbar^2}\rho_0(E_{F2DL}, \lambda) - \left(\frac{\pi n_{z83}}{d_z}\right)^2\right],$$

and

$$\phi_{86}(E_{F2DL}, n_{z83}) = \sum_{r=1}^{s_0} Z_{r,Y}[\phi_{85}(E_{F2DL}, n_{z83})].$$

Combining (1.13) with (8.7)–(8.9), the opto-TPSM in UFs of optoelectronic materials under large magnetic field in accordance with perturbed three and two band models of Kane together with perturbed parabolic energy bands can, respectively, be written as

$$G_0 = \left(\frac{\pi^2 k_B^2 T}{3e}\right)\left[\sum_{n_{z81}=1}^{n_{z81\max}} \left[\phi_{81}\left(E_{F2DL}, n_{z81}\right) + \phi_{82}\left(E_{F2DL}, n_{z81}\right)\right]\right]^{-1} \times \left[\sum_{n_{z81}=1}^{n_{z81\max}} \left[\phi'_{81}(E_{F2DL}, n_{z81}) + \phi'_{82}(E_{F2DL}, n_{z81})\right]\right], \quad (8.10)$$

$$G_0 = \left(\frac{\pi^2 k_B^2 T}{3e}\right)\left[\sum_{n_{z82}=1}^{n_{z82\max}} \left[\phi_{83}(E_{F2DL}, n_{z82}) + \phi_{84}(E_{F2DL}, n_{z82})\right]\right]^{-1} \times \left[\sum_{n_{z82}=1}^{n_{z82\max}} \left[\phi'_{83}\left(E_{F2DL}, n_{z82}\right) + \phi'_{84}\left(E_{F2DL}, n_{z82}\right)\right]\right], \quad (8.11)$$

and

$$G_0 = \left(\frac{\pi^2 k_B^2 T}{3e}\right)\left[\sum_{n_{z83}=1}^{n_{z83\max}} \left[\phi_{85}\left(E_{F2DL}, n_{z83}\right) + \phi_{86}\left(E_{F2DL}, n_{z83}\right)\right]\right]^{-1} \times \left[\sum_{n_{z83}=1}^{n_{z83\max}} \left[\phi'_{85}(E_{F2DL}, n_{z83}) + \phi'_{86}(E_{F2DL}, n_{z83})\right]\right]. \quad (8.12)$$

8.2.2 Optothermoelectric Power in Quantum Wires of Optoelectronic Materials Under Large Magnetic Field

The dispersion relations of the 1D electrons in QWs of optoelectronic materials in the presence of light waves can be expressed from (8.4)–(8.6) as

$$k_y^2 = \frac{2m^*\beta_0(E,\lambda)}{\hbar^2} - \left(\frac{\pi n_{z81}}{d_z}\right)^2 - \left(\frac{\pi n_{x81}}{d_x}\right)^2, \quad (8.13)$$

$$k_y^2 = \frac{2m^*\tau_0(E,\lambda)}{\hbar^2} - \left(\frac{\pi n_{z82}}{d_z}\right)^2 - \left(\frac{\pi n_{x82}}{d_x}\right)^2, \quad (8.14)$$

$$k_y^2 = \frac{2m^*\rho_0(E,\lambda)}{\hbar^2} - \left(\frac{\pi n_{z83}}{d_z}\right)^2 - \left(\frac{\pi n_{x83}}{d_x}\right)^2, \quad (8.15)$$

where $n_{x8J}(J = 1, 2, 3)$ is the size quantum number.

The electron concentration per unit length in the presence of light waves in this case is given by

$$n_0=\frac{2g_{\mathrm{v}}\sqrt{2m^*}}{\pi\hbar}\sum_{n_{x81}=1}^{n_{x81\max}}\sum_{n_{z81}=1}^{n_{z81\max}}\left[\phi_{87}\left(E_{\mathrm{F1DL}},n_{x81},n_{z81}\right)+\phi_{88}\left(E_{\mathrm{F1DL}},n_{x81},n_{z81}\right)\right], \tag{8.16}$$

$$n_0=\frac{2g_{\mathrm{v}}\sqrt{2m^*}}{\pi\hbar}\sum_{n_{x82}=1}^{n_{x82\max}}\sum_{n_{z82}=1}^{n_{z82\max}}\left[\phi_{89}\left(E_{\mathrm{F1DL}},n_{x82},n_{z82}\right)+\phi_{810}\left(E_{\mathrm{F1DL}},n_{x82},n_{z82}\right)\right], \tag{8.17}$$

$$n_0=\frac{2g_{\mathrm{v}}\sqrt{2m^*}}{\pi\hbar}\sum_{n_{x83}=1}^{n_{x83\max}}\sum_{n_{z83}=1}^{n_{z83\max}}\left[\phi_{811}\left(E_{\mathrm{F1DL}},n_{x83},n_{z83}\right)+\phi_{812}\left(E_{\mathrm{F1DL}},n_{x83},n_{z83}\right)\right], \tag{8.18}$$

where E_{F1DL} is the Fermi energy in QWs in the presence of light waves as measured from the edge of the conduction band in the vertically upward direction in the absence of any quantization,

$$\phi_{87}\left(E_{\mathrm{F1DL}},n_{x81},n_{z81}\right)=\left[\beta_0\left(E_{\mathrm{F1DL}},\lambda\right)-G_{81}\left(n_{x81},n_{z81}\right)\right]^{1/2},$$

$$G_{8i}\left(n_{x8i},n_{z8i}\right)=\frac{\hbar^2}{2m^*}\left[\left(\frac{\pi n_{x8i}}{d_x}\right)^2+\left(\frac{\pi n_{z8i}}{d_z}\right)^2\right],$$

$$\phi_{88}\left(E_{\mathrm{F1DL}},n_{x81},n_{z81}\right)=\sum_{r=1}^{s_0}Z_{r,Y}\left[\phi_{87}\left(E_{\mathrm{F1DL}},n_{x81},n_{z81}\right)\right],\quad Y=1\mathrm{DL},$$

$$\phi_{89}\left(E_{\mathrm{F1DL}},n_{x82},n_{z82}\right)=\left[\tau_0(E_{\mathrm{F1DL}},\lambda)-G_{82}(n_{x82},n_{z82})\right]^{1/2},$$

$$\phi_{810}\left(E_{\mathrm{F1DL}},n_{x82},n_{z82}\right)=\sum_{r=1}^{s}Z_{r,Y}\left[\phi_{89}\left(E_{\mathrm{F1DL}},n_{x82},n_{z82}\right)\right],$$

$$\phi_{811}(E_{\mathrm{F1DL}},n_{x83},n_{z83})=\left[\rho_0(E_{\mathrm{F1DL}},\lambda)-G_{83}(n_{x83},n_{z83})\right]^{1/2},\ \text{and}$$

$$\phi_{812}\left(E_{\mathrm{F1DL}},n_{x83},n_{z83}\right)=\sum_{r=1}^{s}Z_{r,Y}\left[\phi_{811}\left(E_{\mathrm{F1DL}},n_{x83},n_{z83}\right)\right].$$

Combining (1.13) with (8.16)–(8.18), the opto-TPSM in QWs of optoelectronic materials under large magnetic field in accordance with perturbed three and two band models of Kane together with perturbed parabolic energy bands can, respectively, be written as

$$G_0=\left(\frac{\pi^2k_{\mathrm{B}}^2T}{3e}\right)\left[\sum_{n_{x81}=1}^{n_{x81\max}}\sum_{n_{z81}=1}^{n_{z81\max}}\left[\phi_{87}(E_{\mathrm{F1DL}},n_{x81},n_{z81})+\phi_{88}\left(E_{\mathrm{F1DL}},n_{x81},n_{z81}\right)\right]\right]^{-1}$$
$$\times\left[\sum_{n_{x81}=1}^{n_{x81\max}}\sum_{n_{z81}=1}^{n_{z81\max}}\left[\phi_{87}'\left(E_{\mathrm{F1DL}},n_{x81},n_{z81}\right)+\phi_{88}'\left(E_{\mathrm{F1DL}},n_{x81},n_{z81}\right)\right]\right], \tag{8.19}$$

$$G_0 = \left(\frac{\pi^2 k_B^2 T}{3e}\right) \left[\sum_{n_{x82}=1}^{n_{x82\max}} \sum_{n_{z82}=1}^{n_{z82\max}} \left[\phi_{89}(E_{\mathrm{F1DL}}, n_{x82}, n_{z82}) + \phi_{810}(E_{\mathrm{F1DL}}, n_{x82}, n_{z82})\right]\right]^{-1}$$
$$\times \left[\sum_{n_{x82}=1}^{n_{x82\max}} \sum_{n_{z82}=1}^{n_{z82\max}} \left[\phi'_{89}(E_{\mathrm{F1DL}}, n_{x82}, n_{z82}) + \phi'_{810}(E_{\mathrm{F1DL}}, n_{x82}, n_{z82})\right]\right], \quad (8.20)$$

and

$$G_0 = \left(\frac{\pi^2 k_B^2 T}{3e}\right) \left[\sum_{n_{x83}=1}^{n_{x83\max}} \sum_{n_{z83}=1}^{n_{z83\max}} \left[\phi_{811}(E_{\mathrm{F1DL}}, n_{x83}, n_{z83}) + \phi_{812}(E_{\mathrm{F1DL}}, n_{x83}, n_{z83})\right]\right]^{-1}$$
$$\times \left[\sum_{n_{x83}=1}^{n_{x83\max}} \sum_{n_{z83}=1}^{n_{z83\max}} \left[\phi'_{811}(E_{\mathrm{F1DL}}, n_{x83}, n_{z83}) + \phi'_{812}(E_{\mathrm{F1DL}}, n_{x83}, n_{z83})\right]\right]. \quad (8.21)$$

8.3 Results and Discussion

Using (8.7); (8.10) and (8.8); (8.11) and (8.9); (8.12) for the perturbed three and the two band model of Kane and the perturbed parabolic energy bands together with the energy band constants from Table 1.1, the TPSM in the presence of light waves is shown from Figs. 8.1 to 8.16 as functions of film thickness, electron concentration, light intensity and wavelength for the ultrathin films of InSb, GaAs, $Hg_{1-x}Cd_xTe$, and $In_{1-x}Ga_xAs_yP_{1-y}$, respectively, at 4.2 K. Figure 8.1 exhibits the variation of TPSM for ultrathin films of InSb as a function of film thickness in which $n_0 = 10^{16}\,\mathrm{m}^{-2}$, $I_0 = 0.1\,\mathrm{W\,m}^{-2}$, and $\lambda = 400\,\mathrm{nm}$. The magnitude of the TPSM increases with the increase of well thickness in quantized steps. The influence of different energy band constants for various materials changes the numerical value of the TPSM for all cases as evident from Figs. 8.5, 8.9, and 8.13, respectively. With the increase of temperature, the TPSM is expected to show rather nonsmooth variation over well thickness. It appears from the said figures that the TPSM exhibits largest value for $In_{1-x}Ga_xAs_yP_{1-y}$ and smallest for InSb. In Figs. 8.2, 8.6, 8.10, and 8.14, we have plotted the TPSM as a function of electron concentration for $d_z = 20\,\mathrm{nm}$, $I_0 = 0.1\,\mathrm{W\,m}^{-2}$, and $\lambda = 400\,\mathrm{nm}$ for all the aforesaid cases. The TPSM in the presence of external photoexcitation decreases with the increase in the electron concentration in an oscillatory way. Figures 8.3, 8.7, 8.11, and 8.15 exhibit the variation of the TPSM as a function of light intensity for $n_0 = 10^{16}\mathrm{m}^{-2}$, $d_z = 20\,\mathrm{nm}$, and $\lambda = 400\,\mathrm{nm}$. InSb and $Hg_{1-x}Cd_xTe$ exhibits a smooth increase variation with the intensity. GaAs and $In_{1-x}Ga_xAs_yP_{1-y}$ signatures an oscillatory behavior at the initial stage. It should be noted that the occurrence of the next subbands strongly depends on the materials constants, although they follow the same respective energy dispersion law. In Figs. 8.4, 8.8, 8.12, and 8.16, we have plotted the TPSM as a function of wavelength for $n_0 = 10^{16}\mathrm{m}^{-2}$, $d_z = 20\,\mathrm{nm}$, and $I_0 = 0.1\,\mathrm{W\,m}^{-2}$. The materials are exposed from red to violet radiation for the possible effect of variation of the TPSM for such optoelectronic materials. It appears that the same increases with increasing wavelength and since the quantum limit has been used, the oscillations are absent in this case.

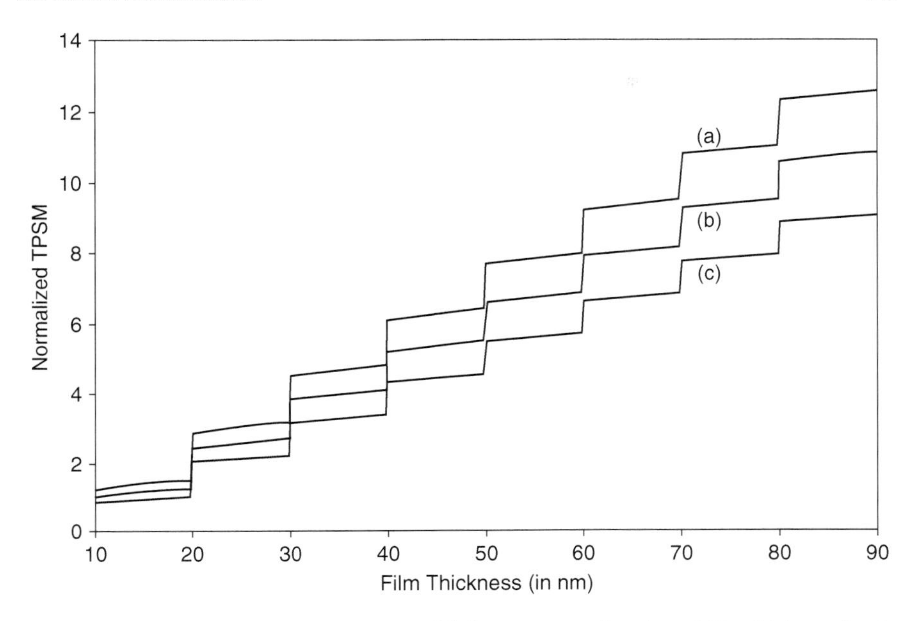

Fig. 8.1 Plot of the TPSM in the presence of light waves as a function of film thickness for the UFs of InSb in accordance with the perturbed (**a**) three and (**b**) two band models of Kane together with the perturbed (**c**) parabolic energy bands

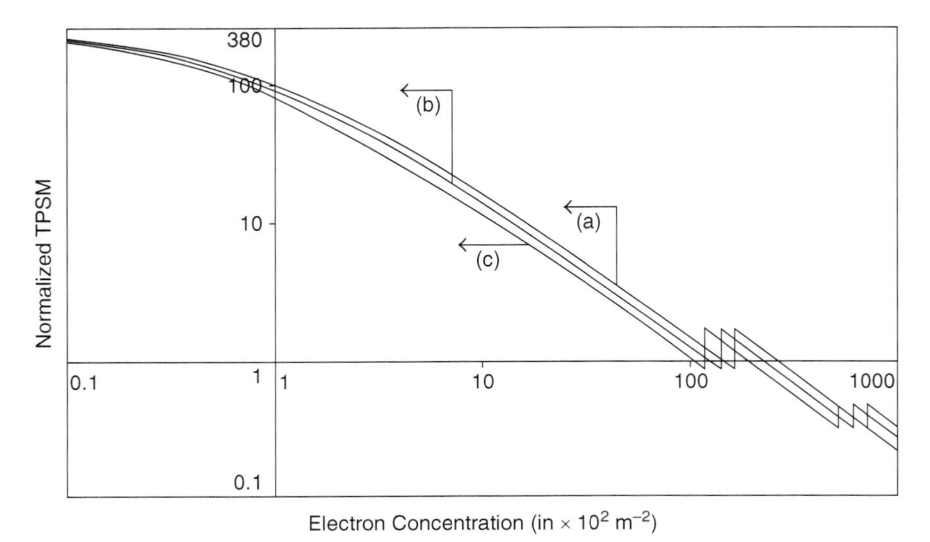

Fig. 8.2 Plot of the TPSM as a function of electron concentration for the UFs of InSb for all cases of Fig. 8.1 in the presence of external photoexcitation

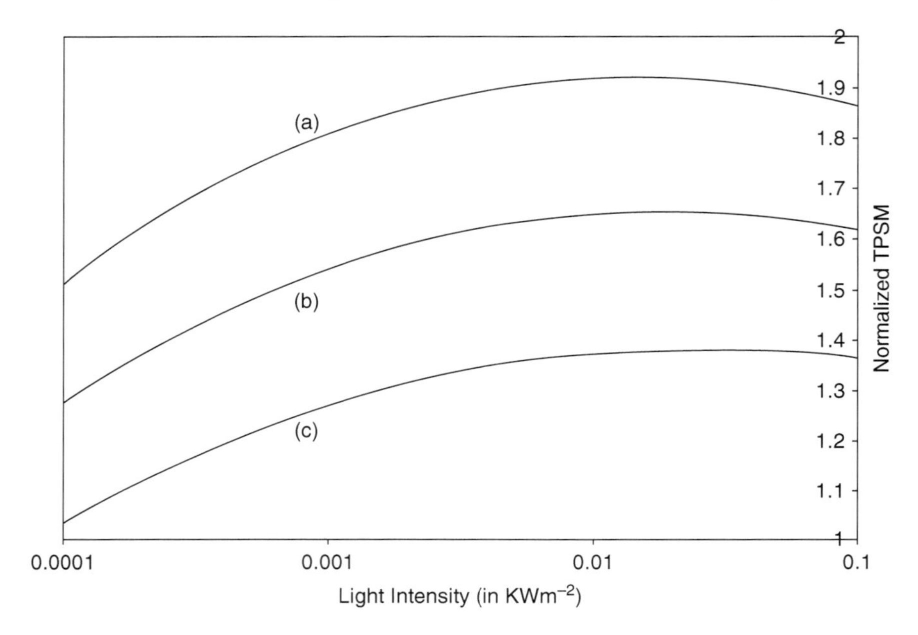

Fig. 8.3 Plot of the TPSM as a function of light intensity for the UFs of InSb for all cases of Fig. 8.1 in the presence of external photoexcitation

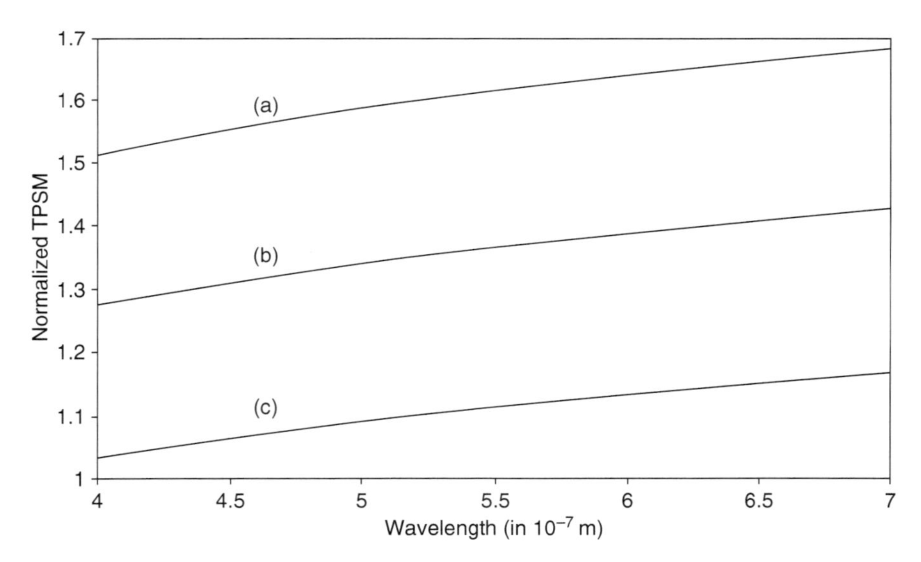

Fig. 8.4 Plot of the TPSM as a function of wavelength for the UFs of InSb for all cases of Fig. 8.1 in the presence of external photoexcitation

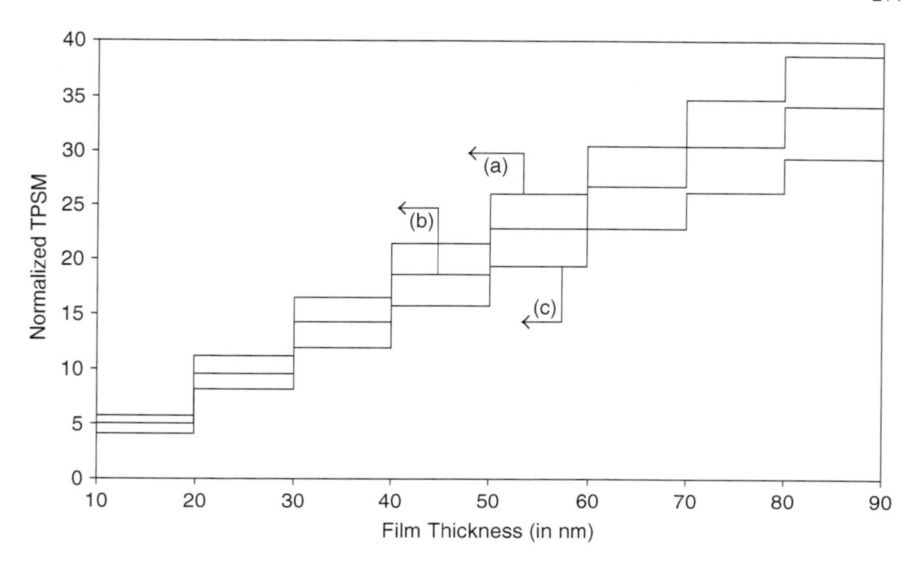

Fig. 8.5 Plot of the TPSM in the presence of light waves as a function of film thickness for the UFs of GaAs for all cases of Fig. 8.1

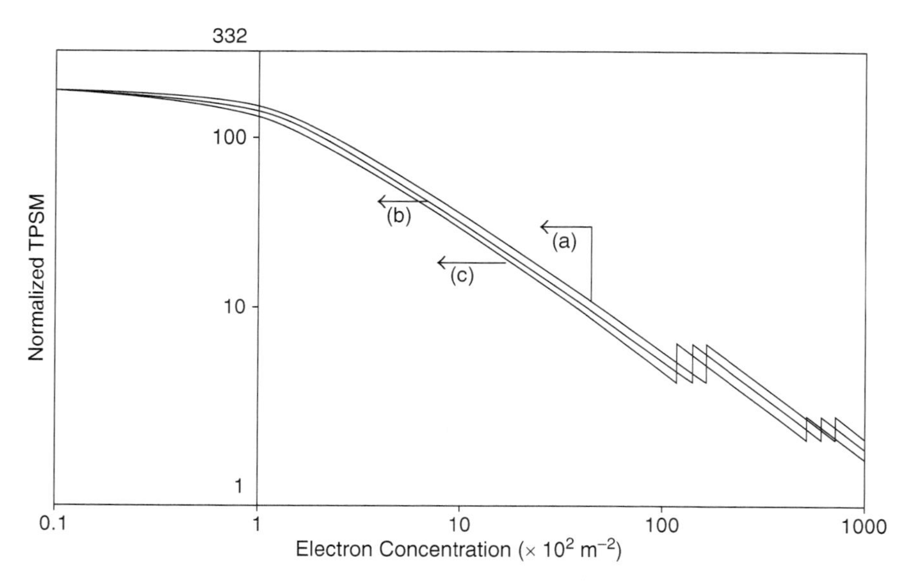

Fig. 8.6 Plot of the TPSM as a function of electron concentration for the UFs of GaAs for all cases of Fig. 8.5 in the presence of external photoexcitation

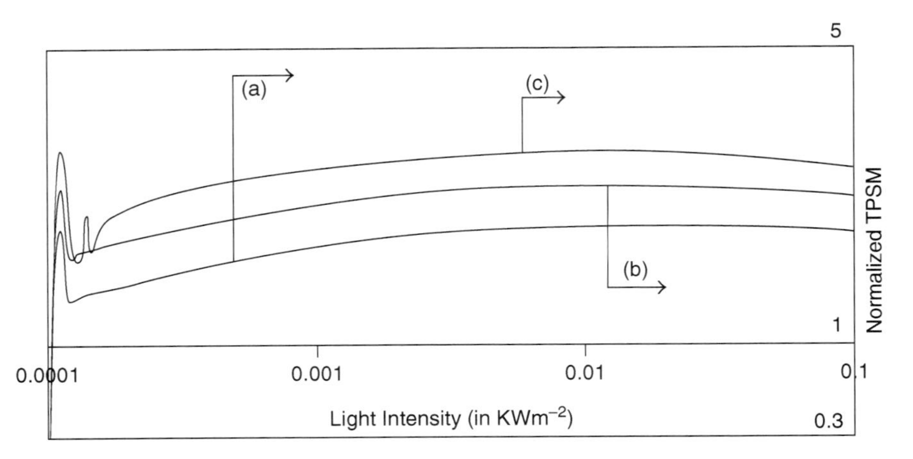

Fig. 8.7 Plot of the TPSM as a function of light intensity for the UFs of GaAs for all cases of Fig. 8.5 in the presence of external photoexcitation

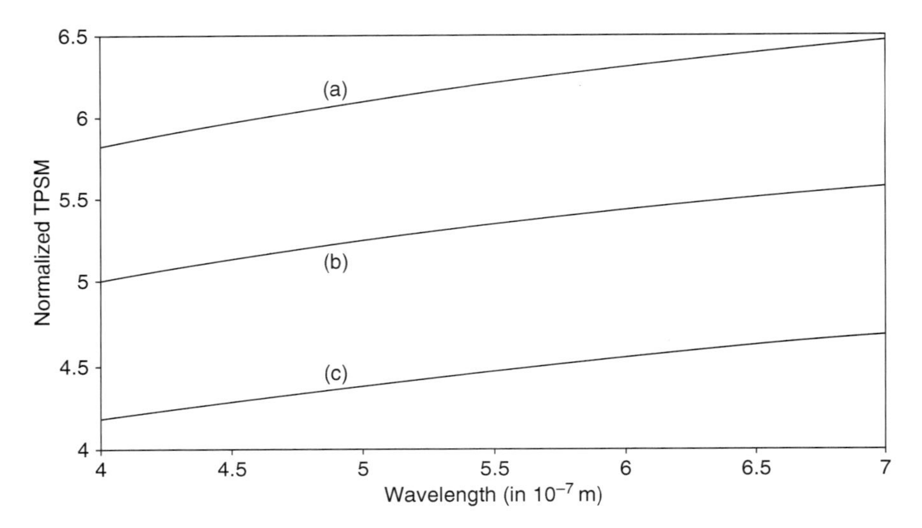

Fig. 8.8 Plot of the TPSM as a function of wavelength for the UFs of GaAs for all cases of Fig. 8.5 in the presence of external photoexcitation

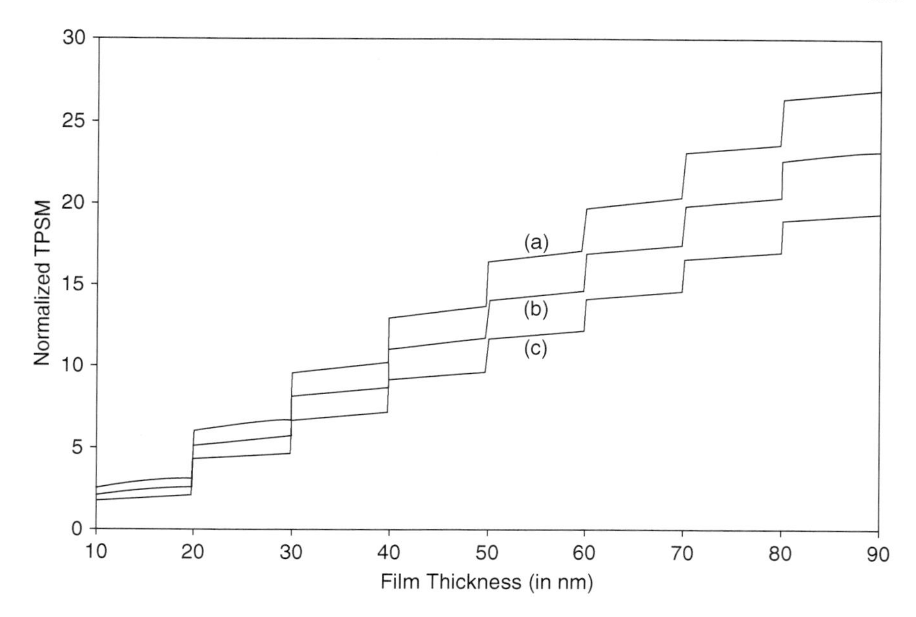

Fig. 8.9 Plot of the TPSM in the presence of light waves as a function of film thickness for the UFs of $Hg_{1-x}Cd_xTe$ for all cases of Fig. 8.1

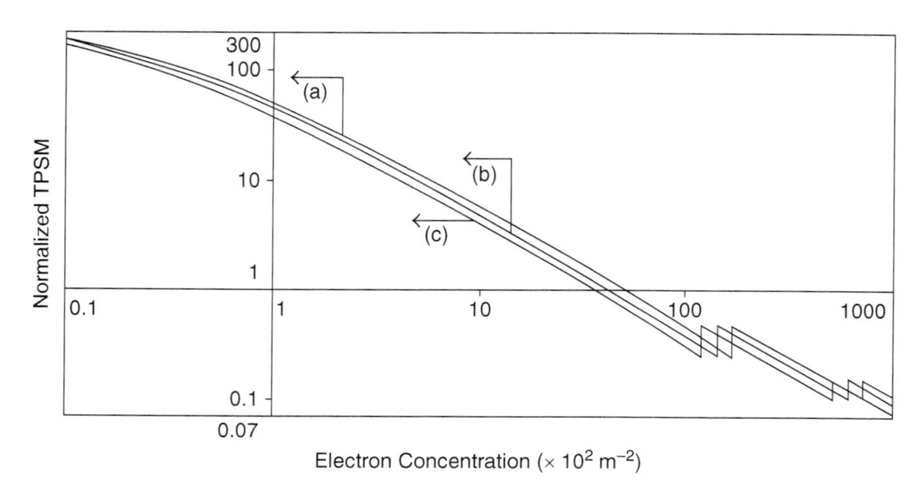

Fig. 8.10 Plot of the TPSM as a function of electron concentration for the UFs of $Hg_{1-x}Cd_xTe$ for all cases of Fig. 8.9 in the presence of external photoexcitation

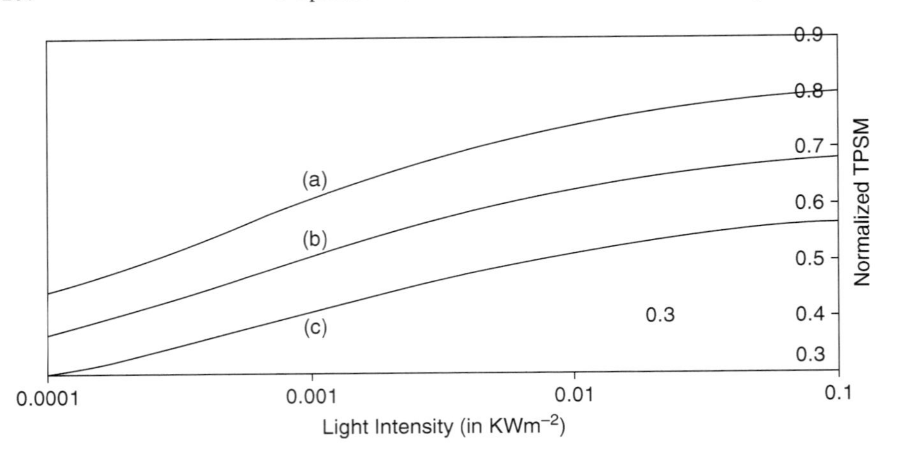

Fig. 8.11 Plot of the TPSM as a function of light intensity for the UFs of $Hg_{1-x}Cd_xTe$ for all cases of Fig. 8.9 in the presence of external photoexcitation

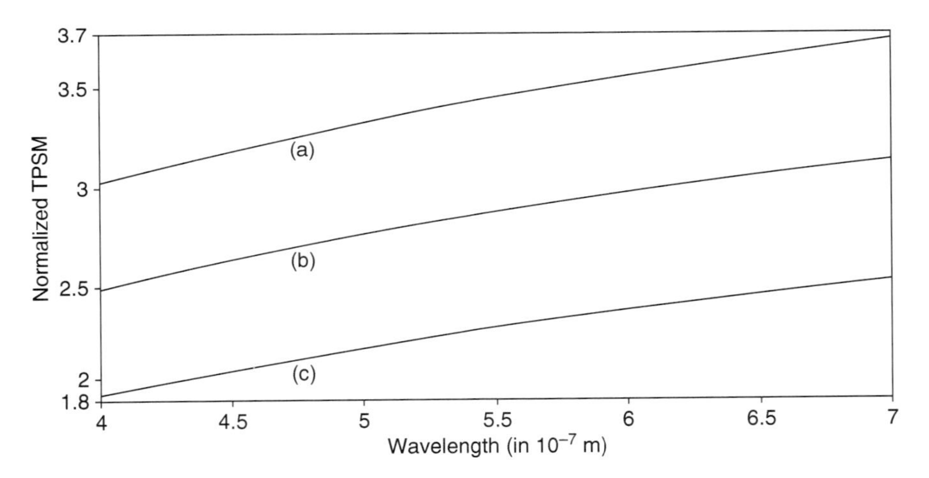

Fig. 8.12 Plot of the TPSM as a function of wavelength for the UFs of $Hg_{1-x}Cd_xTe$ for all cases of Fig. 8.9 in the presence of external photoexcitation

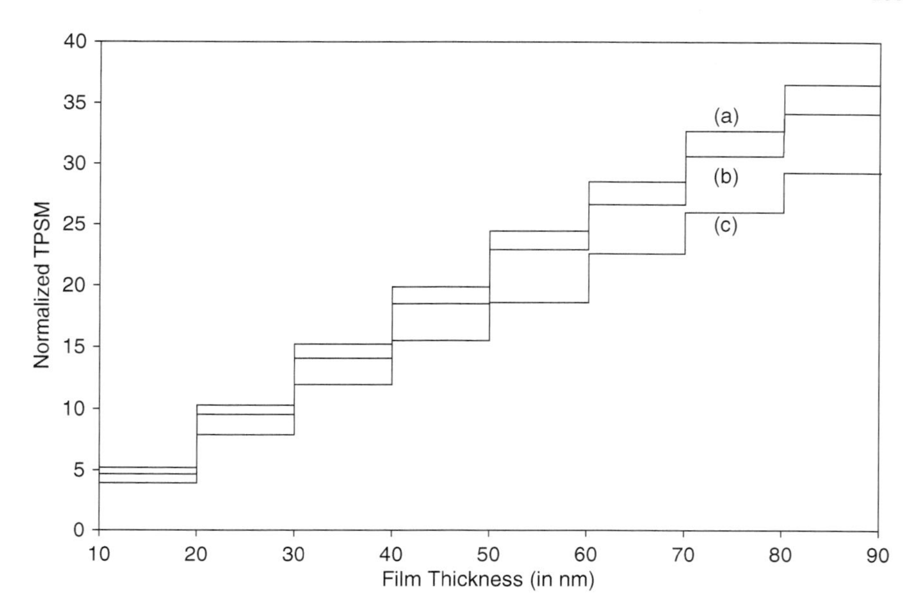

Fig. 8.13 Plot of the TPSM as function of film thickness for the UFs of $In_{1-x}Ga_xAs_{1-y}P_y$ for all cases of Fig. 8.1 in the presence of external photoexcitation

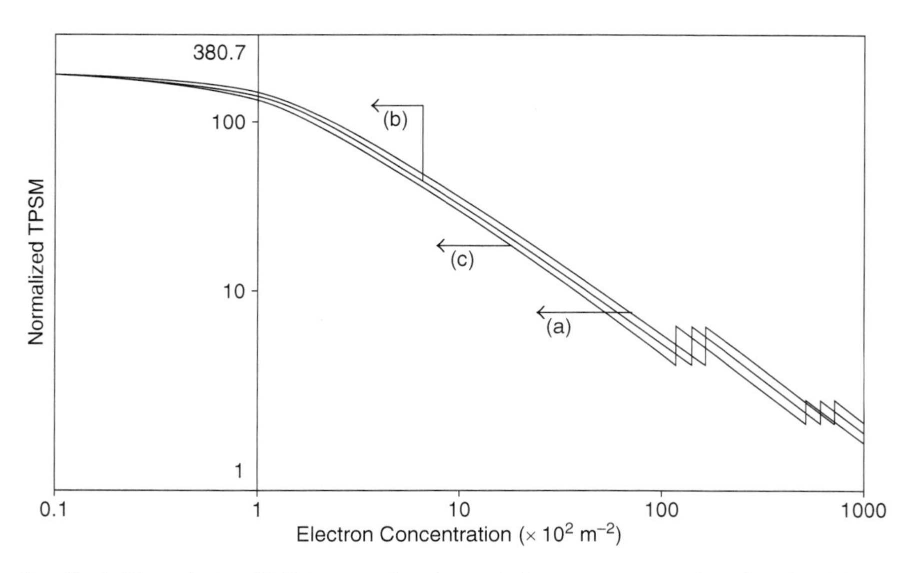

Fig. 8.14 Plot of the TPSM as a function of electron concentration for the UFs of $In_{1-x}Ga_xAs_{1-y}P_y$ for all cases of Fig. 8.13 in the presence of external photoexcitation

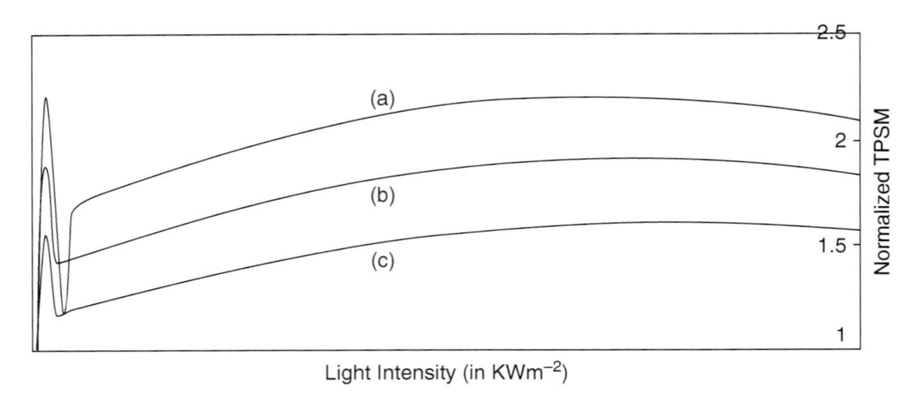

Fig. 8.15 Plot of the TPSM as a function of light intensity for the UFs of $In_{1-x}Ga_xAs_{1-y}P_y$ for all cases of Fig. 8.13 in the presence of external photoexcitation

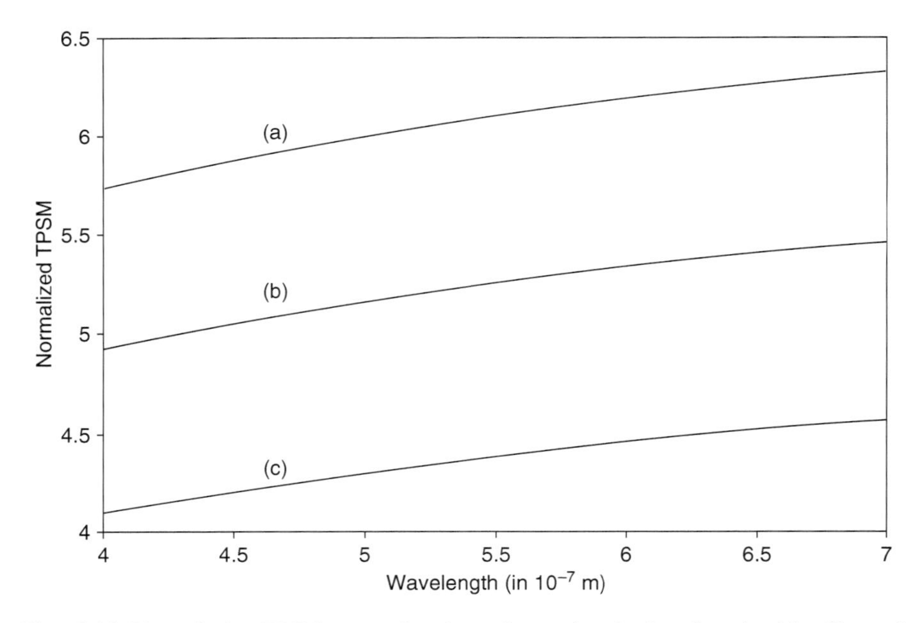

Fig. 8.16 Plot of the TPSM as a function of wavelength for the ultrathin films of $In_{1-x}Ga_xAs_{1-y}P_y$ for all cases of Fig. 8.9 in the presence of external photoexcitation

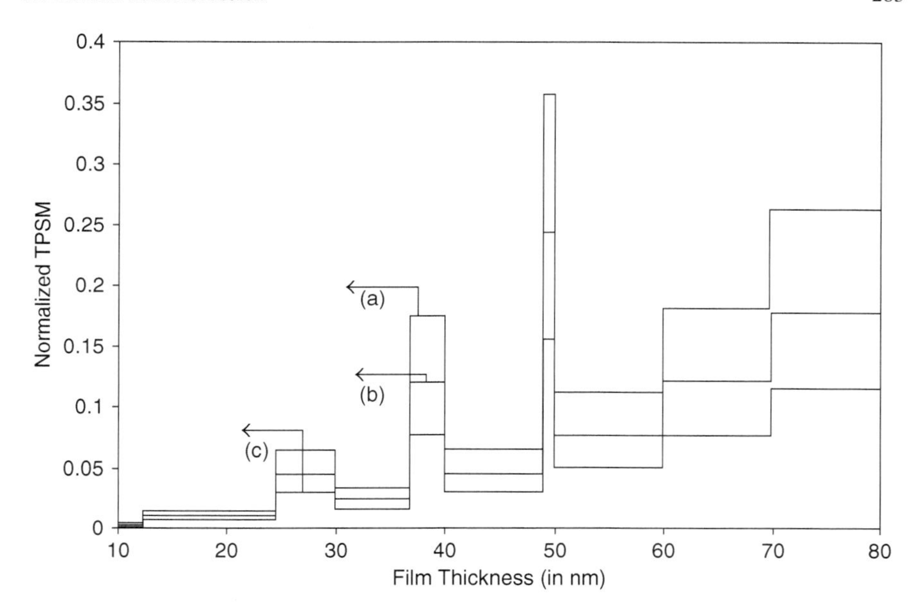

Fig. 8.17 Plot of the TPSM as a function of film thickness for the QWs of InSb for all cases of Fig. 8.1 in the presence of external photoexcitation

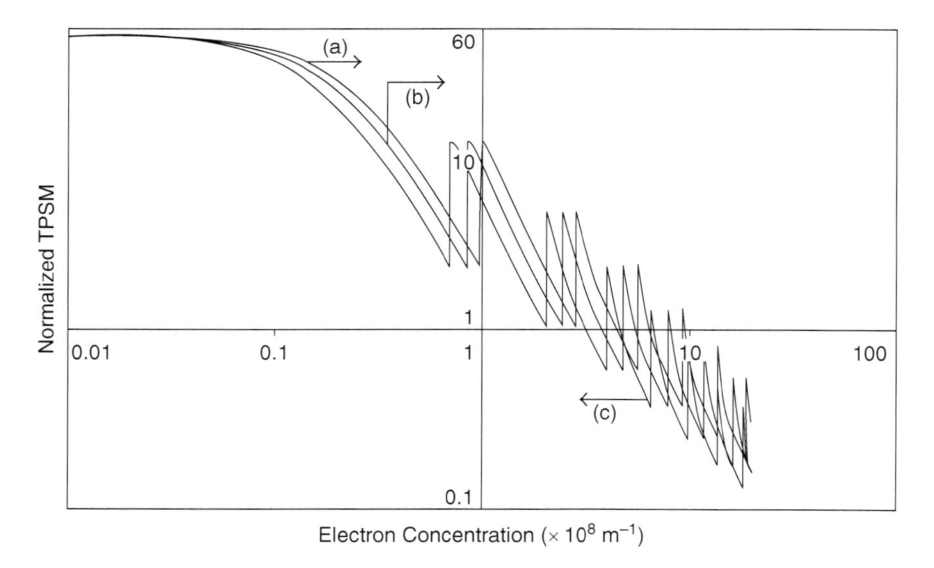

Fig. 8.18 Plot of the TPSM as a function of electron concentration for the quantum wires of InSb for all cases of Fig. 8.17 in the presence of external photoexcitation

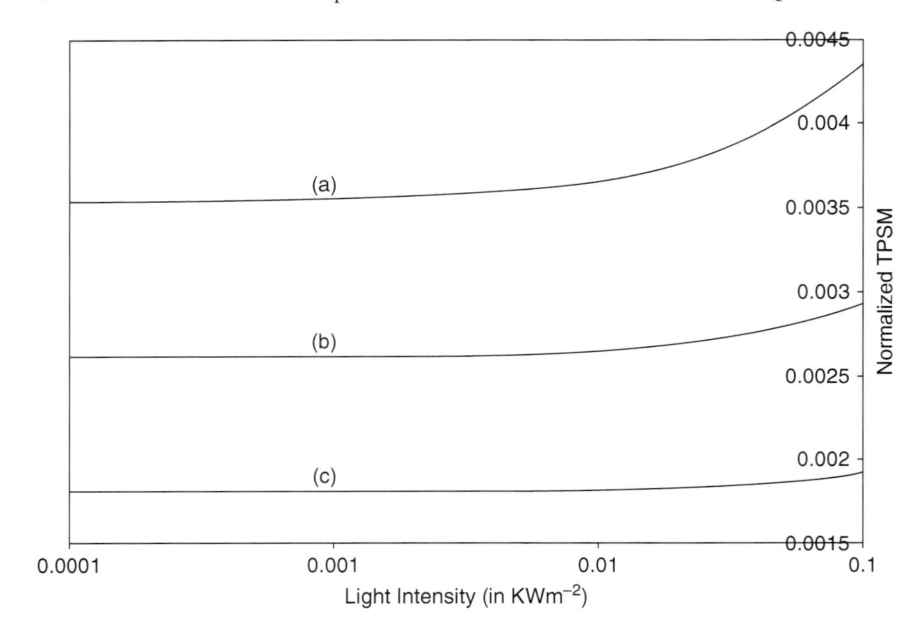

Fig. 8.19 Plot of the TPSM as a function of light intensity for the quantum wires of InSb for all cases of Fig. 8.17 in the presence of external photoexcitation

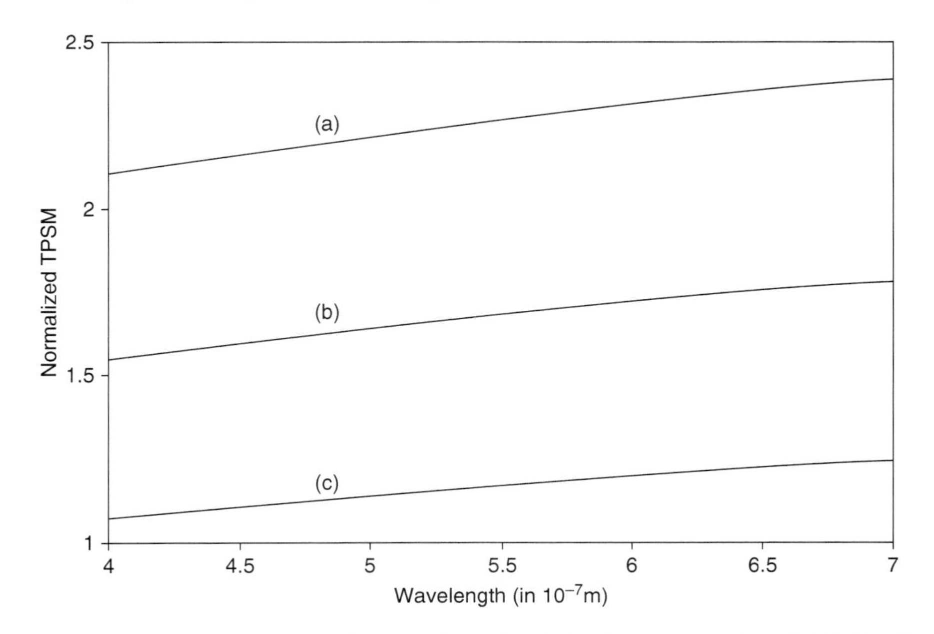

Fig. 8.20 Plot of the TPSM as a function of wavelength for the quantum wires of InSb for all cases of Fig. 8.17 in the presence of external photoexcitation

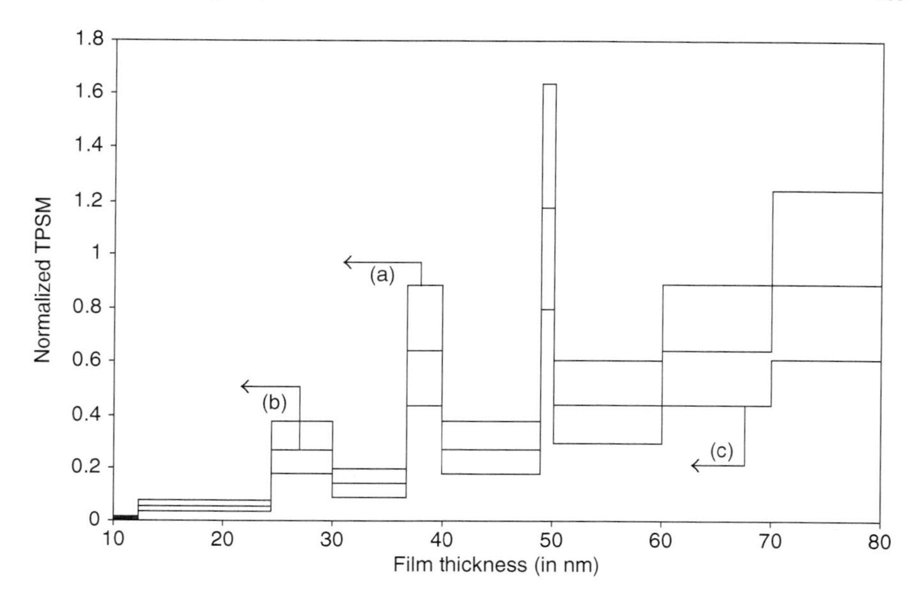

Fig. 8.21 Plot of the TPSM as a function of film thickness for the quantum wires of GaAs for all cases of Fig. 8.1 in the presence of external photoexcitation

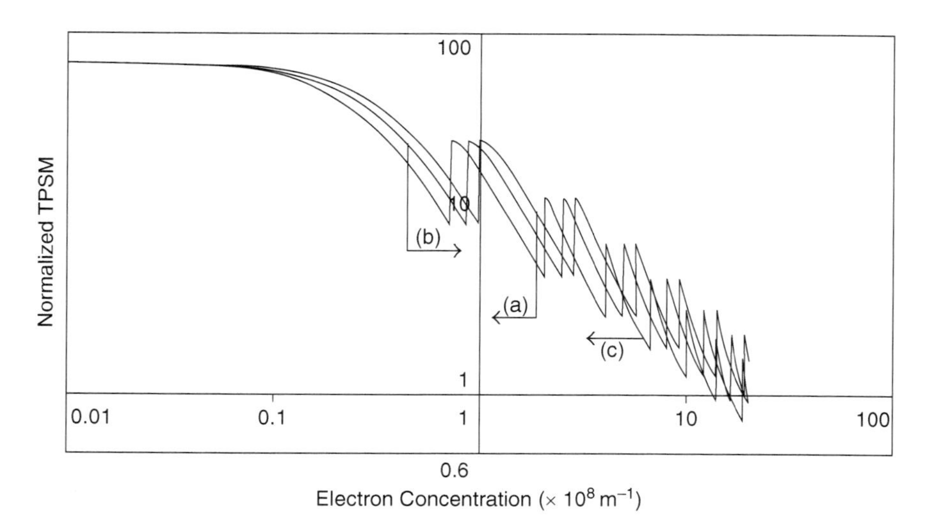

Fig. 8.22 Plot of the TPSM as a function of electron concentration for the quantum wires of GaAs for all cases of Fig. 8.21 in the presence of external photoexcitation

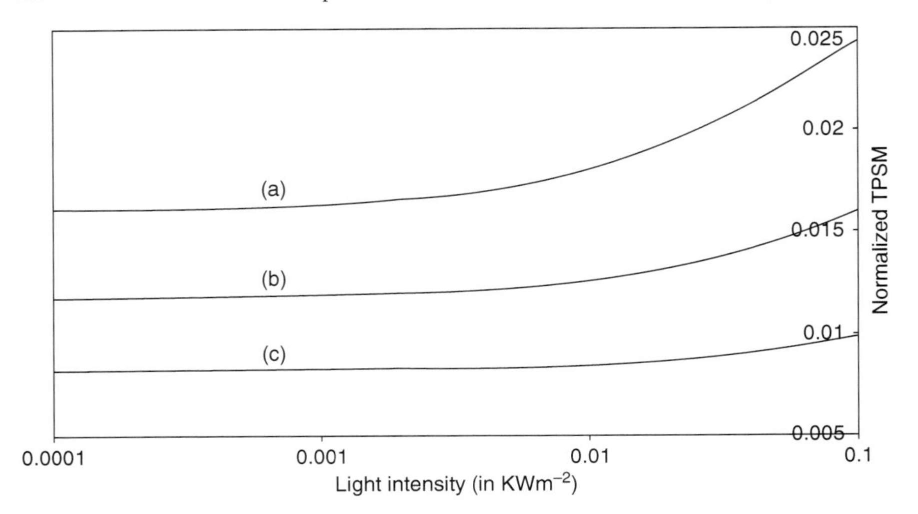

Fig. 8.23 Plot of the TPSM as a function of light intensity for the quantum wires of GaAs for all cases of Fig. 8.21 in the presence of external photoexcitation

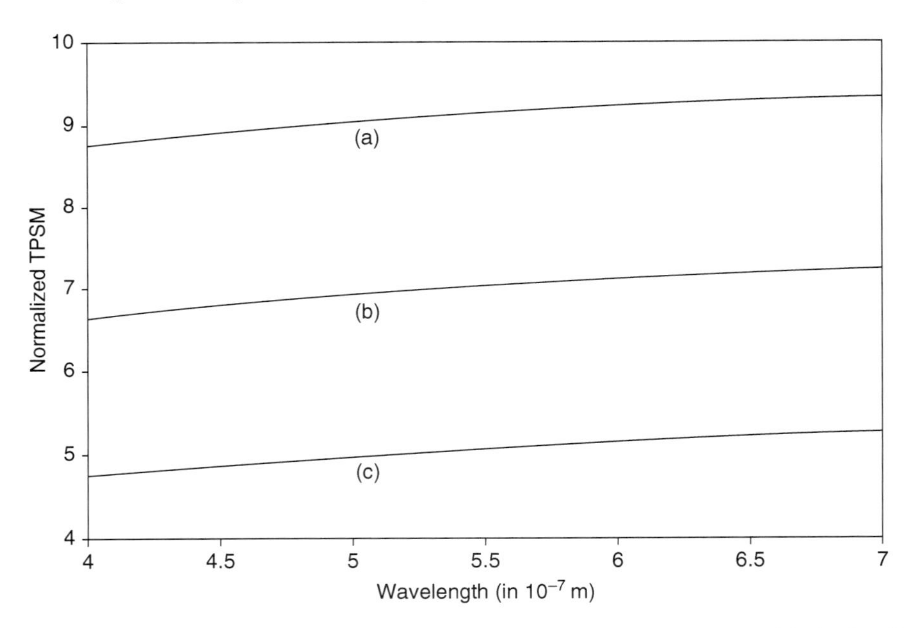

Fig. 8.24 Plot of the TPSM as a function of wavelength for the quantum wires of GaAs for all cases of Fig. 8.21 in the presence of external photoexcitation

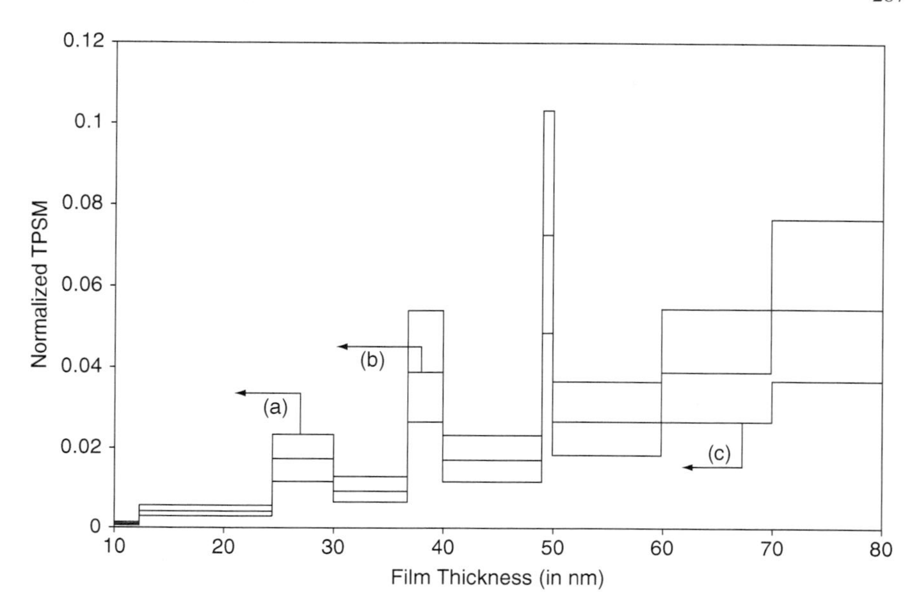

Fig. 8.25 Plot of the TPSM as a function of film thickness for the quantum wires of $Hg_{1-x}Cd_xTe$ for all cases of Fig. 8.1 in the presence of external photoexcitation

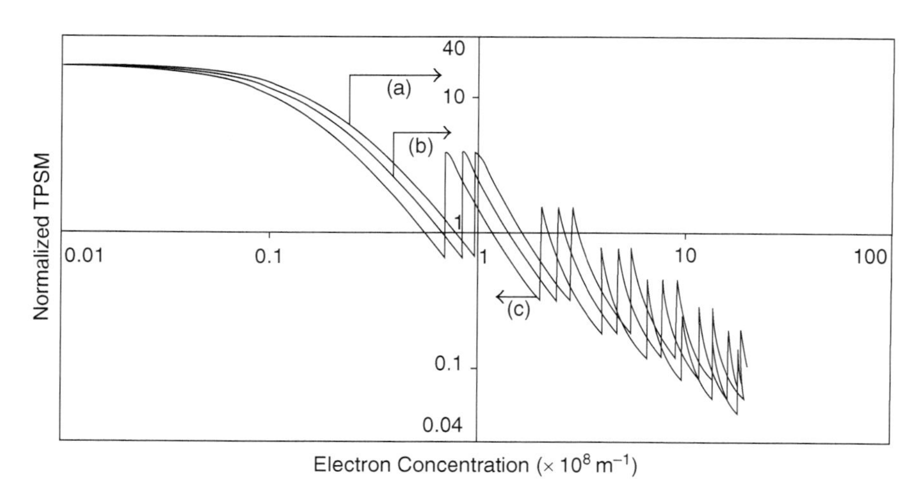

Fig. 8.26 Plot of the TPSM as a function of electron concentration for all cases of Fig. 8.25 in the presence of external photoexcitation

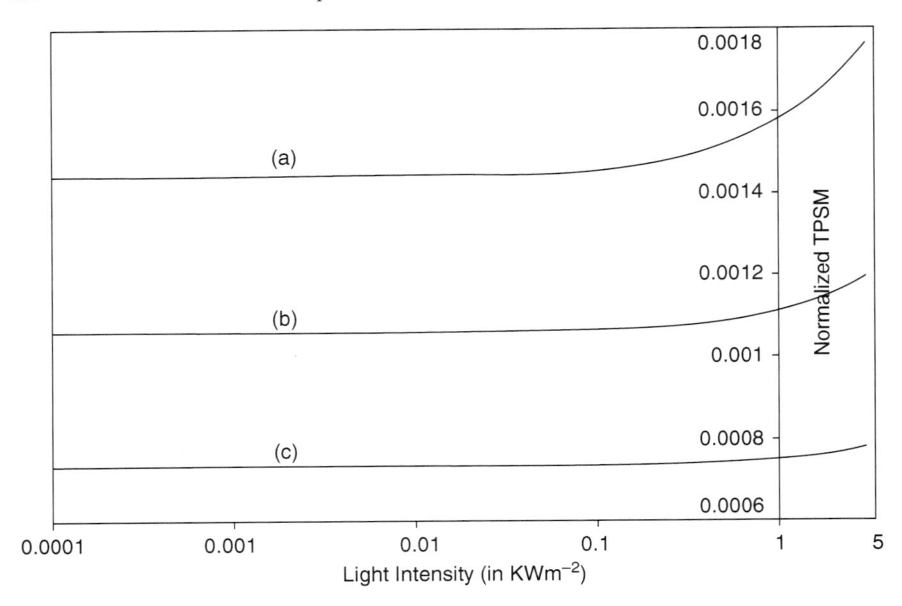

Fig. 8.27 Plot of the TPSM as a function of light intensity for the quantum wires of $Hg_{1-x}Cd_xTe$ for all cases of Fig. 8.25 in the presence of external photoexcitation

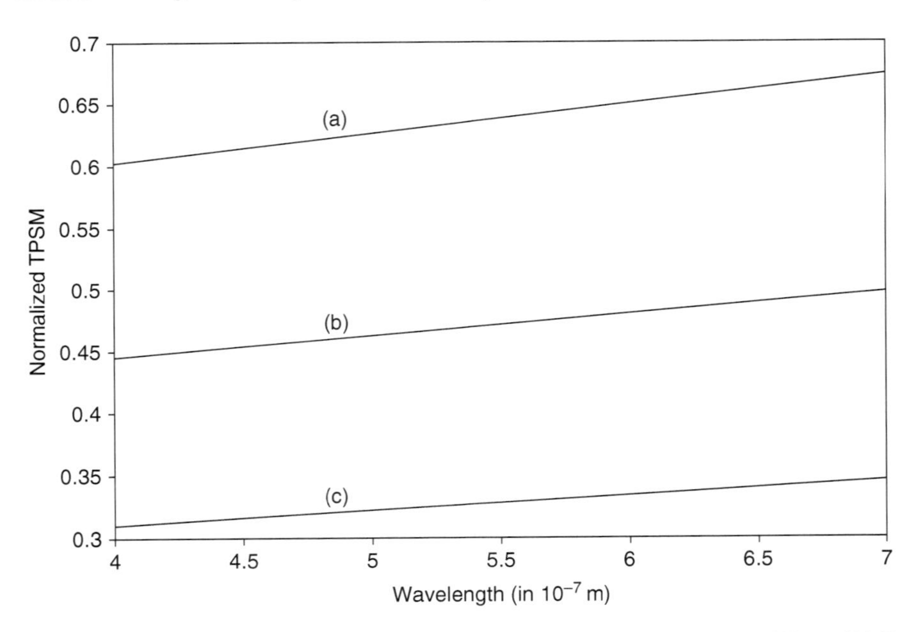

Fig. 8.28 Plot of the TPSM as a function of wavelength for the quantum wires of $Hg_{1-x}Cd_xTe$ for all cases of Fig. 8.25 in the presence of external photoexcitation

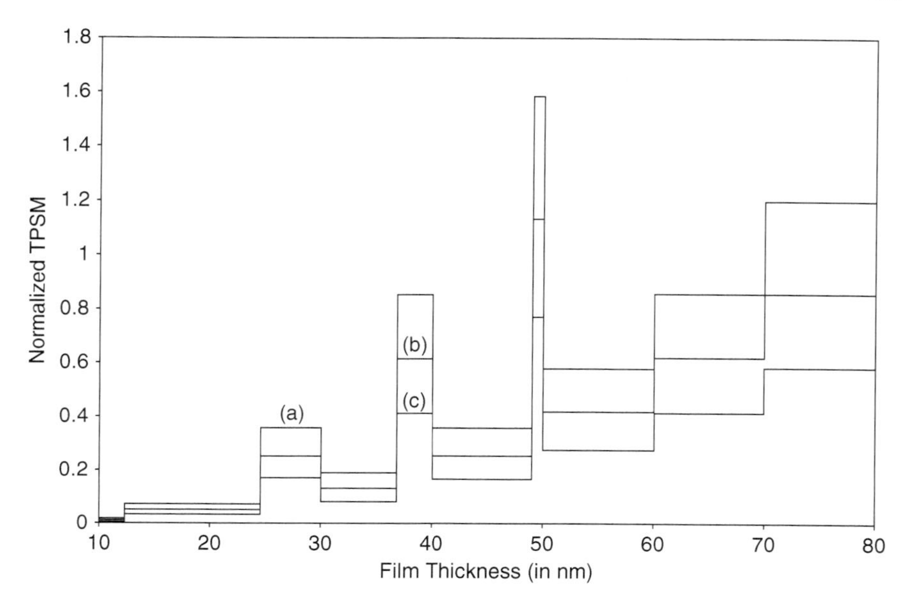

Fig. 8.29 Plot of the TPSM as a function of film thickness for the quantum wires of $In_{1-x}Ga_xAs_{1-y}P_y$ for all cases of Fig. 8.1 in the presence of external photoexcitation

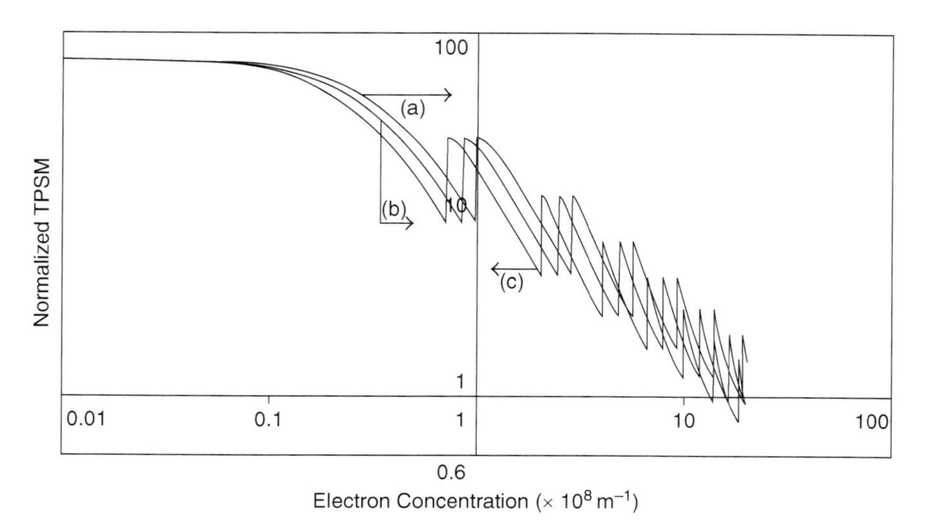

Fig. 8.30 Plot of the TPSM as a function of electron concentration for the quantum wires of $In_{1-x}Ga_xAs_{1-y}P_y$ for all cases of Fig.8.29 in the presence of external photoexcitation

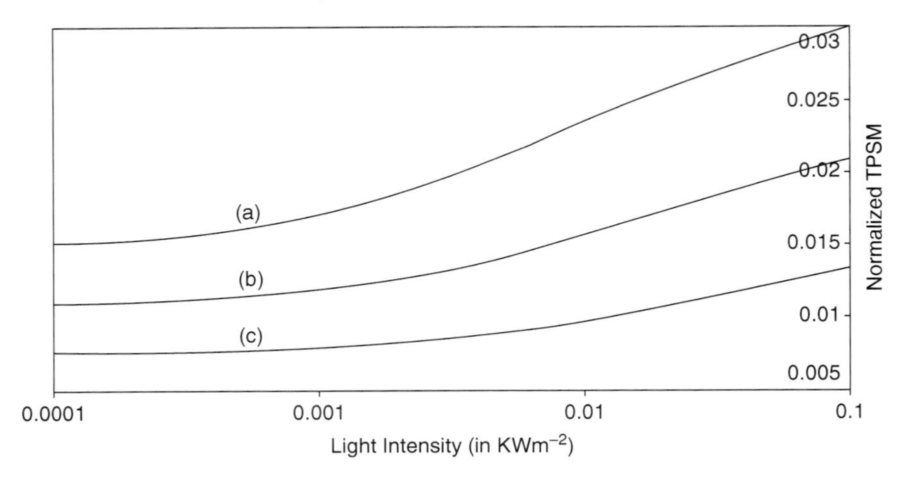

Fig. 8.31 Plot of the TPSM as a function of light intensity for the quantum wires of $In_{1-x}Ga_xAs_{1-y}P_y$ for all cases of Fig. 8.29 in the presence of external photoexcitation

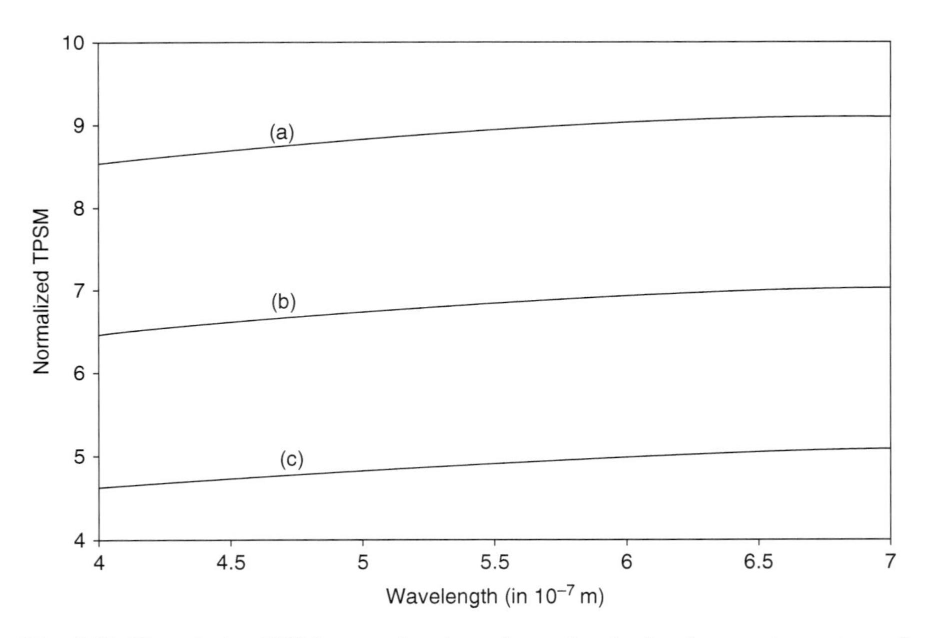

Fig. 8.32 Plot of the TPSM as a function of wavelength for the quantum wires of $In_{1-x}Ga_xAs_{1-y}P_y$ for all cases of Fig. 8.29 in the presence of external photoexcitation

Using (8.16); (8.19) and (8.17); (8.20) and (8.18); (8.21), we have plotted the TPSM from Figs. 8.17 to 8.31 for quantum wires of InSb, GaAs, $Hg_{1-x}Cd_xTe$, and $In_{1-x}Ga_xAs_{1-y}P_y$ as functions of film thickness, electron concentration, light intensity, and wavelength. Figure 8.17 shows the variation of TPSM for QWs of InSb with film thickness for $d_x = 30\,\text{nm}$, $n_0 = 20 \times 10^8\text{m}^{-1}$, $I_0 = 0.1\,\text{W m}^{-2}$, and $\lambda = 400\,\text{nm}$.

Figures 8.21, 8.25, and 8.29 exhibit the said variation for the other materials. It appears from the said figures that for QWs, the TPSM exhibits rectangular variations with the film thickness and exhibits composite oscillations due to the intermingling between the two quantum numbers along two directions. In Figs. 8.18, 8.22, 8.26, and 8.30, we have presented the variation of the TPSM as a function of electron concentration for $n_x = 1$, $d_x = 50\,\text{nm}$, $I_0 = 0.1\,\text{W m}^{-2}$, and $\lambda = 400\,\text{nm}$. From Figs. 8.18, 8.22, 8.26, and 8.30, it appears that TPSM decreases with increasing electron concentration in oscillatory manners.

For large values of electron concentration per unit length, the TPSM exhibits very quick oscillatory decrement whereas for small values of the carrier degeneracy, the TPSM shows the converging tendencies. Figures 8.19, 8.23, 8.27, and 8.31 exhibit the variation of the TPSM for quantum wires of such materials with the light intensity for $n_x = n_y = 1$, $d_x = 10\,\text{nm}$, $d_z = 30\,\text{nm}$, $n_{2\text{DL}} = 1 \times 10^8\,\text{m}^{-1}$, and $I_0 = 0.1\,\text{W m}^{-2}$, respectively.

It may be noted that the TPSM increases with increasing light intensity and due to both quantum limits, the variation is again nonoscillatory. Figures 8.20, 8.24, 8.28, and 8.32 show the variation of the TPSM of the quantum wires of the aforementioned materials with wavelength for $n_x = n_y = 1$, $d_x = 10\,\text{nm}$, $d_z = 30\,\text{nm}$, $n_0 = 20 \times 10^8\,\text{m}^{-1}$, $I_0 = 0.1\,\text{W m}^{-2}$, and $\lambda = 400\,\text{nm}$. Because of the quantum limits, the TPSM changes slowly with wavelength in a nonoscillatory manner, although the wide difference in the TPSM versus λ plot for the perturbed two and three band models of Kane together with parabolic energy bands exhibits the influence of energy band models. The interested readers may perform the numerical computations to obtain oscillatory plots for better assessment and joy of understanding.

For the purpose of condensed presentation, the carrier concentration and the corresponding TPSM for this chapter have been presented in Table 8.1.

8.4 Open Research Problem

(R8.1) Investigate all the appropriate open research problems of Chap. 2 in the presence of arbitrary external photoexcitation and strain, respectively.

Table 8.1 The carrier statistics and the thermoelectric power under large magnetic field for ultrathin films and quantum wires of optoelectronic materials in the presence of external photoexcitation

Type of materials	Carrier statistics	TPSM
1. Ultrathin films of optoelectronic materials under large magnetic field	$$n_0 = \left(\frac{m^* g_v}{\pi \hbar^2}\right) \sum_{n_{z81}=1}^{n_{z81\max}} [\phi_{81}(E_{F2DL}, n_{z81}) + \phi_{82}(E_{F2DL}, n_{z81})] \quad (8.7)$$	On the basis of (8.7), $$G_0 = \left(\frac{\pi^2 k_B^2 T}{3e}\right)\left[\sum_{n_{z81}=1}^{n_{z81\max}} [\phi_{81}(E_{F2DL}, n_{z81}) + \phi_{82}(E_{F2DL}, n_{z81})]\right]^{-1} \left[\sum_{n_{z81}=1}^{n_{z81\max}} [\phi'_{81}(E_{F2DL}, n_{z81}) + \phi'_{82}(E_{F2DL}, n_{z81})]\right] \quad (8.10)$$
	$$n_0 = \left(\frac{m^* g_v}{\pi \hbar^2}\right) \sum_{n_{z82}=1}^{n_{z82\max}} [\phi_{83}(E_{F2DL}, n_{z82}) + \phi_{84}(E_{F2DL}, n_{z82})] \quad (8.8)$$	On the basis of (8.8), $$G_0 = \left(\frac{\pi^2 k_B^2 T}{3e}\right)\left[\sum_{n_{z82}=1}^{n_{z82\max}} [\phi_{83}(E_{F2DL}, n_{z82}) + \phi_{84}(E_{F2DL}, n_{z82})]\right]^{-1} \left[\sum_{n_{z82}=1}^{n_{z82\max}} [\phi'_{83}(E_{F2DL}, n_{z82}) + \phi'_{84}(E_{F2DL}, n_{z82})]\right] \quad (8.11)$$
	$$n_0 = \left(\frac{m^* g_v}{\pi \hbar^2}\right) \sum_{n_{z83}=1}^{n_{z83\max}} [\phi_{85}(E_{F2DL}, n_{z83}) + \phi_{86}(E_{F2DL}, n_{z83})] \quad (8.9)$$	On the basis of (8.9), $$G_0 = \left(\frac{\pi^2 k_B^2 T}{3e}\right)\left[\sum_{n_{z83}=1}^{n_{z83\max}} [\phi_{85}(E_{F2DL}, n_{z83}) + \phi_{86}(E_{F2DL}, n_{z83})]\right]^{-1} \left[\sum_{n_{z83}=1}^{n_{z83\max}} [\phi'_{85}(E_{F2DL}, n_{z83}) + \phi'_{86}(E_{F2DL}, n_{z83})]\right] \quad (8.12)$$

2. Quantum wires of opto-electronic materials under large magnetic field

$$n_0 = \frac{2g_v\sqrt{2m^*}}{\pi\hbar} \sum_{n_{x81}=1}^{n_{x81\max}} \sum_{n_{z81}=1}^{n_{z81\max}} [\phi_{87}(E_{F1DL}, n_{x81}, n_{z81}) + \phi_{88}(E_{F1DL}, n_{x81}, n_{z81})] \tag{8.16}$$

On the basis of (8.16),

$$G_0 = \left(\frac{\pi^2 k_B^2 T}{3e}\right) \left[\sum_{n_{x81}=1}^{n_{x81\max}} \sum_{n_{z81}=1}^{n_{z81\max}} [\phi_{87}(E_{F1DL}, n_{x81}, n_{z81}) + \phi_{88}(E_{F1DL}, n_{x81}, n_{z81})]\right]^{-1} \left[\sum_{n_{x81}=1}^{n_{x81\max}} \sum_{n_{z81}=1}^{n_{z81\max}} [\phi'_{87}(E_{F1DL}, n_{x81}, n_{z81}) + \phi'_{88}(E_{F1DL}, n_{x81}, n_{z81})]\right] \tag{8.19}$$

$$n_0 = \frac{2g_v\sqrt{2m^*}}{\pi\hbar} \sum_{n_{x82}=1}^{n_{x82\max}} \sum_{n_{z82}=1}^{n_{z82\max}} [\phi_{89}(E_{F1DL}, n_{x82}, n_{z82}) + \phi_{810}(E_{F1DL}, n_{x82}, n_{z82})] \tag{8.17}$$

On the basis of (8.17),

$$G_0 = \left(\frac{\pi^2 k_B^2 T}{3e}\right) \left[\sum_{n_{x82}=1}^{n_{x82\max}} \sum_{n_{z82}=1}^{n_{z82\max}} [\phi_{89}(E_{F1DL}, n_{x82}, n_{z82}) + \phi_{810}(E_{F1DL}, n_{x82}, n_{z82})]\right]^{-1} \left[\sum_{n_{x82}=1}^{n_{x82\max}} \sum_{n_{z82}=1}^{n_{z82\max}} [\phi'_{89}(E_{F1DL}, n_{x82}, n_{z82}) + \phi'_{810}(E_{F1DL}, n_{x82}, n_{z82})]\right] \tag{8.20}$$

$$n_0 = \frac{2g_v\sqrt{2m^*}}{\pi\hbar} \sum_{n_{x83}=1}^{n_{x83\max}} \sum_{n_{z83}=1}^{n_{z83\max}} [\phi_{811}(E_{F1DL}, n_{x83}, n_{z83}) + \phi_{812}(E_{F1DL}, n_{x83}, n_{z83})] \tag{8.18}$$

On the basis of (8.18),

$$G_0 = \left(\frac{\pi^2 k_B^2 T}{3e}\right) \left[\sum_{n_{x83}=1}^{n_{x83\max}} \sum_{n_{z83}=1}^{n_{z83\max}} [\phi_{811}(E_{F1DL}, n_{x83}, n_{z83}) + \phi_{812}(E_{F1DL}, n_{x83}, n_{z83})]\right]^{-1} \left[\sum_{n_{x83}=1}^{n_{x83\max}} \sum_{n_{z83}=1}^{n_{z83\max}} [\phi'_{811}(E_{F1DL}, n_{x83}, n_{z83}) + \phi'_{812}(E_{F1DL}, n_{x83}, n_{z83})]\right] \tag{8.21}$$

References

1. P.K. Basu, Theory of Optical Processes in Semiconductors, Bulk and Microstructures (Oxford University Press, Oxford, 1997)
2. K.P. Ghatak, S. Bhattacharya, J. Appl. Phys. **102**, 073704, (2007)
3. K.P. Ghatak, S. Bhattacharya, S.K. Biswas, A. De, A.K. Dasgupta, Phys. Scr. **75**, 820, (2007)
4. K.P. Ghatak, S. Bhattacharya, D. De, S. Pahari, A. Dey, A.K. Dasgupta, S.N. Biswas, J. Comput. Theor. Nanosci. **5**, 1345 (2008)
5. P.Y. Lu, C.H. Wung, C.M. Williams, S.N.G. Chu, C.M. Stiles, Appl. Phys. Letts. **49**, 1372 (1986)
6. N.R. Taskar, I.B. Bhat, K.K. Prat, D. Terry, H. Ehasani, S.K. Ghandhi, J. Vac. Sci. Tech. **7A**, 281 (1989)
7. F. Koch, Springer Series in Solid States Sciences, vol. 53 (Springer, Germany, 1984), pp. 20
8. L.R. Tomasetta, H.D. Law, R.C. Eden, I. Reyhimy, K. Nakano, IEEE J. Quant. Electron. **14**, 800 (1978)
9. T. Yamato, K. Sakai, S. Akiba, Y. Suematsu, IEEE J. Quant. Electron. **14**, 95 (1978)
10. T.P. Pearsall, B.I. Miller, R.J. Capik, Appl. Phys. Letts. **28**, 499 (1976)
11. M.A. Washington, R.E. Nahory, M.A. Pollack, E.D. Beeke, Appl. Phys. Letts. **33**, 854 (1978)
12. M.I. Timmons, S.M. Bedair, R.J. Markunas, J.A. Hutchby, Proceedings of the 16th IEEE Photovoltaic Specialist Conference (IEEE, San Diego, California, (1982), p. 666
13. E. Haga, H. Kimura, J. Phys. Soc. Jpn. **18**, 777, (1963)
14. E. Haga, H. Kimura, J. Phys. Soc. Jpn. **19**, 471, (1964)
15. K.P. Ghatak, S. Bhattacharya, D. De, Einstein Relation in Compound Semiconductors and Their Nanostructures, Springer Series in Materials Science, vol. 116 (Springer, Germany, 2008)

Chapter 9
Optothermoelectric Power in Quantum Dots of Optoelectronic Materials Under Large Magnetic Field

9.1 Introduction

In this chapter, the TPSM in the presence of light in QDs of optoelectronic materials under large magnetic field is investigated. Section 9.2.1 of theoretical background contains the study of the thermoelectric power under large magnetic field in the presence of external photoexcitation in QDs of optoelectronic materials, whose bulk conduction electrons are defined by the dispersion relations as given by (8.1), (8.2), and (8.3), respectively. Sections 9.3 and 9.4 include results and discussion and open research problems, respectively.

9.2 Theoretical Background

9.2.1 Magnetothermopower in Quantum Dots of Optoelectronic Materials

The dispersion relations of the electrons in QDs of optoelectronic materials in the presence of light waves can, respectively, be expressed from (8.1), (8.2), and (8.3) as [1]

$$\frac{2m^*\beta_0(E_{Q1},\lambda)}{\hbar^2} = H_{91}(n_{x91},n_{y91},n_{z91}), \tag{9.1}$$

$$\frac{2m^*\tau_0(E_{Q2},\lambda)}{\hbar^2} = H_{92}(n_{x92},n_{y92},n_{z92}), \tag{9.2}$$

$$\frac{2m^*\rho_0(E_{Q3},\lambda)}{\hbar^2} = H_{93}(n_{x93},n_{y93},n_{z93}), \tag{9.3}$$

where E_{Qi} is the totally quantized energy and

$$H_{9i}(n_{x9i}, n_{y9i}, n_{z9i}) = \left(\frac{\pi n_{x9i}}{d_x}\right)^2 + \left(\frac{\pi n_{y9i}}{d_y}\right)^2 + \left(\frac{\pi n_{z9i}}{d_z}\right)^2.$$

The electron concentration per unit volume can, in general, be written as

$$n_0 = \left(\frac{2g_v}{d_x d_y d_z}\right) \sum_{n_{x9i}=1}^{n_{x9i_{\max}}} \sum_{n_{y9i}=1}^{n_{y9i_{\max}}} \sum_{n_{z9i}=1}^{n_{z9i_{\max}}} F_{-1}(\eta_{9i0D}), \tag{9.4}$$

where $\eta_{9i0D} = (E_{F_0DL} - E_{\theta i})/k_B T$ and E_{F_0DL} is the Fermi energy in QDs in the presence of light waves as measured from the edge of the conduction band in the vertically upward direction in the absence of any quantization.

Combining (1.13) and (9.4), the opto-TPSM in this case assumes the form

$$G_0 = \left(\frac{\pi^2 k_B}{3e}\right) \left[\sum_{n_{x9i}=1}^{n_{x9i_{\max}}} \sum_{n_{y9i}=1}^{n_{y9i_{\max}}} \sum_{n_{z9i}=1}^{n_{z9i_{\max}}} F_{-1}(\eta_{9i0D})\right]^{-1} \times \left[\sum_{n_{x9i}=1}^{n_{x9i_{\max}}} \sum_{n_{y9i}=1}^{n_{y9i_{\max}}} \sum_{n_{z9i}=1}^{n_{z9i_{\max}}} F_{-2}(\eta_{9i0D})\right]. \tag{9.5}$$

It is interesting to note under the condition of carrier nondegeneracy, (9.5) gets simplified into the well-known form of classical TPSM equation as given in the Preface and becomes a band structure invariant physical quantity.

9.3 Results and Discussion

Using (9.4) and (9.5), we have plotted the TPSM for QDs of $In_{1-x}Ga_xAs_yP_{1-y}$ as functions of well thickness, electron concentration, and wavelength for perturbed three and two band models of Kane together with perturbed parabolic energy bands in Figs. 9.1–9.3, respectively. Figure 9.1 exhibits the variation of the normalized TPM as a function of well thickness for $d_y = 30$ nm, $d_z = 40$ nm, $n_0 = 1\times10^{23}\,\mathrm{m}^{-3}$, $I_0 = 0.1\,\mathrm{W\,m}^{-2}$, and $\lambda = 400$ nm.

It appears from Fig. 9.1 that the TPSM increases with increasing film thickness in oscillatory ways for the perturbed three and two band models of Kane and also perturbed parabolic energy bands in the presence of external light waves. It appears that instead of spikes, trapezoidal variations occur during quantum jumps and the length and breadth of the trapezoids are totally dependent on energy band constants. The influence of spin–orbit splitting constant is to enhance the value of the normalized TPSM in the whole range of the thicknesses considered so far as Fig. 9.1 is considered. The nonparabolicity band also increases the TPSM as compared with

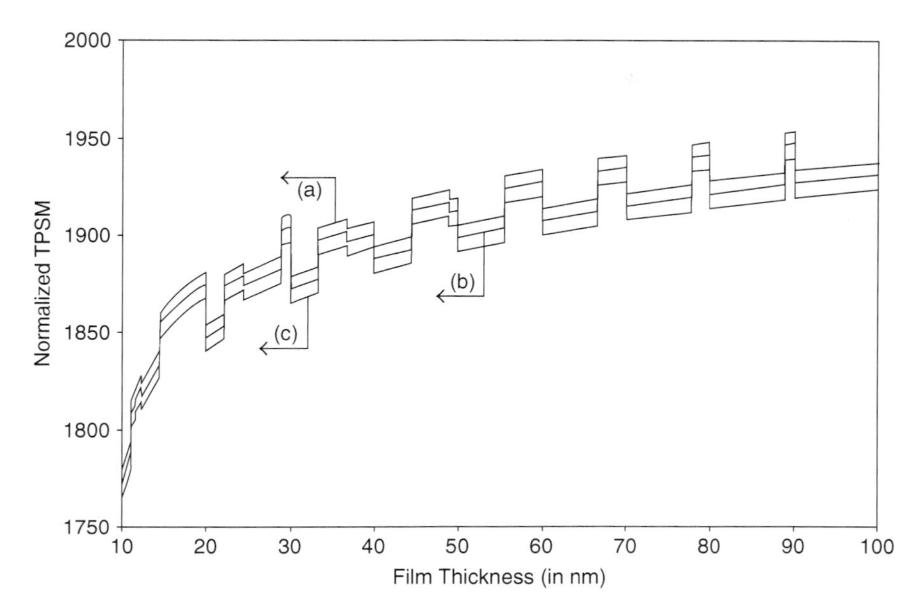

Fig. 9.1 Plot of the TPSM as a function of film thickness for QDs of $In_{1-x}Ga_xAs_{1-y}P_y$ for perturbed (**a**) the three (**b**) and the two band models of Kane together with the (**c**) parabolic model in the presence of external photoexcitation

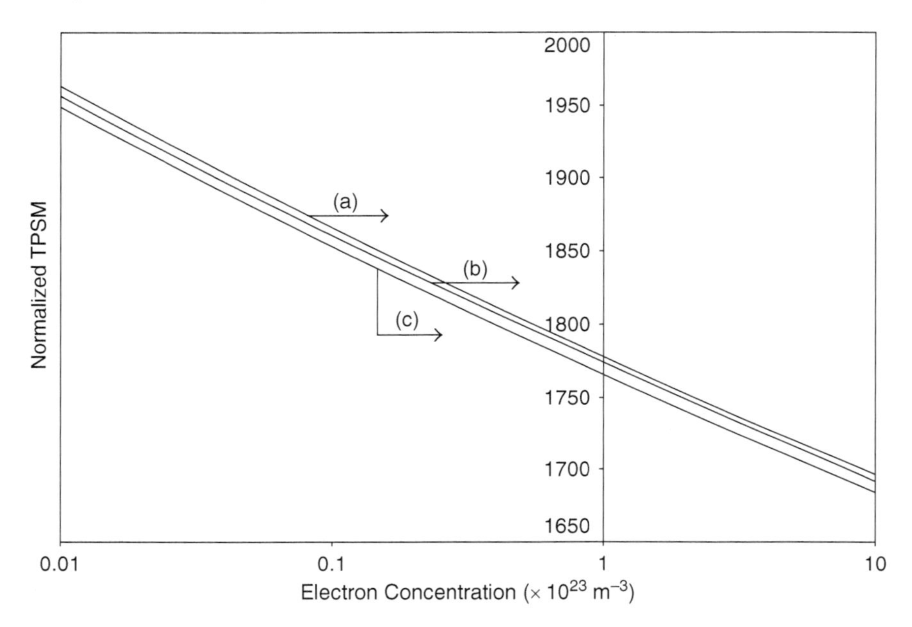

Fig. 9.2 Plot of the TPSM as a function of electron concentration for QDs of $In_{1-x}Ga_xAs_{1-y}P_y$ for all the cases of Fig. 9.1 in the quantum limit in the presence of external photoexcitation

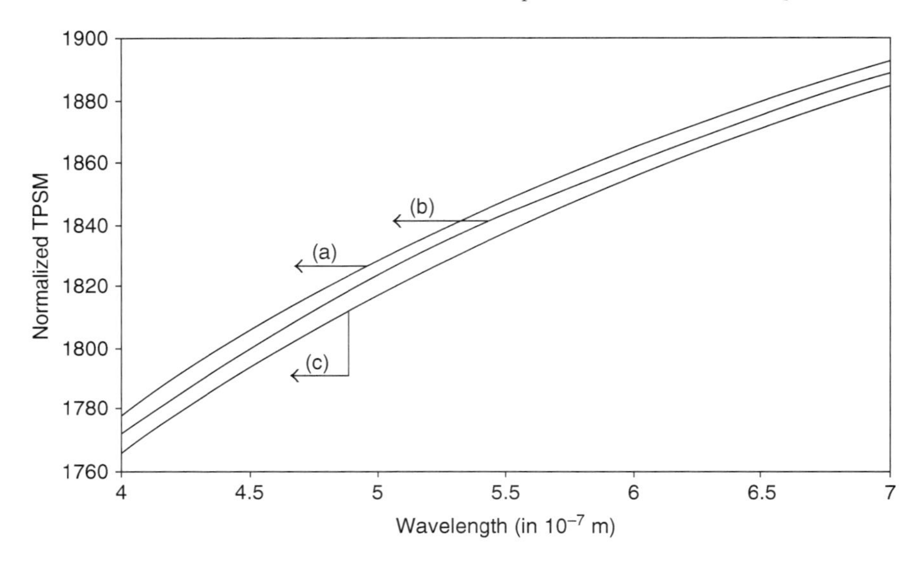

Fig. 9.3 Plot of the TPSM as function of wavelength for QDs of $In_{1-x}Ga_xAs_{1-y}P_y$ for all the cases of Fig. 9.1 in the quantum limit in the presence of external photoexcitation

Table 9.1 The carrier statistics and the thermoelectric power under large magnetic field for quantum dots of optoelectronic materials in the presence of external photoexcitation

Type of materials	Carrier statistics	TPSM
Quantum dots of opto-electronic materials	$n_0 = \left(\frac{2g_v}{d_x d_y d_z}\right) \sum_{n_{x9i}=1}^{n_{x9i_{max}}} \sum_{n_{y9i}=1}^{n_{y9i_{max}}} \sum_{n_{z9i}=1}^{n_{z9i_{max}}} F_{-1}(\eta_{9i0D})$ (9.4)	On the basis of (9.4), $G_0 = \left(\frac{\pi^2 k_B}{3e}\right) \times \left[\sum_{n_{x9i}=1}^{n_{x9i_{max}}} \sum_{n_{y9i}=1}^{n_{y9i_{max}}} \sum_{n_{z9i}=1}^{n_{z9i_{max}}} F_{-1}(\eta_{9i0D})\right]^{-1} \times \left[\sum_{n_{x9i}=1}^{n_{x9i_{max}}} \sum_{n_{y9i}=1}^{n_{y9i_{max}}} \sum_{n_{z9i}=1}^{n_{z9i_{max}}} F_{-2}(\eta_{9i0D})\right]$ (9.5)

the corresponding perturbed parabolic energy bands. The composite oscillations in TPSM as observed in Fig. 9.1 is not only due to the strong correlation among the size quantum numbers but also due to the selection rules through which an electron in a energy level corresponding to a fixed values of the size quantum numbers jumps to another allowed energy level which is again specified by another set of size quantum numbers.

It may be noted that for the purpose of simplified numerical computation, in the rest of the figures, the computer programming have been performed on the basis of electric quantum limit conditions in the three directions of the wave-vector space of the electrons. From Fig. 9.2, it appears that the TPSM decreases sharply in a nonoscillatory way with increasing concentration. From 9.3, we observe that the TPSM increases with increasing wavelength for all types of band models. For the purpose of condensed presentation, the carrier concentration and the corresponding TPSM for this chapter have been presented in Table 9.1.

9.4 Open Research Problem

(R9.1) Investigate the DTP, PTP, and Z in the presence of arbitrary external photoexcitation and strain for wedge shaped, cylindrical, ellipsoidal, conical, triangular, circular, parabolic rotational, and parabolic cylindrical quantum dots of all the materials as discussed in Chap. 1.

Reference

1. K.P. Ghatak, S. Bhattacharya, D. De, in *Einstein Relation in Compound Semiconductors and Their Nanostructures*. Springer Series in Materials Science, vol 116 (Springer, Germany, 2008)

Chapter 10
Optothermoelectric Power in Quantum-Confined Semiconductor Superlattices of Optoelectronic Materials Under Large Magnetic Field

10.1 Introduction

In Chaps. 3, 4, and 7, the thermoelectric power under strong magnetic field has been studied from SLs having various band structures assuming that the band structures of the constituent materials are invariant quantities in the presence of external photoexcitation. In this chapter, this assumption has been removed and in Sect. 10.2.1, an attempt is made to study the optothermoelectric power under large magnetic field in QW effective mass SL of optoelectronic materials. In Sect. 10.2.2, the same in QD effective mass SLs of optoelectronic materials has been investigated. In Sect. 10.2.3, an attempt is made to investigate the optothermoelectric power under large magnetic field in QW SLs with graded interfaces of optoelectronic materials. In Sect. 10.2.4, the same in QD SLs with graded interfaces of optoelectronic materials has been studied. Sections 10.3 and 10.4 contain the results and discussion and open research problems, respectively.

10.2 Theoretical Background

10.2.1 Magnetothermopower in III–V Quantum Wire Effective Mass Superlattices

The electron dispersion relation in this case is given by [1]

$$k_x^2 = \left[\frac{1}{L_0^2}\left[\cos^{-1} f_{101}(E,n_y,n_z)\right]^2 - \left(\frac{\pi n_y}{d_y}\right)^2 - \left(\frac{\pi n_z}{d_z}\right)^2\right], \tag{10.1}$$

where $f_{101}(E,n_y,n_z) = [a_1 \cos[a_0 g_{101}(E,n_y,n_z) + b_0 h_{101}(E,n_y,n_z)] - a_2 \cos [a_0 g_{101}(E,n_y,n_z) - b_0 h_{101}(E,n_y,n_z)]$,

$$g_{101}\left(E,n_y,n_z\right)=\left[\frac{2m_1^*}{\hbar^2}\beta_0\left(E,\lambda,E_{g01},\Delta_1\right)-\left[\left(\frac{\pi n_y}{d_y}\right)^2+\left(\frac{\pi n_z}{d_z}\right)^2\right]\right]^{1/2},$$

and

$$h_{101}\left(E,n_y,n_z\right)=\left[\frac{2m_2^*}{\hbar^2}\beta_0\left(E,\lambda,E_{g02},\Delta_2\right)-\left[\left(\frac{\pi n_y}{d_y}\right)^2+\left(\frac{\pi n_z}{d_z}\right)^2\right]\right]^{1/2}.$$

The electron concentration per unit length is given by

$$n_0=\frac{2g_v}{\pi}\sum_{n_y=1}^{n_{y\max}}\sum_{n_z=1}^{n_{z\max}}\left[K_{101}\left(E_{FQWSLEML},n_y,n_z\right)+K_{102}\left(E_{FQWSLEML},n_y,n_z\right)\right], \tag{10.2}$$

where

$$K_{101}\left(E_{FQWSLEML},n_y,n_z\right)=\left[\frac{1}{L_0^2}\left[\cos^{-1}f_{101}\left(E_{FQWSLEML},n_y,n_z\right)\right]^2-\left(\frac{\pi n_y}{d_y}\right)^2-\left(\frac{\pi n_z}{d_z}\right)^2\right]^{1/2},$$

$E_{FQWSLEML}$ is the Fermi energy in the present case,

$$K_{102}\left(E_{FQWSLEML},n_y,n_z\right)=\sum_{r=1}^{s}Z_{r,Y}\left[K_{101}\left(E_{FQWSLEML},n_y,n_z\right)\right],$$

and $Y=$ QWSLEML. For the perturbed two band model of Kane and that of parabolic energy bands, the term $\beta_0\left(E,\lambda,E_{g0i},\Delta_i\right)$ should be replaced by $\tau_0\left(E,\lambda,E_{g0i}\right)$ and $\rho_0\left(E,\lambda,E_{g0i}\right)$, respectively. The basic forms of (8.9) and (8.10) remain unchanged.

Combining (1.13) and (10.2), the optothermoelectric power under large magnetic field in this case is given by

$$G_0=\left(\frac{\pi^2k_B^2T}{3e}\right)\left[\sum_{n_y=1}^{n_{y\max}}\sum_{n_z=1}^{n_{z\max}}\left[K_{101}\left(E_{FQWSLEML},n_y,n_z\right)+K_{102}\left(E_{FQWSLEML},n_y,n_z\right)\right]\right]^{-1}\left[\sum_{n_y=1}^{n_{y\max}}\sum_{n_z=1}^{n_{z\max}}\left[K'_{101}\left(E_{FQWSLEML},n_y,n_z\right)+K'_{101}\left(E_{FQWSLEML},n_y,n_z\right)\right]\right]. \tag{10.3}$$

10.2.2 Magnetothermopower in III–V Quantum Dot Effective Mass Superlattices

The electron energy spectrum in this case is given by

$$\left(\frac{\pi n_x}{d_x}\right)^2 = \left[\frac{1}{L_0^2}\left[\cos^{-1} f_{101}\left(E, n_y, n_z\right)\big|_{E=E_{102}}\right]^2 - \left(\frac{\pi n_y}{d_y}\right)^2 - \left(\frac{\pi n_z}{d_z}\right)^2\right], \tag{10.4}$$

where E_{102} is the totally quantized electron energy in this case.

The electron concentration per unit volume is given by

$$n_0 = \frac{2g_v}{d_x d_y d_z} \sum_{n_x=1}^{n_{x\max}} \sum_{n_y=1}^{n_{y\max}} \sum_{n_z=1}^{n_{z\max}} F_{-1}\left(\eta_{102}\right), \tag{10.5}$$

where

$$\eta_{102} = \frac{E_{\text{FQDSLEML}} - E_{102}}{k_B T},$$

in which E_{FQDSLEML} is the Fermi energy in the present case.

Combining (1.13) and (9.5), the optothermoelectric power under large magnetic field in this case is given by

$$G_0 = \frac{\pi^2 k_B}{3e}\left[\sum_{n_x=1}^{n_{x\max}} \sum_{n_y=1}^{n_{y\max}} \sum_{n_z=1}^{n_{z\max}} F_{-1}\left(\eta_{102}\right)\right]^{-1}\left[\sum_{n_x=1}^{n_{x\max}} \sum_{n_y=1}^{n_{y\max}} \sum_{n_z=1}^{n_{z\max}} F_{-2}\left(\eta_{102}\right)\right]. \tag{10.6}$$

10.2.3 Magnetothermopower in III–V Quantum Wire Superlattices with Graded Interfaces

The electron energy spectrum in this case is given by

$$k_z^2 = \left[\frac{1}{L_0^2}\left[\cos^{-1}\left[\frac{1}{2}\phi_{101}\left(E, n_x, n_y\right)\right]\right]^2 - \left(\frac{\pi n_x}{d_x}\right)^2 - \left(\frac{\pi n_y}{d_y}\right)^2\right], \tag{10.7}$$

where

$$\phi_{101}\left(E, n_x, n_y\right) = \Big[2\cosh\left\{\beta_{101}\left(E, n_x, n_y\right)\right\}\cos\left\{\gamma_{101}\left(E, n_x, n_y\right)\right\}$$
$$+ \varepsilon_{101}\left(E, n_x, n_y\right)\sinh\left\{\beta_{101}\left(E, n_x, n_y\right)\right\}\sin\left\{\gamma_{101}\left(E, n_x, n_y\right)\right\}$$

$$
\begin{aligned}
&+\Delta_0\left[\left(\frac{K_{101}^2(E,n_x,n_y)}{K_{102}(E,n_x,n_y)}-3K_{102}(E,n_x,n_y)\right)\right.\\
&\times\cosh\{\beta_{101}(E,n_x,n_y)\}\sin\{\gamma_{101}(E,n_x,n_y)\}\\
&+\left(3K_{101}(E,n_x,n_y)-\frac{K_{102}^2(E,n_x,n_y)}{K_{101}(E,n_x,n_y)}\right)\\
&\times\sinh\{\beta_{101}(E,n_x,n_y)\}\cos\{\gamma_{101}(E,n_x,n_y)\}]\\
&+\Delta_0[2\{K_{101}^2(E,n_x,n_y)-K_{102}^2(E,n_x,n_y)\}\\
&\times\cosh\{\beta_{101}(E,n_x,n_y)\}\cos\{\gamma_{101}(E,n_x,n_y)\}\\
&+\frac{1}{12}\left[\frac{5K_{102}^2(E,n_x,n_y)}{K_{101}(E,n_x,n_y)}+\frac{5K_{101}^2(E,n_x,n_y)}{K_{102}(E,n_x,n_y)}\right.\\
&\left.-34K_{101}(E,n_x,n_y)K_{102}(E,n_x,n_y)\right]\\
&\left.\times\sinh\{\beta_{101}(E,n_x,n_y)\}\sin\{\gamma_{101}(E,n_x,n_y)\}]\right],\\
&\beta_{101}(E,n_x,n_y)=K_{101}(E,n_x,n_y)(a_0-\Delta_0),
\end{aligned}
$$

$$
K_{101}(E,n_x,n_y)=\left[\left(\frac{\pi n_x}{d_x}\right)^2+\left(\frac{\pi n_y}{d_y}\right)^2-\frac{2m_2^*}{\hbar^2}\beta_0(E-V_0,\lambda,E_{g02},\Delta_2)\right]^{1/2},
$$

$$
\gamma_{101}(E,n_x,n_y)=K_{102}(E,n_x,n_y)(b_0-\Delta_0),
$$

$$
K_{102}(E,n_x,n_y)=\left[-\left(\frac{\pi n_x}{d_x}\right)^2-\left(\frac{\pi n_y}{d_y}\right)^2+\frac{2m_1^*}{\hbar^2}\beta_0(E,\lambda,E_{g01},\Delta_1)\right]^{1/2},
$$

and

$$
\varepsilon_{101}(E,n_x,n_y)=\left[\frac{K_{101}(E,n_x,n_y)}{K_{102}(E,n_x,n_y)}-\frac{K_{102}(E,n_x,n_y)}{K_{101}(E,n_x,n_y)}\right].
$$

The electron concentration per unit length is given by

$$
n_0=\frac{2g_v}{\pi}\sum_{n_x=1}^{n_{x\max}}\sum_{n_y=1}^{n_{y\max}}[\overline{K}_{101}(E_{\mathrm{FQWSLGIL}},n_x,n_y)+\overline{K}_{102}(E_{\mathrm{FQWSLGIL}},n_x,n_y)],
\tag{10.8}
$$

where

$$
\bar{K}_{101}(E_{\mathrm{FQWSLGIL}},n_x,n_y)=\left[\frac{1}{L_0^2}\left[\cos^{-1}\left[\frac{1}{2}\phi_{101}(E_{\mathrm{FQWSLGIL}},n_x,n_y)\right]\right]^2\right.
\left.-\left(\frac{\pi n_x}{d_x}\right)^2-\left(\frac{\pi n_y}{d_y}\right)^2\right]^{1/2},
$$

E_{FQWSLGIL} is the Fermi energy in this case, $\overline{K}_{102}\left(E_{\mathrm{FQWSLGIL}}, n_x, n_y\right) = \sum_{r=1}^{s_0} Z_{r,Y}\left[\overline{K}_{101}\left(E_{\mathrm{FQWSLGIL}}, n_x, n_y\right)\right]$ and $Y = \mathrm{QWSLGIL}$.

Combining (1.13) and (9.8), the optothermoelectric power under large magnetic field in this case is given by

$$G_0 = \left(\frac{\pi^2 k_B^2 T}{3e}\right)\left[\sum_{n_x=1}^{n_{x\max}} \sum_{n_y=1}^{n_{y\max}} \left[\bar{K}_{101}\left(E_{\mathrm{FQWSLGIL}}, n_x, n_y\right) + \bar{K}_{102}\left(E_{\mathrm{FQWSLGIL}}, n_x, n_y\right)\right]\right]^{-1}$$

$$\left[\sum_{n_x=1}^{n_{x\max}} \sum_{n_y=1}^{n_{y\max}} \left[\bar{K}'_{101}\left(E_{\mathrm{FQWSLGIL}}, n_x, n_y\right) + \bar{K}'_{102}\left(E_{\mathrm{FQWSLGIL}}, n_x, n_y\right)\right]\right]. \quad (10.9)$$

10.2.4 *Magnetothermopower in III–V Quantum Dot Superlattices with Graded Interfaces*

The electron energy spectrum in this case is given by

$$\left(\frac{\pi n_z}{d_z}\right)^2 = \left[\frac{1}{L_0^2}\left[\cos^{-1}\left[\frac{1}{2}\phi_{101}\left(E, n_x, n_y\right)\right]\Big|_{E=E_{103}}\right]^2 - \left(\frac{\pi n_x}{d_x}\right)^2 - \left(\frac{\pi n_y}{d_y}\right)^2\right], \quad (10.10)$$

where E_{103} is the totally quantized energy in this case.

The electron concentration per unit volume is given by

$$n_0 = \frac{2g_v}{d_x d_y d_z} \sum_{n_x=1}^{n_{x\max}} \sum_{n_y=1}^{n_{y\max}} \sum_{n_z=1}^{n_{z\max}} F_{-1}\left(\eta_{103}\right), \quad (10.11)$$

where $\eta_{103} = \frac{E_{\mathrm{FQDSLGIL}} - E_{103}}{k_B T}$, in which E_{FQDSLGIL} is the Fermi energy in this case.

Combining (1.13) and (10.11), the optothermoelectric power under large magnetic field in this case is given by

$$G_0 = \frac{\pi^2 k_B}{3e}\left[\sum_{n_x=1}^{n_{x\max}} \sum_{n_y=1}^{n_{y\max}} \sum_{n_z=1}^{n_{z\max}} F_{-1}\left(\eta_{103}\right)\right]^{-1}\left[\sum_{n_x=1}^{n_{x\max}} \sum_{n_y=1}^{n_{y\max}} \sum_{n_z=1}^{n_{z\max}} F_{-2}\left(\eta_{103}\right)\right] \quad (10.12)$$

10.3 Results and Discussion

Using (10.2), (10.3), and Table 1.1, we have plotted the normalized TPSM as functions of well thickness, concentration, and light intensity in quantum wires of $Al_{0.8}Ga_{0.2}As/GaAs$ effective mass superlattices in the presence of external photoexcitation in Figs. 10.1–10.3 at $T = 4.2$ K. For Figs. 10.1 and 10.2, we have taken the lowest occupied subband along the transverse direction. This removes the multiple fluctuations in the TPSM as mentioned in the earlier chapter. From Fig. 10.1 (taking $d_y = 30$ nm, $n_0 = 1 \times 10^8 \text{m}^{-1}$, $I_0 = 0.01 \text{ KWm}^{-2}$, and $\lambda = 400$ nm), it is observed that the numerical values of the TPSM are the greatest for the perturbed three band Kane model representation of the bulk dispersion relations of effective mass superlattices of optoelectronic materials and the least for the perturbed parabolic band representation of the same in the presence of external photoexcitation. From Fig. 10.2, it can be inferred that the TPSM decreases with increasing electron concentration showing oscillations for large values of the electron concentration. After the electron degeneracy of 10^8 m^{-1}, the TPSM decreases very quickly with increasing electron concentration exhibiting large oscillatory spikes.

From Fig. 10.3, we observe that the TPSM exhibits slow increment with increasing light intensity and we have used only the lowest occupancy levels along both the directions. Using (10.5) and (10.6), we have plotted in Figs. 10.4–10.6 the variations of the normalized TPSM of AlGaAs/GaAs quantum dot effective mass superlattices at the lowest energy levels along all the three directions as functions of film thickness, electron concentration, and light intensity, respectively. From Fig. 10.4, it appears that the TPSM quantum dot effective mass superlattices increases with

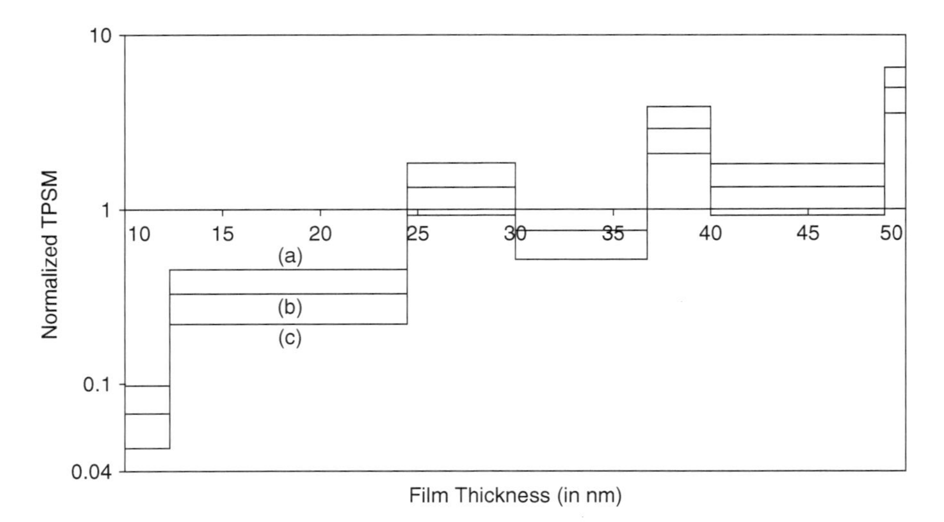

Fig. 10.1 Plot of the TPSM as a function of film thickness for AlGaAs/GaAs quantum wire effective mass superlattices for the perturbed (**a**) three and the (**b**) two band model of Kane together with the (**c**) parabolic model in the presence of external photoexcitation

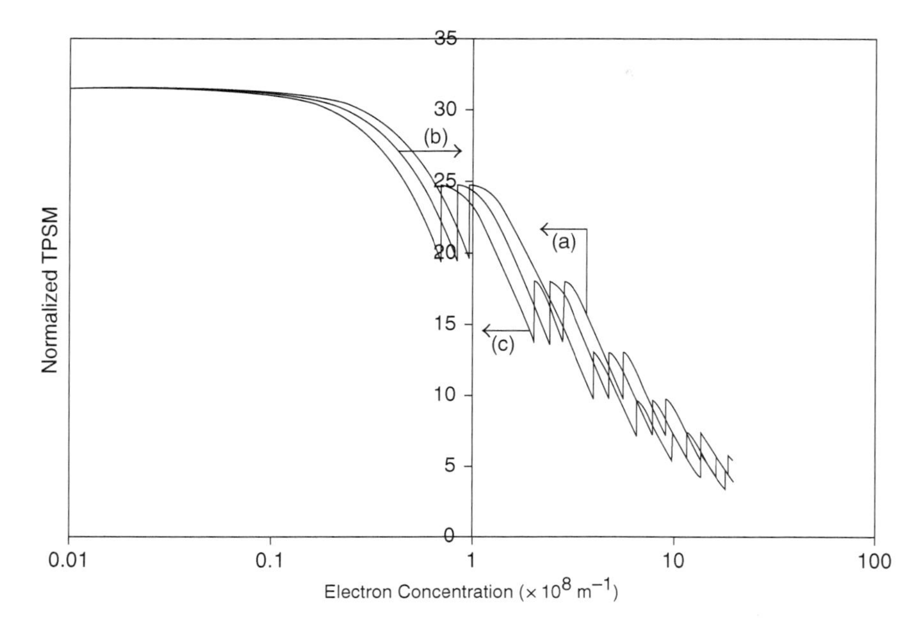

Fig. 10.2 Plot of the TPSM as a function of electron concentration for AlGaAs/GaAs quantum wire effective mass superlattices for all the cases of Fig. 10.1 in the presence of external photoexcitation

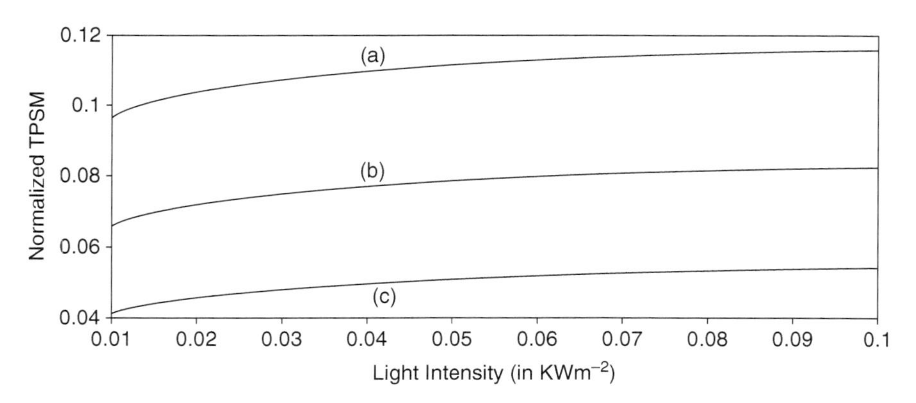

Fig. 10.3 Plot of the TPSM as a function of light intensity for AlGaAs/GaAs quantum wire effective mass superlattices for all the cases of Fig. 10.1 in the presence of external photoexcitation

increasing film thickness in nonoscillatory manners due to quantum limits. Besides for low values of film thickness the TPSM in the said quantized structures for the perturbed (a) three and the (b) two band model of Kane together with the (c) parabolic model in the presence of external photoexcitation converges to a single value. From Fig. 10.5, it is observed that the TPSM decreases sharply with

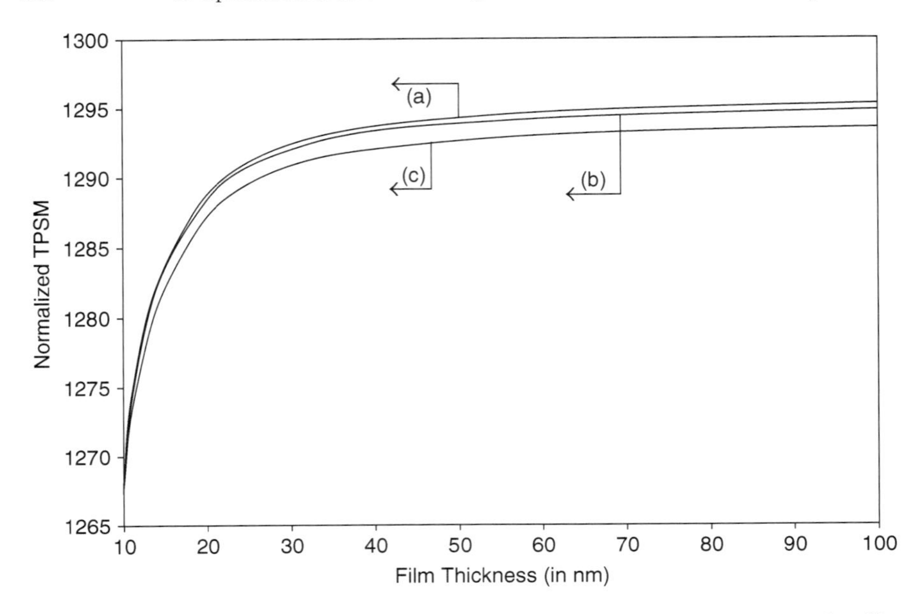

Fig. 10.4 Plot of the TPSM as a function of film thickness for AlGaAs/GaAs quantum dot effective mass superlattices for the perturbed (**a**) three and the (**b**) two band model of Kane together with the (**c**) parabolic model in the presence of external photoexcitation

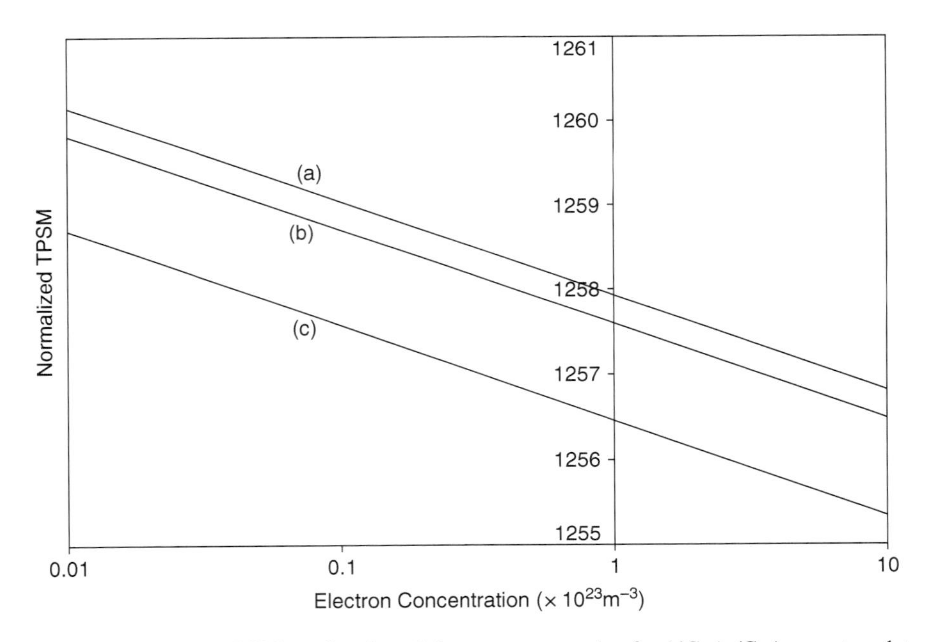

Fig. 10.5 Plot of the TPSM as a function of electron concentration for AlGaAs/GaAs quantum dot effective mass superlattices for all the cases of Fig. 10.4 in the presence of external photoexcitation

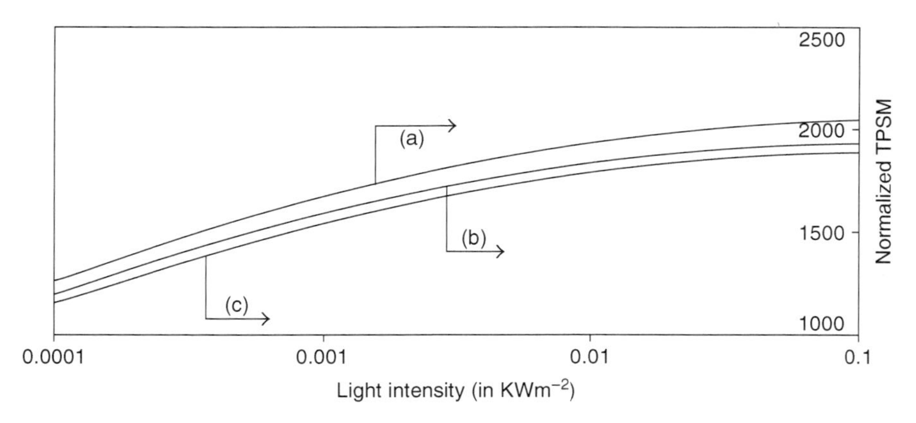

Fig. 10.6 Plot of the TPSM as a function of light intensity for AlGaAs/GaAs quantum dot effective mass superlattices for all the cases of Fig. 10.4 in the presence of external photoexcitation

increasing electron concentration. The difference in the numerical values of the TPSM for the perturbed three band model of Kane and that of perturbed parabolic energy bands becomes large in the presence of photoexcitation. From Fig. 10.6, we can infer that the TPSM increases with increasing light intensity. Furthermore in the whole range of the light intensity, the numerical values of the TPSM for all types of perturbed band models differ widely with respect to the starting value and the ending value of the intensity of light as considered here.

Using (10.8) and (10.9), in Figs. 10.7–10.9, we have plotted the TPSM for AlGaAs/GaAs quantum wire superlattices with graded interfaces at different occupancy levels as functions of film thickness, electron concentration, and light intensity, respectively. From Fig. 10.7 (taking $d_y = 10\,\text{nm}$, $n_0 = 20 \times 10^8\,\text{m}^{-1}$, $I_0 = 0.01\,\text{KWm}^{-2}$, and $\lambda = 400\,\text{nm}$), it appears that the TPSM increases with increasing film thickness showing quantum jumps at intervals determined by the energy band constants and the basic quantum effects involved in it. Besides, the quantum jumps occur at the specified values of the film thickness determined by the selection rules and is independent of the energy band structures although this basic concept totally controls the numerical values of the TPSM, which is maximum for the perturbed three band model of Kane and minimum for the perturbed parabolic presentation of the same.

From Fig. 10.9, it is observed that the TPSM increases slowly with increasing light intensity. From all the figures, it can be stated in general that the inclusion of the subband level creates the singularity in TPSM and other transport properties of the material concerned, due to which the oscillations are observed experimentally at low temperatures, where the quantum effects become prominent. Using (10.11) and (10.12), we have plotted in Figs. 10.10 and 10.11 the normalized TPSM for quantum dot superlattices with graded interfaces of $Al_{0.8}Ga_{0.2}As/GaAs$ as functions of well thickness and wavelength, respectively. From Fig. 10.10, it appears that the TPSM increases with increasing film thickness in an oscillatory way with different

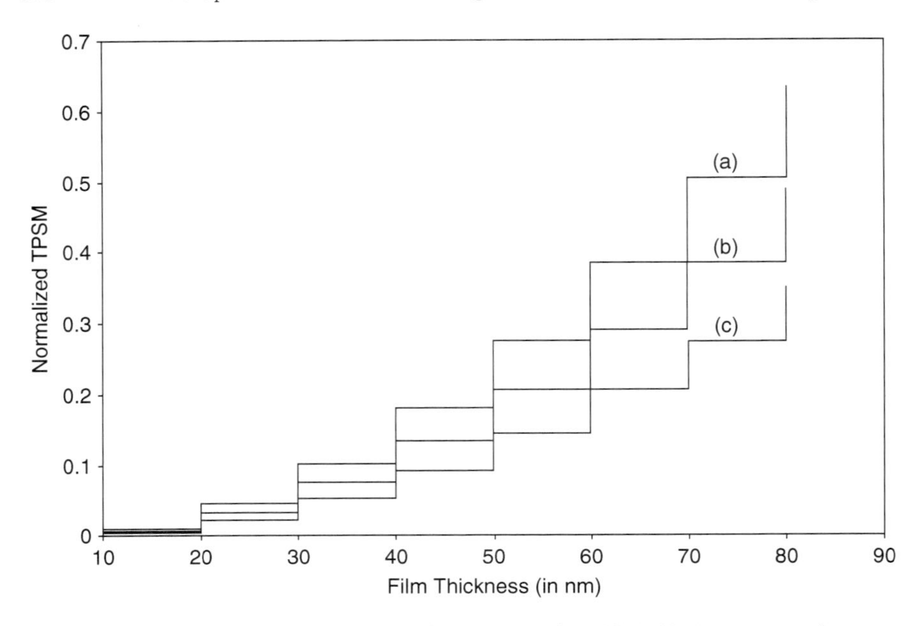

Fig. 10.7 Plot of the TPSM as a function of film thickness for AlGaAs/GaAs quantum wire superlattices with graded interfaces for the perturbed (**a**) three and the (**b**) two band model of Kane together with the (**c**) parabolic model in the presence of external photoexcitation

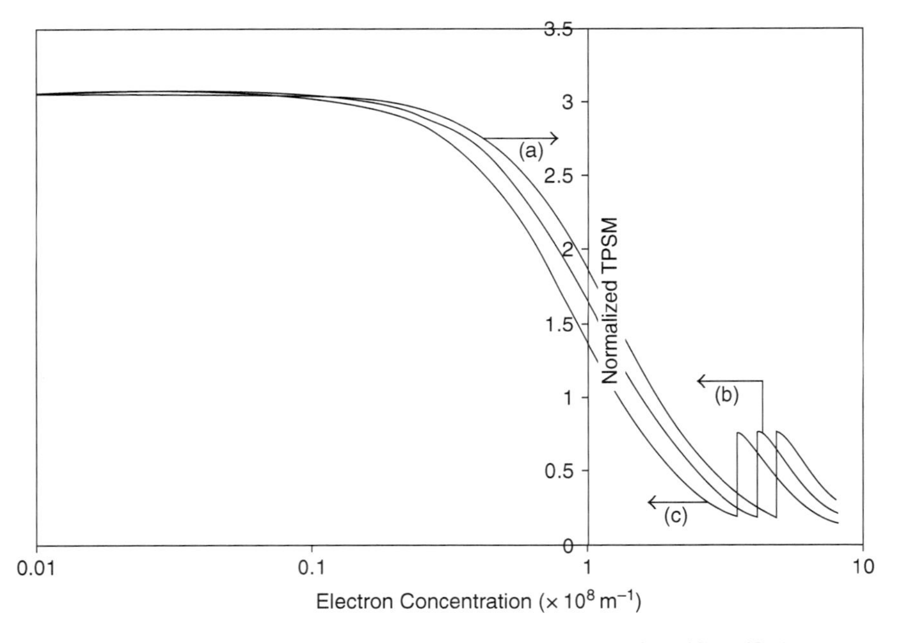

Fig. 10.8 Plot of the TPSM as a function of electron concentration for AlGaAs/GaAs quantum wire superlattices with graded interfaces for all the cases of Fig. 10.7 in the presence of external photoexcitation

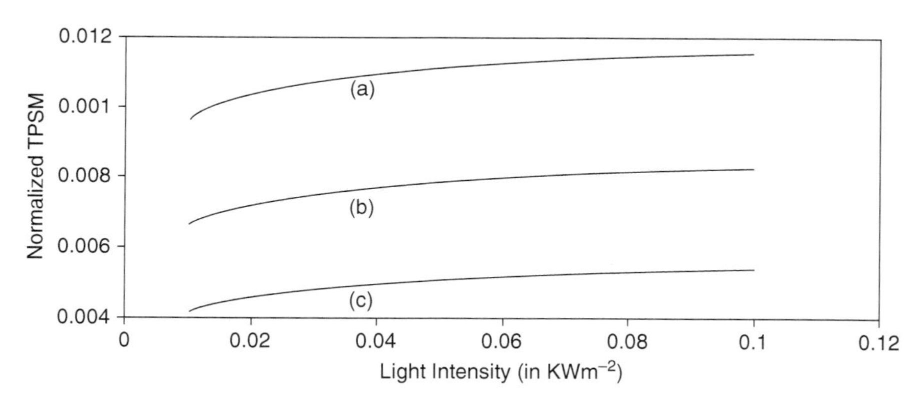

Fig. 10.9 Plot of the TPSM as a function of light intensity for AlGaAs/GaAs quantum wire superlattices with graded interfaces for all the cases of Fig. 10.7 in the presence of external photoexcitation

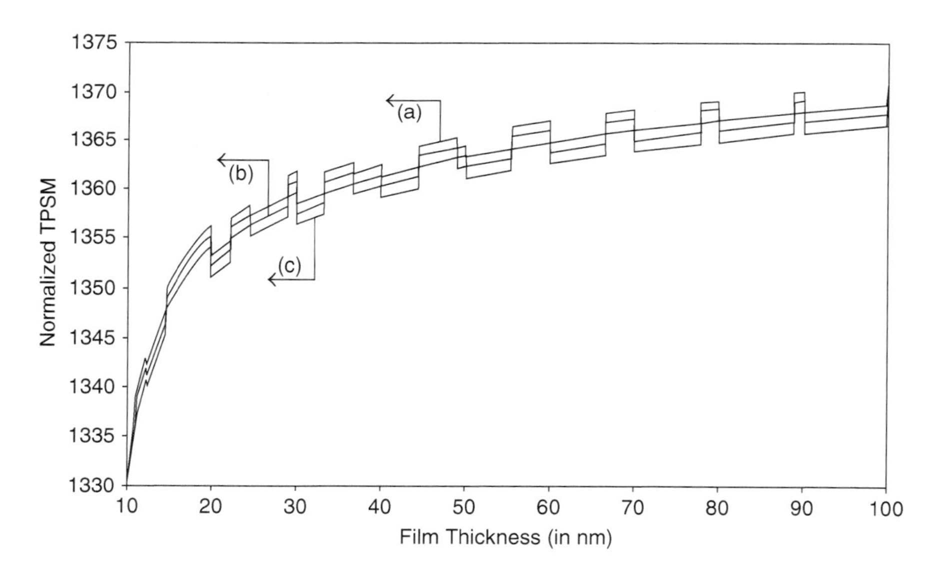

Fig. 10.10 Plot of the TPSM as a function of film thickness for AlGaAs/GaAs quantum dot superlattices with graded interfaces for the perturbed (**a**) three and the (**b**) two band model of Kane together with the (**c**) parabolic model in the presence of external photoexcitation

numerical magnitudes as compared with the corresponding variations with quantum wire superlattices. Besides, Fig. 10.11 states the fact that the TPSM increases with increasing wavelength in a nonoscillatory way due to quantum limit consideration. For the purpose of condensed presentation, the carrier concentration and the corresponding TPSM for this chapter have been presented in Table 10.1.

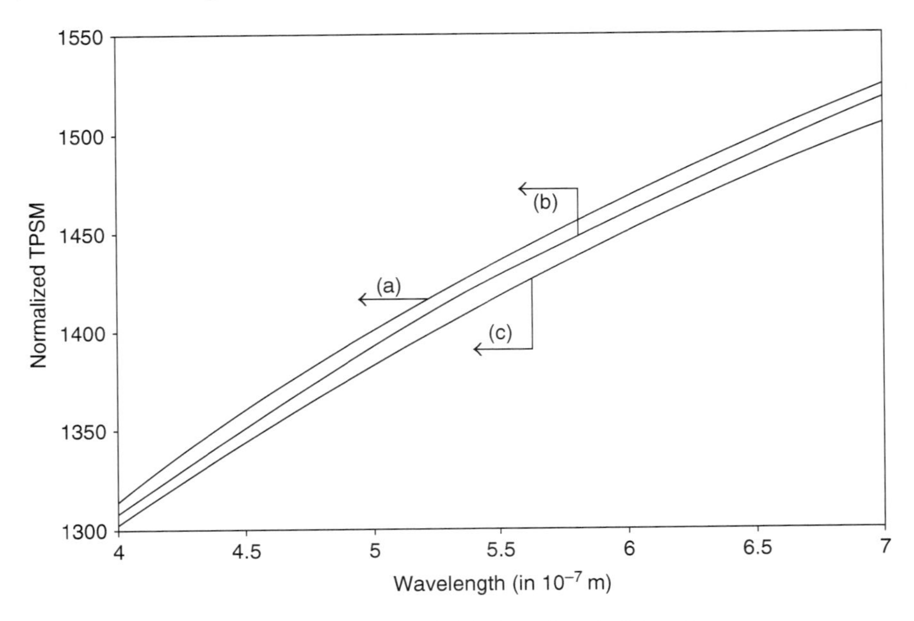

Fig. 10.11 Plot of the TPSM as a function of light intensity of AlGaAs/GaAs quantum dot superlattices with graded interfaces for all the cases of Fig. 10.10 in the presence of external photoexcitation

10.4 Open Research Problems

(R10.1) Investigate the DTP, PTP, and Z in the absence of magnetic field in quantum-confined III–V, II–VI, IV–VI, HgTe/CdTe superlattices with graded interfaces and effective mass superlattices together with short period, strained layer, random, Fibonacci, polytype, and sawtooth superlattices under arbitrarily oriented photoexcitation.

(R10.2) Investigate the DTP, PTP, and Z in the presence of arbitrarily oriented photoexcitation and large magnetic field, respectively, for all the cases of R10.1.

(R10.3) Investigate the DTP, PTP, and Z in the presence of arbitrarily oriented photoexcitation and nonquantizing nonunifoelectric field, respectively, for all the cases of R10.1.

(R10.4) Investigate the DTP, PTP, and Z in the presence of arbitrarily oriented photoexcitation and nonquantizing alternating electric field, respectively, for all the cases of R10.1.

(R10.5) Investigate the DTP, PTP, and Z in the presence of arbitrarily oriented photoexcitation and crossed electric and quantizing magnetic fields, respectively, for all the cases of R10.1.

(R10.6) Investigate the DTP, PTP, and Z from heavily doped quantum-confined superlattices for all the problems of R10.1.

Table 10.1 The carrier statistics and the optomagnetothermoelectric power in quantum-confined semiconductor superlattices of optoelectronic materials

Type of materials	Carrier statistics	TPSM
1. III–V Quantum wire effective mass superlattices	$n_0 = \frac{2g_v}{\pi} \sum_{n_y=1}^{n_{y\max}} \sum_{n_z=1}^{n_{z\max}} [K_{91}(E_{\mathrm{FQWSLEML}}, n_y, n_z) + K_{92}(E_{\mathrm{FQWSLEML}}, n_y, n_z)]$ (10.2)	On the basis of (10.2), $G_0 = \left(\frac{\pi^2 k_B^2 T}{3e}\right) \left[\sum_{n_y=1}^{n_{y\max}} \sum_{n_z=1}^{n_{z\max}} [K_{91}(E_{\mathrm{FQWSLEML}}, n_y, n_z) + K_{92}(E_{\mathrm{FQWSLEML}}, n_y, n_z)]\right]^{-1} \left[\sum_{n_y=1}^{n_{y\max}} \sum_{n_z=1}^{n_{z\max}} [K'_{91}(E_{\mathrm{FQWSLEML}}, n_y, n_z) + K'_{91}(E_{\mathrm{FQWSLEML}}, n_y, n_z)]\right]$ (10.3)
2. III–V Quantum dot effective mass superlattices	$n_0 = \frac{2g_v}{d_x d_y d_z} \sum_{n_x=1}^{n_{x\max}} \sum_{n_y=1}^{n_{y\max}} \sum_{n_z=1}^{n_{z\max}} F_{-1}(\eta_{92})$ (10.5)	On the basis of (10.5), $G_0 = \frac{\pi^2 k_B}{3e} \left[\sum_{n_x=1}^{n_{x\max}} \sum_{n_y=1}^{n_{y\max}} \sum_{n_z=1}^{n_{z\max}} F_{-1}(\eta_{92})\right]^{-1} \left[\sum_{n_x=1}^{n_{x\max}} \sum_{n_y=1}^{n_{y\max}} \sum_{n_z=1}^{n_{z\max}} F_{-2}(\eta_{92})\right]$ (10.6)

(continued)

Table 10.1 (Continued)

Type of materials	Carrier statistics	TPSM
3. III–V Quantum wire superlattices with graded interfaces	$n_0 = \frac{2g_v}{\pi} \sum_{n_x=1}^{n_{x\max}} \sum_{n_y=1}^{n_{y\max}} [\bar{K}_{91}(E_{FQWSLGI}, n_x, n_y) + \bar{K}_{92}(E_{FQWSLGI}, n_x, n_y)]$ (10.8)	On the basis of (10.8), $G_0 = \left(\frac{\pi^2 k_B^2 T}{3e}\right) \left[\sum_{n_x=1}^{n_{x\max}} \sum_{n_y=1}^{n_{y\max}} [\bar{K}_{91}(E_{FQWSLGI}, n_x, n_y) + \bar{K}_{92}(E_{FQWSLGI}, n_x, n_y)]\right]^{-1} \left[\sum_{n_x=1}^{n_{x\max}} \sum_{n_y=1}^{n_{y\max}} [\bar{K}_{91}(E_{FQWSLGI}, n_x, n_y) + \bar{K}_{92}(E_{FQWSLGI}, n_x, n_y)]\right]$ (10.9)
4. III–V Quantum dot superlattices with graded interfaces	$n_0 = \frac{2g_v}{d_x d_y d_z} \sum_{n_x=1}^{n_{x\max}} \sum_{n_y=1}^{n_{y\max}} \sum_{n_z=1}^{n_{z\max}} F_{-1}(\eta_{93})$ (10.11)	On the basis of (10.11), $G_0 = \frac{\pi^2 k_B}{3e} \left[\sum_{n_x=1}^{n_{x\max}} \sum_{n_y=1}^{n_{y\max}} \sum_{n_z=1}^{n_{z\max}} F_{-1}(\eta_{93})\right]^{-1} \left[\sum_{n_x=1}^{n_{x\max}} \sum_{n_y=1}^{n_{y\max}} \sum_{n_z=1}^{n_{z\max}} F_{-2}(\eta_{93})\right]$ (10.12)

(R10.7) Investigate the DTP, PTP, and Z in the presence of arbitrarily oriented photoexcitation and quantizing magnetic field, respectively, for all the cases of R10.6.

(R10.8) Investigate the DTP, PTP, and Z in the presence of arbitrarily oriented photoexcitation and nonquantizing nonunifoelectric field, respectively, for all the cases of R10.6.

(R10.9) Investigate the DTP, PTP, and Z in the presence of arbitrarily oriented photoexcitation and nonquantizing alternating electric field, respectively, for all the cases of R10.6.

(R10.10) Investigate the DTP, PTP, and Z in the presence of arbitrarily oriented photoexcitation and crossed electric and quantizing magnetic fields, respectively, for all the cases of R10.6.

Reference

1. K.P. Ghatak, S. Bhattacharya, D. De, *Photoemission from Optoelectronic Materials and Their Nanostructures*, Springer Series in Nanostructure Science and Technology (Springer, New York, USA, 2009)

Part IV
Thermoelectric Power Under Magnetic Quantization in Macro and Micro-optoelectronic Materials in the Presence of Light Waves

Chapter 11
Optothermoelectric Power in Macro-Optoelectronic Materials Under Magnetic Quantization

11.1 Introduction

This chapter investigates optothermoelectric power in macro-optoelectronic materials under magnetic quantization. Section 11.2.1 investigates the magnetothermopower in the presence of external photoexcitation for bulk specimens of optoelectronic materials whose conduction electrons obey the dispersion relations as given by (8.1)–(8.3). Sections 11.3 and 11.4 contain results and discussion and open research problems, respectively.

11.2 Theoretical Background

11.2.1 Magnetothermopower in Optoelectronic Materials

Using (8.1)–(8.3), the magnetodispersion relations, in the absence of electron spin, for optoelectronic materials in the presence of photoexcitation, whose unperturbed conduction electrons obey the three and two band models of Kane, together with parabolic energy bands, are given by [1]

$$\beta_0(E,\lambda) = \left(n+\frac{1}{2}\right)\hbar\omega_0 + \frac{\hbar^2 k_z^2}{2m^*}, \tag{11.1}$$

$$\tau_0(E,\lambda) = \left(n+\frac{1}{2}\right)\hbar\omega_0 + \frac{\hbar^2 k_z^2}{2m^*}, \tag{11.2}$$

$$\rho_0(E,\lambda) = \left(n+\frac{1}{2}\right)\hbar\omega_0 + \frac{\hbar^2 k_z^2}{2m^*}. \tag{11.3}$$

The density-of-states function per subband for (10.1), (10.2), and (10.3) can, respectively, be written as

$$N_{\mathrm{B}}'(E,\lambda) = \frac{g_{\mathrm{v}}|e|\sqrt{2m^*}}{2\pi^2\hbar^2}\left[\{\beta_0(E,\lambda)\}'\left\{\beta_0(E,\lambda) - \left(n+\frac{1}{2}\right)\hbar\omega_0\right\}^{-1/2}\right], \tag{11.4}$$

$$N_{\mathrm{B}}'(E,\lambda) = \frac{g_{\mathrm{v}}|e|\sqrt{2m^*}}{2\pi^2\hbar^2}\left[\{\tau_0(E,\lambda)\}'\left\{\tau_0(E,\lambda) - \left(n+\frac{1}{2}\right)\hbar\omega_0\right\}^{-1/2}\right], \tag{11.5}$$

$$N_{\mathrm{B}}'(E,\lambda) = \frac{g_{\mathrm{v}}|e|\sqrt{2m^*}}{2\pi^2\hbar^2}\left[\{\rho_0(E,\lambda)\}'\left\{\rho_0(E,\lambda) - \left(n+\frac{1}{2}\right)\hbar\omega_0\right\}^{-1/2}\right]. \tag{11.6}$$

It appears then that evaluation of the optothermoelectric power requires the expression of electron statistics per unit volume which can, respectively, be expressed as

$$n_0 = \frac{g_{\mathrm{v}}eB\sqrt{2m^*}}{\pi^2\hbar^2}\sum_{n=0}^{n_{\max}}\left[M_{101}(E_{\mathrm{FBL}},B,\lambda) + N_{101}(E_{\mathrm{FBL}},B,\lambda)\right], \tag{11.7}$$

$$n_0 = \frac{g_{\mathrm{v}}eB\sqrt{2m^*}}{\pi^2\hbar^2}\sum_{n=0}^{n_{\max}}\left[M_{102}(E_{\mathrm{FBL}},B,\lambda) + N_{102}(E_{\mathrm{FBL}},B,\lambda)\right], \tag{11.8}$$

$$n_0 = \frac{g_{\mathrm{v}}eB\sqrt{2m^*}}{\pi^2\hbar^2}\sum_{n=0}^{n_{\max}}\left[M_{103}(E_{\mathrm{FBL}},B,\lambda) + N_{103}(E_{\mathrm{FBL}},B,\lambda)\right], \tag{11.9}$$

where

$$M_{101}(E_{\mathrm{FBL}},B,\lambda) \equiv \left[\beta_0(E_{\mathrm{FBL}},\lambda) - \left(n+\frac{1}{2}\right)\hbar\omega_0\right]^{1/2},$$

E_{FBL} is the Fermi energy in this case, $N_{101}(E_{\mathrm{FBL}},B,\lambda) \equiv \sum_{r=1}^{s} Z_{r,Y}M_{101}(E_{\mathrm{FBL}},B,\lambda)$, $Y = BL$,

$$M_{102}(E_{\mathrm{FBL}},B,\lambda) \equiv \left[\tau_0(E_{\mathrm{FBL}},\lambda) - \left(n+\frac{1}{2}\right)\hbar\omega_0\right]^{1/2},$$

$$N_{102}(E_{\mathrm{FBL}},B,\lambda) \equiv \sum_{r=1}^{s} Z_{r,Y}M_{102}(E_{\mathrm{FBL}},B,\lambda),$$

$$M_{103}(E_{\mathrm{FBL}},B,\lambda) \equiv \left[\rho_0(E_{\mathrm{FBL}},\lambda) - \left(n+\frac{1}{2}\right)\hbar\omega_0\right]^{1/2},$$

and $N_{103}(E_{\mathrm{FBL}},B,\lambda) \equiv \sum_{r=1}^{s} Z_{y,Y}M_{103}(E_{\mathrm{FBL}},B,\lambda)$.

Combining (1.13) with (11.7), (11.8), and (11.9), the optothermoelectric power in macro-optoelectronic materials in the presence of a quantizing magnetic field B can, respectively, be written as

$$G_0 = \left(\frac{\pi^2 k_B^2 T}{3e}\right)\left[\sum_{n=0}^{n_{\max}} \left[M_{101}(E_{FBL}, B, \lambda) + N_{101}(E_{FBL}, B, \lambda)\right]\right]^{-1} \times \left[\sum_{n=0}^{n_{\max}} \left[M'_{101}(E_{FBL}, B, \lambda) + N'_{101}(E_{FBL}, B, \lambda)\right]\right], \quad (11.10)$$

$$G_0 = \left(\frac{\pi^2 k_B^2 T}{3e}\right)\left[\sum_{n=0}^{n_{\max}} \left[M_{102}(E_{FBL}, B, \lambda) + N_{102}(E_{FBL}, B, \lambda)\right]\right]^{-1} \times \left[\sum_{n=0}^{n_{\max}} \left[M'_{102}(E_{FBL}, B, \lambda) + N'_{102}(E_{FBL}, B, \lambda)\right]\right], \quad (11.11)$$

and

$$G_0 = \frac{\pi^2 k_B^2 T}{3e}\left[\sum_{n=0}^{n_{\max}} \left[M_{103}(E_{FBL}, B, \lambda) + N_{103}(E_{FBL}, B, \lambda)\right]\right]^{-1} \times \left[\sum_{n=0}^{n_{\max}} \left[M'_{103}(E_{FBL}, B, \lambda) + N'_{103}(E_{FBL}, B, \lambda)\right]\right]. \quad (11.12)$$

11.3 Results and Discussion

Using (11.7); (11.10) and (11.8); (11.11) and (11.9); (11.12); and using Table 11.1, the normalized TPSM

$$\left((3G_0 e)/(\pi^2 k_B^2 T)\right).$$

for n-InAs in accordance with the perturbed three and two band models of Kane together with perturbed parabolic energy bands has been drawn at $T = 4.2$ K as functions of inverse quantizing magnetic field, carrier degeneracy, light intensity, and wavelength in Figs. 11.1–11.4, respectively. The aforementioned plots for n-InSb and n-HgCdTe have been shown in Figs. 11.5–11.8 and 11.9–11.12, respectively. Figure 11.13 exhibits the variation of the normalized TPSM as a function of alloy composition for the aforementioned band models for n-HgCdTe, and Figs. 11.14–11.18 exhibit the normalized TPSM for n-InGaAsP as functions of inverse quantizing magnetic field, carrier degeneracy, light intensity, wavelength, and alloy composition, respectively. From Figs. 11.1, 11.5, 11.9, and 11.14 it appears that normalized TPSM is an oscillatory function of $1/B$. The oscillatory

Table 11.1 The carrier statistics and the optothermoelectric power in macro-optoelectronic materials under magnetic quantization

Type of materials	Carrier statistics	TPSM
Optoelectronic materials	$n_0 = \frac{g_v eB\sqrt{2m^*}}{\pi^2\hbar^2} \sum_{n=0}^{n_{\max}} [M_{101}(E_{\mathrm{FBL}}, B, \lambda) + N_{101}(E_{\mathrm{FBL}}, B, \lambda)]$ (11.7)	On the basis of (11.7), $G_0 = \left(\frac{\pi^2 k_B^2 T}{3e}\right)\left[\sum_{n=0}^{n_{\max}} [M_{101}(E_{\mathrm{FBL}}, B, \lambda) + N_{101}(E_{\mathrm{FBL}}, B, \lambda)]\right]^{-1} \times \left[\sum_{n=0}^{n_{\max}} [M'_{101}(E_{\mathrm{FBL}}, B, \lambda) + N'_{101}(E_{\mathrm{FBL}}, B, \lambda)]\right]$ (11.10)
	$n_0 = \frac{g_v eB\sqrt{2m^*}}{\pi^2\hbar^2} \sum_{n=0}^{n_{\max}} [M_{102}(E_{\mathrm{FBL}}, B, \lambda) + N_{102}(E_{\mathrm{FBL}}, B, \lambda)]$ (11.8)	On the basis of (11.8), $G_0 = \left(\frac{\pi^2 k_B^2 T}{3e}\right)\left[\sum_{n=0}^{n_{\max}} [M_{102}(E_{\mathrm{FBL}}, B, \lambda) + N_{102}(E_{\mathrm{FBL}}, B, \lambda)]\right]^{-1} \times \left[\sum_{n=0}^{n_{\max}} [M'_{102}(E_{\mathrm{FBL}}, B, \lambda) + N'_{102}(E_{\mathrm{FBL}}, B, \lambda)]\right]$ (11.11)
	$n_0 = \frac{g_v eB\sqrt{2m^*}}{\pi^2\hbar^2} \sum_{n=0}^{n_{\max}} [M_{103}(E_{\mathrm{FBL}}, B, \lambda) + N_{103}(E_{\mathrm{FBL}}, B, \lambda)]$ (11.9)	On the basis of (11.9), $G_0 = \frac{\pi^2 k_B^2 T}{3e}\left[\sum_{n=0}^{n_{\max}} [M_{103}(E_{\mathrm{FBL}}, B, \lambda) + N_{103}(E_{\mathrm{FBL}}, B, \lambda)]\right]^{-1} \times \left[\sum_{n=0}^{n_{\max}} [M'_{103}(E_{\mathrm{FBL}}, B, \lambda) + N'_{103}(E_{\mathrm{FBL}}, B, \lambda)]\right]$ (11.12)

dependence is due to the well-known Subhnikov–de Hass effect, and in the presence of external quantizing magnetic field, the TPSM oscillates in a periodic manner which has been already mentioned in the previous chapters. In the presence of external photoexcitation, it appears that there is a tendency of increase in the Fermi energy of the system with an increase in either of the intensity or the wavelength under strong quantizing magnetic field, which reduces the magnitude of the TPSM. It may be noted that in the presence of light, the SdH periodicity has been preserved by the system. Figures 11.2, 11.6, 11.10, and 11.15 exhibit the fact that the TPSM decreases with increasing carrier degeneracy in an oscillatory way. The spikes are due to the reorganization of the carriers in the Landau levels.

Figures 11.3, 11.7, 11.11, and 11.16 (taking $B = 10\,\text{T}$, $n_0 = 10^{23}\text{m}^{-3}$, $I_0 = 10^{-5}\,\text{kW m}^{-2}$, and $\lambda = 400\,\text{nm}$) exhibit the variation of the TPSM as function of the light intensity for the said materials and it appears that the TPSM increases with the increase in intensity for all types of band models. From Figs. 11.4, 11.8, 11.12, and 11.16 we observe that TPSM increases with increasing wavelength and the wide difference among the band models reflects the fact that the influence of energy band constants are prominent. From Figs. 11.13 and 11.18, it appears that the TPSM increases with the increasing alloy composition for both ternary and quaternary materials. All the figures from 11.9 to 11.12 and 11.14 to 11.17 have been plotted

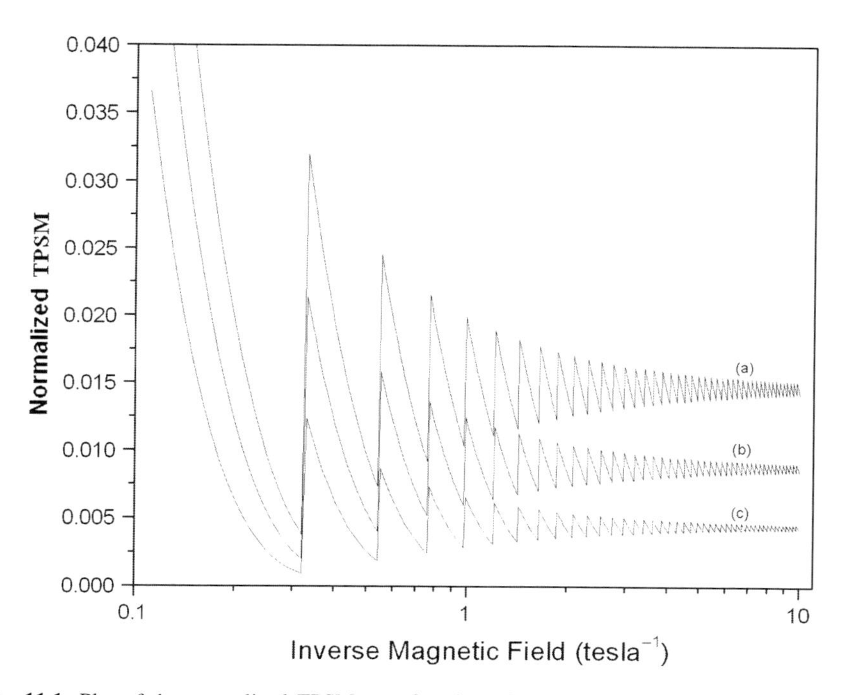

Fig. 11.1 Plot of the normalized TPSM as a function of inverse quantizing magnetic field for n-InAs in accordance with the perturbed (**a**) three and the (**b**) two band model of Kane together with the (**c**) perturbed parabolic energy bands in the presence of external photoexcitation

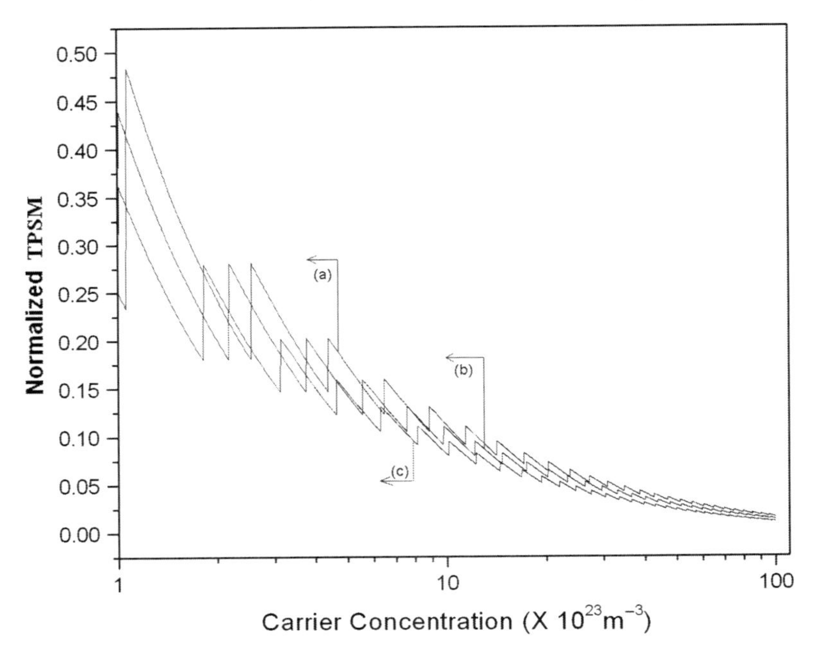

Fig. 11.2 Plot of the normalized TPSM as a function of carrier concentration for n-InAs for all the cases of Fig. 11.1 in the presence of external photoexcitation

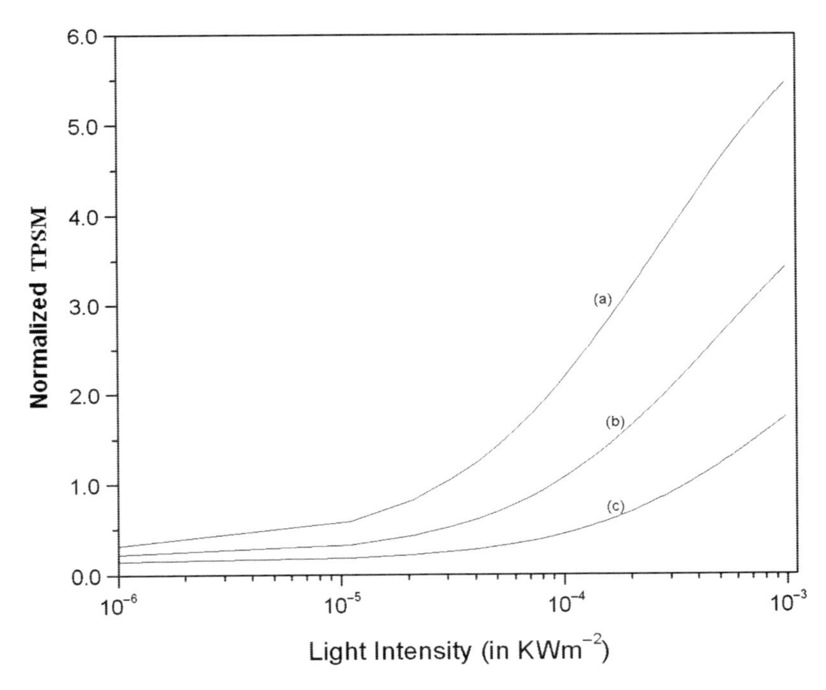

Fig. 11.3 Plot of the normalized TPSM as a function of light intensity for n-InAs for all the cases of Fig. 11.1 in the presence of external photoexcitation

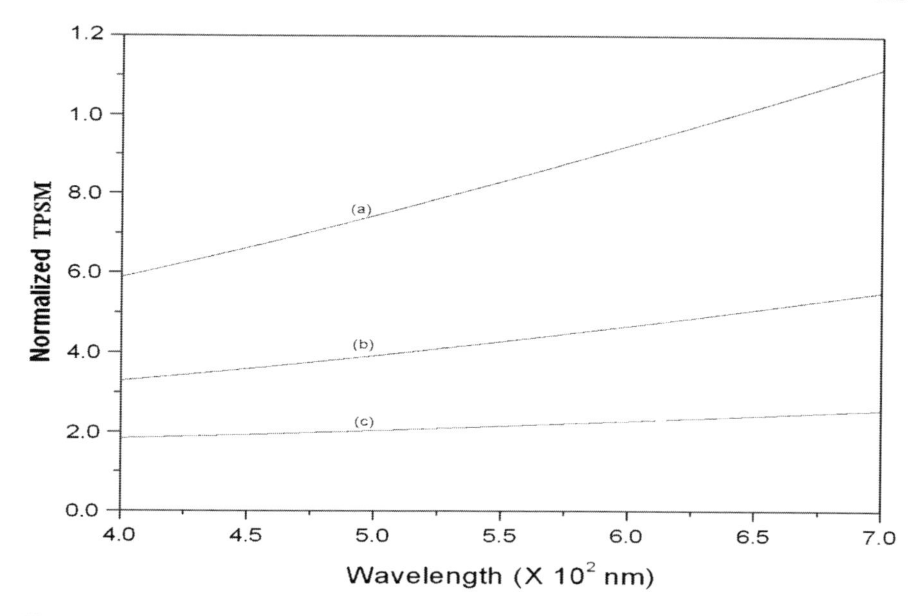

Fig. 11.4 Plot of the normalized TPSM as a function of wavelength for n-InAs for all the cases of Fig. 11.1

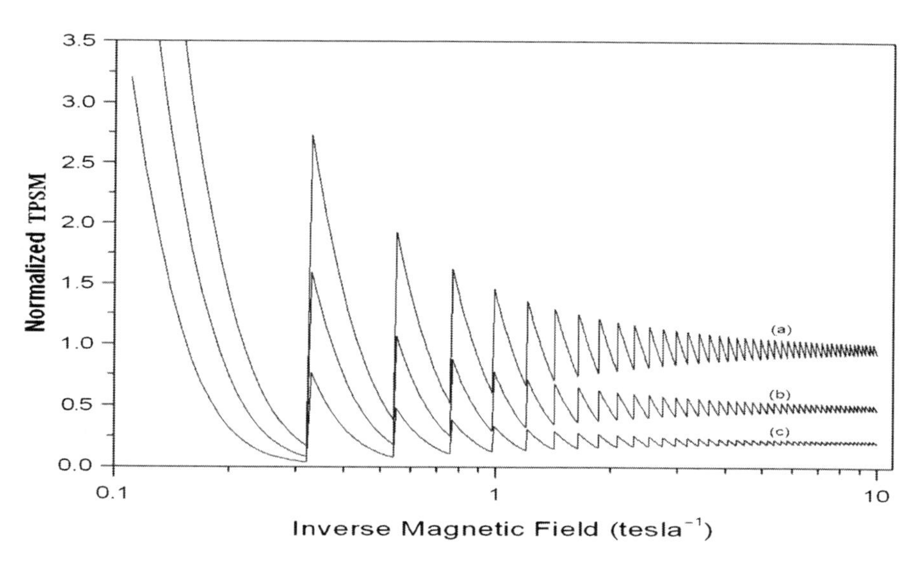

Fig. 11.5 Plot of the normalized TPSM as a function of $1/B$ for n-InSb for all cases of Fig. 11.1

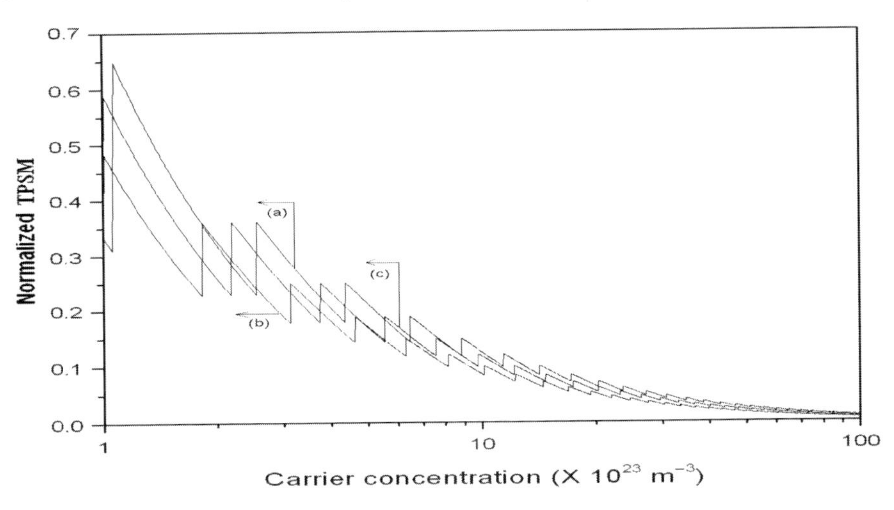

Fig. 11.6 Plot of the normalized TPSM as a function of n_0 for n-InSb for all the cases of Fig. 11.1

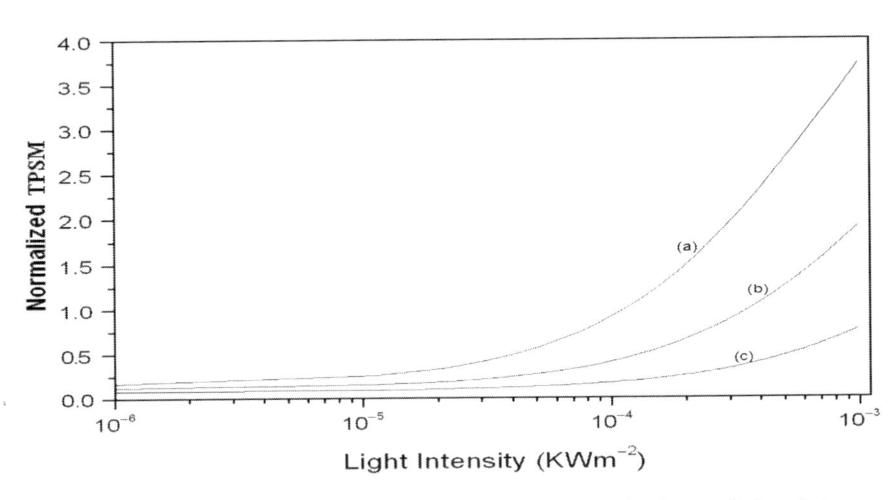

Fig. 11.7 Plot of the normalized TPSM as a function of light intensity for n-InSb for all the cases of Fig. 11.1 in the presence of external photoexcitation

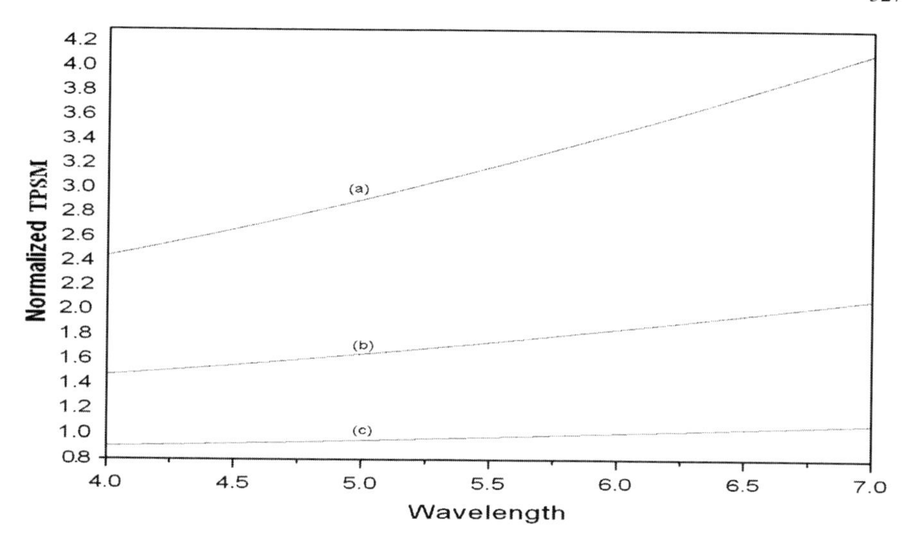

Fig. 11.8 Plot of the normalized TPSM as a function of wavelength for n-InSb for all the cases of Fig. 11.1 in the presence of external photoexcitation

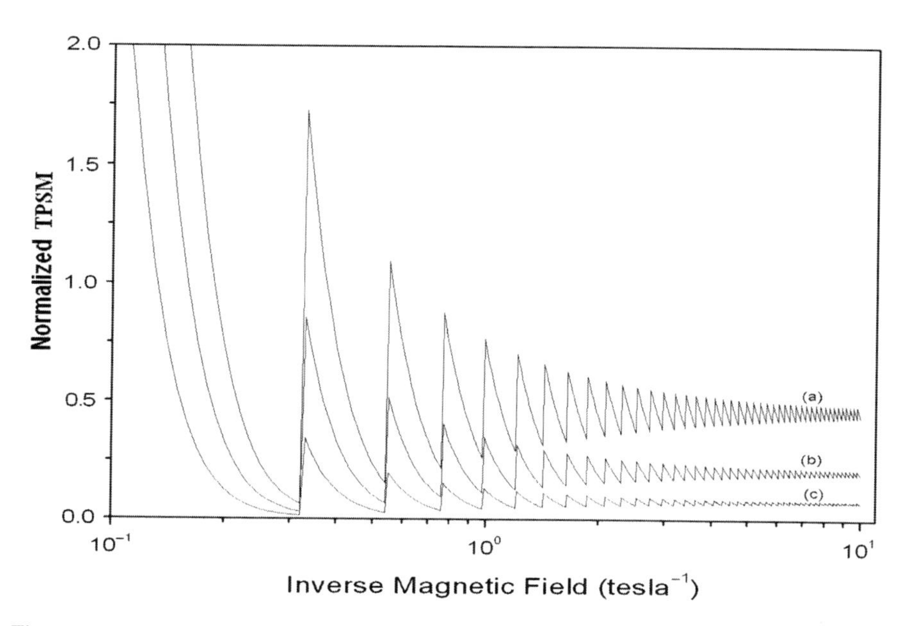

Fig. 11.9 Plot of the normalized TPSM as a function of $1/B$ for n-HgCdTe for all cases of Fig. 11.1

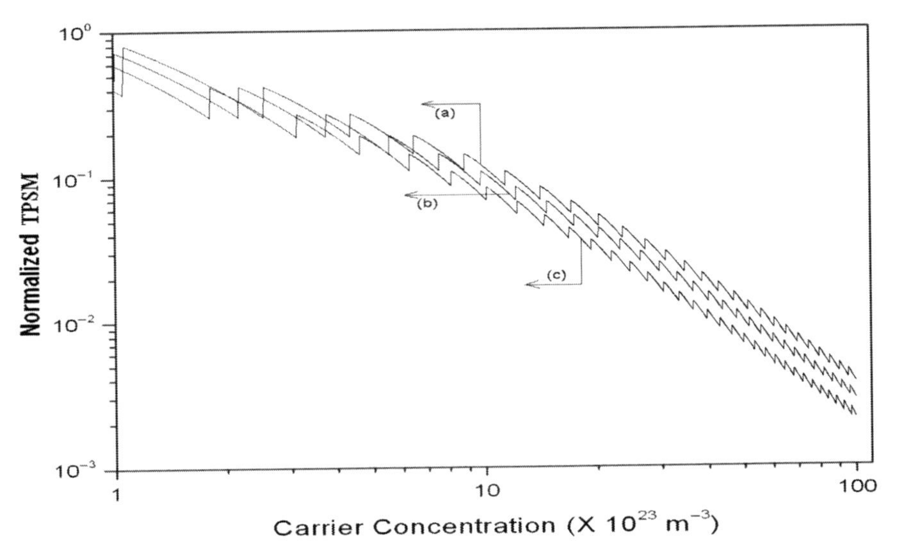

Fig. 11.10 Plot of the normalized TPSM as a function of n_0 for n-HgCdTe for all the cases of Fig. 11.1

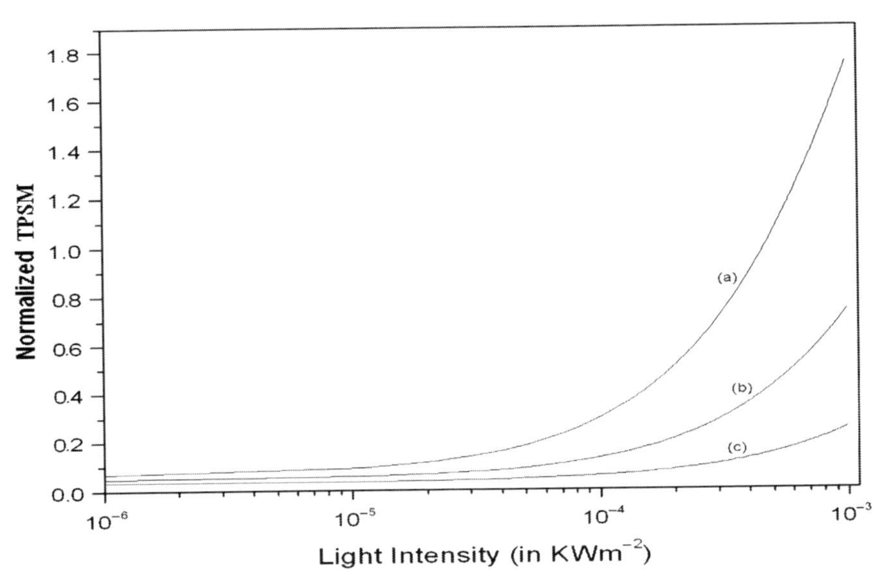

Fig. 11.11 Plot of the normalized TPSM as a function of I for n-HgCdTe for all the cases of Fig. 11.1

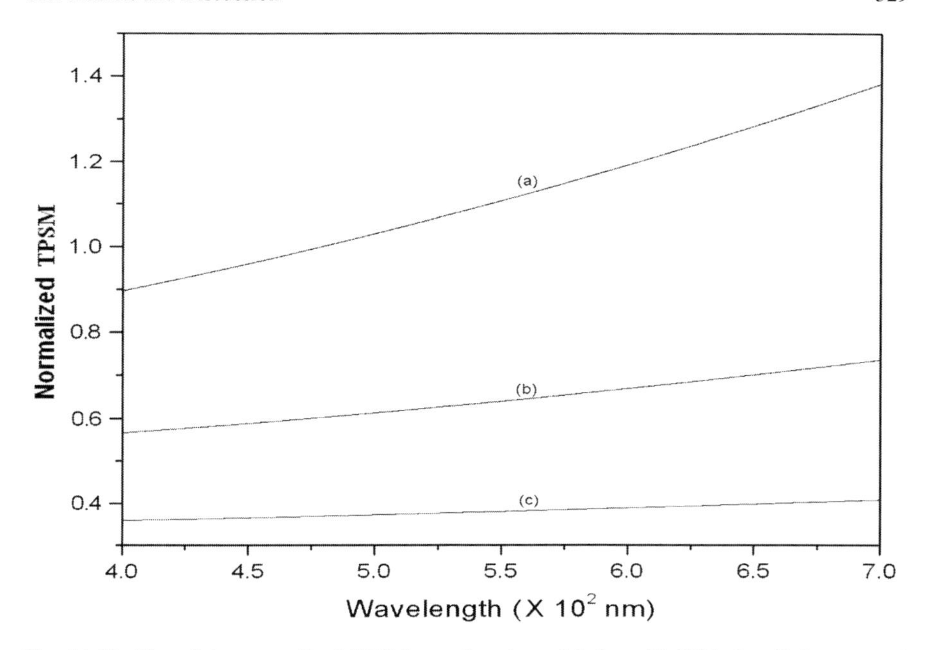

Fig. 11.12 Plot of the normalized TPSM as a function of λ for n-HgCdTe for all the cases of Fig. 11.1

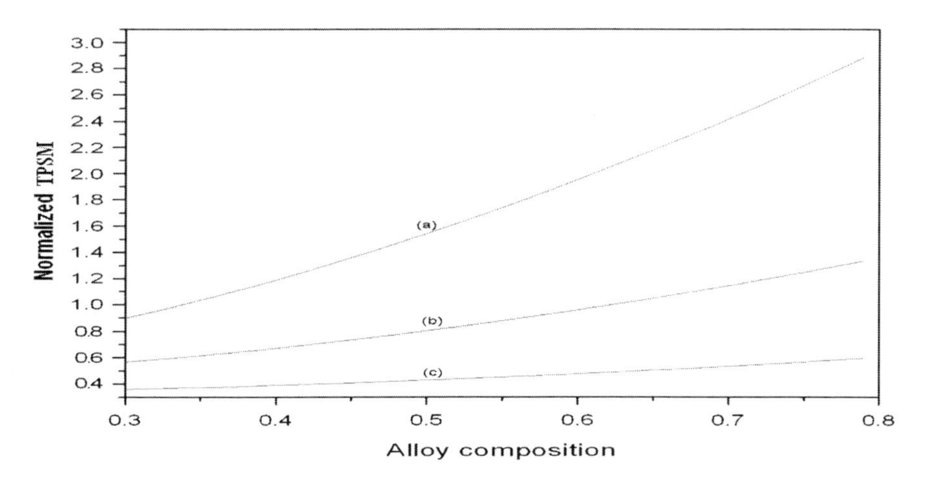

Fig. 11.13 Plot of the normalized TPSM as a function of alloy composition for n-HgCdTe all the cases of Fig. 11.1 in the presence of external photoexcitation

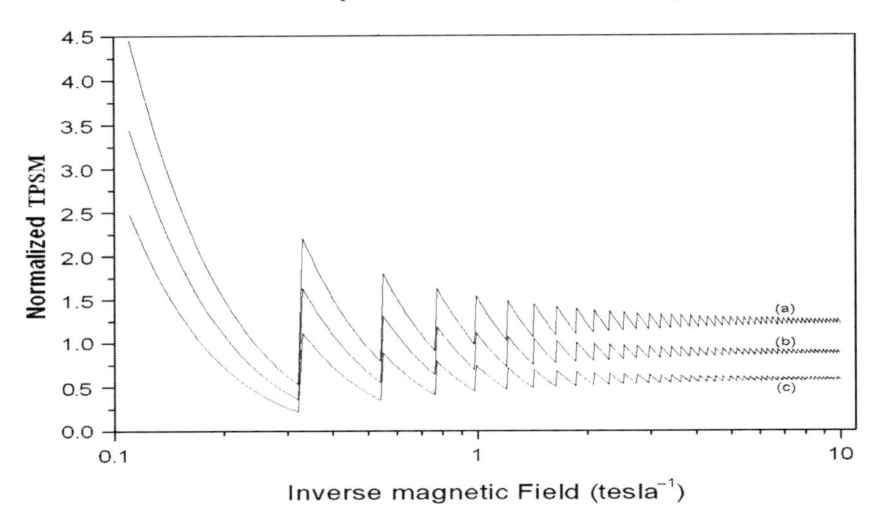

Fig. 11.14 Plot of the normalized TPSM as a function of $1/B$ for n-InGaAsP for all cases of Fig. 11.1

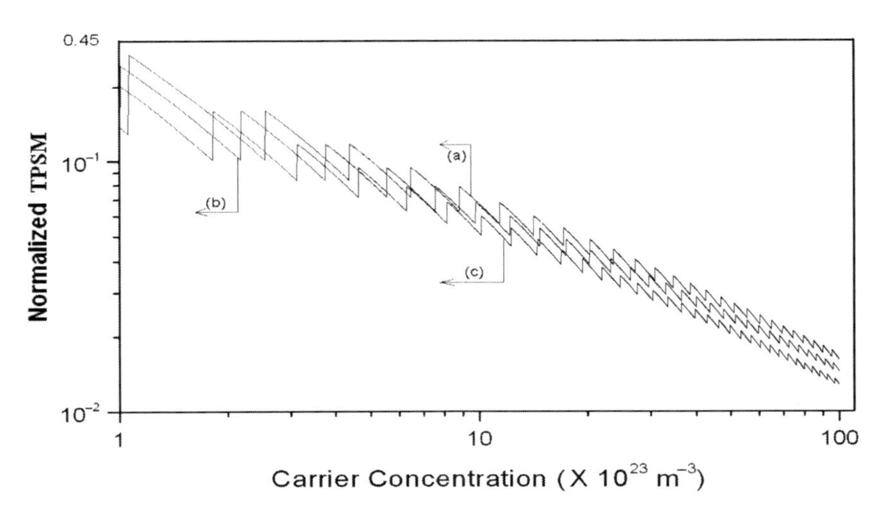

Fig. 11.15 Plot of the normalized TPSM as a function of n_0 for n-InGaAsP for all cases of Fig. 11.1

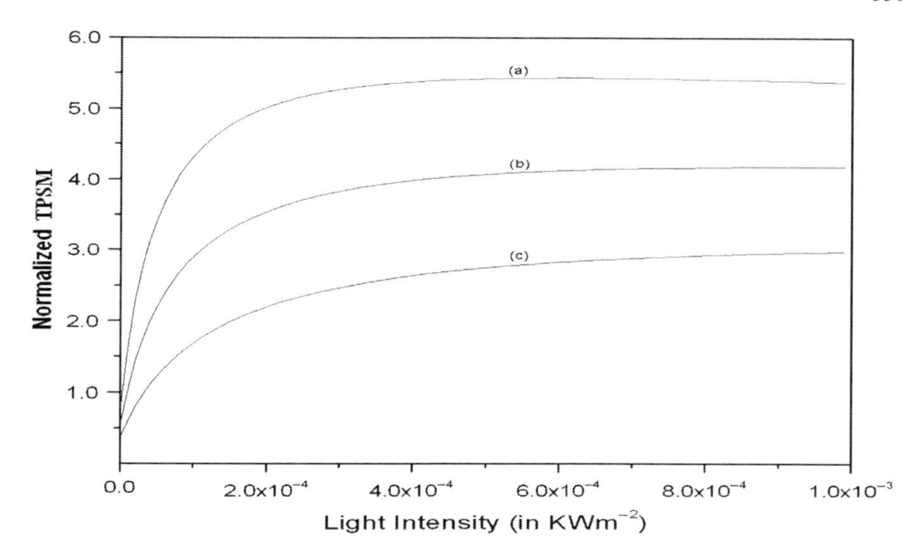

Fig. 11.16 Plot of the normalized TPSM as a function of I for n-InGaAsP for all the cases of Fig. 11.1

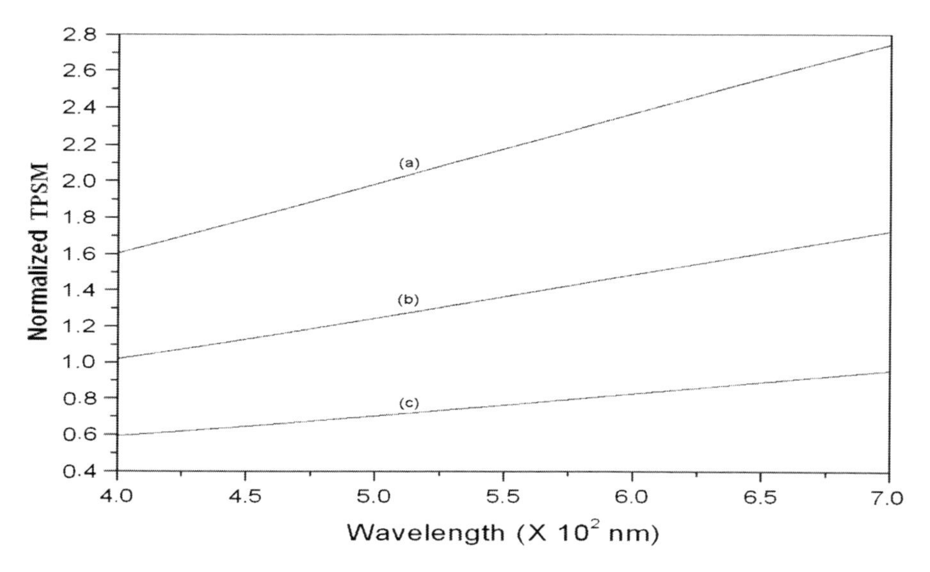

Fig. 11.17 Plot of the normalized TPSM as a function of λ for n-InGaAsP for all the cases of Fig. 11.1

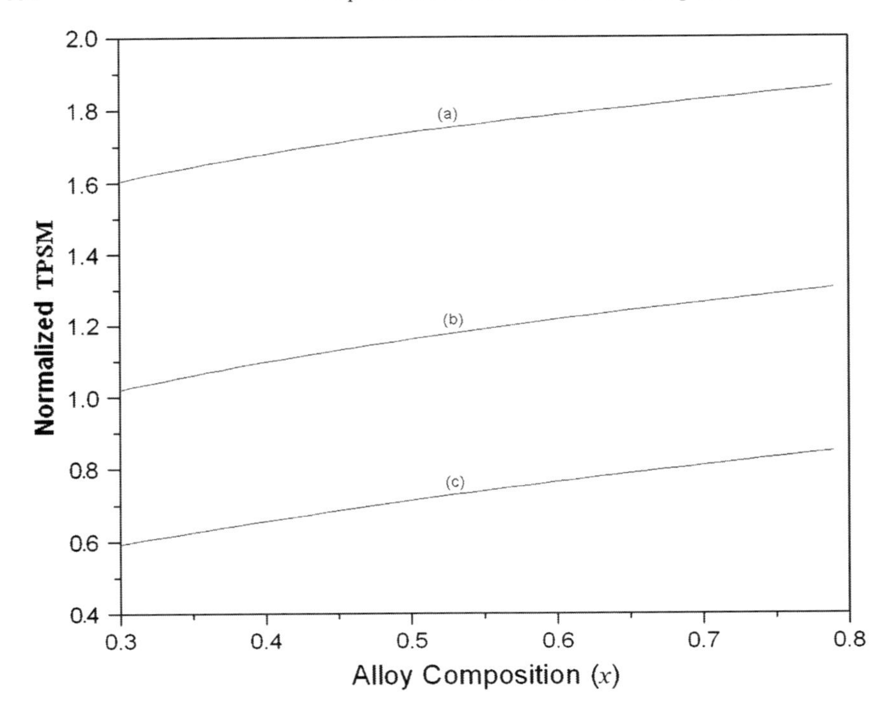

Fig. 11.18 Plot of the normalized TPSM as a function of wavelength for n-InGaAsP for all the cases of Fig. 11.1 in the presence of external photoexcitation

for $x = 0.3$. From all the figures, it may be noted that under the present physical conditions, n-InSb exhibits the maximum TPSM as compared with n-$Hg_{1-x}Cd_sTe$, In $In_xGa_xAs_yP_{1-y}$, and n-InAs, respectively. For the purpose of condensed presentation, the carrier concentration and the corresponding TPSM for this chapter have been presented in Table 11.1.

11.4 Open Research Problem

(R11.1) Investigate the DTP, PTP, and Z for all the appropriate problems of Chap. 5 in the presence of arbitrarily oriented external photoexcitation.

Reference

1. K.P. Ghatak, S. Bhattacharya, D. De, *Einstein Relation in Compound Semiconductors and Their Nanostructures*, Springer Series in Materials Science, vol 116 (Springer, Germany, 2008)

Chapter 12
Optothermoelectric Power in Ultrathin Films of Optoelectronic Materials Under Magnetic Quantization

12.1 Introduction

In this chapter, we shall study optothermoelectric power in UFs of optoelectronic materials in the presence of a quantizing magnetic field in Sect. 12.2.1 of theoretical background. Sections 12.3 and 12.4 contain results and discussion and open research problems, respectively.

12.2 Theoretical Background

12.2.1 Magnetothermopower in Ultrathin Films of Optoelectronic Materials

Using (10.1)–(10.3), the magnetodispersion relations, in the absence of electron spin, for UFs of optoelectronic materials in the presence of photoexcitation, whose unperturbed conduction electrons obey the three and two band models of Kane, together with parabolic energy bands, are given by [1]

$$\beta_0 \left(E_{121}, \lambda\right) = \left(n + \frac{1}{2}\right)\hbar\omega_0 + \frac{\hbar^2}{2m^*}\left(\frac{\pi n_z}{d_z}\right)^2, \tag{12.1}$$

$$\tau_0 \left(E_{122}, \lambda\right) = \left(n + \frac{1}{2}\right)\hbar\omega_0 + \frac{\hbar^2}{2m^*}\left(\frac{\pi n_z}{d_z}\right)^2, \tag{12.2}$$

$$\rho_0 \left(E_{123}, \lambda\right) = \left(n + \frac{1}{2}\right)\hbar\omega_0 + \frac{\hbar^2}{2m^*}\left(\frac{\pi n_z}{d_z}\right)^2, \tag{12.3}$$

where E_{121}, E_{122}, and E_{123} are the totally quantized energies in their respective cases.

The electron statistics per unit area can be expressed as

$$n_0 = \frac{g_v eB}{\pi\hbar} \sum_{n=0}^{n_{\max}} \sum_{n_z=1}^{n_{z\max}} F_{-1}(\eta_{121}), \tag{12.4}$$

$$n_0 = \frac{g_v eB}{\pi\hbar} \sum_{n=0}^{n_{\max}} \sum_{n_z=1}^{n_{z\max}} F_{-1}(\eta_{122}), \tag{12.5}$$

$$n_0 = \frac{g_v eB}{\pi\hbar} \sum_{n=0}^{n_{\max}} \sum_{n_z=1}^{n_{z\max}} F_{-1}(\eta_{123}), \tag{12.6}$$

where $\eta_{121} = (E_{F2DBL} - E_{121})/k_B T$, E_{F2DBL} is the Fermi energy in the present case, $\eta_{122} = (E_{F2DBL} - E_{122})/k_B T$, and $\eta_{123} = (E_{F2DBL} - E_{123})/k_B T$.

Therefore, combining (12.4), (12.5), and (12.6) with (1.13), the optothermoelectric power, in the absence of electron spin, for UFs of optoelectronic materials in the presence of photoexcitation, whose unperturbed conduction electrons obey the three and two band models of Kane, together with parabolic energy bands, are, respectively, given by

$$G_0 = \frac{\pi^2 k_B}{3e} \left[\sum_{n=0}^{n_{\max}} \sum_{n_z=1}^{n_{z\max}} F_{-1}(\eta_{121}) \right]^{-1} \left[\sum_{n=0}^{n_{\max}} \sum_{n_z=1}^{n_{z\max}} F_{-2}(\eta_{121}) \right], \tag{12.7}$$

$$G_0 = \frac{\pi^2 k_B}{3e} \left[\sum_{n=0}^{n_{\max}} \sum_{n_z=1}^{n_{z\max}} F_{-1}(\eta_{122}) \right]^{-1} \left[\sum_{n=0}^{n_{\max}} \sum_{n_z=1}^{n_{z\max}} F_{-2}(\eta_{122}) \right], \tag{12.8}$$

and

$$G_0 = \frac{\pi^2 k_B}{3e} \left[\sum_{n=0}^{n_{\max}} \sum_{n_z=1}^{n_{z\max}} F_{-1}(\eta_{123}) \right]^{-1} \left[\sum_{n=0}^{n_{\max}} \sum_{n_z=1}^{n_{z\max}} F_{-2}(\eta_{123}) \right]. \tag{12.9}$$

12.3 Results and Discussion

Combining (12.4)–(12.6) and (12.7)–(12.9) and using Table 1.1, the normalized TPSM

$$\left((3G_0 e)/(\pi^2 k_B^2 T)\right).$$

for n-InAs in accordance with the perturbed three and two band models of Kane together with perturbed parabolic energy bands has been drawn at $T = 4.2\,\mathrm{K}$ as functions of inverse quantizing magnetic field, film thickness, carrier degeneracy, and wavelength in Figs. 12.1–12.4, respectively. Figure 12.1 exhibits the variation of the TPSM as a function of inverse quantizing magnetic field for $n_z = 1$ and the oscillations are prominent for relatively larger values of quantizing magnetic field. From Fig. 12.2, it appears that due to the simultaneous occupancy of both the size quantized and magnetic subbands the dual oscillations in the TPSM appear in the present system. The higher spikes are due to the change in subbands due to the size quantization, whereas the magnetic quantization creates the lower peaks.

Figure 12.3 exhibits the variation of the TPSM with the carrier concentration at both the quantum limits, where $n_z = 1$ and $n = 0$ and it appears that the TPSM decreases for relatively large values of the carrier concentration. Figure 12.4 shows the dependence of TPSM on the wavelength in this case and it appears that for both variations of the quantum numbers, the TPSM exhibits sharp oscillatory spikes for relatively low values of the wavelength as considered in the said plot. For the purpose of condensed presentation, the carrier concentration and the corresponding TPSM for this chapter have been presented in Table 12.1.

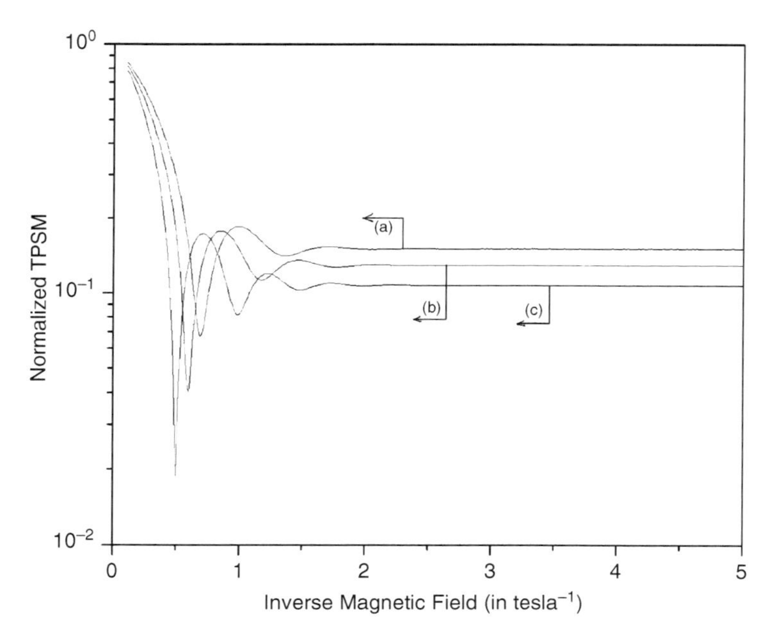

Fig. 12.1 Plot of the normalized TPSM as a function of inverse quantizing magnetic field for UFs of n-InSb in accordance with the perturbed (**a**) three and the (**b**) two band model of Kane together with the (**c**) parabolic model in the presence of external photoexcitation

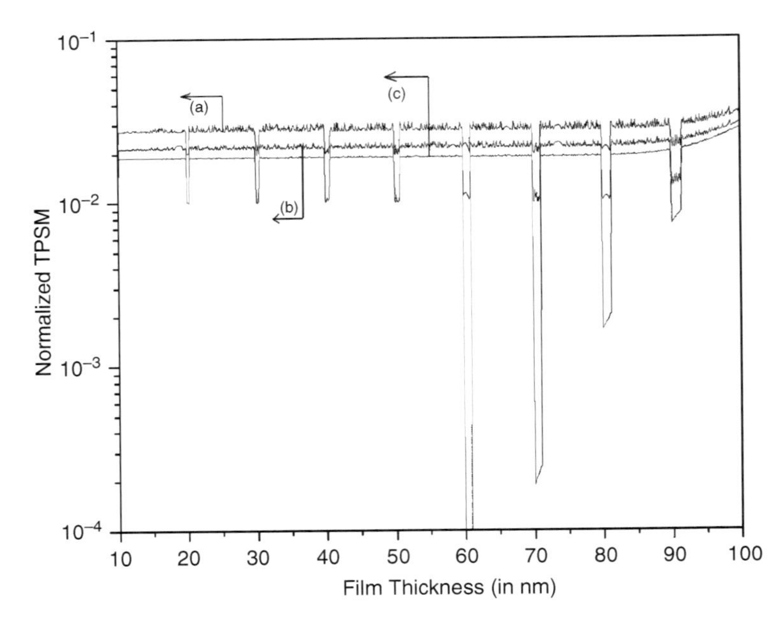

Fig. 12.2 Plot of the normalized TPSM as a function of film thickness for UFs of n-InSb for all cases of Fig. 12.1 in the presence of external photoexcitation

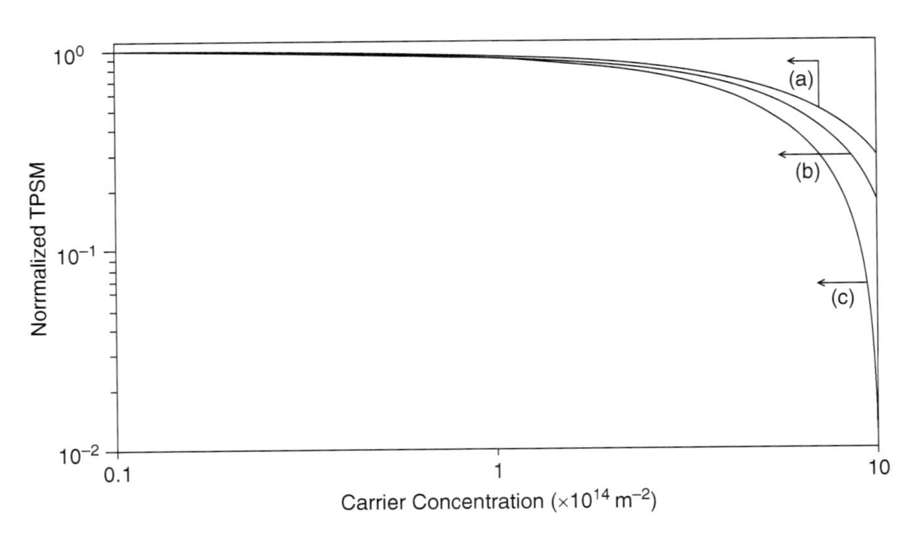

Fig. 12.3 Plot of the normalized TPSM as a function of carrier degeneracy for UFs of n-InSb for all cases of Fig. 12.1 in the presence of external photoexcitation

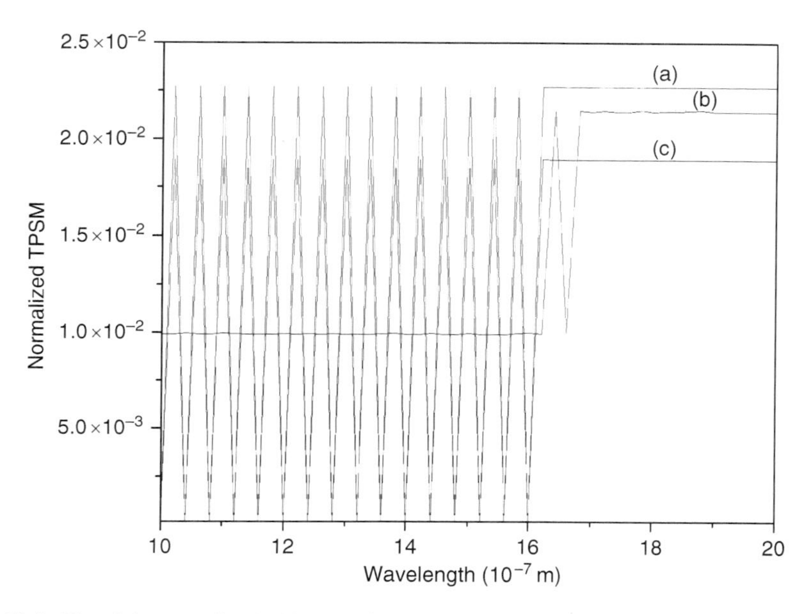

Fig. 12.4 Plot of the normalized TPSM as a function of wavelength for UFs of n-InSb for all cases of Fig. 12.1 in the presence of external photoexcitation

Table 12.1 The carrier statistics and the optothermoelectric power in ultrathin films of optoelectronic materials under magnetic quantization

Type of materials	Carrier statistics	TPSM
Optoelectronic materials	$$n_0 = \frac{g_v eB}{\pi\hbar} \sum_{n=0}^{n_{max}} \sum_{n_z=1}^{n_{zmax}} F_{-1}(\eta_{121}) \quad (12.4)$$	$$G_0 = \frac{\pi^2 k_B}{3e} \left[\sum_{n=0}^{n_{max}} \sum_{n_z=1}^{n_{zmax}} F_{-1}(\eta_{121}) \right]^{-1} \times \left[\sum_{n=0}^{n_{max}} \sum_{n_z=1}^{n_{zmax}} F_{-2}(\eta_{121}) \right] \quad (12.7)$$
	$$n_0 = \frac{g_v eB}{\pi\hbar} \sum_{n=0}^{n_{max}} \sum_{n_z=1}^{n_{zmax}} F_{-1}(\eta_{122}) \quad (12.5)$$	$$G_0 = \frac{\pi^2 k_B}{3e} \left[\sum_{n=0}^{n_{max}} \sum_{n_z=1}^{n_{zmax}} F_{-1}(\eta_{122}) \right]^{-1} \times \left[\sum_{n=0}^{n_{max}} \sum_{n_z=1}^{n_{zmax}} F_{-2}(\eta_{122}) \right] \quad (12.8)$$
	$$n_0 = \frac{g_v eB}{\pi\hbar} \sum_{n=0}^{n_{max}} \sum_{n_z=1}^{n_{zmax}} F_{-1}(\eta_{123}) \quad (12.6)$$	$$G_0 = \frac{\pi^2 k_B}{3e} \left[\sum_{n=0}^{n_{max}} \sum_{n_z=1}^{n_{zmax}} F_{-1}(\eta_{123}) \right]^{-1} \times \left[\sum_{n=0}^{n_{max}} \sum_{n_z=1}^{n_{zmax}} F_{-2}(\eta_{123}) \right] \quad (12.9)$$

12.4 Open Research Problem

(R12.1) Investigate the DTP, PTP, and Z for all the appropriate problems of Chap. 6 in the presence of arbitrarily oriented external photoexcitation.

Reference

1. K.P. Ghatak, S. Bhattacharya, D. De, *Einstein Relation in Compound Semiconductors and Their Nanostructures*, Springer Series in Materials Science, vol 116 (Springer, Germany, 2008)

Chapter 13
Optothermoelectric Power in Superlattices of Optoelectronic Materials Under Magnetic Quantization

13.1 Introduction

In this chapter, the optothermoelectric power in III–V quantum well-effective mass superlattices and III–V quantum well superlattices with graded interfaces in the presence of a quantizing magnetic field have been studied in Sects. 13.2.1 and 13.2.2, respectively. Sections 13.3 and 13.4 contain, respectively, the results and discussion and open research problems.

13.2 Theoretical Background

13.2.1 Magnetothermopower in III–V Quantum Well-Effective Mass Superlattices

The electron energy spectrum in this case can be expressed following [1] as

$$k_x^2 = \left[\frac{1}{L_0^2}\left[\cos^{-1} f_{130}\left(E, k_y, k_z\right)\right]^2 - k_\perp^2\right], \tag{13.1}$$

where $f_{130}\left(E, k_y, k_z\right) = [a_1 \cos\left[a_0 g_{130}\left(E, k_\perp\right) + b_0 h_{130}\left(E, k_\perp\right)\right] - a_2 \cos[a_0 g_{130} \left(E, k_\perp\right) - b_0 h_{130}\left(E, k_\perp\right)]]$, $g_{130}\left(E, k_\perp\right) = \left[\frac{2m_1^*}{\hbar^2}\beta_0\left(E, \lambda, E_{g_{01}}, \Delta_1\right) - k_\perp^2\right]^{1/2}$, and $h_{130}\left(E, k_\perp\right) = \left[\frac{2m_2^*}{\hbar^2}\beta_0\left(E, \lambda, E_{g_{02}}, \Delta_2\right) - k_\perp^2\right]^{1/2}$. When the unperturbed bulk dispersion law of the constituent materials is defined by the two band model of Kane, (13.1) remains as it is and only $\beta_0\left(E, \lambda, E_{g_{0i}}, \Delta_i\right)$ is being replaced by $\tau_0\left(E, \lambda, E_{g_{0i}}\right)$ and again when the unperturbed bulk dispersion law of the constituent materials is defined by parabolic energy bands, the (13.1) remains as it is and only $\beta_0\left(E, \lambda, E_{g_{0i}}, \Delta_i\right)$ should be replaced by $\rho_0\left(E, \lambda, E_{g_{0i}}\right)$.

In the presence of a quantizing magnetic field along x-direction, the electron dispersion relation in quantum well-effective mass superlattices is given by

$$\left(\frac{\pi n_x}{d_x}\right)^2 = \frac{1}{L_0^2}\left[\cos^{-1}\overline{f}_{130}(E,n)\big|_{E=E_{130}}\right]^2 - \frac{2eB}{\hbar}\left(n+\frac{1}{2}\right), \qquad (13.2)$$

where E_{130} is the totally quantized energy,

$$\overline{f}_{130}(E,n) = \Big[a_1\cos\left[a_0\overline{g}_{130}(E,n) + b_0\overline{h}_{130}(E,n)\right] - a_2\cos\left[a_0\overline{g}_{130}(E,n) - b_0\overline{h}_{130}(E,n)\right]\Big],$$

$$\overline{g}_{130}(E,n) = \left[\frac{2m_1^*}{\hbar^2}\beta_0\left(E,\lambda,E_{g01},\Delta_1\right) - \frac{2|e|B}{\hbar}\left(n+\frac{1}{2}\right)\right]^{1/2},$$

and

$$\overline{h}_{130}(E,n) = \left[\frac{2m_2^*}{\hbar^2}\beta_0\left(E,\lambda,E_{g02},\Delta_2\right) - \frac{2|e|B}{\hbar}\left(n+\frac{1}{2}\right)\right]^{1/2}.$$

The electron concentration per unit area is given by

$$n_0 = \left(\frac{g_v eB}{\pi\hbar}\right)\sum_{n=0}^{n_{\max}}\sum_{n_x=1}^{n_{x\max}} F_{-1}(\eta_{130}), \qquad (13.3)$$

where

$$\eta_{130} = \frac{E_{FBQWSLEML} - E_{130}}{k_B T}$$

and $E_{FBQWSLEML}$ is the Fermi energy in this case.

When the unperturbed bulk dispersion relation of the constituent materials are defined by the two band model of Kane, all the above pertinent equations remain unchanged, where $\beta_0\left(E,\lambda,E_{g0i},\Delta_i\right)$ is to be replaced by $\tau_0\left(E,\lambda,E_{g0i}\right)$. For perturbed parabolic bulk dispersion relation of constituent materials in this case $\beta_0\left(E,\lambda,E_{g0i},\Delta_i\right)$ should be $\rho_0\left(E,\lambda,E_{g0i}\right)$.

Combining (1.13) and (13.3), the optothermoelectric power in this case assumes the form

$$G_0 = \left(\frac{\pi^2 k_B}{3e}\right)\left[\sum_{n=0}^{n_{\max}}\sum_{n_x=1}^{n_{x\max}} F_{-1}(\eta_{130})\right]^{-1}\left[\sum_{n=0}^{n_{\max}}\sum_{n_x=1}^{n_{x\max}} F_{-2}(\eta_{130})\right]. \qquad (13.4)$$

13.2.2 Magnetothermopower in III–V Quantum Well Superlattices with Graded Interfaces

The electron energy spectrum in this case can be expressed following [1] as

$$k_z^2 = \frac{1}{L_0^2}\left[\cos^{-1}\left[\frac{1}{2}\phi_{131}(E,k_s)\right]\right]^2 - k_s^2, \tag{13.5}$$

where

$$\begin{aligned}
&\phi_{131}(E,k_s) = \\
&\Bigg[2\cosh\{\beta_{131}(E,k_s)\}\cos\{\gamma_{131}(E,k_s)\} \\
&+\varepsilon_{131}(E,k_s)\sinh\{\beta_{131}(E,k_s)\}\sin\{\gamma_{131}(E,k_s)\} \\
&+\Delta_0\Bigg[\left(\frac{K_{131}^2(E,k_s)}{K_{132}(E,k_s)} - 3K_{132}(E,k_s)\right)\cosh\{\beta_{131}(E,k_s)\}\sin\{\gamma_{131}(E,k_s)\} \\
&+\left(3K_{131}(E,k_s) - \frac{K_{132}^2(E,k_s)}{K_{131}(E,k_s)}\right)\sinh\{\beta_{131}(E,k_s)\}\cos\{\gamma_{131}(E,k_s)\}\Bigg] \\
&+\Delta_0\Bigg[2\left\{K_{131}^2(E,k_s) - K_{132}^2(E,k_s)\right\}\cosh\{\beta_{131}(E,k_s)\}\cos\{\gamma_{131}(E,k_s)\} \\
&+\frac{1}{12}\left[\frac{5K_{132}^2(E,k_s)}{K_{131}(E,k_s)}\right] + \frac{5K_{131}^2(E,k_s)}{K_{132}(E,k_s)} - 34K_{131}(E,k_s)K_{132}(E,k_s)] \\
&\times\sinh\{\beta_{131}(E,k_s)\}\sin\{\gamma_{131}(E,k_s)\}\Bigg]\Bigg],
\end{aligned}$$

$$\beta_{131}(E,k_s) = K_{131}(E,k_s)(a_0 - \Delta_0),$$

$$K_{131}(E,k_s) = \left[k_s^2 - \frac{2m_2^*}{\hbar^2}\beta_0\left(E - V_0,\lambda,E_{g02},\Delta_2\right)\right]^{1/2},$$

$$k_s^2 = k_x^2 + k_y^2,\ V_0 = \left|E_{g02} - E_{g01}\right|,$$

$$\gamma_{131}(E,k_s) = K_{132}(E,k_s)(b_0 - \Delta_0),$$

$$K_{132}(E,k_s) = \left[\frac{2m_1^*}{\hbar^2}\beta_0\left(E,\lambda,E_{g01},\Delta_1\right) - k_s^2\right]^{1/2},$$

and

$$\varepsilon_{131}(E,k_s) \equiv \left[\frac{K_{131}(E,k_s)}{K_{132}(E,k_s)} - \frac{K_{132}(E,k_s)}{K_{131}(E,k_s)}\right].$$

In the presence of a quantizing magnetic field B along z-direction, the magnetodispersion relation assumes the form

$$k_z^2 = \frac{1}{L_0^2}\left[\cos^{-1}\left[\frac{1}{2}\overline{\Phi}_{131}(E,n)\right]\right]^2 - \frac{2|e|B}{\hbar}\left(n+\frac{1}{2}\right), \tag{13.6}$$

where

$$\begin{aligned}
\overline{\varphi}_{131}(E,n) = \Bigg[& 2\cosh\left\{\overline{\beta}_{131}(E,n)\right\}\cos\left\{\overline{\gamma}_{131}(E,n)\right\} + \overline{\varepsilon}_{131}(E,n) \\
& \times \sinh\left\{\overline{\beta}_{131}(E,n)\right\}\sin\left\{\overline{\gamma}_{131}(E,n)\right\} \\
& + \Delta_0\Bigg[\left(\frac{\overline{K}_{131}^2(E,n)}{\overline{K}_{132}(E,n)} - 3\overline{K}_{132}(E,n)\right)\cosh\left\{\overline{\beta}_{131}(E,n)\right\} \\
& \times \sin\left\{\overline{\gamma}_{131}(E,n)\right\} \\
& + \left(3\overline{K}_{131}(E,n) - \frac{\overline{K}_{132}^2(E,n)}{\overline{K}_{131}(E,n)}\right)\sinh\left\{\overline{\beta}_{131}(E,n)\right\} \\
& \times \cos\left\{\overline{\gamma}_{131}(E,n)\right\}\Bigg] \\
& + \Delta_0\Bigg[2\left\{\overline{K}_{131}^2(E,n) - \overline{K}_{132}^2(E,n)\right\}\cosh\left\{\overline{\beta}_{131}(E,n)\right\} \\
& \times \cos\left\{\overline{\gamma}_{131}(E,n)\right\} + \frac{1}{12}\Bigg[\frac{5\overline{K}_{132}^2(E,n)}{\overline{K}_{131}(E,n)} \\
& + \frac{5\overline{K}_{131}^2(E,n)}{\overline{K}_{132}(E,n)} - 34\overline{K}_{131}(E,n)\overline{K}_{132}(E,n)\Bigg] \\
& \times \sinh\left\{\overline{\beta}_{131}(E,n)\right\}\sin\left\{\overline{\gamma}_{131}(E,n)\right\}\Bigg]\Bigg],
\end{aligned}$$

$$\begin{aligned}
\overline{\beta}_{131}(E,n) &= \overline{K}_{131}(E,n)(a_0 - \Delta_0), \\
\overline{K}_{131}(E,n) &= \left[\frac{2|e|B}{\hbar}\left(n+\frac{1}{2}\right) - \frac{2m_2^*}{\hbar^2}\beta_0\left(E - V_0, \lambda, E_{g02}, \Delta_2\right)\right]^{1/2}, \\
\overline{\gamma}_{131}(E,n) &= \overline{K}_{132}(E,n)(b_0 - \Delta_0), \\
\overline{K}_{132}(E,n) &= \left[\frac{2m_1^*}{\hbar^2}\beta_0\left(E, \lambda, E_{g01}, \Delta_1\right) - \frac{2|e|B}{\hbar}\left(n+\frac{1}{2}\right)\right]^{1/2},
\end{aligned}$$

and

$$\bar{\varepsilon}_{131}(E,n) = \left[\frac{\overline{K}_{131}(E,n)}{\overline{K}_{132}(E,n)} - \frac{\overline{K}_{132}(E,n)}{\overline{K}_{131}(E,n)}\right].$$

In quantum well superlattices with graded interfaces the magnetodispersion law assumes the form

$$\left(\frac{\pi n_z}{d_z}\right)^2 = \frac{1}{L_0^2}\left[\cos^{-1}\left[\frac{1}{2}\bar{\phi}_{131}(E,n)\right]\big|_{E=E_{131}}\right]^2 - \frac{2|e|B}{\hbar}\left(n+\frac{1}{2}\right), \quad (13.7)$$

where E_{131} is the quantized energy in this case.

The electron concentration per unit area is given by

$$n_0 = \frac{g_v eB}{\pi\hbar}\sum_{n_z=1}^{n_{z\max}}\sum_{n=0}^{n_{\max}} F_{-1}(\eta_{131}), \quad (13.8)$$

where $\eta_{131} = \frac{E_{FQWSLGIL} - E_{131}}{k_B T}$, in which $E_{FQWSLGIL}$ is the Fermi energy in this case.

When the unperturbed bulk dispersion relation of the constituent materials are defined by the two band model of Kane, all the above pertinent equations remain unchanged, where $\beta_0(E,\lambda,E_{g_{0i}},\Delta_i)$ is to be replaced by $\tau_0(E,\lambda,E_{g_{0i}})$. For perturbed parabolic bulk dispersion relation of constituent materials in this case $\beta_0(E,\lambda,E_{g_{0i}},\Delta_i)$ should be $\rho_0(E,\lambda,E_{g_{0i}})$.

Combining (1.13) and (13.8), the optothermoelectric power in this case assumes the form

$$G_0 = \frac{\pi^2 k_B}{3e}\left[\sum_{n_z=1}^{n_{z\max}}\sum_{n=0}^{n_{\max}} F_{-1}(\eta_{131})\right]^{-1}\left[\sum_{n_z=1}^{n_{z\max}}\sum_{n=0}^{n_{\max}} F_{-2}(\eta_{131})\right]. \quad (13.9)$$

13.3 Results and Discussion

Using (13.3); (13.4) and (13.8); (13.9), we have plotted the variation of the normalized opto-TPSM $\left(G_0\big/\left(\frac{\pi^2 k_B}{3e}\right)\right)$ under magnetic quantization for quantum well-effective mass and graded interface GaAs/$Ga_{1-x}Al_x$As superlattices as functions of the inverse magnetic field, carrier concentration, and wavelength as shown in Figs. 13.1–13.3, respectively. Figure 13.1 exhibits the variation of the TPSM with the inverse quantizing magnetic field under size quantum limit and it appears that the TPSM exhibits oscillatory behavior for relatively large values of the quantizing magnetic field both types of superlattices. The said figure also exhibits the fact for relatively small values of magnetic field the oscillation becomes damped and the numerical value of the TPSM for effective mass superlattices is greater as compared

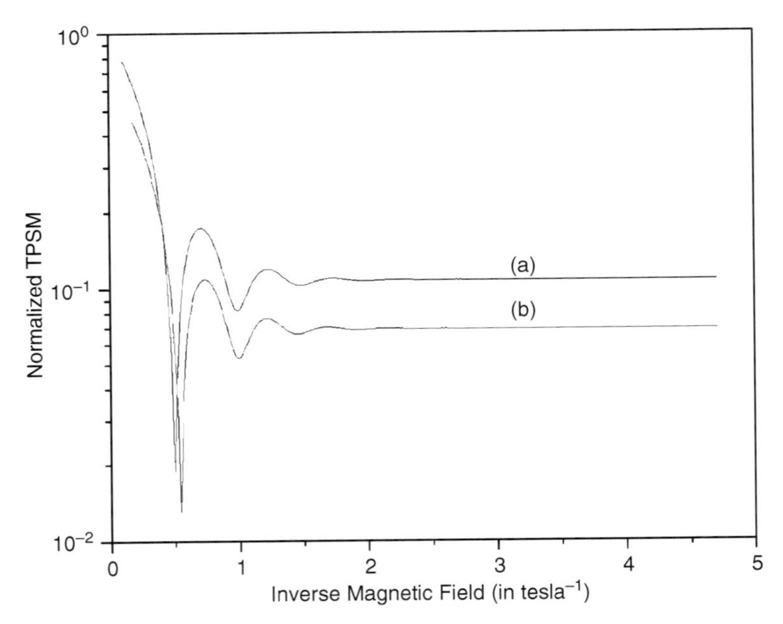

Fig. 13.1 Plot of the normalized TPSM as a function of inverse magnetic field for QW (**a**) effective mass (**b**) graded interface $GaAs/Ga_{1-x}Al_xAs$ superlattice in the presence of external photoexcitation in accordance with the perturbed parabolic energy band model

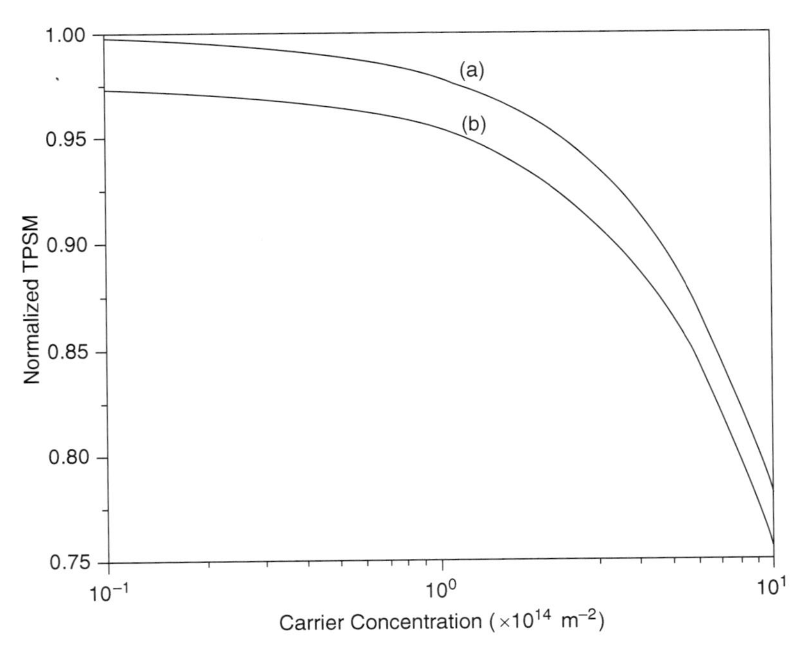

Fig. 13.2 Plot of the normalized TPSM as a function of carrier concentration for QW (**a**) effective mass (**b**) graded interface $GaAs/Ga_{1-x}Al_xAs$ superlattice in the presence of external photoexcitation in accordance with the perturbed parabolic energy band model

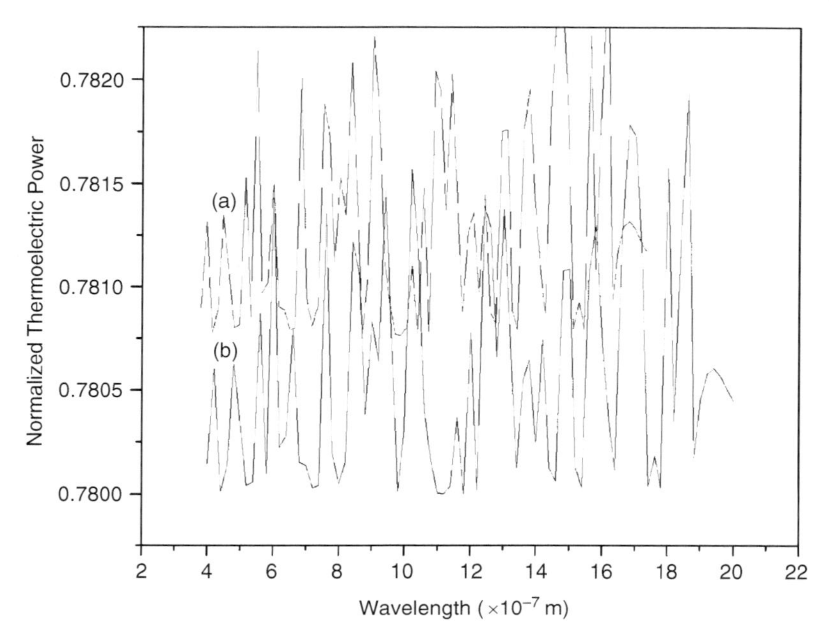

Fig. 13.3 Plot of the normalized TPSM as a function of wavelength for QW (**a**) effective mass (**b**) graded interface GaAs/$Ga_{1-x}Al_xAs$ superlattice in the presence of external photoexcitation in accordance with the perturbed parabolic energy band model

with the superlattices of optoelectronic materials with graded interfaces in the presence of external photoexcitation. Figure 13.2 exhibits the plot of the normalized TPSM as a function of carrier concentration for QW (a) effective mass (b) graded interface GaAs/$Ga_{1-x}Al_xAs$ superlattice in the presence of external photoexcitation in accordance with the perturbed parabolic energy band model.

From Fig. 13.2, we observe that the normalized TPSM decreases with increasing carrier concentration. For low values of carrier concentration the numerical values of the Opto-TPSM for quantum well-effective mass superlattice is greater as compared with quantum well superlattices with graded interfaces under magnetic quantization, where as for high values of carrier degeneracy they exhibit the converging tendency. Figure 13.3 exhibits the fact that the TPSM exhibits spiky oscillations over the wavelength as considered. For the purpose of simplified presentation, we have considered only the perturbed parabolic energy bands and have not presented the corresponding variation of the TPSM for the perturbed three and two band models of Kane. Nevertheless, one can form a rough idea about the magnitude and rate of variation of the TPSM, which can be estimated from the curves as presented in this chapter. The readers can perform the intricate numerical computation involved for the consideration of the perturbed three and two band models of Kane in this context. Finally, we can write that although the TPSM from SLs has been investigated in Chaps. 3, 4, 6, 10, and 13, still one can easily infer that how little is presented and how much more there is yet to be investigated in the research field of diffusion thermoelectric power, phonon-drag thermoelectric power, and thermoelectric figure-of-merit for quantum-confined SLs having different band structures in general which is the signature of the coexistence of new physics, advanced mathematics combined with the inner fire for performing innovative researches in this context from the young scientists since like Kikoin [2], we firmly believe that "A young scientist is no good if his teacher learns nothing from him and gives his teacher nothing to be proud of." As usual, we present for the last time a condensed presentation concerning the carrier concentration and the corresponding TPSM for this chapter in Table13.1.

13.4 Open Research Problems

(R13.1) Investigate the DTP, PTP, and Z in the absence of external magnetic field and including all types of scattering mechanisms for III–V, II–VI, IV–VI, and HgTe/CdTe quantum-confined superlattices with graded interfaces and also the effective mass superlattices of the aforementioned materials in the presence of arbitrarily oriented external photoexcitation.

(R13.2) Investigate the DTP, PTP, and Z in the absence of magnetic field by considering all types of scattering mechanisms for strained layer, random, short period and Fibonacci, polytype and saw-tooth quantum-confined superlattices in the presence of arbitrarily oriented external photoexcitation.

(R13.3) Investigate the DTP, PTP, and Z in the presence of an arbitrarily oriented quantizing magnetic field considering the effects of spin and broadening by considering all types of scattering mechanisms for (R13.1) and (R13.2) under an arbitrarily oriented (a) nonuniform electric field and (b) alternating electric field, respectively, in the presence of arbitrarily oriented external photoexcitation.

(R13.4) Investigate the DTP, PTP, and Z by considering all types of scattering mechanisms for (R13.1) and (R13.2) under an arbitrarily oriented alternating magnetic field by including broadening and the electron spin, respectively, in the presence of arbitrarily oriented external photoexcitation.

(R13.5) Investigate the DTP, PTP, and Z by considering all types of scattering mechanisms for (R13.1) and (R13.2) under an arbitrarily oriented quantizing alternating magnetic field and crossed alternating electric field by including broadening and the electron spin, respectively, in the presence of arbitrarily oriented external photoexcitation.

(R13.6) Investigate the DTP, PTP, and Z by considering all types of scattering mechanisms for (R13.1) and (R13.2) under an arbitrarily oriented alternating quantizing magnetic field and crossed alternating nonuniform electric field by including broadening and the electron spin, respectively, in the presence of arbitrarily oriented external photoexcitation.

Table 13.1 The carrier statistics and the optothermoelectric power in superlattices of optoelectronic materials under magnetic quantization

Type of materials	Carrier statistics	TPSM
III–V Quantum well-effective mass superlattices	$n_0 = \left(\frac{g_v eB}{\pi\hbar}\right)\sum_{n=0}^{n_{max}}\sum_{n_x=1}^{n_{x max}} F_{-1}(\eta_{130})$ (13.3)	On the basis of (13.3) $G_0 = \left(\frac{\pi^2 k_B}{3e}\right)\left[\sum_{n=0}^{n_{max}}\sum_{n_x=1}^{n_{x max}} F_{-1}(\eta_{130})\right]^{-1}\left[\sum_{n=0}^{n_{max}}\sum_{n_x=1}^{n_{x max}} F_{-2}(\eta_{130})\right]$ (13.4)
III–V Quantum well superlattices with graded interfaces	$n_0 = \left(\frac{g_v eB}{\pi\hbar}\right)\sum_{n_z=1}^{n_{z max}}\sum_{n=0}^{n_{max}} F_{-1}(\eta_{131})$ (13.9)	$G_0 = \frac{\pi^2 k_B}{3e}\left[\sum_{n_z=1}^{n_{z max}}\sum_{n=0}^{n_{max}} F_{-1}(\eta_{131})\right]^{-1}\left[\sum_{n_z=1}^{n_{z max}}\sum_{n=0}^{n_{max}} F_{-2}(\eta_{131})\right]$ (13.9)

(R13.7) Investigate the DTP, PTP, and Z in the absence of magnetic field for all types of quantum-confined superlattices under exponential, Kane, Halperin, Lax, and Bonch-Bruevich band tails [3], respectively, in the presence of arbitrarily oriented external photoexcitation.

(R13.8) Investigate the DTP, PTP, and Z in the presence of quantizing magnetic field by incorporating spin and broadening for the problem as defined in (R13.7) under an arbitrarily oriented (a) nonuniform electric field and (b) alternating electric field, respectively, in the presence of arbitrarily oriented external photoexcitation.

(R13.9) Investigate the DTP, PTP, and Z for the problem as defined in (R13.7) under an arbitrarily oriented alternating quantizing magnetic field by incorporating broadening and the electron spin, respectively, in the presence of arbitrarily oriented external photoexcitation.

(R13.10) Investigate the DTP, PTP, and Z for the problem as defined in (R13.7) under an arbitrarily oriented alternating quantizing magnetic field and crossed alternating electric field by incorporating broadening and the electron spin, respectively, in the presence of arbitrarily oriented external photoexcitation.

(R13.11) Introducing new theoretical formalisms, investigate all the problems of this chapter in the presence of hot electron effects.

(R13.12) Investigate the influence of deep traps and surface states separately for all the appropriate problems of this chapter after proper modifications.

References

1. K.P. Ghatak, S. Bhattacharya, D. De, *Einstein Relation in Compound Semiconductors and Their Nanostructures*, Springer Series in Materials Science, vol 116 (Springer-Verlag, Germany, 2008)
2. I.K. Kikoin, *Science for Everyone: Encounters with Physicists and Physics* (Mir Publishers, Russia, 1989), p. 154
3. B.R. Nag, *Electron Transport in Compound Semiconductors*, Springer Series in Solid-State Sciences, vol 11 (Springer-Verlag, Germany, 1980)

Chapter 14
Applications and Brief Review of Experimental Results

14.1 Introduction

In this book, we have discussed many aspects of TPSM based on the dispersion relations of the nanostructures of different technologically important materials having different band structures in the presence of 1D, 2D, and 3D confinements of the wave-vector space of the charge carriers, respectively. In this chapter, we discuss few applications in this context in Sect. 14.2 and we shall also present a very brief review of the experimental investigations in Sect. 14.3 which is a sea in itself. Section 14.4 contains the single experimental open research problem.

14.2 Applications

The investigations as presented in this monograph find eight different applications in the realm of modern electronic devices.

14.2.1 Effective Electron Mass

The effective mass of the carriers in different materials, being connected with the mobility, is known to be one of the most important physical quantities used for the analysis of the semiconductor devices under different operating conditions and different areas of materials science in general for the investigations of different physical properties [1]. It must be noted that among the various definitions of the effective electron mass [2], it is the effective momentum mass that should be regarded as the basic quantity [3]. This is due to the fact that it is this mass which appears in the description of transport phenomena and all other properties of the conduction electrons of the semiconductors having arbitrary dispersion laws [4]. It is the effective momentum mass which enters in various transport coefficients and plays the most dominant role in explaining the experimental results under different scattering mechanisms [5–8]. The carrier degeneracy in semiconductors influences the effective mass when it is energy dependent. Under degenerate conditions, only the electrons at the Fermi surface of n-type semiconductors participate in the conduction process and, hence, the effective momentum mass (EMM) of the electrons corresponding to the Fermi level would be of interest in electron transport under such conditions. The Fermi energy is again determined by the carrier energy spectrum and the carrier concentration and therefore these two features would determine the dependence of the EMM in degenerate materials on the degree of carrier degeneracy. In recent years, the EMM in such materials under different external conditions has been studied extensively [9–26]. It has, therefore, different values in different materials and varies with electron concentration, with the magnitude of the reciprocal quantizing magnetic field under magnetic quantization, with the quantizing electric field as in inversion layers, with the nanothickness as in quantum wells and quantum well wires, and with superlattice period as in the quantum-confined superlattices having various carrier energy spectra.

The expression of the EMM in the ith direction is given by

$$m_i^* (E_{\mathrm{F}}) = \hbar^2 \left[k_{i_0} \left(\frac{\partial k_{i_0}}{\partial E} \right) \right] \Bigg|_{E=E_{\mathrm{F}}}, \tag{14.1}$$

where $i_0 = x, y$, and z.

From the different chapters of this monograph, the EMM can be formulated by using the respective dispersion relation and their concentration dependence can be studied from the expressions of carrier statistics as formulated for different materials in this monograph. In many cases, in addition to Fermi energy and other system constraints, the effective mass will depend on the quantum numbers depending on particular band structure under different physical conditions.

14.2.2 Debye Screening Length

It is well known that the Debye screening length (DSL) of the carriers in the semiconductors is a fundamental quantity, characterizing the screening of the Coulomb field of the ionized impurity centers by the free carriers. It affects many special features of the modern semiconductor devices, the carrier mobility under different mechanisms of scattering, and the carrier plasmas in semiconductors [27–45]. The DSL (L_{D}) can, in general, be written as [32–45]

$$L_{\mathrm{D}} = \left(\frac{|e|^2}{\varepsilon_{\mathrm{sc}}} \frac{\partial n_0}{\partial E_{\mathrm{F}}} \right)^{-1/2}, \tag{14.2}$$

where n_0 and E_{F} are applicable for bulk samples. Equation (1.13) is also valid for bulk materials in the presence of a classically large magnetic field [46]. Using (14.2) and (1.13), one obtains

$$L_{\mathrm{D}} = \left(3\,|e|^3\,n_0 G_0 \Big/ \varepsilon_{\mathrm{sc}}\pi^2 k_{\mathrm{B}}^2 T\right)^{-1/2}. \tag{14.3}$$

Therefore, we can experimentally determine L_{D} by knowing the experimental curve of G_0 versus n_0 at a fixed temperature.

14.2.3 Carrier Contribution to the Elastic Constants

The knowledge of the carrier contribution to the elastic constants is useful in studying the mechanical properties of the materials and has been investigated in the literature [47–69]. The electronic contribution to the second- and third-order elastic constants can be written as [47–69]

$$\Delta C_{44} = -\frac{\left(\overline{G_0}\right)^2}{9}\frac{\partial n_0}{\partial E_{\mathrm{F}}} \tag{14.4}$$

and

$$\Delta C_{456} = \frac{\left(\overline{G_0}\right)^3}{27}\frac{\partial^2 n_0}{\partial E_{\mathrm{F}}^2}, \tag{14.5}$$

where $\overline{G_0}$ is the deformation potential constant. Thus, using (1.13), (14.4), and (14.5), we can write

$$\Delta C_{44} = \left[-n_0\left(\overline{G_0}\right)^2 |e|\, G_0 \Big/ \left(3\pi^2 k_{\mathrm{B}}^2 T\right)\right] \tag{14.6}$$

and

$$\Delta C_{456} = \left(n_0\,|e|\left(\overline{G_0}\right)^3 G_0^2 \Big/ \left(3\pi^4 k_{\mathrm{B}}^3 T\right)\right)\left(1 + \frac{n_0}{G_0}\frac{\partial G_0}{\partial n_0}\right). \tag{14.7}$$

Thus, again the experimental graph of G_0 versus n_0 allows us to determine the electronic contribution to the elastic constants for materials having arbitrary spectra.

14.2.4 Diffusivity–Mobility Ratio

The diffusivity (D) to mobility (μ) ratio (DMR) of the carriers in semiconductor devices is known to be very useful [70] since the diffusion constant (a quantity often used in device analysis but whose exact experimental determination is rather difficult) can be obtained from this ratio by knowing the experimental values of the mobility. In addition, it is more accurate than any of the individual relation for the diffusivity or the mobility, which are the two widely used quantities of carrier transport of modern nanostructured materials and devices. The classical DMR equation is valid for both types of carriers. In its conventional form, it appears that the DMR increases linearly with the temperature T being independent of the carrier concentration. This relation holds only under the condition of carrier nondegeneracy, although its validity has been suggested erroneously for degenerate materials [71]. The performance of the electron devices at the device terminals and the speed of operation of modern switching transistors are significantly influenced by the degree of carrier degeneracy present in these devices [72]. The simplest way of analyzing them under degenerate condition is to use the appropriate DMR to express the performance of the devices at the device terminals and the switching speed in terms of the carrier concentration [72].

It is well known from the fundamental work of Landsberg [73–75] that the DMR for electronic materials having degenerate electron concentration is essentially determined by their respective energy band structures. It has therefore different values in different materials and varies with the doping, with the magnitude of the reciprocal quantizing magnetic field under magnetic quantization, with the quantizing electric field as in inversion layers, with the nanothickness as in quantum wells and quantum well wires, and with superlattice period as in the quantum-confined superlattices of small gap semiconductors with graded interfaces having various carrier energy spectra. This relation is useful for semiconductor homostructures [76,77], semiconductor–semiconductor heterostructures [78, 79], metal–semiconductor heterostructures [80–84], and insulator–semiconductor heterostructures [85–88]. It can, in general, be proved that for bulk specimens the DMR is given by [89]

$$\frac{D}{\mu} = \left(\frac{n_0}{|e|}\right) \Big/ \left(\frac{\mathrm{d}n_0}{\mathrm{d}E_{\mathrm{F}}}\right) \tag{14.8}$$

Combining (1.13) with (14.8), we get

$$\frac{D}{\mu} = \left(\frac{\pi^2 k_{\mathrm{B}}^2 T}{3\,|e|^2\,G_0}\right) \tag{14.9}$$

Thus, the DMR for degenerate materials can be determined by knowing the experimental values of G_0. Since G_0 decreases with increasing n_0, from (14.9), one can infer that for constant temperature, the DMR increases with increasing carrier degeneracy which exhibits the compatibility test. Equation (14.9) is independent of the dimensions of quantum confinement. We should note that the present analysis is not valid for totally k-space quantized systems such as quantum dots, magnetoinversion and accumulation layers, magneto size quantization, magneto nipis, quantum dot superlattices, and quantum well superlattices under magnetic quantization. Under the said conditions, the electron motion is possible in the broadened levels. The experimental results of G for degenerate materials will provide an experimental check on the DMR and also a technique for probing the band structure of degenerate compounds having arbitrary dispersion laws.

14.2.5 Diffusion Coefficient of the Minority Carriers

This particular coefficient in quantum-confined lasers can be expressed [89] as

$$D_i/D_0 = \mathrm{d}E_{\mathrm{F}i}/\mathrm{d}E_{\mathrm{F}} \tag{14.10}$$

where D_i and D_0 are the diffusion coefficients of the minority carriers both in the presence and in the absence of quantum confinements and $E_{\mathrm{F}i}$ and E_{F} are the Fermi energies in the respective cases. It appears then that the formulation of the above ratio requires a relation between $E_{\mathrm{F}i}$ and E_{F}, which, in turn, is determined by the appropriate carrier statistics. Thus, our present study plays an important role in determining the diffusion coefficients of the minority carriers of quantum-confined lasers with materials having arbitrary band structures. In this context, it may be noted that the investigation of the optical excitation of the optoelectronic materials leads to the study of the ambipolar diffusion coefficients in which the present results contribute significantly.

14.2.6 Nonlinear Optical Response

The nonlinear response from the optical excitation of the free carriers is given by [90]

$$Z_0 = \frac{-e^2}{\omega^2\hbar^2}\int_0^\infty \left(k_x\frac{\partial k_x}{\partial E}\right)^{-1} f(E)N(E)\mathrm{d}E \tag{14.11}$$

where ω is the optical angular frequency, $f(E)$ is the Fermi–Dirac occupation probability factor, $N(E)$ is the density-of-states function. From the various E–k relations of different materials under different physical conditions, we can formulate the expression of $N(E)$ and from band structure we can derive the term $(k_x(\partial k_x/\partial E))$ and thus by using the density-of-states function as formulated, we can study the Z_0 for all types of materials as considered in this monograph.

14.2.7 Third-Order Nonlinear Optical Susceptibility

This particular susceptibility can be written as [91]

$$\chi_{\mathrm{NP}}(\omega_1,\omega_2,\omega_3) = \frac{n_0 e^4\langle\varepsilon^4\rangle}{24\omega_1\omega_2\omega_3(\omega_1+\omega_2+\omega_3)\hbar^4} \tag{14.12}$$

where $n_0\langle\varepsilon^4\rangle = \int_0^\infty (\partial^4 E/\partial k_z^4)N(E)f(E)\mathrm{d}E$ and the other notations are defined in [91]. The term $\left(\partial^4 E/\partial k_z^4\right)$ can be formulated by using the dispersion relations of different materials as given in appropriate sections of this monograph. Thus, one can investigate the $\chi_{\mathrm{NP}}(\omega_1,\omega_2,\omega_3)$ for all materials as considered in this book.

14.2.8 Generalized Raman Gain

The generalized Raman gain in optoelectronic materials can be expressed as [92]

$$R_{\mathrm{G}} = \overline{I}\left(\frac{16\pi^2 c^2}{\hbar\omega\rho g\omega_s^2 n_s n_p}\right)\left(\frac{\Gamma_\rho}{\Gamma}\right)\left(\left(\frac{e^2}{mc^2}\right)^2 m^2 R^2\right) \tag{14.13}$$

where $\overline{I} = \sum\limits_{n.t_z}[f_0(n, k_z \uparrow) - f_0(n, k_z \downarrow)]$, $f_0(n, k_z \uparrow)$ is the Fermi factor for spin-up Landau levels, $f_0(n, k_z \downarrow)$ is the Fermi factor for spin down Landau levels, n is the Landau quantum number, and the other notations are defined in [92]. It appears then the formulation of R_G is determined by the appropriate derivation of I which in turn requires the magnetodispersion relations. By using the formulas of Chaps. 5–7 and 11–13, the band structure is derived in the said chapters. R_G can, in general, be investigated.

14.3 Brief Review of Experimental Works

The experimental aspect of thermoelectrics is extremely wide and it is virtually impossible to even highlight the major developments in a chapter. For the purpose of condensed presentation, the experimental aspects of thermoelectrics and the related topics for different technologically important bulk samples with different band structures are given in Sect. 14.3.1 and the same for nanostructured materials have been discussed in Sect. 14.3.2.

14.3.1 Bulk Samples

Using (A.1) and (A.2) and the energy band constants of n-6-Cd_3As_2 [93–97] ($|E_{\mathrm{g}}| = 0.095\,\mathrm{eV}$, $\Delta_{||} = 0.27\,\mathrm{eV}$, $\Delta_{\perp} = 0.25\,\mathrm{eV}$, $m_{||}^* = 0.00697\,m_0$, $m_{\perp}^* = 0.013933\,m_0$, $\delta = 0.085\,\mathrm{eV}$, $g_{\mathrm{v}} = 1$, and $\varepsilon_{\mathrm{sc}} = 16\varepsilon_0$), the plot of TPSM as a function of electron concentration in bulk specimens of n$-Cd_3As_2$ (which is an example of tetragonal material, the conduction electrons of which obey the generalized energy–wave-vector dispersion relation for nonlinear optical compounds as formulated in (1.2) of Chap. 1) is shown in Fig. 14.1, where the circular points exhibit the experimental result [98]. It appears from Fig. 14.1 that the TPSM in bulk specimens of n-type cadmium arsenide decreases with increasing electron concentration in the whole range of the carrier degeneracy as considered here and the theoretical plot is in good agreement with the experimental data as given in [98]. It is worth remarking to note that the generalized theoretical formulation of the TPSM for different materials, defined by the respective carrier energy spectrum, as formulated in Appendix A

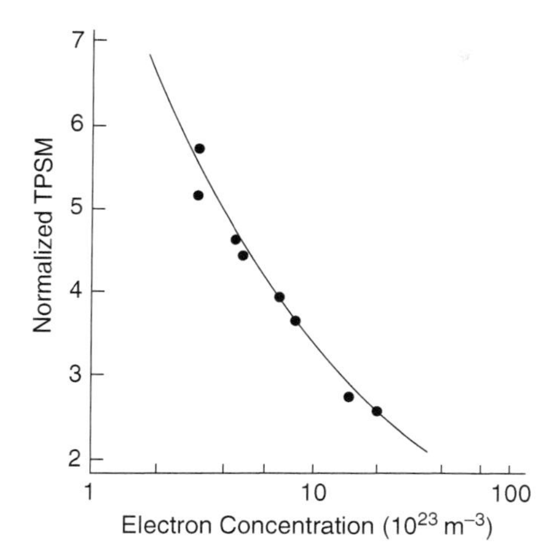

Fig. 14.1 The *solid line* exhibits the plot of the normalized TPSM in bulk specimens of Cd_3As_2 as a function of electron concentration and the *circular points* show the experimental results

together with the open research problem as given there and the consequent experimental verification in each case will constitute very important experimental study in this particular arena of thermoelectrics.

It may be noted that the study of the silicides as a prominent thermoelectric material was reported by Nikitin [99]. The thermoelectric power of ruthenium sesquisilicide [100], higher manganese silicide [101], chromium disilicide [102], cobalt monosilicide, and iron disilicide [103–105] has been investigated in the literature. In conclusion, thermoelectric silicides are emerging materials for low-cost, eco-friendly, and effective thermoelectric generators. The clathrate structure (taking Ge clathrate as an example) can be modeled as a derivative of four-coordinated diamond lattice structure of Ge. The clathrate compounds can be synthesized and thereby they are of interest for further research in the domain of thermoelectrics [106–110]. The promising thermoelectric properties of these compounds indicate a new category of thermoelectric materials for researchers and Rowe and his group [111] measured the thermoelectric power of type 1 clathrate compounds. These results are very important for thermoelectric applications and represent a high figure-of-merit value of the basically nonoptimized materials. The promising properties of these important compounds rightly indicate a new category of thermoelectric materials for future research [112].

With the discovery of $NaCo_2O_4$ as a potential candidate of thermoelectric materials in 1997 [113, 114], the oxide materials have received the warm entry in the arena of thermoelectrics in general since they have many advantages such as nontoxicity, thermal stability, high oxidation resistance, etc. Cobalt-oxide-based layer-structured crystals including $NaCo_2O_4$, $Ca_2Co_2O_5$, and their derivative compounds have been fabricated as p-type compounds having fairly high thermoelectric performance [115, 116]. The thermoelectric power of Na_xCoO_2 single crystals is $100\,\mu V/K$ at

300 K and the origin of this large thermopower has been explained on the basis of strongly correlated systems [117, 118]. It may be noted that the thermoelectric power of calcium cobalt oxides is 150 μV/K [119]. Nanostructure controlled through nanoblock integration is a promising path for the development of new oxide thermoelectric which is itself very important field of research [119].

In recent years, there has been considerable interest in studying the thermoelectric properties of various organic materials [120, 121] and the magnitude of the thermoelectric power for some organic semiconductors may reach up to 2 mV/K [122, 123] The quasione-dimensional organic crystals represent a completely new class of organic crystals and the one dimensionality of electronic structures results from good planarity of π-conjugated organic molecules. The model of these types of organic materials with high numerical magnitude of the thermoelectric power has been described in the literature [124]. It has been shown analytically that under certain constraint in some quasi1D materials, the value of the figure-of-merit reaches up to 20 and the dependence of the same on the carrier concentration and other external variables has also been studied [125].

Besides, there has been considerable interest in studying the temperature dependence of the thermoelectric properties of the functionally graded materials [126]. The thermoelectric power of the p-type functionally graded $(Bi_2Te_3)_{1-x-y}$ $(Sb2Te3)_x(Sb2Se3)_y$ specimen increases with increasing temperature after attaining a maximum value exhibits the decreasing tendency in the whole range of temperature considered. The determination of temperature dependence of thermoelectric properties and conversion efficiency of functionally graded materials under operating conditions present significant difficulties related to nonhomogeneity of the composition, and variable transport properties along the length of the sample considered. A number of methods have been developed to study the temperature profile, optimum carrier concentration distribution, and conversion efficiency of functionally graded thermoelectric materials. It has been experimentally realized that the conversion efficiency of functionally graded PbTe- and Bi_2Te_3-based specimens can be increased by at least 10% in comparison with that of materials where the distribution of the carrier concentration is uniform It may be noted that although the field of thermoelectrics advances very rapidly and the important role of heavily doped semiconductors as good thermoelectric materials has been accepted, but still it appears from the literature that there lies enough scope for the study of various thermoelectric properties of heavily doped semiconductors based on Boltzmann's transport equation which, in turn, is based on the energy–wavevector dispersion relation of the carriers. The future of thermoelectrics is based on new materials and consequently solid state chemistry must play very prominent role in this field together with the fact that the optimization of new materials is connected with multidimensional parameter space.

Extensive investigations exhibit the fact that filled skutterudites are promising novel thermoelectric materials [127–130]. Among skutterudites, there are compounds that have exceeded figure-of-merit value by 1 and they are of current interest for thermoelectric power generation applications in various industrial operations. Skutterudites are remarkable in that they achieve optimum performance in a given range of carrier concentration as that of metals and semimetals. Fleurial et al. [131] measured the high temperature thermopower of several $CeFe_{4-x}Co_xSb_{12}$ and from that data it appears that the thermoelectric power of $CeFe_4Sb_{12}$ (59 μV/K) is the least, whereas for Co_4Sb_{12}, the same power is more than twice and exactly 138 μV/K. In recent years, it has been suggested [132], based on their band structure calculations, that the n-type $La(Ru_{1-x}Rh_x)_4Sb_{12}$ compounds might be exceptionally good thermoelectrics. Filled skutterudites represent a rich and fertile ground and by making use of filled skutterudites in the upper stages of segmented unicouples, it might be possible to achieve efficiencies of thermoelectric power conversion devices approaching 15%. Realizing such efficiencies would make thermoelectricity very appealing to a vast number of industrial applications where there is great need for electrical power or an ample amount of waste heat available [133].

The extensive recent studies [134–142] of ternary and multinary half-Heusler phases have revealed many thermoelectric properties in this class of bandgap intermetallic compounds. At relatively high and low temperature, most of the half-Heusler alloys exhibit semiconducting and semimetallic behaviors, respectively. In ferromagnetic half-Heusler phases, the crossover from semiconducting zone to semimetallic regime happens in a particular region where the magnetism becomes prominent. In accordance with the formulation of energy band structures, the ferromagnetism is of the itinerant and highly spin-polarized type. Large thermopower and moderate resistivity, with the former attributable to the existence of relatively large values of the mass of the carriers, are measured at and above ambient temperature. The thermoelectric figure-of-merit, which is found to reach $\sim$0.6 at 800 K, underscores the potential of half-Heusler alloys as a new class of prospective thermoelectric materials above ambient temperature. The value of the thermoelectric figure-of-merit is found to be further enhanced with the reduction of the lattice contribution to thermal conductivity. Unlike other thermoelectric materials, half-Heusler alloys exhibit low carrier mobility. Thus, attempts made to reduce the lattice thermal conductivity are seen to have relatively less effect on the power factor. The measured properties are found to be sensitive to annealing conditions, with the latter presumably determining the underlying crystallographic order [143].

In the world of materials science, quasicrystals form an important group of relatively new materials [144] for exhibiting high values of mechanical strength, hardness [145] and corrosion resistance [146], and low thermal conductivity. More than hundred quasicrystalline systems have been experimentally realized and all possesses 5-, 8-, 10-, or 12-fold classically forbidden symmetries. The quasicrystals possess favorable values of electrical resistivity for thermoelectrics and the resistivity and thermoelectric power can be appropriately changed by varying the composition without sacrificing the low values of thermal conductivity. Thermopower values in quasicrystals vary with the quality and composition of the crystal. Thermopower in AlPdMn quasicrystals is seen to vary substantially with composition and temperature. For AlPdMn, the thermopower is as large as +75 μV/K for 70–22.5–7.5 composition, whereas for the 70.5–22.5–7 composition, the thermopower in the same material is very small and negative [147]. The thermopower has been observed to vary with quality and composition. Thermopower in AlCuFe changes from plus to minus 20 μV/K with composition and also varies with changes in annealing together with the fact that maximum room temperature value of thermopower becomes 55 μV/K [148]. It is important to note that a very distinct difference in the values of the thermopower has been observed even though there is no apparent change in the density-of-states function. It may be noted that the worldwide concern regarding the harmful effect of global warming and the consequent recognition that the thermoelectric technology offers important solutions of converting waste heat into electrical power has resulted the commercial availability of modules to be designed for new generation [100].

14.3.2 Nanostructured Materials

The film thickness dependencies of the thermoelectric power of n-PbTe/p-SnTe/n-PbTe heterostructures on the SnTe quantum well width at fixed PbTe barrier layer thicknesses were studied by the Dresselhaus group [149]. It was established that the thickness variation of the thermoelectric power, at room temperature is distinctly nonmonotonic. In this work, the Dresselhaus group has assumed that the basic carrier dispersion relation for such IV–VI compounds obeys Kane's dispersion law. This behavior is attributed to the manifestation of size quantization of the hole gas in the SnTe quantum wells between n-PbTe barriers. The experimental value of the oscillation period and the position of extrema points are in good agreement with the results of the theoretical calculations, taking into account a finite barrier height. Dresselhaus [150] also theoretically predicted that the thermoelectric power for superlattice nanowires oscillates with increase in the Fermi energy and shows strong oscillations near the minigaps. The superlattice nanowires may be tailored to exhibit n- or p-type properties, using the same dopants (e.g., electron donors) by carefully controlling the Fermi energy or the dopant concentration. More importantly, they also concluded that the thermoelectric power extrema of superlattice nanowires have substantially larger magnitudes for Fermi energies near the minigaps with only slightly reduced electrical conductivity compared to alloy nanowires, which is a direct consequence of the unique potential profile in the transport direction.

Recently, the Catteni group [151] has shown that the thermoelectric power of thin film platinum thermocouples is related to the bulk thermopower as a linear function of film thickness (t) when the electron mean free path (l) is very less compared with the film thickness. They also predicted that for such conditions, quantum size effect (QSE) is negligible. For $t < l$, QSEs become relevant and they showed that this nonlinear $1/t$ behavior is due to QSE. The experiments on the thermopower of a double GaAs quantum well have been studied in the temperature range 0.3–4.2 K as a function of voltage applied to a top gate by Smith et al. [152]. The group has calculated the thermopower based on a model of two independent two-dimensional electron gases (2DEGs) connected in parallel. They found that the thermoelectric power exhibits a T^5 dependence at low temperatures instead of the standard T^4

expected for single GaAs quantum wells. Also, for the lowest densities examined, the local-field correction enhances the magnitude of the calculated thermoelectric power by over a factor of 2, in good agreement with experiment. As a check on the model of two independent 2DEGs, they have also calculated that the thermoelectric power at resonance taking into account interwell coupling and the theoretical values were in good agreement with those obtained for uncoupled wells. This confirms the experimental result that the thermoelectric power is insensitive to the resonance condition.

Detailed results on the thermoelectric power of four high-mobility GaAs–Ga_{1-x} Al_xAs quantum well heterojunctions at magnetic fields up to 20 T has been extensively carried out by the Fletcher and Ploog group [153]. They showed that in the presence of a magnetic field, the thermoelectric power behaves qualitatively as expected from the predictions of the diffusion model exhibit large magnitudes in particular, in the magnetic quantum limit and the thermopower exhibited a magnitude of -50 mV/K instead of the expected $-120\,\mu V/K$. The fractional quantum Hall effect (FQHE) is not visible in the thermopower, which makes it possible to trace the sensitivity to the FQHE to a particular component of the thermoelectric tensor.

The Shakouri group [154] proposed a tall barrier HgCdTe superlattice structure that can achieve a large effective thermoelectric figure of merit $ZT_{\max} \sim 3$ at cryogenic temperatures. On the basis of the Boltzmann transport equation and taking into account the quantum mechanical electron transmission, they showed that the thermopower can be increased significantly at low temperatures with the use of nonplanar barriers as the thermal spreading of the electron density is tightened around the Fermi level. This provided a better asymmetric differential conductivity around the Fermi level close to the top of the barrier and, consequently, a high thermoelectric power factor is produced resulting in a large *ZT*. The group proposed improved thermoelectric properties in heterostructure's thermionic emission coolers at low temperatures for HgCdTe superlattices for low temperature cooling applications. They came out with the conclusion that since CdTe and HgTe have virtually the same lattice constants, the Hg_xCd_{1-x} Te system permits a wide range of energy gaps by alloying. Tall barrier HgCdTe superlattices, when doped appropriately, exhibited more than two orders of magnitude improvement in *ZT* ($ZT \sim 3$ at 100 K). According to their observation, the main reason for the low value of *ZT* in HgCdTe bulk is that it has a very low effective mass and a single conduction band. Thermionic emissions in heterostructures loosen up the high effective mass requirement since the improvement in the thermoelectric power is achieved through the induced asymmetric differential conductivity by a potential barrier. At a low temperature, the thermal spreading of the electron density is narrower around the Fermi level and the potential barrier generates a larger asymmetric differential conductivity around the Fermi level close to the top of the barrier, which resulted in a higher value of the thermoelectric power.

Assuming an ellipsoidal parabolic energy dispersion relation, Broido and Reinecke [155] gave a quantitative theoretical description of the power factor (P) for thermoelectric transport in superlattices and have made calculations for PbTe and GaAs quantum well and quantum wire superlattices. These calculations include

(1) 3D superlattice band structure used in (2) a multisubband inelastic Boltzmann equation for carrier transport. They have shown that these two features are needed for a quantitative treatment of thermoelectric transport in superlattice systems. They found that a strong dependence of P on orientation occurs for both PbTe quantum well and quantum wire superlattice systems. It results from the anisotropic multivalley bulk band structure, which causes the effective masses for each valley to depend on the choice of confinement direction and lifts the valley degeneracy along all but special directions. For PbTe quantum well superlattices, they found that the increased carrier scattering rates that occur with increasing confinement cause the power factor to remain near the bulk value for all barrier heights, a result that contrasts strongly with the large enhancements in P predicted from calculations employing the constant relaxation time approximation. For both PbTe and GaAs quantum wire superlattices, only modest increases in P are seen for a wide range of realistic potential offsets. These results made them to suggest that the features presented in this case are to be expected for all semiconductor superlattice systems. We suggest here that significant enhancements in P can be achieved only by eliminating the parasitic effect of heat transport through the barrier material, which might be achieved, for example, in freestanding wire systems.

The Fletcher [156] group reported experiments based on the parabolic energy bands on the effect of a parallel magnetic field on the thermopower of a double quantum well GaAs–Al0.67Ga0.33As over the field range 0–7 T at temperatures of 0.3–4 K. The main feature of interest was the effect on the thermopower of the anticrossing of the dispersion curves of the electrons in each well. At low temperatures, where diffusion effects dominate the thermopower, the results are generally in accordance with the theoretical predictions, though the magnitude of the observed effects is much smaller due to impurity and thermal broadening of the electronic energy levels. In this regime they showed that the thermopower results can be quantitatively related to the derivative of the observed resistivity with respect to magnetic field. At high temperatures, where phonon drag is dominant, the behavior of the thermopower at the anticrossing closely resembles that of the resistivity. They concluded that as the field increases and the resonance condition is largely destroyed, the smooth background variation of the thermopower is found to be much less dependent on the magnetic field than is the case for the resistivity. They thus confirmed that phonon drag thermopower is almost independent of tunneling at resonance, and that the diffusion thermopower is very similar for the electrons in the two wells.

At this point, it is to be noted that the experimental determination of the thermoelectric power with respect to the alloy composition for low-dimensional ternary and quaternary compounds is not available in detail in the literature to the best of the knowledge of the authors. However, to get some idea about how thermoelectric power varies with alloy composition for bulk HgCdTe, Sofo et al. [157] argued that the general result is that the best figure of merit is obtained for low doping levels of bulk HgCdTe, of the order of $10^{15}\ \mathrm{cm}^{-3}$ or lower, and the composition in the range of 0.11–0.14, depending on temperature. At 300 K they obtained a maximum value of $ZT \approx 0.33$ at a composition $x \approx 0.13$. They thus concluded by considering that the materials used in industry nowadays have a figure of merit of 1, the calculated values for HgCdTe suggest that it is not a promising material for thermoelectric devices.

PbSeTe-based quantum dot superlattice structures grown by molecular beam epitaxy have been experimentally investigated by Herman et al. [158] for applications in thermoelectrics. They demonstrated improved cooling values relative to the conventional bulk $(Bi, Sb)_2(Se, Te)_3$ thermoelectric materials using an n-type film in a one-leg thermoelectric device test setup, which cooled the cold junction 43.7 K below the room temperature hot junction temperature of 299.7 K. The typical device consists of a substrate-free, bulk-like (typically 0.1 mm in thickness, 10 mm in width, and 5 mm in length) slab of nanostructured PbSeTe/PbTe as the n-type leg and a metal wire as the p-type leg.

Thus, we observe from all the important experimental evidences that the thermoelectric power as experimentally achieved for both the bulk and low-dimensional materials depends on the carrier scattering, whether it be acoustic phonon, ionized, alloy, etc. The values of the thermoelectric power obtained from these data for the respective materials as discussed in this chapter find extreme usefulness in designing thermoelectric coolers or space applications if the material is subjected to follow diffusive transport regime. It should be noted that we were unable to show the experimental curves of the thermoelectric power for the low-dimensional $CoSb_3$ and other skutterudite group due its nonavailability in the literature. The diffusive thermoelectric properties of such group have been extensively studied both experimentally and theoretically in past few years [159–168] for both degenerate and nondegenerate bulk materials. This formulation can be achieved following the fundamental work of Zawadzki [24] that the thermoelectric power for degenerate bulk and low-dimensional systems having different band structures can be determined without considering the scattering mechanism under the application of a large magnetic field [25, 46] and is essentially determined by their respective carrier dispersion laws.

In this book, we have studied the thermoelectric power under strong magnetic field in quantum-confined nonlinear optical, III–V, II–VI, GaP, Ge, $PtSb_2$, zero-gap, stressed, bismuth, GaSb, IV–VI, $Pb_{1-x}Ge_xTe$, graphite, tellurium, II–V, cadmium and zinc diphosphides, Bi_2Te_3, antimony, III–V, II–VI, IV–VI, and HgTe/CdTe quantum well superlattices with graded interfaces under magnetic quantization, III–V, II–VI, IV–VI, and HgTe/CdTe effective mass superlattices under magnetic quantization, the QDs of the aforementioned superlattices, quantum-confined effective mass superlattices, and superlattices of optoelectronic materials with graded interfaces on the basis of appropriate carrier energy spectra. It is also interesting to note that although we have considered a plethora of materials having different band structures and the thermoelectric power under strong magnetic field in such quantized structures theoretically, the corresponding experimental studies in this pin-pointed topic have relatively been less investigated. Thus, the detailed experimental works are needed for an in-depth study of the thermoelectric power under strong magnetic field in such nanostructured materials as functions of externally controllable quantities which, in turn, will add new physical phenomenon in the regime of nanostructured thermoelectric and related topics. In this context, we believe that the identification of open research problems is one of the biggest problems in research. We hope that the scientists will carry out new experimental researches, not only in the said directions, but also in many other interdisciplinary aspects to add new knowledge and concepts in the experimental portion of this particular important area of nanostructured thermodynamics.

14.4 Open Research Problem

The single open research problem of this chapter is provided with the solitary endeavor of stimulating the ever spawning creativity of ace scientists.

(R14.1) Investigate experimentally the DTP, PTP, and Z for all the systems and the appropriate research problems of this monograph.

References

1. S.J. Adachi, J. Appl. Phys. **58**, R11 (1985)
2. R. Dornhaus, G. Nimtz, *Springer Tracts in Modern Physics*, vol 78 (Berlin-Heidelberg, Springer, 1976), p. 1
3. W. Zawadzki, *Handbook of Semiconductor Physics*, vol 1, ed. by W. Paul (Amsterdam, North Holland, 1982) p. 719
4. I.M. Tsidilkovski, Cand. Thesis Leningrad University SSR (1955)
5. F.G. Bass, I.M. Tsidilkovski, Ivz. Acad. Nauk Azerb SSR **10**, 3 (1966)
6. I.M. Tsidilkovski, *Band Structures of Semiconductors* (Pergamon Press, London 1982)
7. B Mitra, K.P Ghatak, Phys. Scr. **40**, 776 (1989)
8. S.K. Biswas, A.R. Ghatak, A. Neogi, A. Sharma, S. Bhattacharya, K.P. Ghatak, Physica E **36**, 163 (2007)
9. P.K. Charkaborty, G.C. Dutta, K.P. Ghatak, Phys. Scr. **68**, 368 (2003)
10. K.P. Ghatak, S.N. Biswas, Nonlinear Opt. Quant. Opts. **4**, 347, (1993)
11. A.N. Chakravarti, A.K. Choudhury, K.P. Ghatak, S. Ghosh, and A. Dhar, Appl. Phys. **25**, 105 (1981)
12. K.P. Ghatak, M. Mondal, Z. F. Physik B **B69**, 471 (1988)
13. M. Mondal and K.P. Ghatak, Phys. Letts. **131 A**, 529, (1988)
14. K.P. Ghatak, A. Ghoshal, B. Mitra, Nouvo Cimento **14D**, 903 (1992)
15. B. Mitra, A. Ghoshal, K.P. Ghatak, Nouvo Cimento D **12D**, 891 (1990)
16. K.P. Ghatak, S.N. Biswas, Nonlinear Opt. Quant. Opts., **12** 83, (1995)
17. B. Mitra, K.P. Ghatak, Solid State Electron. **32**, 177 (1989)
18. K.P. Ghatak, S.N. Biswas, Proc. SPIE **1484**, 149 (1991)
19. M. Mondal, K.P. Ghatak, *Graphite Intercalation Compounds: Science and Applications, MRS Proceedings*, ed. by M. Endo, M.S. Dresselhaus, G. Dresselhaus, MRS Fall Meeting, **EA** 16, 173, (1988)
20. M. Mondal, N. Chattapadhyay, K.P. Ghatak, J. Low Temp. Phys. **66**, 131 (1987)
21. A. N. Chakravarti, K.P. Ghatak, K.K. Ghosh, S. Ghosh, A. Dhar, Z. Physik B. **47**, 149 (1982)
22. V.K. Arora, H. Jeafarian, Phys. Rev. B. **13**, 4457 (1976)
23. M. Singh, P.R. Wallace, S.D. Jog, J.J. Erushanov, J. Phys. Chem. Solids **45**, 409 (1984)
24. W. Zawadski, Adv. Phys. **23**, 435 (1974)
25. K.P. Ghatak, M. Mondal, Z. fur Nature A **41A**, 881 (1986)
26. T. Ando, A.H. Fowler, F. Stern, Rev. Modern Phys. **54**, 437 (1982)

27. K.P. Ghatak, B. De, M. Mondal, S.N. Biswas, Epitaxial Heterostructures, MRS Symposium Proceedings Spring Meeting **198**, 327 (1990)
28. K.P. Ghatak, S.N. Biswas, Long wave length semiconductor devices, materials and processes MRS symposium proceedings, MRS spring meeting **216**, 465 (1990)
29. K.P. Ghatak, B. De, Modern perspective on thermoelectrics and related materials, MRS symposium proceedings, spring meeting **234**, 55 (1991)
30. Modern perspective on thermoelectrics and related materials, MRS symposium proceedings, spring meeting **234**, 59 (1991)
31. K.P. Ghatak, Proceedings of SPIE, USA, fiber optic and laser sensors IX **1584**, 435 (1992)
32. R.B. Dingle, Phil. Mag. **46**, 813 (1955)
33. D. Redfield, M.A. Afromowitz, Phil. Mag. **18**, 831 (1969)
34. H.C. Cassey, F. Stern, J. Appl. Phys. **47**, 631 (1976)
35. M. Mondal, K.P. Ghatak, Phys. Lett. **102A**, 54 (1984)
36. P.K. Chakraborty, G.C. Datta and K.P. Ghatak, Physica Scripta, **68**, 368 (2003)
37. B. Mitra, D.K. Basu, B. Nag and K.P. Ghatak, Nonlinear Optics **17**, 171 (1997)
38. K.P. Ghatak, S. Bhattacharya J. Appl. Phys. **102**, 073704 (2007)
39. K.P. Ghatak, S. Bhattacharya, H. Saikia, D. Baruah, A. Saikia, K.M. Singh, A. Ali, S.N. Mitra, P.K. Bose, A. Sinha, J. Comput. Theor. Nanosci. **3**, 727 (2006)
40. E.O. Kane, Solid State Electron. **8**, 3 (1985)
41. T. Ando, A.H. Fowler, F. Stern, Rev. Mod. Phys. **54**, 437 (1982)
42. P.K. Basu, *Optical Processes in Semiconductors* (Oxford University Press, 2001)
43. A.N. Chakravarti, D. Mukherjee, Phys. Lett. **53A**, 403 (1975)
44. A.N. Chakravarti, S. Swaminathan, Phys. Stat. Sol. (a) **23**, K191 (1974)
45. A.N. Chakravarti, Phys. Stat. Sol (a) **25**, K 105 (1974)
46. G.P. Chuiko, Sov. Phys. Semi. **19**, 1279 (1985)
47. A.K. Sreedhar, S.C. Gupta, Phys. Rev. B **5**, 3160 (1972)
48. R.W. Keyes, IBM. J. Res. Dev. **5**, 266 (1961)
49. R.W. Keyes, Solid State Phys. **20**, 37 (1967)
50. S. Bhattacharya, S. Chowdhury, K.P. Ghatak, J. Comput. Theor. Nanosci. **3**, 423 (2006)
51. S. Choudhury, L.J. Singh, K.P. Ghatak, Physica B **365**, 5 (2005)
52. L.J. Singh, S. Choudhary, A. Mallik, K.P. Ghatak, J. Comput. Theor. Nanosci. **2**, 287, (2005)
53. K.P. Ghatak, J.Y. Siddiqui, B. Nag, Phys. Lett. A **282**, 428 (2001)
54. K.P. Ghatak, J.P. Banerjee, B. Nag, J. Appl. Phys. **83**, 1420 (1998)
55. B. Nag, K.P. Ghatak, Nonlinear Opt. **19**, 1 (1998)
56. K.P. Ghatak, B. Nag, Nanostruct. Mater. **10**, 923 (1998)
57. B. Nag, K.P. Ghatak, J. Phys. Chem. Sol. **58**, 427 (1997)
58. K.P. Ghatak, D.K. Basu, B. Nag, J. Phys. Chem. Solids **58**, 133, (1997)
59. K.P. Ghatak, J.P. Banerjee, B. Goswami, B. Nag, Nonlinear Opt. Quant. Opt. **16**, 241 (1996)
60. K.P. Ghatak, J.P. Banerjee, D. Bhattacharyya, B. Nag, Nanotechnology **7**, 110 (1996)
61. K.P. Ghatak, J.P. Banerjee, M. Mitra, B. Nag, Nonlinear Opt. **17**, 193 (1996)
62. B. Nag, K.P. Ghatak, Phys. Scr. **54**, 657 (1996)
63. K.P. Ghatak, B. Mitra, Phys. Scr., **46**, 182 (1992)
64. K.P. Ghatak, Int. J. Electron. **71**, 239 (1991)
65. K.P. Ghatak, B. De, S.N. Biswas, M. Mondal, Mechanical behavior of materials and structures in microelectronics, MRS Symposium Proceedings, Spring Meeting, **2216**, 191, (1991)
66. K.P. Ghatak, B. De, MRS Symp. Proc. **226**, 191 (1991)
67. K.P. Ghatak, B. Nag, G. Majumdar, MRS Symp. Proc. **379**, 109 (1995)
68. K.P. Ghatak, S. Bhattacharyya, S. Pahari, S.N. Mitra, P.K. Bose, J. Phys. Chem. Solids **70**, 122 (2009)
69. D. Baruah, S. Choudhury, K.M. Singh, K.P. Ghatak, J. Phys. Conf. Series **61**, 80 (2007)
70. H. Kroemer, IEEE Trans Electron. Devices **25**, 850 (1978)
71. R.W. Lade, Proc. IEEE. **51**, 743 (1964)
72. S.N. Mohammed, J. Phys. C.: Solid State Phys. **13**, 2685 (1980)
73. P.T. Landsberg, Eur. J. Phys. **2**, 213 (1981)
74. P.T. Landsberg, Proc. R. Soc. A **213**, 226 (1952).

75. S.A. Hope, G. Feat, P.T. Landsberg, J. Phys. A Math. Gen. **14**, 2377 (1981)
76. C.H. Wang, A. Neugroschel, IEEE Electron. Dev. Lett. **ED-11**, 576 (1990)
77. I.-Y. Leu, A. Neugroschel, C.H. Wang, A. Neugroschel, IEEE Trans. Electron. Dev. **ED-40**, 1872 (1993)
78. F. Stengel, S.N. Mohammad, H. Morkoç, J. Appl. Phys. **80**, 3031 (1996)
79. H.J. Pan, W.C. Wang, K.B. Thai, C.C. Cheng, K.H. Yu, K.W. Lin, C.Z. Wu, W.C. Liu, Semicond. Sci. Technol. **15**, 1101 (2000)
80. S.N. Mohammad, J. Appl. Phys. **95**, 4856 (2004)
81. V.K. Arora, Appl. Phys. Lett. **80**, 3763 (2002)
82. S.N. Mohammad, J. Appl. Phys. **95**, 7940 (2004)
83. S.N. Mohammad, Phil. Mag. **84**, 2559 (2004)
84. S.N. Mohammad, J. Appl. Phys. **97**, 063703 (2005)
85. S.G. Dmitriev, Yu. V. Markin, Semiconductors **34**, 931 (2000)
86. M. Tao, D. Park, S.N. Mohammad, D. Li, A.E. Botchkerav, H. Morkoç, Phil. Mag. B **73**, 723 (1996)
87. D.G. Park, M. Tao, D. Li, A.E. Botchkarev, Z. Fan, S.N. Mohammad, H. Morkoç, J. Vac. Sci. Technol. B **14**, 2674 (1996)
88. Z. Chen, D.G. Park, S.N. Mohammad, H. Morkoç, Appl. Phys. Lett. **69**, 230 (1996)
89. K.P. Ghatak, S. Bhattacharya, D. De, *Einstein Relation in Compound Semiconductors and Their Nanostructures*, Springer Series in Materials Science, Vol. 116 (Springer-Verlag, Germany, 2008)
90. A.S. Filipchenko, I.G. Lang, D.N. Nasledov, S.T. Pavlov, L.N. Radaikine, Phys. Stat. Sol. (b) **66**, 417 (1974)
91. M. Wegener, *Extreme Nonlinear Optics* (Springer-Verlag, Germany, 2005)
92. B.S. Wherreff, W. Wolland, C.R. Pidgeon, R.B. Dennis, S.D. Smith, *Proceedings of the 12th International Conference of the Physics of the Semiconductors*, ed. by M.H. Pilkahn, R.G. Tenbner (Staffgard, 1978), p. 793
93. O. Madelung, *Semiconductors: Data handbook*, 3rd Ed. Springer (2004)
94. M. Krieehbaum, P. Kocevar, H. Pascher, G. Bauer, IEEE QE **24**, 1727 (1988)
95. G.P. Chuiko, Sov. Phys. Semiconduct. **19**, 1279 (1985)
96. M.J. Gelten, C.V.M. VanEs, F.A.P. Blom, J.W.F. Jongencelen, Solid State Commun., **33**, 833 (1980)
97. A.A. El-Shazly, H.S. Soliman, H.E.A. El-Sayed, D.A.A. El-Hady, J. Vac. **47**, 53 (1996)
98. E.A. Arushanov, A.F. Knyazev, A.N. Natepov, S.T. Radautsan, Sov. Phys. Semiconduct. 15, 828 (1981) (This paper contains the experimental results of TPSM in bulk specimens of Cd_3As_2 and the less sophisticated theory in accordance with Bodnar model, whereas more generalized analysis has been presented here)
99. E.N. Nikitin, Zhurnal Tekhnicheskoj Fiziki **28**, 23 (1958)
100. D. Souptel, G. Behr, L. Ivanenko, H. Vinzelberg, J. Schumann, J. Cryst. Growth **244**, 296 (2002)
101. B.K. Voronov, L.D. Dudkin, N.N. Trusova, Krystallografia **12**, 519, 1967
102. B.K. Voronov, L.D. Dudkin, N.N. Trusova, in *Khimicheskaya Svyaz v Poluprovodnikah* (Nauka i Tekhnika, Russia, 1969), p. 291
103. V.I. Kaidanov, L.S. Lyakhina, V.A. Tselishchev, B.K. Voronov, N.N. Trusova, L.D. Dudkin, Sov. Phys. Solid State **1**, 2481 (1967)
104. A. Heinrich, G. Behr, H. Griessmann, *Proceedings ICT'97. 16th International Conference on Thermoelectrics* (IEEE, USA, 1997), p. 287
105. M.I. Fedorov, V.K. Zaitsev, I.S. Eremin, N.F. Kartenko, P.P. Konstantinov, V.V. Popov, M. Kurisu, G. Nakamoto, T. Souma, *Proceedings of the 20th International Conference on Thermoelectrics (ICT 2001)* (IEEE, USA, 2001), p. 214
106. A.A. Demkov, O.F. Sankey, K.E. Schmidt, G.B. Adams, M. O'Keeffe, Phys. Rev. B **50**, 17001 (1995)
107. J. Dong, O.F. Sankey, J. Phys. Condens. Matter **11**, 6129 (1999)
108. J. Dong, O.F. Sankey, G. Kern, Phys. Rev. B **60**, 950 (1999)

109. B.B. Iverson, A.E.C. Palmqvist, D.E. Cox, G.S. Nolas, G.D. Stucky, N.P. Blake, H. Metiu, J. Solid State Chem. **149**, 455 (1999)
110. R. Kröner, K. Peters, H.G. vo Schnering, R. Nesper, Z. Kristallogr. 213, 675 (1998)
111. V.L. Kuznetsov, L.A. Kuznetsova, A.E. Kaliazin, D.M. Rowe, *J. Appl. Phys.* **87**, 7871 (2000)
112. G.S. Nolas, G.A. Slack, S.B. Schujman, in *Semiconductors and Semimetals*, vol 69, Chapter 6 (Academic Press, USA, 2001)
113. I. Terasaki, Y. Sasago, K. Uchinokura, Phys. Rev. B **56**, R12685 (1997)
114. H. Yakabe, K. Kikuchi, I. Terasaki, Y. Sasago, K. Uchinokura, *Proceedings of the 16th International Conference on Thermoelectrics* (Germany, 1997), p. 523
115. R. Funahashi, I. Matsubara, H. Ikuta, T. Takeuchi, U. Mizutani, S. Sodeoka, Jpn. J. Appl. Phys **39**, L1127 (2000)
116. M. Ohtaki, Y. Nojiri, E. Maeda, *Proceedings of the 19th International Conference on Thermoelectrics* (U.K., 2000), p. 190
117. A. Satake, H. Tanaka, T. Ohkawa, T. Fujii, I. Terasaki, J. Appl. Phys. **96**, 931 (2004)
118. W. Koshibae, K. Tsutsui, S. Maekawa, Phys. Rev. B **62**, 6869 (2000)
119. K. Koumoto, I. Terasaki, T. Kajitani, M. Ohtaki, R. Funahashi, in *Thermoelectrics Handbook: Macro to Nano*, ed. by D. M. Rowe, Chapter 35, (Taylor and Francis Group, USA, 2006)
120. A. Casian, J. Thermoelectr. **3**, 5 (1996)
121. M. Pope, C.E. Swenberg, *Electronic Processes in Organic Crystals and Polymers*, 2nd ed. (Oxford University Press, Oxford, 1999)
122. M. Zabrowska, High Temp. High Press. **17**, 215 (1985)
123. H. Meier, *Organic Semiconductors* (Chemie, Weinheim, 1974)
124. A. Casian V. Dusciac, Iu. Coropceanu, Phys. Rev. B **66**, 165404-1 (2002)
125. A. Casian, A. Balandin, V. Dusciac, Iu. Coropceanu, Phys. Low-Dim. Struct. **9/10**, 43 (2002)
126. A. Casian, A. Balandin, V. Dusciac, R. Dusciac, *Proceedings of the 21st International Conference on Thermoelectrics* (IEEE, USA, 2003), p. 310
127. G.S. Nolas, D.T. Morelli, T.M. Tritt, Annu. Rev. Mater. Sci **29**, 89 (1999)
128. C. Uher, *Proceedings of the 22nd International Conference on Thermoelectrics* (IEEE, USA, 2003), p. 42
129. D.A. Gajewski, N.R. Dilley, E.D. Bauer, E.J. Freeman, R. Chau, M.B. Maple, D. Mandrus, B.C. Sales, A.H. Lacerda, J. Phys.: Condens. Matter **10**, 6973 (1998)
130. E. Bauer, A. Galatanu, H. Michor, G. Hilscher, P. Rogl, P. Boulet, H. Noel, Eur. Phys. J **B14**, 483 (2000)
131. J.-P. Fleurial, A. Borshchevsky, T. Caillat, D.T. Morelli, G.P. Meisner, *Proceedings of the 15th International Conference on Thermoelectrics*. IEEE Catalog 96TH8169 (Piscataway, NJ, 1996), p. 91
132. M. Fornari, D.J. Singh, Appl. Phys. Lett. **74**, 3666 (1999)
133. C. Uher, in *Semiconductors and Semimetals*, vol 69, Chapter 5 (Academic Press, USA, 2001)
134. C. Uher, J. Yang, G. P. Meisner, in: *Proceedings of the ICT '99* (1999), p. 56
135. C. Uher, J. Yang, S. Hu, MRS Syrup. Proc. **545**, 247 (1999)
136. D. Young, K. Mastronardi, P. Khalifah, C.C. Wang, R.J. Cava, A.P. Ramirez, Appl. Phys. Lett. **74**, 3999 (1999)
137. S. Sportouch, P. Larson, M. Bastea, P. Brazis, J. Ireland, C.R. Kannenwurf, S.D. Mahanti, C. Uher, M.G. Kanatzidis, MRS Syrup. Proc. **545**, 421 (1999)
138. K. Mastronardi, D. Young, C.C. Wang, P. Khalifah, R.J. Cava, A.P. Ramirez, Appl. Phys. Lett. **74**, 1415 (1999)
139. H. Hohl, A.R. Ramirez, W. Kaefer, K. Fess, Ch. Thurner, Ch. Kloc, E. Bucher, in *Thermoelectric Materials – New Directions and Approaches*, ed. by T.M. Tritt, M. Kanatzidis, H.B. Lyon, Jr., G.D. Mahan, MRS Symp. Proc. **478**, 109 (1997)
140. V.M. Browning, S.J. Poon, T.M. Tritt, A.L. Pope, S. Bhattacharya, P. Volkov, J.G. Song, V. Ponnambalam, A.C. Ehrlich, in *Thermoelectric Materials 1998 – The Next Generation Materials for Small-Scale Refrigeration and Power Generation Applications*, ed. by T.M. Tritt, M. Kanatzidis, H.B. Lyon, Jr., G.D. Mahan, MRS Symp. Proc. **545**, 403 (1999)
141. B.A. Cook, J.L. Harringa, Z.S. Tan, W.A. Jesser, in: *Proceedings of the ICT '96: 15th International Conference on Thermoelectrics,* IEEE Catalog No. 96TH8169 (1996), p. 122

142. B.A. Cook, G.P. Meisner, J. Yang, C. Uher, *Proceedings of the ICT' 99* (1999), p. 64
143. S.J. Poon, in Semiconductors and Semimetals, vol. 70, Chapter 2 (Academic Press, USA, 2001)
144. D. Shechtman, I. Blech, D. Gratias, J.W. Cahn, Phys. Rev. Lett. 53, 1951 (1984)
145. S.S. Kang, J.M. Dubois, J. von Stebut, J. Mater. Res **8**, 2471 (1993)
146. Q. Guo, S.J. Poon, Phys. Rev. B **54**, 12793 (1996)
147. F. Giroud, T. Grenet, C. Berger, P. Lindquist, C. Gignoux, G. Fourcaudot, Czech. J. Phys. **46**, 2709 (1996)
148. S.J. Poon, F.S. Pierce, Q. Guo, P. Volkov. in *Proceedings of the 6th International Conference on Quasicrystals (ICQ6)*, ed. by S. Takeuchi, T. Fujiwara (World Scientific, Singapore, 1997)
149. E.I. Rogacheva, O.N. Nashchekina, A.V. Meriuts, S.G. Lyubchenko M.S. Dresselhaus, G. Dresselhaus, Appl. Phys. Lett. **86**, 063103 (2005)
150. Y.-M. Lin, M.S. Dresselhaus, Phys. Rev. B **68**, 075304 (2003)
151. M. Cattani, M.C. Salvadori, A.R. Vaz, F.S. Teixeira, I.G. Brown, J. Appl. Phys. **100**, 114905 (2006)
152. T. Smith, M. Tsaousidou, R. Fletcher, P.T. Coleridge, Z.R. Wasilewski, Y. Feng, Phys. Rev. B. **67**, 155328 (2003).
153. R. Fletcher, J.C. Maan, K. Ploog, G. Weimann, Phys. Rev. B. **33**, 7122 (1986)
154. D. Vashaee, A. Shakouri, Appl. Phys. Lett. **88**, 132110 (2006)
155. D.A. Broido, T.L. Reinecke, Phys. Rev. B **64**, 045324 (2001)
156. R. Fletcher, T. Smith, M. Tsaousidou, P.T. Coleridge, Z.R. Wasilewski, Y. Feng, Phys. Rev. B, **70**, 155333 (2004)
157. J.O. Sofo, G.D. Mahan, J. Baars, J. Appl. Phys. **76**, 2249 (1994)
158. T.C. Harman, P.J. Taylor, M.P. Walsh, B.E. LaForge, Science **297**, 2229 (2002)
159. T. Caillat, A. Borshchevsky, J.-P. Fleurial, J. Appl. Phys. **80**, 4442 (1996)
160. E. Arushanov, K. Fess, W. Kaefer, Ch. Kloc, E. Bucher, Phys. Rev. B. **56**, 1911 (1997)
161. G.A. Slack, V.G. Tsoukala, J. Appl. Phys. **76**, 1665 (1994)
162. M.S. Brandt, P. Herbst, H. Angerer, O. Ambacher, M. Stutzmann, Phys. Rev. B. **58**, 7786 (1998)
163. R.C. Mallik, J.-Y. Jung, S.-C. Ur, I.-H. Kim, Met Mater Int **14**, 2 223 (2008)
164. H. Anno, K. Matsubara, Y. Notohara, T. Sakakibara, H. Tashiro, J. Appl. Phys. **86**, 3780 (1999)
165. V.L. Kuznetsov, L.A. Kuznetsova, D.M. Rowe, J. Phys. Condens. Matter **15**, 5035 (2003)
166. M. Puyet, B. Lenoir, A. Dauscher, M. Dehmas, C. Stiewe, E. Müller, J. Appl. Phys. **95**, 4852 (2004)
167. Z. He, C. Stiewe, D. Platzek, G. Karpinski, E. Müller, S. Li, M. Toprak, M. Muhammed, J. Appl. Phys., **101**, 053713 (2007)
168. K.T. Wojciechowski, J. Tobola, J. Leszczynski, J. Alloys Compd. **361**, 19 (2003)

Chapter 15
Conclusion and Future Research

This monograph represents the combined effort of a research team over the span of more than 20 years and deals with the thermoelectric power under strong magnetic field in various types of ultrathin films, quantum wires, quantum dots, effective mass superlattices, and superlattices with graded interfaces under different physical conditions which, in turn, generate pin-pointed knowledge regarding thermoelectric power in various nanostructured materials having different band structures. The experimental data of G_0 in this context are not available in the literature although the in-depth experimental investigations are extremely important to uncover the underlying physics and mathematics. The TPSM is basically temperature-induced thermodynamic phenomena and we have formulated the simplified expressions of G_0 for three-dimensional quantized systems together with the fact that our investigations are based on the simplified $k.p$ formalism of solid-state science without incorporating the advanced field theoretic techniques. In spite of such constraints, the wonderful role of band structure behind the curtain, which generates, in turn, new concepts are really astonishing and are discussed throughout the text.

We further present a bouquet of open research problems to our esteemed readers in this particular area of nanostructured thermodynamics which is a sea in itself.

(R15.1) Investigate the DTP, PTP, and Z in the presence of a quantizing magnetic field under exponential, Kane, Halperin, Lax, and Bonch-Bruevich band tails [1] for all the problems of this monograph of all the materials, whose unperturbed carrier energy spectra are defined in Chap. 1 by considering all types of scattering mechanisms including spin and broadening effects.

(R15.2) Investigate all the appropriate problems after proper modifications introducing new theoretical formalisms for the problems as defined in (R15.1) for negative refractive index, macromolecular, nitride, and organic materials by considering all types of scattering mechanisms.

(R15.3) Investigate all the appropriate problems of this monograph for all types of quantum-confined p-InSb, p-CuCl, and semiconductors having diamond structure valence bands, whose dispersion relations of the carriers in bulk materials are given by Cunningham [2], Yekimov et al. [3], and Roman and Ewald [4], respectively, by considering all types of scattering mechanisms.

(R15.4) Investigate the influence of defect traps and surface states separately on the thermoelectric power, carrier diffusion thermopower, and thermoelectric figure-of-merit for all the appropriate problems of all the chapters after proper modifications by considering all types of scattering mechanisms.

(R15.5) Investigate the DTP, PTP, and Z under the condition of nonequilibrium of the carrier states for all the appropriate problems of this monograph.

(R15.6) Investigate the DTP, PTP, and Z for all the appropriate problems of this monograph for the corresponding p-type materials.

(R15.7) Investigate the DTP, PTP, and Z for all the appropriate problems of this monograph for all the materials under mixed conduction in the presence of strain.

(R15.8) Investigate the DTP, PTP, and Z for all the appropriate problems of this monograph for all the materials in the presence of hot electron effects.

(R15.9) Investigate the DTP, PTP, and Z for all the appropriate problems of this monograph for all the materials for nonlinear charge transport.

(R15.10) Investigate the DTP, PTP, and Z for all the appropriate problems of this monograph for all the materials in the presence of strain in an arbitrary direction.

(R15.11) Investigate Benedicks [5] thermoelectric power for all the appropriate problems of this monograph for all the materials in the presence of strain in an arbitrary direction.

(R15.12) Investigate all the appropriate problems of this monograph for $Bi_2Te_{3-x}Se_x$ and $Bi_{2-x}Sb_xTe_3$, respectively in the presence of strain.

(R15.13) Investigate all the appropriate problems of this monograph for all types of skutterudites in the presence of strain.

(R15.14) Investigate all the appropriate problems of this monograph for semiconductor clathrates in the presence of strain.

(R15.15) Investigate all the appropriate problems of this monograph for quasicrystalline materials in the presence of strain.

(R15.16) Investigate all the appropriate problems of this monograph for strongly correlated electron systems in the presence of strain.

(R15.17) Investigate Z for all the appropriate problems of this monograph for all types of transition metal silicides in the presence of strain.

(R15.18) Investigate Z for all the appropriate problems of this monograph for all types of electrically conducting organic materials in the presence of strain.

(R15.19) Investigate Z for all the appropriate problems of this monograph for all types of functionally graded materials in the presence of strain.

(R15.20) Investigate the upper limit of the thermoelectric figure-of-merit for all the appropriate problems of this monograph both in the presence of strain.

(R15.21) Investigate all the appropriate problems of this chapter in the presence of arbitrarily oriented photon field and strain.

(R15.22) Investigate all the appropriate problems of this monograph for paramagnetic semiconductors in the presence of strain.

(R15.23) Investigate all the appropriate problems of this monograph for Boron Carbides in the presence of strain.

(R15.24) Investigate all the appropriate problems of this monograph for all types of Argyrodites in the presence of strain.

(R15.25) Investigate all the appropriate problems of this monograph for layered cobalt oxides and complex chalcogenide compounds in the presence of strain.

(R15.26) Investigate all the appropriate problems of this monograph for all types of nanotubes in the presence of strain.

(R15.27) Investigate all the appropriate problems of this monograph for various types of half-Heusler compounds in the presence of strain.

(R15.28) Investigate all the appropriate problems of this monograph for various types of pentatellurides in the presence of strain.

(R15.29) Investigate all the appropriate problems of this monograph for Bi_2Te_3–Sb_2Te_3 superlattices in the presence of strain.

(R15.30) Investigate the influence of temperature-dependent energy band constants for all the appropriate problems of this monograph.

(R15.31) Investigate the relation of Z for all the materials with the corresponding thermoelectric generator.

(R15.32) Investigate the ambipolar thermodiffusion of the carriers for all the materials as discussed in this monograph in the presence of strain.

(R15.33) Investigate the thermal diffusivity for all the appropriate problems of this monograph in the presence of strain.

(R15.34) Investigate Z for $Ag_{(1-x)}Cu_{(x)}TlTe$ for different appropriate physical conditions as discussed in this monograph in the presence of strain.

(R15.35) Investigate Z for p-type SiGe under different appropriate physical conditions as discussed in this monograph in the presence of strain.

(R15.36) Investigate Z for different metallic alloys under different appropriate physical conditions as discussed in this monograph in the presence of strain.

(R15.37) Investigate Z for different intermetallic compounds under different appropriate physical conditions as discussed in this monograph in the presence of strain.

(R15.38) Investigate Z for GaN under different appropriate physical conditions as discussed in this monograph in the presence of strain.

(R15.39) Investigate Z for different disordered conductors under different appropriate physical conditions as discussed in this monograph in the presence of strain.

(R15.40) Investigate Z for various semimetals under different appropriate physical conditions as discussed in this monograph in the presence of strain.

(R15.41) (a) Investigate the DTP, PTP, and Z for all the problems of this monograph in the presence of many body effects and strain. (b) Investigate the influence of the localization of carriers for all the appropriate problems of this monograph. (c) Investigate all the problems of this monograph by removing all the physical and mathematical approximations and establishing the respective appropriate uniqueness conditions.

Total 150 open research problems have been presented in this monograph and we sincerely believe that our esteemed readers will not only solve these condensed and challenging research problems but also will generate new concepts, both theoretical and experimental. The thermoelectric power in general is the consequence of the solution of empirically well-known and well-adjusted Boltzmann transport equation (BTE) and all the assumptions behind BTE are also applicable to thermoelectric power. The formulation of all types of scattering mechanisms for all types of materials as we discussed in this book, which will also be needed in the formulation of the thermoelectric power in the respective cases is, in general, a Herculean task. Such investigations covering the total materials spectrum of modern materials science require huge amount of creative insight. It may also be noted that the last open research problem, namely, (R15.41) alone is sufficient to draw the attention of the first-order creative minds with a strong enthusiasm for mathematics from diverse fields. This particular approach will metamorphose you into time-tested and experienced scientists bubbling with creativity much more original than that of us, although there is no hide-bound prescription for creativity. One should remember that the great men think alike which can be exemplified by the harmony of the philosophical frequency of the creatively prolific mathematician Godfrey Harold Hardy [6] tells us "in his roll-call of mathematicians: 'Galois died at twenty-one, Abel at twenty-seven, Ramanujan at thirty-three, Riemann at forty... I do not know an instance of a major mathematical advance initiated by a man past fifty'." with the physicist extraordinaire Nobel Laureate Lev Davidovich Landau, who is also famous for the anecdote "What, so young and already so unknown?" [7]. We wish to induce the passion for research activity in you, since we eternally like to hope that you are the right person to carry forward the lineage of this subject for further enhancement and supersede us. And just for this reason you will enjoy to innovate in the real sense of the term the altogether new physics and the related mathematics behind the screen of this pin-pointed research topic. We believe that physics should march ahead by leaps and bounds by creative young minds and we greet your appearance in the present research scenario in lieu of us. In the mean time, our research interest has also been shifted and we are leaving the legacy of this wonderful research arena of materials science in general on your first-order ingenuity.

References

1. B.R. Nag, *Electron Transport in Compound Semiconductors*, Springer Series in Solid State Sciences, vol 11 (Springer, Germany, 1980)
2. R.W. Cunningham, Phys. Rev. **167**, 761 (1968)
3. A.I. Yekimov, A.A. Onushchenko, A.G. Plyukhin Al, L. Efros, J. Expt. Theor. Phys. **88**, 1490 (1985)
4. B.J. Roman, A.W. Ewald, Phys. Rev. B **5,** 3914 (1972)
5. M.C. Benedicks, Acad. Sci. Comptes Rendus **165**, 391 (1917)
6. G.H. Hardy, *A mathematician's Apology* (Cambridge University Press, Cambridge, 1990), p. 37
7. A. Livanova, *Landau: A Great Physicist and Teacher*, In the Preface, pp. viii by Sir Rudolf Peierls (Pergamon, Oxford, 1985)

Appendix A

For the purpose of complete and condensed presentation, we present the brief formulation of TPSM for bulk specimens of nonlinear optical and Cd_3As_2, III–V, GaP, II–VI, Bi_2Te_3, stressed materials, IV–VI, n-Ge, $PtSb_2$, n-GaSb, n-Te, and bismuth in accordance with the appropriate band models in the following 12 sections. This appendix ends with the last set of open research problems.

A.1 Nonlinear Optical Materials and Cd_3As_2

The electron concentration of bulk specimens in this case can be expressed following [1] as

$$n_0 = g_v \left(3\pi^2\right)^{-1} \left[M_{1a}\left(E_F\right) + N_{1a}\left(E_F\right)\right], \tag{A.1}$$

where

$$M_{1a}(E_{F_b}) \equiv \left[\frac{\left[\gamma\left(E_{F_b}\right)\right]^{\frac{3}{2}}}{f_1\left(E_{F_b}\right)\sqrt{f_2\left(E_{F_b}\right)}}\right],$$

E_{F_b} is the Fermi energy as measured from the edge of the conduction band in the vertically upward direction in the absence of any quantization

$$N_{1a}(E_{F_b}) \equiv \sum_{r=1}^{s} Z_{1a}(r) M_{1a}(E_{F_b})$$

and

$$Z_{1a}(r) \equiv \left[2\left(k_B T\right)^{2r}\left(1 - 2^{1-2r}\right)\xi\left(2r\right)\right]\left[\frac{\partial^{2r}}{\partial E_{F_b}^{2r}}\right].$$

Using (A.1) and (1.13), the TPSM in this case is given by

$$G_0 = \frac{\pi^2 k_B^2 T}{3e}\left[M'_{1a}\left(E_{F_b}\right) + N'_{1a}\left(E_{F_b}\right)\right]\left[M_{1a}\left(E_{F_b}\right) + N_{1a}\left(E_{F_b}\right)\right]^{-1} \tag{A.2}$$

A.2 III–V Materials

A.2.1 Three Band Model of Kane

In accordance with this model, the electron concentration can be expressed as

$$n_0 = \frac{g_\mathrm{v}}{3\pi^2}\left(\frac{2m^*}{\hbar^2}\right)^{3/2}\left[\bar{M}_A\left(E_{\mathrm{F}_b}\right) + \bar{N}_A\left(E_{\mathrm{F}_b}\right)\right], \tag{A.3}$$

where

$$\bar{M}_A\left(E_{\mathrm{F}_b}\right) = \left[\frac{E_{\mathrm{F}_b}\left(E_{\mathrm{F}_b}+E_\mathrm{g}\right)\left(E_{\mathrm{F}_b}+E_\mathrm{g}+\Delta\right)\left(E_\mathrm{g}+\frac{2}{3}\Delta\right)}{E_\mathrm{g}\left(E_\mathrm{g}+\Delta\right)\left(E_{\mathrm{F}_b}+E_\mathrm{g}+\frac{2}{3}\Delta\right)}\right]^{3/2}$$

and

$$\bar{N}_A\left(E_{\mathrm{F}_b}\right) = \sum_{r=1}^{s} 2\left(k_\mathrm{B}T\right)^{2r}\left(1-2^{1-2r}\right)\zeta(2r)\frac{\partial^{2r}}{\partial E_{\mathrm{F}_b}^{2r}}\left[\bar{M}_A\left(E_{\mathrm{F}_b}\right)\right].$$

Using (A.3) and (1.13), the TPSM in this case can be written as

$$G_0 = \left(\frac{\pi^2 k_\mathrm{B}^2 T}{3e}\right)\left[\frac{\left(\bar{M}_A\left(E_{\mathrm{F}_b}\right)\right)' + \left(\bar{N}_A\left(E_{\mathrm{F}_b}\right)\right)'}{\bar{M}_A\left(E_{\mathrm{F}_b}\right) + \bar{N}_A\left(E_{\mathrm{F}_b}\right)}\right]. \tag{A.4}$$

A.2.2 Two Band Model of Kane

The electron concentration for this model assumes the form [1]

$$n_0 = N_{A_c}\left[F_{\frac{1}{2}}(\eta) + \frac{15\alpha k_\mathrm{B}T}{4}F_{\frac{3}{2}}(\eta)\right], \tag{A.5}$$

where

$$N_{A_c} = 2g_\mathrm{v}\left(\frac{2\pi m^* k_\mathrm{B}T}{h^2}\right)^{3/2}$$

and

$$\eta = \frac{E_{\mathrm{F}_b}}{k_\mathrm{B}T}$$

Using (A.5) and (1.13), the TPSM in this case can be written as

$$G_0 = \left(\frac{\pi^2 k_B}{3e}\right)\left[\frac{F_{-\frac{1}{2}}(\eta) + \frac{15\alpha k_B T}{4} F_{\frac{1}{2}}(\eta)}{F_{\frac{1}{2}}(\eta) + \frac{15\alpha k_B T}{4} F_{\frac{3}{2}}(\eta)}\right]. \tag{A.6}$$

A.2.3 Parabolic Energy Bands

The electron concentration and the TPSM in this case can, respectively, be written from (A.5) and (A.6) as

$$n_0 = N_{A_c}\left(F_{\frac{1}{2}}(\eta)\right) \tag{A.7}$$

and

$$G_0 = \left(\frac{\pi^2 k_B}{3e}\right)\left[\frac{F_{-\frac{1}{2}}(\eta)}{F_{\frac{1}{2}}(\eta)}\right]. \tag{A.8}$$

A.2.4 The Model of Stillman Et al.

The expression of electron concentration in this case can be written as

$$n_0 = \frac{g_v}{3\pi^2}\left(\frac{2m^*}{\hbar^2}\right)^{3/2}\left[M_{A_{10}}\left(E_{F_b}\right) + N_{A_{10}}\left(E_{F_b}\right)\right], \tag{A.9}$$

where

$$M_{A_{10}}\left(E_{F_b}\right) = \left[I_{11}(E_{F_b})\right]^{3/2}$$

and

$$N_{A_{10}}\left(E_{F_b}\right) = \sum_{r=1}^{s} 2\left(k_B T\right)^{2r}\left(1 - 2^{1-2r}\right)\zeta(2r)\frac{\partial^{2r}}{\partial E_{F_b}^{2r}}\left[M_{A_{10}}\left(E_{F_b}\right)\right].$$

Using (A.9) and (1.13), the TPSM can be expressed as

$$G_0 = \left(\frac{\pi^2 k_B^2 T}{3e}\right)\left[\frac{M'_{A_{10}}\left(E_{F_b}\right) + N'_{A_{10}}\left(E_{F_b}\right)}{M_{A_{10}}\left(E_{F_b}\right) + N_{A_{10}}\left(E_{F_b}\right)}\right]. \tag{A.10}$$

A.2.5 The Model of Palik Et al.

In accordance with this model, the electron concentration can be expressed as

$$n_0 = \frac{g_v}{3\pi^2}\left(\frac{2m^*}{\hbar^2}\right)^{3/2}\left[\bar{M}_{12A_b}\left(E_{F_b}\right) + \bar{N}_{12A_b}\left(E_{F_b}\right)\right], \tag{A.11}$$

where

$$\bar{M}_{12A_b}\left(E_{F_b}\right) = \left[I_{12}(E_{F_b})\right]^{3/2}$$

and

$$\bar{N}_{12A_b}\left(E_{F_b}\right) = \sum_{r=1}^{s} 2\left(k_B T\right)^{2r}\left(1 - 2^{1-2r}\right)\zeta(2r)\frac{\partial^{2r}}{\partial E_{F_b}^{2r}}\left[\bar{M}_{12A_b}\left(E_{F_b}\right)\right].$$

Using (A.11) and (1.13), the TPSM in this case can be written as

$$G_0 = \left(\frac{\pi^2 k_B^2 T}{3e}\right)\left[\frac{\left(\bar{M}_{12A_b}\left(E_{F_b}\right)\right)' + \left(\bar{N}_{12A_b}\left(E_{F_b}\right)\right)'}{\bar{M}_{12A_b}\left(E_{F_b}\right) + \bar{N}_{12A_b}\left(E_{F_b}\right)}\right]. \tag{A.12}$$

A.2.6 Model of Johnson and Dicley

The expressions of the electron concentration and the TPSM for this model are given by

$$n_0 = \frac{g_v}{3\pi^2}\left[M_{13A_b}\left(E_{F_b}\right) + N_{13A_b}\left(E_{F_b}\right)\right] \tag{A.13}$$

and

$$G_0 = \left(\frac{\pi^2 k_B^2 T}{3e}\right)\left[\frac{M'_{13A_b}\left(E_{F_b}\right) + N'_{13A_b}\left(E_{F_b}\right)}{M_{13A_b}\left(E_{F_b}\right) + N_{13A_b}\left(E_{F_b}\right)}\right], \tag{A.14}$$

where

$$M_{13A_b}\left(E_{F_b}\right) = \left[\bar{e}_8\left(E_{F_b}\right)\right]^{3/2},$$

$$N_{13A_b}\left(E_{F_b}\right) = \sum_{r=1}^{s} 2\left(k_B T\right)^{2r}\left(1 - 2^{1-2r}\right)\zeta(2r)\frac{\partial^{2r}}{\partial E_{F_b}^{2r}}\left[M_{13A_b}\left(E_{F_b}\right)\right],$$

$$\bar{e}_8\left(E_{F_b}\right) = \left[\left(E_{g0} + 2E_{F_b}\right)e_7 + \frac{E_{g0}^2}{4}e_8\left(E_{F_b}\right) - \left[E_{g0}^2\left[e_7^2 + \frac{E_{g0}^2}{16}e_8^2\left(E_{F_b}\right)\right.\right.\right.$$
$$\left.\left.\left. E_{F_b}e_7 e_8\left(E_{F_b}\right) + \frac{e_7 E_{g0}}{2}e_8\left(E_{F_b}\right)\right]\right]^{1/2}\right]\cdot\left(2e_7^2\right)^{-1},$$

$$e_7 = \frac{\hbar^2}{2}\left[\frac{1}{m^*} - \frac{1}{m_0}\right],$$

$$e_8\left(E_{\mathrm{F}_b}\right)=\frac{2\hbar^2\phi_{A_1}\left(E_{\mathrm{F}_b}\right)}{E_{g_0}m^*},$$

$$\phi_{A_1}\left(E_{\mathrm{F}_b}\right)=\frac{\left(E_{g_0}+\Delta\right)\left(E_{\mathrm{F}_b}+E_{g_0}+\frac{2}{3}\Delta\right)}{\left(E_{g_0}+\frac{2}{3}\Delta\right)\left(E_{\mathrm{F}_b}+E_{g_0}+\Delta\right)}.$$

A.3 n-Type Gallium Phosphide

In this case, the electron concentration and the TPSM can, respectively, be written as

$$n_0=\frac{2g_{\mathrm{v}}}{4\pi^2}\left[M_{A_1}\left(E_{\mathrm{F}_b}\right)+N_{A_1}\left(E_{\mathrm{F}_b}\right)\right], \tag{A.15}$$

$$G_0=\left(\frac{\pi^2k_{\mathrm{B}}^2T}{3e}\right)\left[\frac{M'_{A_1}\left(E_{\mathrm{F}_b}\right)+N'_{A_1}\left(E_{\mathrm{F}_b}\right)}{M_{A_1}\left(E_{\mathrm{F}_b}\right)+N_{A_1}\left(E_{\mathrm{F}_b}\right)}\right], \tag{A.16}$$

where

$$\begin{aligned}M_{A_1}\left(E_{\mathrm{F}_b}\right)=&\left[\left(t_{A_1}\right).\left(E_{\mathrm{F}_b}\right)\theta_-\left(E_{\mathrm{F}_b}\right)+t_{A_2}\theta_-\left(E_{\mathrm{F}_b}\right)-\frac{t_{A_3}\left(\theta_-\left(E_{\mathrm{F}_b}\right)\right)^3}{3}\right.\\&-\frac{t_{A_4}\theta_-\left(E_{\mathrm{F}_b}\right)}{2}\left[\left(\theta_-\left(E_{\mathrm{F}_b}\right)\right)^2+t_{A_5}\left(E_{\mathrm{F}_b}\right)\right]^{\frac{1}{2}}+\frac{t_{A_4}t_{A_5}\left(E_{\mathrm{F}_b}\right)}{2}\\&\left.\ln\left|\frac{\theta_-\left(E_{\mathrm{F}_b}\right)+\sqrt{\left(\theta_-\left(E_{\mathrm{F}_b}\right)\right)^2+t_{A_5}\left(E_{\mathrm{F}_b}\right)}}{\sqrt{t_{A_5}\left(E_{\mathrm{F}_b}\right)}}\right|\right],\end{aligned}$$

$$t_{A_1}=\frac{1}{a},\quad a=\left(\frac{\hbar^2}{2m_\perp^*}+A\frac{\hbar^2}{2m_\parallel^*}\right),\quad b=\frac{\hbar^2}{2m_\parallel^*},\quad c=\frac{\hbar^2k_0^2}{m_\parallel^{*2}},$$

$$D=\left|V_G\right|^2,\quad t_{A_2}=\left[\frac{g_1}{2a^2}\right],\quad t_{A_3}=\left(\frac{b}{a}\right),\quad t_{A_4}=\left(\frac{\sqrt{g_3}}{2a^2}\right),$$

$$g_1=(2aD-c),\quad g_2=\left[4a^2b^2+c^2-4acD\right],\quad g_3=\left[4abc+4a^2c\right]$$

$$t_{A_5}\left(E_{\mathrm{F}_b}\right)=\left[\frac{g_2-(4ac).\left(E_{\mathrm{F}_b}\right)}{g_3}\right],$$

$$t_{A_6}=\left(t_{A_4}^2+2t_{A_2}t_{A_3}\right),\quad t_{A_7}=\left(2t_{A_1}t_{A_3}\right),$$

$$t_{A_8}=\left[t_{A_4}^4+4t_{A_4}^2t_{A_2}t_{A_3}+\left(4t_{A_3}^2t_{A_4}^2g_2/g_3\right)\right],$$

$$t_{A_9}=\left[4t_{A_1}t_{A_3}t_{A_4}^2+8t_{A_1}t_{A_2}t_{A_3}^2-\left(16t_{A_3}^2t_{A_4}^2ac/g_3\right)\right],$$

and

$$\theta_-\left(E_{\mathrm{F}_b}\right) = \left(t_{A_3}.\sqrt{2}\right)^{-1}\left[t_{A_6} + \left(E_{\mathrm{F}_b}\right).(t_{A7}) - \left(t_{A8} + (t_{A9}).\left(E_{\mathrm{F}_b}\right)\right)^{1/2}\right]$$

A.4 II–VI Materials

The expressions of electron concentration and the TPSM for II–VI materials assume the forms

$$n_0 = \frac{1}{2}\left(\frac{k_\mathrm{B}T}{\pi B_0}\right)^{3/2}\left(\frac{B_0}{A_0}\right)\left[F_{\frac{1}{2}}(\eta) + \frac{C_0^2}{2A_0k_\mathrm{B}T}F_{\frac{-1}{2}}(\eta)\right], \tag{A.17}$$

$$G_0 = \left(\frac{\pi^2 k_\mathrm{B}}{3e}\right)\left[\frac{F_{-1/2}(\eta) + \left(C_0^2/2A_0k_\mathrm{B}T\right)F_{-3/2}(\eta)}{F_{1/2}(\eta) + \left(C_0^2/2A_0k_\mathrm{B}T\right)F_{-1/2}(\eta)}\right]. \tag{A.18}$$

A.5 Bismuth Telluride

In this case, electron concentration and TPSM are given by (A5) and (A6), respectively, where

$$N_{A_c} = 2\left(\frac{2\pi m_0 k_\mathrm{B}T}{h^2}\right)^{3/2} g_\mathrm{v}\left[\bar{\bar{\alpha}}_{11}\alpha_{22}\alpha_{33} - 4\bar{\bar{\alpha}}_{11}(\alpha_{23})^2\right]^{-1/2}.$$

A.6 Stressed Materials

In this case, electron concentration and TPSM assume the forms

$$n_0 = g_\mathrm{v}\left(3\pi^2\right)^{-1}\left[M_{A_2}\left(E_{\mathrm{F}_b}\right) + N_{A_2}\left(E_{\mathrm{F}_b}\right)\right], \tag{A.19}$$

$$G_0 = \left(\frac{\pi^2 k_\mathrm{B}^2 T}{3e}\right)\left[\frac{M'_{A_2}\left(E_{\mathrm{F}_b}\right) + N'_{A_2}\left(E_{\mathrm{F}_b}\right)}{M_{A_2}\left(E_{\mathrm{F}_b}\right) + N_{A_2}\left(E_{\mathrm{F}_b}\right)}\right], \tag{A.20}$$

where $M_{A_2}\left(E_{\mathrm{F}_b}\right) = \left[a^*\left(E_{\mathrm{F}_b}\right)b^*\left(E_{\mathrm{F}_b}\right)c^*\left(E_{\mathrm{F}_b}\right)\right]$ and

$$N_{A_2}\left(E_{\mathrm{F}_b}\right) = \sum_{r=1}^{s} 2(k_\mathrm{B}T)^{2r}\left(1 - 2^{1-2r}\right)\zeta(2r)\frac{\partial^{2r}}{\partial E_{\mathrm{F}_b}^{2r}}\left[M_{A_2}\left(E_{\mathrm{F}_b}\right)\right]$$

A.7 IV–VI Semiconductors

A.7.1 Bangert and Kästner Model

In this case, electron concentration and theTPSM can, respectively, be expressed as

$$n_0 = \left(\frac{g_v}{3\pi^2}\right)\left[M_{A_3}\left(E_{F_b}\right) + N_{A_3}\left(E_{F_b}\right)\right] \tag{A.21}$$

and

$$G_0 = \left(\frac{\pi^2 k_B^2 T}{3e}\right)\left[\frac{M'_{A_3}\left(E_{F_b}\right) + N'_{A_3}\left(E_{F_b}\right)}{M_{A_3}\left(E_{F_b}\right) + N_{A_3}\left(E_{F_b}\right)}\right], \tag{A.22}$$

where

$$M_{A_3}\left(E_{F_b}\right) = \left[\tau_A\left(E_{F_b}\right)\right]^{3/2}\left[\bar{F}_1\left(E_{F_b}\right)\sqrt{\bar{F}_2\left(E_{F_b}\right)}\right]^{-1},$$

$\tau_A\left(E_{F_b}\right) = 2E_{F_b}$, and

$$N_{A_3}\left(E_{F_b}\right) = \sum_{r=1}^{s} 2\left(k_B T\right)^{2r}\left(1 - 2^{1-2r}\right)\zeta(2r)\frac{\partial^{2r}}{\partial E_{F_b}^{2r}}\left[M_{A_3}\left(E_{F_b}\right)\right].$$

A.7.2 Cohen Model

In this case, electron concentration and the TPSM can, respectively, be written as

$$n_0 = \left(\frac{g_v\sqrt{m_1 m_3}}{\pi^2\hbar}\right)\left[M_{A_3}\left(E_{F_b}\right) + N_{A_3}\left(E_{F_b}\right)\right], \tag{A.23}$$

$$G_0 = \left(\frac{\pi^2 k_B^2 T}{3e}\right)\left[\frac{M'_{A_3}\left(E_{F_b}\right) + N'_{A_3}\left(E_{F_b}\right)}{M_{A_3}\left(E_{F_b}\right) + N_{A_3}\left(E_{F_b}\right)}\right], \tag{A.24}$$

where

$$M_{A_3}\left(E_{F_b}\right) = \tau_{A_1}\left(E_{F_b}\right)\left[E_{F_b}\left(1 + \alpha E_{F_b}\right) - \frac{\tau_{A_1}^4\left(E_{F_b}\right)}{20 m_2 m'_2} + \frac{\alpha E_{F_b}\tau_{A_1}^2\left(E_{F_b}\right)}{6m'_2} - \frac{\tau_{A_1}^2\left(E_{F_b}\right)\left(1 + \alpha E_{F_b}\right)}{6m_2}\right],$$

$$\tau_{A_1}\left(E_{F_b}\right)=\left[\frac{\alpha}{2m_2m_2'}\right]^{-1/2}\left[-\left[\frac{1+\alpha E_{F_b}}{2m_2}-\frac{\alpha E_{F_b}}{2m_2'}\right]\right.$$
$$\left.+\left[\left[\frac{1+\alpha E_{F_b}}{2m_2}-\frac{\alpha E_{F_b}}{2m_2'}\right]^2+\frac{\alpha E_{F_b}\left(1+\alpha E_{F_b}\right)}{m_2m_2'}\right]^{1/2}\right]^{1/2},$$

and

$$N_{A_3}\left(E_{F_b}\right)=\sum_{r=1}^{s}2\left(k_BT\right)^{2r}\left(1-2^{1-2r}\right)\zeta\left(2r\right)\frac{\partial^{2r}}{\partial E_{F_b}^{2r}}\left[M_{A_3}\left(E_{F_b}\right)\right].$$

A.7.3 Dimmock Model

In this case, electron concentration and the TPSM assume the forms

$$n_0=\left(\frac{g_v}{2\pi^2}\right)\left[M_{A_4}\left(E_{F_b}\right)+N_{A_4}\left(E_{F_b}\right)\right], \tag{A.25}$$

$$G_0=\left(\frac{\pi^2k_B^2T}{3e}\right)\left[\frac{M'_{A_4}\left(E_{F_b}\right)+N'_{A_4}\left(E_{F_b}\right)}{M_{A_4}\left(E_{F_b}\right)+N_{A_4}\left(E_{F_b}\right)}\right], \tag{A.26}$$

where

$$M_{A_4}\left(E_{F_b}\right)=\left[\alpha_5J_{A_1}\left(E_{F_b}\right)-\alpha_3\left(E_{F_b}\right)\bar{\tau}_{A_1}\left(E_{F_b}\right)-\frac{\alpha_4}{3}\left[\bar{\tau}_{A_1}\left(E_{F_b}\right)\right]^3\right],$$

$$\alpha_5=\left[\frac{2m_t^+m_t^-}{\alpha\hbar^2}\omega_{A_1}\right],$$

$$\omega_{A_1}=\left[\frac{\alpha^2}{16}\left[\frac{1}{m_t^-m_l^+}+\frac{1}{m_l^-m_t^+}\right]^2-\frac{\alpha^2}{4m_l^+m_t^-m_l^-m_t^+}\right],$$

$$J_{A_1}\left(E_{F_b}\right)=\frac{A_A\left(E_{F_b}\right)}{3}\left[-\left(A_A^2\left(E_{F_b}\right)+B_A^2\left(E_{F_b}\right)\right)E\left(\lambda,q\right)\right.$$
$$\left.+2B_A^2\left(E_{F_b}\right)F\left(\lambda,q\right)\right]+\frac{\bar{\tau}_{A_1}\left(E_{F_b}\right)}{3}\left[\left(\bar{\tau}_{A_1}\left(E_{F_b}\right)\right)^2+A_A^2\left(E_{F_b}\right)\right.$$
$$\left.+2B_A^2\left(E_{F_b}\right)\right]\left[A_A^2\left(E_{F_b}\right)+\bar{\tau}_{A_1}^2\left(E_{F_b}\right)\right]^{1/2}\left[B_A^2\left(E_{F_b}\right)+\bar{\tau}_{A_1}^2\left(E_{F_b}\right)\right]^{-1/2},$$

$$\lambda=\tan^{-1}\frac{\bar{\tau}_A\left(E_{F_b}\right)}{B_A\left(E_{F_b}\right)},$$

$$q = \left[\frac{\sqrt{A_A^2\left(E_{\mathrm{F}_b}\right) - B_A^2\left(E_{\mathrm{F}_b}\right)}}{A_A\left(E_{\mathrm{F}_b}\right)} \right],$$

$$A_A\left(E_{\mathrm{F}_b}\right) = \left[\tau_{A_2}\left(E_{\mathrm{F}_b}\right) + \sqrt{\tau_{A_2}^2\left(E_{\mathrm{F}_b}\right) - 4\tau_{A_3}\left(E_{\mathrm{F}_b}\right)}\right]^{1/2} \Big/ \sqrt{2},$$

$$B_A\left(E_{\mathrm{F}_b}\right) = \left[\tau_{A_2}\left(E_{\mathrm{F}_b}\right) - \sqrt{\tau_{A_2}^2\left(E_{\mathrm{F}_b}\right) - 4\tau_{A_3}\left(E_{\mathrm{F}_b}\right)}\right]^{1/2} \Big/ \sqrt{2},$$

$$\tau_{A_2}\left(E_{\mathrm{F}_b}\right) = \frac{\omega_{A_2}\left(E_{\mathrm{F}_b}\right)}{\omega_{A_1}^2},$$

$$\tau_{A_3}\left(E_{\mathrm{F}_b}\right) = \frac{\omega_{A_3}\left(E_{\mathrm{F}_b}\right)}{\omega_{A_1}^2},$$

$$\omega_{A_2}\left(E_{\mathrm{F}_b}\right) = \left[\frac{\alpha}{2}\left[\frac{1}{2m_t^*} - \frac{\alpha.E_{\mathrm{F}_b}}{2m_t^+} + \frac{1+\alpha.E_{\mathrm{F}_b}}{2m_t^-}\right].\left[\frac{1}{m_t^- m_l^+} + \frac{1}{m_l^- m_t^+}\right] - \frac{\alpha}{m_t^+ m_t^-}\left[\frac{1}{2m_l^*} + \frac{\alpha.E_{\mathrm{F}_b}}{2m_l^+} + \frac{1+\alpha.E_{\mathrm{F}_b}}{2m_l^-}\right]\right],$$

$$\omega_{A_3}\left(E_{\mathrm{F}_b}\right) = \left[\frac{\alpha.E_{\mathrm{F}_b}\left(1+\alpha.E_{\mathrm{F}_b}\right)}{m_t^+ m_t^-} + \left[\frac{1}{2m_t^*} - \frac{\alpha.E_{\mathrm{F}_b}}{2m_t^+} + \frac{1+\alpha.E_{\mathrm{F}_b}}{2m_t^-}\right]^2\right],$$

$$\alpha_2\left(E_{\mathrm{F}_b}\right) = \left[\frac{1}{2m_t^*} - \frac{\alpha.E_{\mathrm{F}_b}}{2m_t^+} + \frac{1+\alpha.E_{\mathrm{F}_b}}{2m_t^-}\right],$$

$$\alpha_3 = \frac{\alpha\hbar^2}{4}\left[\frac{1}{m_t^- m_l^+} + \frac{1}{m_l^- m_t^+}\right],$$

$$\tau_{A_1}\left(E_{\mathrm{F}_b}\right) = \left[\frac{2m_l^+ m_l^-}{\alpha\hbar^2}\right]^{1/2}\left[-\left[\frac{1}{2m_l^*} + \frac{1+\alpha.E_{\mathrm{F}_b}}{m_l^-} - \frac{\alpha.E_{\mathrm{F}_b}}{2m_l^+}\right] + \left[\left[\frac{1}{2m_l^*} + \frac{1+\alpha.E_{\mathrm{F}_b}}{m_l^-} - \frac{\alpha.E_{\mathrm{F}_b}}{2m_l^+}\right]^2 + \frac{\alpha.E_{\mathrm{F}_b}\left(1+\alpha.E_{\mathrm{F}_b}\right)}{m_l^- m_l^+}\right]^{1/2}\right]^{1/2}.$$

$E(\lambda, q) = \int_0^{\lambda} \left[1 - q^2 \sin^2 \alpha\right]^{1/2} d\alpha$ is the complete Elliptic integral of second kind,

$$F(\lambda, q) = \int_0^{\lambda} \frac{d\alpha}{\sqrt{1 - q^2 \sin^2 \alpha}}$$

is the complete Elliptic integral of first kind, and

$$N_{A_4}\left(E_{\mathrm{F}_b}\right) = \sum_{r=1}^{s} 2\left(k_{\mathrm{B}}T\right)^{2r}\left(1 - 2^{1-2r}\right)\zeta(2r)\frac{\partial^{2r}}{\partial E_{\mathrm{F}_b}^{2r}}\left[M_{A_4}\left(E_{\mathrm{F}_b}\right)\right]$$

A.7.4 Foley and Langenberg Model

In this case, electron concentration and theTPSM can, respectively, be expressed as

$$n_0 = \left(\frac{2g_{\mathrm{v}}}{4\pi^2}\right)\left[h_{A_6}\left(E_{\mathrm{F}_b}\right) + h_{A_7}\left(E_{\mathrm{F}_b}\right)\right], \tag{A.27}$$

$$G_0 = \left(\frac{\pi^2 k_{\mathrm{B}}^2 T}{3e}\right)\left[\frac{h'_{A_6}\left(E_{\mathrm{F}_b}\right) + h'_{A_7}\left(E_{\mathrm{F}_b}\right)}{h_{A_6}\left(E_{\mathrm{F}_b}\right) + h_{A_7}\left(E_{\mathrm{F}_b}\right)}\right], \tag{A.28}$$

$$h_{A_6}\left(E_{\mathrm{F}_b}\right) = \left[\frac{1}{3}\delta_{A_5} h_{A_3}^3\left(E_{\mathrm{F}_b}\right) - \delta_{A_4}\left(E_{\mathrm{F}_b}\right) h_{A_3}\left(E_{\mathrm{F}_b}\right) + \delta_{A_{10}} J_{A_6}\left(E_{\mathrm{F}_b}\right)\right],$$

$$\delta_{A_6} = \left[\frac{\hbar^4}{2}\left(\frac{1}{\left(m_{\perp}^{+}\right)^2} - \frac{1}{\left(m_{\perp}^{-}\right)^2}\right)\right]^{-1},$$

$$\delta_{A_4}\left(E_{\mathrm{F}_b}\right) = \delta_{A_6}\left[\frac{\hbar^2}{2m_{\perp}^{-}}\left(E_{\mathrm{g}0} + 2E_{\mathrm{F}_b}\right) + P_{\perp}^2 + \frac{\hbar^2 E_{\mathrm{g}0}}{2m_{\perp}^{+}}\right],$$

$$\delta_{A_7} = \hbar^8\left[\frac{1}{4\left(m_{\perp}^{+} m_{\parallel}^{+}\right)^2} - \frac{1}{2m_{\perp}^{+} m_{\perp}^{-} m_{\parallel}^{+} m_{\parallel}^{-}} + \frac{1}{4\left(m_{\perp}^{-} m_{\parallel}^{-}\right)^2}\right],$$

$$\delta_{A_5} = \delta_{A6}\hbar^4\left[\frac{1}{2m_{\perp}^{-} m_{\parallel}^{-}} - \frac{1}{2m_{\perp}^{+} m_{\parallel}^{+}}\right],$$

$$\delta_{A_8}\left(E_{\mathrm{F}_b}\right) = \left[\frac{\hbar^6 E_{g0}}{2\left(m_\perp^+\right)^2 m_\parallel^-} + \frac{\hbar^6\left(E_{g0}+2E_{\mathrm{F}_b}\right)}{2m_\perp^+ m_\perp^- m_\parallel^+} - \frac{\hbar^6 E_{g0}}{2m_\perp^+ m_\perp^- m_\parallel^-} - \frac{\hbar^4 P_\perp^2}{m_\perp^- m_\parallel^-}\right.$$
$$-\frac{\hbar^6\left(E_{g0}+2E_{\mathrm{F}_b}\right)}{2\left(m_\perp^-\right)^2 m_\parallel^-} - \frac{\hbar^6\left(E_{g0}+2E_{\mathrm{F}_b}\right)}{2m_\parallel^-\left(m_\perp^+\right)^2} - \frac{\hbar^6 E_{g0}}{2m_\parallel^+\left(m_\perp^+\right)^2} - \frac{P_\parallel^2\hbar^4}{\left(m_\perp^+\right)^2}$$
$$\left.+\frac{\hbar^6\left(E_{g0}+2E_{\mathrm{F}_b}\right)}{2m_\parallel^-\left(m_\perp^-\right)^2} + \frac{\hbar^6 E_{g0}}{2m_\parallel^+\left(m_\perp^-\right)^2} + \frac{\hbar^4 P_\parallel^2}{\left(m_\perp^+\right)^2}\right],$$

$$\delta_{A_9}\left(E_{\mathrm{F}_b}\right) = \left[P_\perp^4 + \frac{\hbar^4 E_{g0}^2}{4\left(m_\perp^+\right)^2} + \frac{\hbar^4\left(E_{g0}+2E_{\mathrm{F}_b}\right)}{4\left(m_\perp^-\right)^2} + \frac{E_{g0}\hbar^2 P_\perp^2}{m_\perp^+}\right.$$
$$\left.+\frac{\hbar^2 P_\perp^2\left(E_{g0}+2E_{\mathrm{F}_b}\right)}{m_\perp^-}\right] + \left[\frac{E_{g0}\hbar^4\left(E_{g0}+2E_{\mathrm{F}_b}\right)}{2m_\perp^+ m_\perp^-} + \frac{\hbar^4 E_{\mathrm{F}_b}^2}{\left(m_\perp^+\right)^2}\right.$$
$$\left.+\frac{\hbar^4 E_{\mathrm{F}_b} E_{g0}}{\left(m_\perp^+\right)^2} - \frac{\hbar^4 E_{\mathrm{F}_b}^2}{\left(m_\perp^-\right)^2} - \frac{\hbar^4 E_{\mathrm{F}_b} E_{g0}}{\left(m_\perp^-\right)^2}\right],$$

$$\delta_{A_{10}} = \delta_{A_6}\left(\delta_{A_7}\right)^{1/2},$$

$$\delta_{A_{11}}\left(E_{\mathrm{F}_b}\right) = \left[\delta_{A_8}\left(E_{\mathrm{F}_b}\right)/\delta_{A_7}\right],\ h_{A_2}\left(E_{\mathrm{F}_b}\right) = \frac{\hbar^2}{2}\left[\frac{2E_{\mathrm{F}_b}+E_{g0}}{m_\parallel^-} + \frac{E_{g0}}{m_\parallel^+} + P_\parallel^2\right],$$

$$h_{A_1} = \left[\frac{\hbar^4}{4}\left(\frac{1}{\left(m_\parallel^+\right)^2} - \frac{1}{\left(m_\parallel^-\right)^2}\right)\right],\quad \delta_{A_{12}}\left(E_{\mathrm{F}_b}\right) = \left[\delta_{A_9}/\delta_{A_7}\right]$$

$$h_{A_3}\left(E_{\mathrm{F}_b}\right) = \left(2h_{A_1}\right)^{1/2}\left[\sqrt{h_{A_2}^2\left(E_{\mathrm{F}_b}\right) + 4h_{A_1}E_{\mathrm{F}_b}\left(E_{\mathrm{F}_b}+E_{g0}\right)} - h_{A_2}\left(E_{\mathrm{F}_b}\right)\right]^{1/2},$$

$$J_{A_6}\left(E_{\mathrm{F}_b}\right) = \frac{h_{A_4}\left(E_{\mathrm{F}_b}\right)}{3}\left[-E\left(\lambda_1,q_1\right)\left[h_{A_4}^2\left(E_{\mathrm{F}_b}\right) + h_{A_5}^2\left(E_{\mathrm{F}_b}\right)\right] + 2h_{A_5}^2\left(E_{\mathrm{F}_b}\right)\right.$$
$$\times F\left(\lambda_1,q_1\right) + \frac{h_{A_4}\left(E_{\mathrm{F}_b}\right)}{3}\left[h_{A_3}^2\left(E_{\mathrm{F}_b}\right) + h_{A_4}^2\left(E_{\mathrm{F}_b}\right) + 2h_{A_5}^2\left(E_{\mathrm{F}_b}\right)\right]$$
$$\left[\left(h_{A_4}^2\left(E_{\mathrm{F}_b}\right) + h_{A_5}^2\left(E_{\mathrm{F}_b}\right)\right)\right]/\left[\left(h_{A_5}^2\left(E_{\mathrm{F}_b}\right) + h_{A_3}^2\left(E_{\mathrm{F}_b}\right)\right)\right]\Big]^{1/2},$$

$$\lambda_1 = \tan^{-1}\left[h_{A_3}\left(E_{\mathrm{F}_b}\right)/h_{A_5}\left(E_{\mathrm{F}_b}\right)\right],$$
$$q_1 = \left[\frac{h_{A_4}^2\left(E_{\mathrm{F}_b}\right) - h_{A_5}^2\left(E_{\mathrm{F}_b}\right)}{h_{A_4}\left(E_{F_b}\right)}\right],$$

and

$$h_{A_7}\left(E_{\mathrm{F}_b}\right) = \sum_{r=1}^{s_0} 2\left(k_{\mathrm{B}}T\right)^{2r}\left(1-2^{1-2r}\right)\zeta\left(2r\right)\frac{\partial^{2r}}{\partial E_{\mathrm{F}_b}^{2r}}\left[h_{A_6}\left(E_{\mathrm{F}_b}\right)\right].$$

A.8 n-Ge

A.8.1 Model of Cardona Et al.

The expressions for the electron concentration and the TPSM can be written as

$$n_0 = N_{c0}\left[F_{\frac{1}{2}}(\eta) + \bar{\alpha}_2 F_{\frac{3}{2}}(\eta) - \bar{\alpha}_3 F_{\frac{7}{2}}(\eta)\right], \tag{A.29}$$

$$G_0 = \left(\frac{\pi^2 k_B}{3e}\right)\left[\frac{F_{\frac{-1}{2}}(\eta) + \bar{\alpha}_2 F_{\frac{1}{2}}(\eta) - \bar{\alpha}_3 F_{\frac{5}{2}}(\eta)}{F_{\frac{1}{2}}(\eta) + \bar{\alpha}_2 F_{\frac{3}{2}}(\eta) - \bar{\alpha}_3 F_{\frac{7}{2}}(\eta)}\right], \tag{A.30}$$

where

$$N_{c0} = 2g_v\left(2\pi m_D^* k_B T/h^2\right)^{3/2},$$

$$m_D^* = \left(\left(m_\perp^*\right)^2 m_\parallel^*\right)^{1/3},$$

$$\bar{\alpha}_2 = \frac{45\alpha k_B T}{24},$$

and

$$\bar{\alpha}_3 = \frac{189}{8}\alpha\left(k_B T\right)^2\left(\frac{k_B T\left(m_\parallel^*\right)^2}{\hbar^4}\right).$$

A.8.2 Model of Wang and Ressler

The expressions for the electron concentration and the TPSM assume the forms

$$n_0 = \left(\frac{m_\perp^* g_v}{\pi^2\hbar^2}\right)\left[M_{A_5}\left(E_{F_b}\right) + N_{A_5}\left(E_{F_b}\right)\right], \tag{A.31}$$

$$G_0 = \left(\frac{\pi^2 k_B^2 T}{3e}\right)\left[\frac{M'_{A_5}\left(E_{F_b}\right) + N'_{A_5}\left(E_{F_b}\right)}{M_{A_5}\left(E_{F_b}\right) + N_{A_5}\left(E_{F_b}\right)}\right], \tag{A.32}$$

where

$$M_{A_5}\left(E_{F_b}\right) = \left[\bar{\alpha}_8\rho_{A_1}\left(E_{F_b}\right) - \frac{\bar{\alpha}_9}{3}\rho_{A_1}^3\left(E_{F_b}\right) - \bar{\alpha}_{10}J_{A_2}\left(E_{F_b}\right)\right],$$

$$\bar{\alpha}_4 = \beta_4\left(2m_\perp^*/\hbar^2\right)^2, \quad \beta_4 = 1.4\beta_5,$$

$$\beta_5 = \frac{1}{4}\left(\alpha\hbar^4/\left(m_\perp^*\right)^2\right)\cdot\left(1 - \frac{m_\perp^*}{m_0}\right)^2,$$

$$\bar{\alpha}_5 = \bar{\alpha}_7 \left(4m_\perp^* m_\parallel^* \big/ \hbar^4\right), \quad \bar{\alpha}_7 = 0.8\beta_5, \quad \bar{\alpha}_6 = (0.005\beta_5)\left(2m_\parallel^* \big/ \hbar^2\right)^2,$$

$$\bar{\alpha}_{10} = \left(\frac{1}{2\bar{\alpha}_4}\right) \cdot \left(\frac{\hbar^2}{2m_\parallel^*}\right) \left[\bar{\alpha}_5^2 - 4\bar{\alpha}_4\bar{\alpha}_6\right]^{1/2},$$

$$\bar{\alpha}_{11} = \left(\frac{2m_\parallel^*}{\hbar^2}\right) \left[\frac{4\bar{\alpha}_4 - 2\bar{\alpha}_5}{\bar{\alpha}_5^2 - 4\bar{\alpha}_4\bar{\alpha}_6}\right],$$

$$\bar{\alpha}_{12}\left(E_{F_b}\right) = \left(\frac{2m_\parallel^*}{\hbar^2}\right)^2 \left[\frac{\left(1 - 4\bar{\alpha}_4 E_{F_b}\right)}{\bar{\alpha}_5^2 - 4\bar{\alpha}_4\bar{\alpha}_6}\right],$$

$$\rho_{A_1}\left(E_{F_b}\right) = \frac{1}{\hbar}\left(\frac{m_\parallel^*}{\bar{\alpha}_6}\right)^{1/2} \left[1 - \sqrt{1 - 4\bar{\alpha}_6\left(E_{F_b}\right)}\right]^{1/2},$$

$$\bar{A}_{A_1}^2\left(E_{F_b}\right) = \frac{1}{2}\left[\bar{\alpha}_{11} + \left[\bar{\alpha}_{11}^2 - 4\bar{\alpha}_{12}\left(E_{F_b}\right)\right]^{1/2}\right],$$

$$\bar{B}_{A_1}^2\left(E_{F_b}\right) = \frac{1}{2}\left[\bar{\alpha}_{11} - \left[\bar{\alpha}_{11}^2 - 4\bar{\alpha}_{12}\left(E_{F_b}\right)\right]^{1/2}\right],$$

$$J_{A_2}\left(E_{F_b}\right) = \frac{\bar{A}_{A_1}\left(E_{F_b}\right)}{3}\left[-E\left(\lambda_3, q_3\right)\left[\bar{A}_{A_1}^2\left(E_{F_b}\right) + \bar{B}_{A_1}^2\left(E_{F_b}\right)\right]\right.$$
$$+ 2\bar{B}_{A_1}^2\left(E_{F_b}\right) F\left(\lambda_3, q_3\right)\Big] + \frac{\bar{A}_{A_1}\left(E_{F_b}\right)}{3}\left[\bar{\rho}_{A_1}^2\left(E_{F_b}\right) + \bar{A}_{A_1}^2\left(E_{F_b}\right)\right.$$
$$\left. + 2\bar{B}_{A_1}^2\left(E_{F_b}\right)\right]\left[\frac{\bar{A}_{A_1}^2\left(E_{F_b}\right) + \rho_{A_1}^2\left(E_{F_b}\right)}{\bar{B}_{A_1}^2\left(E_{F_b}\right) + \rho_{A_1}^2\left(E_{F_b}\right)}\right]^{1/2},$$

$$\lambda_3 = \tan^{-1}\frac{\rho_{A_1}\left(E_{F_b}\right)}{\bar{B}_{A_1}\left(E_{F_b}\right)},$$

$$q_3 = \left[\frac{\bar{A}_{A_1}^2\left(E_{F_b}\right) - \bar{B}_{A_1}^2\left(E_{F_b}\right)}{\bar{A}_{A_1}\left(E_{F_b}\right)}\right],$$

$$N_{A_5}\left(E_{F_b}\right) = \sum_{r=1}^{s} 2\left(k_B T\right)^{2r}\left(1 - 2^{1-2r}\right)\zeta(2r)\frac{\partial^{2r}}{\partial E_{F_b}^{2r}}\left[M_{A_5}\left(E_{F_b}\right)\right].$$

A.9 Platinum Antimonide

The expressions for the electron concentration and the TPSM can be written as

$$n_0 = \left(\frac{g_{\mathrm{v}}}{2\pi^2}\right)\left[M_{A6}\left(E_{\mathrm{F}_b}\right) + N_{A6}\left(E_{\mathrm{F}_b}\right)\right], \tag{A.33}$$

$$G_0 = \left(\frac{\pi^2 k_{\mathrm{B}}^2 T}{3e}\right)\left[\frac{M'_{A6}\left(E_{\mathrm{F}_b}\right) + N'_{A6}\left(E_{\mathrm{F}_b}\right)}{M_{A6}\left(E_{\mathrm{F}_b}\right) + N_{A6}\left(E_{\mathrm{F}_b}\right)}\right], \tag{A.34}$$

$$M_{A6}\left(E_{\mathrm{F}_b}\right) = \left[T_{A9}\left(E_{\mathrm{F}_b}\right)\rho_{A2}\left(E_{\mathrm{F}_b}\right) - T_{A10}\left(E_{\mathrm{F}_b}\right)\frac{\rho_{A2}^3\left(E_{\mathrm{F}_b}\right)}{3} - T_{A11} J_{A3}\left(E_{\mathrm{F}_b}\right)\right],$$

$$\begin{aligned}
&T_{A1} = \left[I_1 + \omega_1\omega_3\right], \quad T_{A2}\left(E_{\mathrm{F}_b}\right) = \left[-E_{\mathrm{F}_b}\omega_3 + \omega_1\left(E_{\mathrm{F}_b} + \delta_0\right)\right],\\
&T_{A3} = \left[2I_1 + \omega_2\omega_4 + \omega_3\omega_2\right], \quad T_{A4} = \left[I_1 + \omega_2\omega_4\right],\\
&T_{A5}\left(E_{\mathrm{F}_b}\right) = \omega_2\left(E_{\mathrm{F}_b} + \delta_0\right), \quad T_{A6}\left(E_{\mathrm{F}_b}\right) = \left[E_{\mathrm{F}_b}\left(E_{\mathrm{F}_b} + \delta_0\right) - E_{\mathrm{F}_b}\omega_4\right],\\
&\bar{T}_{A6} = \left[T_{A3}^2 - 4T_{A1}T_{A4}\right], T_{A7}\left(E_{\mathrm{F}_b}\right) = \left[2T_{A3}T_{A2}\left(E_{\mathrm{F}_b}\right) - 4T_{A1}T_{A5}\left(E_{\mathrm{F}_b}\right)\right],\\
&T_{A8}\left(E_{\mathrm{F}_b}\right) = \left[T_{A2}^2\left(E_{\mathrm{F}_b}\right) + 4T_{A1}T_{A6}\left(E_{\mathrm{F}_b}\right)\right],
\end{aligned}$$

$$\begin{aligned}
T_{A9}\left(E_{\mathrm{F}_b}\right) &= \frac{T_{A2}\left(E_{\mathrm{F}_b}\right)}{2T_{A1}},\\
T_{A10} &= \left[T_{A3}/2T_{A1}\right],\\
T_{A11} &= \frac{\sqrt{\bar{T}_{A6}}}{2T_{A1}},
\end{aligned}$$

$$T_{A12}\left(E_{\mathrm{F}_b}\right) = \left[T_{A7}\left(E_{\mathrm{F}_b}\right)/\bar{T}_{A6}\right], \quad T_{A13}\left(E_{\mathrm{F}_b}\right) = T_{A8}\left(E_{\mathrm{F}_b}\right)/\bar{T}_{A6},$$

$$\rho_{A2}\left(E_{\mathrm{F}_b}\right) = \left[\left[T_{A5}\left(E_{\mathrm{F}_b}\right) - \sqrt{T_{A5}^2\left(E_{\mathrm{F}_b}\right) + 4T_{A4}T_{A6}\left(E_{\mathrm{F}_b}\right)}\right]\Big/\left(2T_{A4}\right)\right]^{1/2},$$

$$A_{A3}^2\left(E_{\mathrm{F}_b}\right) = \frac{1}{2}\left[T_{A12}\left(E_{\mathrm{F}_b}\right) + \sqrt{T_{A12}^2\left(E_{\mathrm{F}_b}\right) - 4T_{A13}\left(E_{\mathrm{F}_b}\right)}\right],$$

$$B_{A3}^2\left(E_{\mathrm{F}_b}\right) = \frac{1}{2}\left[T_{A12}\left(E_{\mathrm{F}_b}\right) - \sqrt{T_{A12}^2\left(E_{\mathrm{F}_b}\right) - 4T_{A13}\left(E_{\mathrm{F}_b}\right)}\right],$$

$$\begin{aligned}
J_{A3}\left(E_{\mathrm{F}_b}\right) = {}& \frac{\rho_{A2}\left(E_{\mathrm{F}_b}\right)}{3}\left[\left[A_{A3}^2\left(E_{\mathrm{F}_b}\right) + B_{A3}^2\left(E_{\mathrm{F}_b}\right)\right]E\left(\eta_1, t_1\right) - \left[A_{A3}^2\left(E_{\mathrm{F}_b}\right)\right.\right.\\
&\left.\left. - B_{A3}^2\left(E_{\mathrm{F}_b}\right)\right]F\left(\eta_1, t_1\right)\right] + \frac{\rho_{A2}\left(E_{\mathrm{F}_b}\right)}{3}\left[\left[A_{A3}^2\left(E_{\mathrm{F}_b}\right) - \rho_{A2}^2\left(E_{\mathrm{F}_b}\right)\right]\right.\\
&\left.\left[B_{A3}^2\left(E_{\mathrm{F}_b}\right) - \rho_{A2}^2\left(E_{\mathrm{F}_b}\right)\right]\right]^{1/2},
\end{aligned}$$

$$\eta_1 = \tan^{-1}\left[\rho_{A_2}\left(E_{F_b}\right)/B_{A_3}\left(E_{F_b}\right)\right],$$
$$t_1 = \left[B_{A_3}\left(E_{F_b}\right)/A_{A_3}\left(E_{F_b}\right)\right],$$

and

$$N_{A_6}\left(E_{F_b}\right) = \sum_{r=1}^{s} 2\left(k_B T\right)^{2r}\left(1-2^{1-2r}\right)\zeta(2r)\frac{\partial^{2r}}{\partial E_{F_b}^{2r}}\left[M_{A_6}\left(E_{F_b}\right)\right].$$

A.10 n-GaSb

In accordance with model of Mathur and Jain, the electron concentration and the TPSM can be expressed as

$$n_0 = \frac{g_v}{3\pi^2}\left(\frac{2m^*}{\hbar^2}\right)^{3/2}\left[\delta_{A_2}\left(E_{F_b}\right)+\delta_{A_3}\left(E_{F_b}\right)\right], \tag{A.35}$$

$$G_0 = \left(\frac{\pi^2 k_B^2 T}{3e}\right)\left[\frac{\delta'_{A_2}\left(E_{F_b}\right)+\delta'_{A_3}\left(E_{F_b}\right)}{\delta_{A_2}\left(E_{F_b}\right)+\delta_{A_3}\left(E_{F_b}\right)}\right], \tag{A.36}$$

where

$$\delta_{A_2}\left(E_{F_b}\right) = \left[\delta_{A_1}\left(E_{F_b}\right)\right]^{3/2},$$

$$\delta_{A_1}\left(E_{F_b}\right) = \left[E_{F_b}+E_{g1}-\frac{m^*}{m_0}\frac{E_{g1}}{2}-\left[\left(\frac{E_{g1}}{2}\right)^2+\left[\frac{E_{g1}}{2}\left(1-\frac{m^*}{m_0}\right)\right]^2 + \frac{\left(E_{g1}\right)^2}{2}\left(1-\frac{m^*}{m_0}\right)+E_{F_b}E_{g1}\left(1-\frac{m^*}{m_0}\right)\right]^{1/2}\right],$$

and

$$\delta_{A_3}\left(E_{F_b}\right) = \sum_{r=1}^{s} 2\left(k_B T\right)^{2r}\left(1-2^{1-2r}\right)\zeta(2r)\frac{\partial^{2r}}{\partial E_{F_b}^{2r}}\left[\delta_{A_2}\left(E_{F_b}\right)\right].$$

A.11 n-Te

The electron concentration and TPSM in n-Te in accordance with the model of Bouat et al. can be written as

$$n_0 = \frac{g_v}{3\pi^2}\left[M_{A_9}\left(E_{F_b}\right)+N_{A_9}\left(E_{F_b}\right)\right], \tag{A.37}$$

$$G_0 = \left(\frac{\pi^2 k_B^2 T}{3e}\right)\left[\frac{M'_{A_9}\left(E_{F_b}\right)+N'_{A_9}\left(E_{F_b}\right)}{M_{A_9}\left(E_{F_b}\right)+N_{A_9}\left(E_{F_b}\right)}\right], \tag{A.38}$$

$$M_{A_9}\left(E_{\mathrm{F}_b}\right)=\left[3\psi_5\left(E_{\mathrm{F}_b}\right)\Gamma_3\left(E_{\mathrm{F}_b}\right)-\psi_6\Gamma_3^3\left(E_{\mathrm{F}_b}\right)\right],$$

$$\psi_5\left(E_{\mathrm{F}_b}\right)=\left[\frac{E_{\mathrm{F}_b}}{\psi_2}+\frac{\psi_4^2}{2\psi_2^2}\right],$$

$$\Gamma_3\left(E_{\mathrm{F}_b}\right)=\left[2\psi_1\right]^{-1}\left[\sqrt{\psi_3^2+4\psi_1E_{\mathrm{F}_b}}-\psi_3\right],$$

$$\psi_6=\left(\psi_1/\psi_2\right),$$

$$N_{A_9}\left(E_{\mathrm{F}_b}\right)=\sum_{r=1}^{s}2\left(k_{\mathrm{B}}T\right)^{2r}\left(1-2^{1-2r}\right)\zeta\left(2r\right)\frac{\partial^{2r}}{\partial E_{\mathrm{F}_b}^{2r}}\left[M_{A_9}\left(E_{\mathrm{F}_b}\right)\right],$$

$\psi_1=A_6$, $\psi_2=A_7$, $\psi_3^2=A_8$, and $\psi_4^2=A_9$.

A.12 Bismuth

A.12.1 McClure and Choi Model

The electron concentration and TPSM in Bi in accordance with this model can be written as

$$n_0=\left(\frac{g_{\mathrm{v}}}{4\pi^3}\right)h_{A_8}\left[h_{A_{10}}\left(E_{\mathrm{F}_b}\right)+h_{A_{11}}\left(E_{\mathrm{F}_b}\right)\right], \tag{A.39}$$

$$G_0=\left(\frac{\pi^2k_{\mathrm{B}}^2T}{3e}\right)\left[\frac{h'_{A_{10}}\left(E_{\mathrm{F}_b}\right)+h'_{A_{11}}\left(E_{\mathrm{F}_b}\right)}{h_{A_{10}}\left(E_{\mathrm{F}_b}\right)+h_{A_{11}}\left(E_{\mathrm{F}_b}\right)}\right], \tag{A.40}$$

where

$$h_{A_8}=\frac{4\pi^2\sqrt{m_1m_3}}{\hbar^2\theta_{A_4}},$$

$$\theta_{A_4}=\frac{\alpha\hbar^2}{2m_2},$$

$$\theta_{A_3}=\frac{\alpha\hbar^4}{4m_2m_2'},$$

$$\theta_{A_2}\left(E_{\mathrm{F}_b}\right)=\left(\alpha E_{\mathrm{F}_b}\hbar^4/2m_2\right)\left[1-\frac{m_2}{m_2'}\right],$$

$$\theta_{A_5}^2=\frac{1}{\theta_{A_4}},h_{A_{10}}\left(E_{\mathrm{F}_b}\right)=\left[\frac{h_{A_9}\left(E_{\mathrm{F}_b}\right)}{2\theta_{A_5}}\ln\left|\frac{\theta_{A_5}+\bar{h}_{A_4}\left(E_{\mathrm{F}_b}\right)}{\theta_{A_5}-\bar{h}_{A_4}\left(E_{\mathrm{F}_b}\right)}\right|\right.$$
$$\left.+\left(\theta_{A_5}\left(E_{\mathrm{F}_b}\right)+\theta_{A_3}\theta_{A_5}^2\right)\bar{h}_{A_4}\left(E_{\mathrm{F}_b}\right)+\frac{\theta_{A_3}}{3}\left[\bar{h}_{A_4}\left(E_{\mathrm{F}_b}\right)\right]^3\right],$$

$$h_{A_9}\left(E_{\mathrm{F}_b}\right)=\left[E_{\mathrm{F}_b}\left(1+\alpha E_{\mathrm{F}_b}\right)-\theta_{A_2}\left(E_{\mathrm{F}_b}\right)\theta_{A_5}^2-\theta_{A_3}\theta_{A_5}^4\right],$$

$$\bar{h}_{A_4}\left(E_{\mathrm{F}_b}\right)=\frac{\sqrt{2m_2m_2'}}{\sqrt{\alpha}\hbar^2}\left[\frac{-\alpha E_{\mathrm{F}_b}\hbar^2}{2m_2}\left(1-\frac{m_2}{m_2'}\right)+\left[\frac{\alpha^2E_{\mathrm{F}_b}^2\hbar^4}{4m_2^2}\left(1-\frac{m_2}{m_2'}\right)^2\right.\right.$$
$$\left.\left.+\frac{\alpha E\left(1+\alpha E_{\mathrm{F}_b}\right)\hbar^4}{m_2m_2'}\right]^{1/2}\right]^{1/2},$$

and

$$h_{A_{11}}\left(E_{\mathrm{F}_b}\right)=\sum_{r=1}^{s}2\left(k_{\mathrm{B}}T\right)^{2r}\left(1-2^{1-2r}\right)\zeta\left(2r\right)\frac{\partial^{2r}}{\partial E_{\mathrm{F}_b}^{2r}}\left[h_{A_{10}}\left(E_{\mathrm{F}_b}\right)\right].$$

A.12.2 Hybrid Model

In accordance with Hybrid model, the expressions for n_0 and G_0 are given by

$$n_0=\left(\frac{g_{\mathrm{v}}}{2\pi^2}\right)\left[h_{A_{12}}\left(E_{\mathrm{F}_b}\right)+h_{A_{13}}\left(E_{\mathrm{F}_b}\right)\right],\tag{A.41}$$

$$G_0=\left(\frac{\pi^2k_{\mathrm{B}}^2T}{3e}\right)\left[\frac{h'_{A_{12}}\left(E_{\mathrm{F}_b}\right)+h'_{A_{13}}\left(E_{\mathrm{F}_b}\right)}{h_{A_{12}}\left(E_{\mathrm{F}_b}\right)+h_{A_{13}}\left(E_{\mathrm{F}_b}\right)}\right],\tag{A.42}$$

where

$$h_{A_{12}}\left(E_{\mathrm{F}_b}\right)=\left[E_{\mathrm{F}_b}\left(1+\alpha E_{\mathrm{F}_b}\right)-\frac{L_{A_1}\left(E_{\mathrm{F}_b}\right)\hbar^2I_{A_4}^2\left(E_{\mathrm{F}_b}\right)}{6M_2}-\frac{L_{A_2}\hbar^4I_{A_5}^5\left(E_{\mathrm{F}_b}\right)}{20M_2^2E_{\mathrm{g}0}}\right],$$

$$L_{A_1}\left(E_{\mathrm{F}_b}\right)=\left[1+L_{A_3}+\alpha E_{\mathrm{F}_b}\left(1-L_{A_2}\right)\right],\quad L_{A_3}=M_2/m_2,$$
$$L_{A_2}=M_2/M_2',$$

$$I_{A_4}\left(E_{\mathrm{F}_b}\right)=\left[\frac{L_{A_2}}{2E_{\mathrm{g}0}M_2^2}\right]^{-1/2}\left[\frac{-L_{A_1}\left(E_{\mathrm{F}_b}\right)}{2M_2}+\left[\frac{L_{A_1}^2\left(E_{\mathrm{F}_b}\right)}{4M_2^2}\right.\right.$$
$$\left.\left.+\frac{L_{A_2}E_{\mathrm{F}_b}\left(1+\alpha E_{\mathrm{F}_b}\right)}{4E_{\mathrm{g}0}M_2^2}\right]^{1/2}\right]^{1/2},$$

and

$$h_{A_{13}}\left(E_{\mathrm{F}_b}\right) = \sum_{r=1}^{s} 2\left(k_{\mathrm{B}}T\right)^{2r}\left(1-2^{1-2r}\right)\zeta\left(2r\right)\frac{\partial^{2r}}{\partial E_{\mathrm{F}_b}^{2r}}\left[h_{A_{12}}\left(E_{\mathrm{F}_b}\right)\right].$$

A.12.3 Lax Ellipsoidal Nonparabolic Model

In accordance with this model, the expressions for n_0 and G_0 assume the same forms as given by (A.5) and (A.6), where
$N_{A_c} = 2g_{\mathrm{v}}\left(2\pi m_D^* k_{\mathrm{B}}T/h^2\right)^{3/2}$ and $m_D^* = \left(m_1 m_2 m_3\right)^{1/3}$.

A.12.4 Ellipsoidal Parabolic Model

For this well-known model, the expressions for n_0 and G_0 assume the same forms as given by (A.7) and (A.8) where N_{A_c} and m_D^* are defined above.

A.13 Open Research Problem

(RA.1)

(a) Investigate the TPSM for bulk specimens for all the materials whose carrier energy spectra are described in Chap. 1 excluding the dispersion relations as considered in this Appendix.
(b) Investigate the DTP, PTP, and Z for all the materials whose carrier energy spectra are defined in Chap. 1 by considering all types of scattering mechanisms.
(c) Investigate the TPSM for bulk specimens of all the materials whose carrier energy spectra are described in problem (R1.10) of Chap. 1.
(d) Investigate the DTP, PTP, and Z for all the materials whose carrier energy spectra are defined in problem (R1.10) of Chap. 1.

Reference

1. K.P. Ghatak, S. Bhattacharya, D. De, *Einstein Relation in Compound Semiconductors and Their Nanostructures*, Springer Series in Materials Science, vol 116 (Springer, Germany, 2008)

Subject Index

G

H

I

K

L

M

N

O

P

Q

S

T

V

W

Z

Material Index